AF776009

Evolution of Island Mammals

In memory of Paul Yves Sondaar (1934–2003)

Evolution of Island Mammals

Adaptation and Extinction of Placental Mammals on Islands

Second Edition

Alexandra van der Geer
Naturalis Biodiversity Center, Postbus 9517,
2300 RA Leiden, the Netherlands

George Lyras
Department of Geology and Geo-environment, Panepistimiopolis,
15784 Zografou, Greece

and

John de Vos
Naturalis Biodiversity Center, Postbus 9517,
2300 RA Leiden, the Netherlands

WILEY Blackwell

This second edition first published 2021
© 2021 John Wiley & Sons Ltd

Edition History
First edition © 2010 by John Wiley & Sons Ltd.

All rights reserved. No part of this publication may be reproduced, stored in a retrieval system, or transmitted, in any form or by any means, electronic, mechanical, photocopying, recording or otherwise, except as permitted by law. Advice on how to obtain permission to reuse material from this title is available at http://www.wiley.com/go/permissions.

The right of Alexandra van der Geer, George Lyras and John de Vos to be identified as the authors of this work has been asserted in accordance with law.

Registered Offices
John Wiley & Sons, Inc., 111 River Street, Hoboken, NJ 07030, USA
John Wiley & Sons Ltd, The Atrium, Southern Gate, Chichester, West Sussex, PO19 8SQ, UK

Editorial Office
9600 Garsington Road, Oxford, OX4 2DQ, UK

For details of our global editorial offices, customer services, and more information about Wiley products, visit us at www.wiley.com.

Wiley also publishes its books in a variety of electronic formats and by print-on-demand. Some content that appears in standard print versions of this book may not be available in other formats.

Limit of Liability/Disclaimer of Warranty

While the publisher and authors have used their best efforts in preparing this work, they make no representations or warranties with respect to the accuracy or completeness of the contents of this work and specifically disclaim all warranties, including without limitation any implied warranties of merchantability or fitness for a particular purpose. No warranty may be created or extended by sales representatives, written sales materials or promotional statements for this work. The fact that an organisation, website, or product is referred to in this work as a citation and/or potential source of further information does not mean that the publisher and authors endorse the information or services the organisation, website, or product may provide or recommendations it may make. This work is sold with the understanding that the publisher is not engaged in rendering professional services. The advice and strategies contained herein may not be suitable for your situation. You should consult with a specialist where appropriate. Further, readers should be aware that websites listed in this work may have changed or disappeared between when this work was written and when it is read. Neither the publisher nor authors shall be liable for any loss of profit or any other commercial damages, including but not limited to special, incidental, consequential, or other damages.

Library of Congress Cataloging-in-Publication Data

Names: Geer, Alexandra van der, 1963- author. | Lyras, George | Vos, John de
Title: Evolution of island mammals : adaptation and extinction of placental mammals on islands / Alexandra van der Geer, George Lyras, John de Vos.
Description: Second edition. | Hoboken, NJ : Wiley-Blackwell, 2021. | Includes bibliographical references and index.
Identifiers: LCCN 2020032947 (print) | LCCN 2020032948 (ebook) | ISBN 9781119675730 (hardback) | ISBN 9781119675747 (adobe pdf) | ISBN 9781119675716 (epub)
Subjects: LCSH: Mammals–Evolution. | Island animals–Evolution.
Classification: LCC QL708.5 .E845 2021 (print) | LCC QL708.5 (ebook) | DDC 599–dc23
LC record available at https://lccn.loc.gov/2020032947
LC ebook record available at https://lccn.loc.gov/2020032948

Cover Design: Wiley
Cover Images: Main Image © Haveseen/AdobeStock, Inset images © George Lyras

Set in 10/12pt MeliorLTStd by SPi Global, Chennai, India
Printed and bound by CPI Group (UK) Ltd, Croydon, CR0 4YY

C9781119675730_030421

CONTENTS

PREFACE

This is the second edition of our state-of-the-art reference book about fossil insular mammals. It provides an extensive overview of what is known about their evolution, adaptation, and extinction. Fossil insular mammals often show remarkable and sometimes even bizarre adaptations, such as dwarfism and gigantism. Understanding the processes underlying these adaptations helps us to understand the patterns of evolution, not only those on islands but also those on the mainland and in fragmented habitats.

Many important studies describe the biogeography and ecology of extant insular mammals, but similar works on extinct insular mammals are few. The information on individual taxa and islands is scattered over many journals, often not widely available, or information on the subject is limited to a few paragraphs. Moreover, since the time of the first edition (2010), many new discoveries were made, and novel techniques and methodologies have been implemented in the study of the fossil record. This edition is substantially modified and offers extensive coverage of the latest developments in the field. It is a comprehensive book that offers an updated concise synthesis of available studies. Our overview of fossil insular faunas provides an updated approach to the subject and elaborates on published studies – excellent as some of these in many respects are. We have designed this book as a synthesis of available data somewhat less formal than research papers or systematic revisions. In this way, it will be of use to the many researchers, regardless of speciality, who need a source of data and interpretations about fossil insular mammals, as well as qualified graduate students in palaeontology, zoology, evolutionary biology, and biogeography.

Why is an up-to-date overview of fossil insular mammals important? Our knowledge of island biogeography, ecology, and evolution is limited because it is mainly based on present-day patterns of biodiversity on islands and in fragmented habitats. This means that the available sources are by definition restricted to very short and recent periods of time. To analyse and infer biogeographical patterns that developed over time spans of thousands or even millions of years, fossil data need to be part of our studies.

By ignoring the fact that biodiversity on islands was much larger in the past than it is today and by only taking into account present-day biodiversity on islands, an impoverished, unbalanced view of island biogeography may come into being. For example, the megafauna of Madagascar contained large taxa, such as hippos, elephant birds, and giant lemurs, which are all extinct now. This means that the average body mass of Malagasy taxa is much lower now than during the Late Pleistocene. Furthermore, because of the effects of time, the resulting speciation differs much from what is seen today. Where present-day island taxa often are not smaller than roughly 80% of their mainland ancestor, fossil insular taxa sometimes reduced their body mass to less than 5% as is the case with the dwarf elephant of Sicily (*Palaeoloxodon falconeri*). As a result, islands today provide a poor example of mammal evolution on islands.

The fossil record of the supercontinental islands Australia and South America is excluded from our synthesis. The reason is that their fossil faunas represent balanced or harmonic faunas, thus containing a representative number of elements pertaining to all orders typical for the geographic latitude and altitude with representatives of all trophic levels. In fact, these mega-islands are ecologically more similar to continents than to islands. Their long-term isolation resulted in endemic balanced faunas with their own stamp, not comparable to the endemic faunas we describe in this book. The inclusion of Australia and South America would lead to yet another book on general vertebrate palaeontology. In order to keep the scope of the book more focussed, we excluded Australia and South America and restricted ourselves to placental mammals.

The separate chapters of this and/or the previous edition were reviewed by the following experts in the field, in alphabetical order, for which we are grateful: the late Larry Agenbroad, Athanassios Athanassiou, Chiara Angelone, Laura Bonfiglio, Pere Bover, the late Johanna de Visser, Lawrence Heaney, Christine Hertler, Mark Lomolino, Ross MacPhee, Sandra Olsen, Hidetoshi Ota, Hiroyuki Otsuka, Maria Rita Palombo, Lorenzo Rook, Haruo Saegusa, Ian Tattersal, George Theodorou, Gert van den Bergh, Lars van den Hoek Ostende, and Jan van der Made. We thank every one of them for variously commenting on the draft and final material, supplying answers to queries, and sharing with us their most recent papers.

We are further indebted to the following colleagues (in alphabetical order), who provided us with up-to-date information and relevant papers and with whom we had valuable discussions on various aspects of island mammals throughout the years: Paul Albers, Josep Alcover, Athanassios Athanassiou, Fachroel Aziz, Danilo Balete, Peter Ballmann, Angel Bautista, Rafael Borroto-Páez, Hans Brinkerink, Massimo

Delfino, Eusebio Dizon, Clare Flemming, Nuría García, Steven Goodman, Victoria Herridge, George Iliopoulos, Thomas Ingicco, Josephine Joordens, Tassos Kotsakis, Adrian Lister, Raquel López Antoñanzas, Federica Marcolini, David Mayhew, Hanneke Meijer, Shai Meiri, Gian Luigi Pillola, Pasquale Raia, David Reese, Jelle Reumer, Roberto Rozzi, Marina Sotnikova, William Stanley, Rab Sukamto, Keiichi Takahashi, Bartholomeus van der Geer, Boris Villier, Rebekka Volmer and Gerard Willemsen. Sadly, between the first and second edition, Larry Agenbroad, Johanna (Hannie) de Visser, Danilo (Danny) Balete, David Mayhew, and William (Bill) Stanley passed away. We greatly miss the discussions we had with them on island mammals.

We thank Marieke de Loos, Janet Kamphorst, and Bas van der Geer for a critical reading of parts of earlier versions of the manuscript.

We are grateful to those individuals and organisations who granted permission for the reproduction of copyright material: the derivation of which is indicated in the relevant figure and plate legends, supported by references where relevant. In this respect, we want to thank especially (in alphabetical order) Paul Albers, the American Museum of Natural History Library, Luc Costeur, Thomas Ingicco, Eelco Kruidenier, Ross MacPhee, Naturalis Biodiversity Center (Dubois Archive), Gert van den Bergh, Bartholomeus van der Geer, Jan van der Made, Rebekka Volmer, and the Washington Academy of Sciences. Artistic reconstructions were drawn by Stephen Nash (lemurs); most line drawings and figures were prepared by the authors. Our special thanks go to Rosie Hayden and Andrew Harrison of Wiley-Blackwell, for their guidance and advice during all stages of this project.

Finally, we thank our family members, Kees, Vera, Carmen, Anna, Rita, Zacharo and the late Alexandros, for all their support and patience during the preparation of this book.

PART I

Beyond the Mainland

All over the world, islands were and still are inhabited by unique species, restricted to their own island and found nowhere else. Their ancestors managed to reach the island from the mainland, and once isolated from this mainland with its ecological restrictions, they often evolved spectacular adaptations. In this part, after a general introduction to island studies, a short overview of the history of island studies is given, followed by an overview of what defines islands and island faunas as opposed to the mainland and its faunas.

CHAPTER ONE

Introduction

Evolution of Island Mammals: Adaptation and Extinction of Placental Mammals on Islands, Second Edition.
Alexandra van der Geer, George Lyras and John de Vos.
© 2021 John Wiley & Sons Ltd. Published 2021 by John Wiley & Sons Ltd.

Isolated from the continental landmasses, island species evolve novel and often extreme adaptations to new ecological niches. In this book, the past effects of insularity on mammal lineages are discussed, based on more than 300 fossil insular species, which were endemic to about 80 islands all over the world, and ranging from the Eocene to the Holocene, with a few Cretaceous cases.

The evolution of body size change is at first sight the most spectacular and certainly the best-known effect of ecological release – a shift and decline in the relative importance of interspecific interactions to an increase in the importance of intraspecific interactions. Many large herbivores, like elephants (Plate 1) and hippopotamuses, evolved towards miniature forms on islands, while many small mammals, like rodents and pikas, evolved towards giant forms. Since island faunas are highly disharmonic, with many major groups missing, most or all ecologically relevant competitors are absent. Therefore, the colonising species could expand or even change their ecological niche. In fossil species, this can be deduced from observations such as the evolution of hypsodonty – high-crowned cheek teeth – in herbivores, shifts in prey species in carnivores, fusion and shortening of limb bones, and changes in body proportions. Patterns are not the same everywhere, as we shall see. Islands differ amongst each other, and so do their faunas: these may be balanced (mainland ratio between carnivore and herbivore species), unbalanced (ratio between carnivores and herbivores shifted towards the latter), disharmonic or depauperate (poor taxonomic diversity on higher levels), entirely endemic (all species restricted to the island, not found elsewhere), or mainland-like (hardly different from continental faunas of similar latitude).

The geological time covered in this book ranges from the late Early–Middle Eocene to the Late Pleistocene–Early Holocene with a few instances from the Cretaceous. The earliest faunas, however, have a poor fossil record, and many uncertainties prevail about the level of endemism of the individual taxa. The late Early or Middle Eocene fauna of Sardinia, for example, is known only by two endemic tapirs and an opossum. The fauna of Jamaica of the same period contains an early rhinocerotid and a semi-aquatic sea-cow, but no endemic features have been described. The Early Oligocene sloth of Puerto Rico may even belong to an ancestral mainland fauna. The Miocene insular faunas on the other hand are well documented, such as the Early Miocene faunas of Sardinia and Japan and the Late Miocene faunas of the Balearics, Gargano, and Tuscany. However, the vast majority of fossil insular faunas belong to the Pleistocene period, or the Ice Ages. Especially the Late Pleistocene is known for its wealth of fossil insular mammals in, for example, the West Indies, Madagascar, and the Mediterranean. When relevant to the discussion, recently extinct and still living but endangered insular mammals are included in our synthesis as well.

Not treated separately in this book is the Eocene of Europe at large. During this geological period, Europe was an archipelago, as inferred from stratigraphy and, for example, findings of fossil marine species. These faunas, such as those of Messel in Germany – one of the richest fossil sites in the world, have been extensively covered elsewhere (see further Box 1.1). Excluded as well is the early Miocene St Bathans fauna of New Zealand (16–19 million years ago). Here, fossils of at least two bat species belonging to the extant endemic family of burrowing bats (Mystacinidae) have been retrieved (described by Suzanne Hand and colleagues, 2015). The St Bathans fauna further includes remains of primitive frogs – *Leiopelma*, still extant and unique to New Zealand – a lizard-like sphenodontine reptile (a relative of the extant tuatara), birds (including kiwi, moa, and waterfowl), an armoured stem turtle (possibly related to *Meiolania* of Australia and New Caledonia), geckos, skinks, a crocodile, and a primitive currently unnamed mammal, as summarised by Trevor Worthy and colleagues (2017). This highly endemic and partly 'primitive' fauna confirms the vicariant or ancient dispersal origin of the 'old endemics' of New Zealand and refutes the idea of total drowning of Zealandia – a continental fragment of Gondwana – during the late Oligocene–earliest Miocene (22–25 million years ago) high sea level stand.

BOX 1.1

The European Eocene Archipelago

During the Eocene, much of Europe was under water, forming a kind of archipelago of islands. There was a large Central European Island, which consisted of parts of present-day England, France and Germany, and archipelagos of smaller islands in what is now south-eastern Europe. The islands were inhabited by endemic mammalian faunas – which were different from their contemporaries in North America and Asia – and included various odd-toed ungulates (palaeotheriids and lophiodontids), even-toed ungulates (anoplotheriids and xiphodontids), and hippopotamus-like even-toed ungulates (choeropotamids and cebochoerids). Most of these faunas, particularly those in Western Europe – such as Messel and Quercy – bear a mainland stamp and have been extensively covered by studies on mainland mammals.

The isolation of Europe ended around the Eocene–Oligocene boundary and many European endemic lineages went extinct synchronous with the arrival of new immigrants from Asia. This major faunal turnover, known as the 'Grande Coupure', was first recognised and named by the Swiss palaeontologist and geologist Hans Georg Stehlin (1910). It should be noted, however, that remains of the anthracothere *Prominatherium* – a new arrival from Asia – have been found in Late Eocene coastal marine deposits of Croatia, Romania, and Italy, so dating from before the Grande Coupure proper. According to Laureline Scherler and colleagues (2019), this indicates the existence of a maritime path as there was yet no terrestrial connection to Asia. According to them, *Prominatherium* had a sufficiently aquatic lifestyle to allow its dispersal through the archipelago of Southern Europe, but likely not for crossing the Perialpine and Paratethys seas and settle in Western Europe.

One of the islands in the south-east was the Pontide terrane of Central Anatolia, in what is today Turkey. According to Grégoire Métais and colleagues (2017), the terrane hosted a unique combination of Laurasian and Gondwanan mammals. The mammalian fauna of the island included rhino-like paenungulates (embrithopods), stem marsupials (herpetotheriids, anatoliadelphyid), and an endemic radiation of archaic ungulates (pleuraspidotheriids). Pleuraspidotheriids were also present during the Palaeogene in Western Europe, but already went extinct near the Palaeocene–Eocene boundary. Their survival in the Pontide terrane during the mid-Eocene is considered relictual by Métais and colleagues (2017). They may have arrived through land bridges that existed during the Palaeocene. The other taxa, such as embrithopods and metatherians, likely arrived via overwater dispersal, according to Métais and colleagues (2018). The Pontide terrane remained insular until the end of the Eocene.

Throughout this book, not all islands in the biogeographical sense – a habitat surrounded by inhospitable areas – are included. Our selection has been restricted to islands in the geographical sense: a piece of land surrounded by water. In the geological sense, roughly two main types of islands are further recognised: (i) continental shelf islands, which are islands sitting on a continental plate and may have been (intermittently) connected to the mainland and (ii) oceanic islands, which sit on an oceanic plate and were never connected to the mainland and arose from the sea bottom.

The first type of islands is characterised by a depauperate fauna – consisting of a limited but often representative subset of the continental fauna – with a low degree of endemism. The second type is characterised by a disharmonic fauna – fewer higher order taxa when compared to equivalent patches of nearby mainland – with a high degree of endemism. In the biogeographical sense, some continental shelf islands are like oceanic islands and have a similar fauna. Thus, many gradations exist within the continental shelf islands, with faunas ranging from balanced and harmonic to balanced but depauperate and disharmonic and endemic, mainly depending on degree, type, and duration of isolation.

Part I of this book forms an introduction to island studies, starting with this general introduction, followed by a short overview of the history of island studies and a chapter on the various factors which typify insular faunas, such as distance to the mainland, type of island and area, and the various ways of dispersal to islands.

Part II gives a concise overview of the faunas of individual islands, starting with the Mediterranean Islands, followed by Madagascar, the Indonesian and Japanese islands, the Californian Channel Islands, and ending with the West Indies. In these chapters, a comprehensive treatment is provided of the history of discoveries, biozones, or faunal units and the peculiarities and evolutionary aspects of individual endemic species or lineages. Here, we define a biozone as a stratigraphic layer characterised by one or two taxa, sometimes divided into subzones when an evolution within the taxon can be discerned. A faunal unit (at times also referred to as a *faunal complex*), on the other hand, is a subset of the fauna with the most characteristic elements, sometimes applied when the complete fauna is unknown. For some islands, biozones are defined, for other islands faunal units. Part II thus defines the context of the various insular species, their time span, arrival to the island, and eventual extinction, where known. Non-mammalian vertebrate species – such as birds, tortoises, and frogs – are included where relevant to the discussion.

Part III starts with a short overview of insular mammals with information about their distribution, range, and dispersal to the islands. Insular species are by definition endemic and thus unique to the island, but the area-by-area treatment of Part II obscures the existence of parallelisms. Dwarf elephants are, for example, found all over the northern hemisphere. Discussing them in relation to each other allows us to highlight similarities and dissimilarities. Furthermore, a number of islands are known only for a single endemic species, and these are limited to this part of the book. The second part of Part III discusses overall patterns and trends – including not only the island rule of body size evolution but also other parallel evolutionary processes – as well as a treatment of the processes of speciation on islands. The conceptual context for viewing biogeographical patterns is based on the handbooks by Whittaker and Palacios (2007) and Lomolino, Riddle and Whittaker (2017). Part III ends with a discussion on the possible reasons for the extinction of insular mammals.

References

Hand S.J., Lee D.E., Worthy T.H., Archer M., Worthy J.P., Tennyson A.J.D., Salisbury S.W., Scofield R.P., Mildenhall D.C., Kennedy E.M., Lindqvist J.K. 2015. Miocene fossils reveal ancient roots for New Zealand's endemic *Mystacina* (Chiroptera) and its rainforest habitat. *PLoS ONE*, 10 (6): e0128871.

Lomolino M.V., Riddle B.R., Whittaker R.J., 2017. *Biogeography: Biological Diversity Across Space and Time*, 5th edn. Sinauer Associates, Sunderland, MA.

Métais G., Erdal O., Erturaç K., Beard K.C. 2017. Tarsal morphology of the pleuraspidotheriid mammal *Hilalia* from the middle Eocene of Turkey. *Acta Palaeontologica Polonica*, 62 (1): 173–179.

Métais G., Coster P.M., Kappelman J.W., Licht A., Ocakoğlu F., Taylor M.H., Beard C.K. 2018. Eocene metatherians from Anatolia illuminate the assembly of an island fauna during Deep Time. *PLoS ONE*, 13 (11): e0206181.

Stehlin H.G. 1910. Remarques sur les faunules de Mammifères des couches éocènes et oligocènes du Bassin de Paris. *Bulletin de la Société Géologique de France*, 4 (9): 488–520.

Scherler L., Lihoreau F., Becker D. 2019. To split or not to split *Anthracotherium*? A phylogeny of Anthracotheriinae (Cetartiodactyla: Hippopotamoidea) and its palaeobiogeographical implications. *Zoological Journal of the Linnean Society*, 185 (2): 487–510.

Whittaker R.J., Fernández-Palacios J.M. 2007. *Island Biogeography: Ecology, Evolution, and Conservation*, 2nd edn. Oxford University Press, Oxford.

Worthy T.H., De Pietri V.L., Scofield R.P. 2017. Recent advances in avian palaeobiology in New Zealand with implications for understanding New Zealand's geological, climatic and evolutionary histories. *New Zealand Journal of Zoology*, 44 (3): 177–211.

CHAPTER TWO

History of Island Studies

Evolution of Island Mammals: Adaptation and Extinction of Placental Mammals on Islands, Second Edition.
Alexandra van der Geer, George Lyras and John de Vos.
© 2021 John Wiley & Sons Ltd. Published 2021 by John Wiley & Sons Ltd.

Since Charles Darwin's earliest report on his voyage around the world (1839), his seminal book on natural selection and evolutionary theory (1859), and the many writings of Alfred Russel Wallace on evolution and island life (e.g. 1855, 1858, 1869, 1876, and 1880), generations of naturalists and biogeographers have maintained a keen interest in the nature and evolution of insular biotas. The first published book on islands worldwide, however, predated Darwin by roughly three centuries. In 1528, Benedetto Bordone, an astronomer and cartographer from Padua, Italy, had his *Libro* published in Venice, renamed to *Isolario* in the edition of 1534. The book was an illustrated guide about the then known islands for seafarers. It is divided into three parts, describing, respectively, the islands and peninsulas of the western ocean – now the Atlantic Ocean – the Mediterranean Sea, and the Indian Ocean plus the waters of the Far East. The New World was presented as an island, because, at that time, the correct contours were unknown. Apart from many valuable maps, it contains the earliest known European map of Japan, known to Bordone as the island Ciampagu. For our purpose, his descriptions of the insular biota are interesting, especially those that refer to fossil faunas. For example, in the part on Cyprus ('Cipro') (Figure 2.1), Bordone described a hill at Kyrenia ('Zyrenes'), entirely made of bones of animals and humans. This would lead naturalists in the nineteenth century to investigate these and similar fossiliferous deposits, eventually resulting in the discovery of pygmy hippos and dwarf elephants. Indeed, the lessons were not lost on Darwin and Wallace, who both remarked on the bizarre life forms reported in the fossil record.

Although many early naturalists and travellers deliberately went to the various islands to search for fossils, as given in detail in the relevant chapters of this book, Charles Forsyth Major was one of the first naturalists to seriously make an attempt to compare the fossil faunas from various islands in order to understand the underlying evolutionary principles. His search for fossil island faunas started in 1877 when he was funded by the Italian government to collect fossils on Corsica, Sardinia, and Sicily. In 1886, he began to study fossils from Cyprus, Crete, and Samos, which were partly sent to the British Museum of Natural History, London, and the Geological Museum at Lausanne, Switzerland. These were not the first fossils from Crete for the British Museum, as, previously, fossils had been sent to Richard Owen and Hugh Falconer. At the British Museum, Forsyth Major started to work on the primate collection from Madagascar, both extinct and extant. In 1893, he discovered, amongst others, a new species of extinct giant lemurs (*Megaladapis madagascariensis*) for which he erected a new family (Megaladapidae) and five new species in the

ISOLARIO
DI BENEDETTO BORDONE
Nel qual ſi ragiona di tutte l'Iſole del mondo,
con li lor nomi antichi & moderni, hiſtorie,
fauole, & modi del loro viuere, & in qual
parte del mare ſtanno, & in qual pa-
rallelo & clima giaciono. Ri-
coreto, & di Nuouo
riſtampato.
Con la gionta del Monte del Oro
nouamente ritrouato.
CON IL BREVE DEL PAPA
Et gratia & priuilegio della Illuſtriſſi-
ma Signoria di Venetia co-
me in quelli appare.
M. D. XLVII.

Figure 2.1 Title page of Bordone's *Isolario*, 1547.

genera *Lepilemur* and *Cheirogaleus*. A year later, he undertook an expedition to the island, funded by the Royal Society, the banker Lionel Walter Rothschild, and others. The expedition lasted two years, during which a very large collection of fossils and zoological specimens was gathered:

> 'it may be remarked that the very large collection was obtained under circumstances of great difficulty and danger. The swampy nature of the deposits made the task of excavating very arduous, and the work was frequently interrupted for days at a time through the growing hostility of the natives. Dr Forsyth Major and his companion, M. Robert, are therefore the more to be congratulated that, under such unfavourable conditions, they have added so much to our knowledge of the extinct fauna of Central Madagascar'. [Andrews 1897, 358]

Towards the end of the nineteenth century, a young woman, Dorothea Bate, came to the British Museum to look for a job.

Self-educated, she first started to work in the Bird Room to prepare bird skins, but soon her abilities were recognised. Forsyth Major began to tutor her on fossil island faunas and suggested that she should go to Cyprus, to collect materials for him, fossil as well as extant. He expected her to find fossils of dwarf hippos and dwarf elephants, just as they had been found on Sicily, and Malta. In 1901, she went for the first time and explored the greater part of the island, guided by the earlier descriptions. She discovered 12 new fossiliferous caves and relocated some earlier mentioned sites. Amongst the fossils that she had sent to the British Museum, Forsyth Major recognised fossils of the expected small hippo, which were extremely similar to those depicted by Georges Cuvier (1804) as 'hippopotame petit', literally small hippo. He realised that the material curated in Paris obviously originated from Cyprus, not from southern France as generally thought. A year later, Bate returned to Cyprus and finally found the remains of a dwarf elephant. The similarity in fossil faunas of the Mediterranean islands, containing dwarf hippos, elephants, and giant rodents prompted her to go to Mallorca and Minorca as well, but this turned out a great disappointment. In contrast to what she and Forsyth Major had expected, she found nothing else but bizarre goats, giant dormice, and shrews. It took some years before *Myotragus balearicus*, described by her (1909), was appreciated for what it actually was, a peculiar insular dwarf ruminant with ever-growing incisors like a rodent that had lived for some 5 million years undisturbed on the islands. The picture of fossil island faunas had become more diverse.

The American palaeontologist, William Dillon Matthew (1918), was the first to observe that faunas on oceanic islands are typically unbalanced, which is to say that they only contain a limited and unrepresentative number of elements of the contemporaneous mainland fauna. Later, his fellow countryman George Gaylord Simpson (1940) established the link between the nature of these unbalanced faunas and the means of colonisation of these islands, introducing the term 'sweepstake route', a free translation of Matthew's 'accidents of transportation'. The term, changed into 'sweepstake dispersal', is today widely applied in cases of 'by chance' arrival across large distances over water.

The idea of long-distance overseas dispersal was initially not commonly accepted. A prevalent explanation for the presence of elephants, hippopotamuses, and small mammals on islands was dispersal across ancient but now submerged land bridges. Elephants were believed not to be able to swim (see especially Bourlière, 1970), and thus, these islands had been connected to the nearest mainland by the time of arrival of the ancestors of the insular species. The findings of elephants and hyenas in Sicily were accordingly considered proof of the connection between

Sicily and North Africa, as Hugh Falconer wrote to Darwin on 9 July 1860:

> 'What I want to tell you now is quite a different affair— but one which I am sure will interest you very much. Baron Anca a Sicilian Friend, who followed up my inquiries in the Sicilian caves, has brought over from Sicily two molars of the *Existing African Elephant* and upwards of 20 jaws of the Existing Spotted Hyæna (Hyæna Crocuta), of the Cape—from the Caves! Admiral Smyth laid down "Adventure Bank" a shoal with a narrow channel, between Trapani the Western End of Sicily & Capo Bono—the promontory of Tunis. We can now show that the division of Sicily from the African Continent is quite as late—if not later than the separation of England from France'. (Letter 2863, the Darwin Correspondence Project.)

Despite Matthew (1918) and Simpson (1940), the land bridge idea remained a mainstream explanation for island faunas for decades. The Dutch palaeontologist Dirk Hooijer, for example, explained (Hooijer, 1964) the presence of similar-sized dwarf elephants in Sulawesi, Timor, Flores, and Java by a large landmass that included these islands during the Pleistocene. He referred to this immense landmass as 'Stegoland'. Now we know that the first three islands were never connected to each other or the continent and belong to Wallacea, while Java, on the other hand, was indeed intermittently connected to the mainland and part of the Sunda Shelf.

The Dutch palaeontologist and geologist Paul Yves Sondaar was one of the first to embrace the hypothesis of swimming and rafting as main modes of dispersal to the islands. Sondaar (1977) noticed that the Pleistocene fossil faunas of islands worldwide often consisted of a few genera and species only – mainly elephants, hippopotamuses, deer, and rats – and no terrestrial carnivores and referred to these as *unbalanced endemic island faunas*. He also noticed that, on many islands, fauna with a very small dwarf elephant (pygmy elephants in his terminology) was often replaced at the start of the mid-Pleistocene by fauna with a larger elephant and the presence of Palaeolithic stone tools. He attributed these 'dramatic faunal turnovers' to the arrival of humans, and the absence of extreme dwarfism in these new elephants to the presence of a large carnivore species: humans. Today, the long-distance swimming and rafting abilities of several species (see Chapter 3), including hominins, are widely acknowledged, and sweepstake dispersal is an accepted theory. We now also know that early humans on islands had a wider distribution indeed, not only in maritime Southeast Asia but also in the Mediterranean and elsewhere. In some cases, however, land connections are still the best explanation for the occurrence on islands of unlikely overseas dispersers like rabbits and frogs, or in the case of a species-rich balanced fauna on the island, such as the Late Pleistocene rainforest fauna of Java.

A few early authors, including Piero Leonardi (1954) and Sigfried Kuss (1965), explained the presence of these pygmy herbivores on islands as the result of inbreeding or genetic degeneration in the absence of selective pressure exerted by large terrestrial carnivores. This model was never widely accepted, though from time to time, it is revived, such as in the case of the Late Pleistocene diminutive human of Flores, *Homo floresiensis*, whose small stature and low brain capacity are considered by some as evidence of a pathological condition (see Chapter 13). Leonardi (1954) also provided an alternative theory, according to which the dwarfs had evolved elsewhere and had migrated to the various islands. According to that view, the various dwarf elephants found on Mediterranean islands were considered conspecific with the Sicilian pygmy elephant (*Elephas* [=*Palaeoloxodon*] *falconeri*). This mistake survived well into the mid-70s, with several unrelated island dwarf elephants worldwide being referred to *E. falconeri*, simply based on size.

Body size evolution is not only one of the most fundamental responses to island environments but also the easiest response to quantify and compare. This explains the almost unique focus on body size changes in island studies. In his comprehensive review, J. Bristol Foster (1964) was the first who noted that there are different tendencies in body size evolution amongst taxonomic groups. Whereas insular even-toed ruminants, lagomorphs, and carnivores tend to become smaller, insular rodents and possibly insular marsupials as well tend to become larger. Leigh Van Valen (1973) dubbed this observation a 'rule' (quotation marks his), not directly referring to Foster but indirectly through his own 1970 publication. Later authors, e.g. Lawrence Heaney (1978) and Mark Lomolino (1985), refined this island rule and interpreted the pattern as a graded trend across as well as within taxa, from dwarfism in the larger species to gigantism in the smaller species. The intersection point is then a crude estimate of an 'optimal' body size for a species of a particular design or Bauplan and ecological strategy under optimal conditions, as suggested by Ted Case (1978).

One of the key driving forces for these evolutionary trends is interspecific interactions, or release from those interactions on unbalanced and species-poor islands. Therefore, a general theory of body size evolution on islands also needs to explain the paucity of competitors and predators on those islands, along with the subsequent evolutionary responses of the island's endemics. In a paper published in the journal *Evolution* (1963), and later in a monograph (1967), Robert MacArthur and Edward Wilson developed a quantitative mathematical model to predict the species diversity in a given isolated area. Since then, most island studies use quantitative data instead of purely morphological data with the results presented as plots and graphs. The equilibrium theory of MacArthur and Wilson, based on extant species and populations only, suggests that species diversity on islands is the sum of new arrivals minus the number of extinctions.

Thus, the factors that are used in these quantitative studies to calculate or predict species diversity and body mass of endemic species include surface area of the island, its distance to the mainland, limited resources on the island, decreased inter-species competition, and absence of predation by mammalian carnivores. For example, the greater the distance, or the more difficult the passage or filter, the lower the species richness, and the more prevalent ecological release and tendencies towards body size shifts.

In the past decades, various models have been proposed to explain the underlying processes influencing patterns of diversity and evolution on islands. These models, which are mainly based on studies of living biotas, attempt to explain the arrival, evolution, and extinction in both single islands and archipelagos. Robert Whittaker and José María Fernández-Palacios (2007) provide a comprehensive overview of all these previous works. Application of rules and methods from these and similar works to fossil insular taxa became a new trend in the early 2000s, for example, Pasquale Raia and colleagues (2003) on life-history traits of the dwarf elephant from Sicily; Raia and Shai Meiri (2006) on body size in fossil ungulates and carnivores; Virginie Millien (2006) on the speed of evolution on islands, calculated from the fossil record; Maria Rita Palombo (2007) on evolution of insular elephants from the Mediterranean; and our contributions in the 2010s with quantification of body size evolution in island populations worldwide, extinct, extant as well as introduced (Lyras and colleagues, 2010; Lomolino and colleagues, 2012, 2013; van der Geer and colleagues 2013, 2016, 2017). The number of these studies is increasing, as we will see in the coming chapters, contributing fundamentally to our understanding of evolution on islands in a broader context. The important factor in this new trend is time. Where most island studies are based on recent or extant species only, those occurring within tens of thousands of years at most, more integrative and insightful island studies nowadays include time spans ranging from hundreds of thousands to even millions of years.

References

Andrews C.W. 1897. On some fossil remains of carinate birds from Central Madagascar. *Ibis*, 39 (3): 343–359.

Bate D.M.A. 1909. Preliminary note on a new artiodactyle from Mallorca *Myotragus balearicus*, gen. et sp. nov. *Geological Magazine (Decade 5)*, 6: 385–389.

Bordone B. 1528. *Libro di Benedetto Bordone. Nel qual si ragiona di tutte l'Isole del mondo, con li lor nomi antichi & moderni, historie, favole, & modi del loro vivere.* Niccolo Zoppino, Venice. In *facsimile, Edizioni Aldine, Modena, 1982.*

Bourlière F. 1970. *The Natural History of Mammals*, 3rd edn. Alfred A. Knopf, New York.

Case T.J. 1978. A general explanation for insular body size trends in terrestrial vertebrates. *Ecology*, 59: 1–18.

Cuvier G. 1804. Sur les ossement fossiles d'hippopotame. *Annales du Muséum National d'Histoire Naturelle, Paris*, 5: 99–122.

Darwin Ch. 1839. *Narrative of the Surveying Voyages of His Majesty's Ships Adventure and Beagle between the years 1826 and 1836, Describing Their Examination of the Southern Shores of South America, and the Beagle's Circumnavigation of the Globe. Journal and Remarks 1832–1836.* London: Henry Colburn.

Darwin Ch. 1859. *On the Origin of Species by Means of Natural Selection, or the Preservation of Favoured Races in the Struggle for Life.* John Murray, London.

Forsyth Major C.I. 1893. On *Megaladapis madagascariensis*, an extinct gigantic lemuroid from Madagascar. *Proceedings of the Royal Society*, 54: 176–179.

Forsyth Major C.I. 1901. Exhibition of, and remarks upon, the skull of a new fossil mammal (*Enhydrictis galictoides*). *Proceedings of the Zoological Society of London*, 1901 (2): 625–628.

Foster J.B. 1964. Evolution of mammals on islands. *Nature*, 202: 234–235.

Heaney L.R. 1978. Island area and body size of insular mammals: Evidence from the tri-colored squirrel (*Callosciurus prevosti*) of Southeast Asia. *Evolution*, 32 (1): 29–44.

Hooijer D.A. 1964. Pleistocene vertebrates from Celebes XII. Notes on pygmy Stegodonts. *Zoologische Mededelingen*, 40: 37–44.

Kuss S.E. 1965. Eine pleistozäne Säugetierfauna der Insel Kreta. *Berichte der Naturforschenden Gesellschaft zu Freiburg im Breisgau*, 55 (2): 271–348.

Leonardi P. 1954. Les Mammifères nains du Pleistocène méditerranéen. *Annales de Paleontologie*, 40: 189–201.

Lomolino M.V. 1985. Body size of mammals on islands: The island rule reexamined. *American Naturalist*, 125: 310–316.

Lomolino M.V., Sax D.F., Palombo M.R., van der Geer A.A.E. 2012. Of mice and mammoths: Evaluations of causal explanations for body size evolution in insular mammals. *Journal of Biogeography*, 39: 842–854.

Lomolino M.V., van der Geer A.A.E., Lyras G.A., Palombo M.R., Sax D., Rozzi R. 2013. Of mice and mammoths: Generality and antiquity of the island rule. *Journal of Biogeography*, 40: 1427–1439.

Lyras G.A., van der Geer A., Rook L. 2010. Body size of insular carnivores: Evidence from the fossil record. *Journal of Biogeography*, 37: 1007–1021.

MacArthur R.H., Wilson E.O. 1963. An equilibrium theory of insular zoogeography. *Evolution*, 17: 373–387.

MacArthur R.H., Wilson E.O. 1967. *The Theory of Island Biogeography*. Monographs on Population Biology, Princeton University Press, Princeton.

Matthew W.D. 1918. Affinities and origin of the Antillean mammals. *Bulletin of the Geological Society of America*, 29: 657–666.
Millien V. 2006. Morphological evolution is accelerated among island mammals. *PLoS Biology*, 4 (10): e321.
Palombo M.R. 2007. How can endemic proboscideans help us understand the "island rule"? A case study of Mediterranean Islands. *Quaternary International*, 169–170: 105–124.
Raia P., Barbera C., Conte M. 2003. The fast life of a dwarfed giant. *Evolutionary Ecology* 17: 293–312.
Raia P., Meiri S. 2006. The island rule in large mammals: Paleontology meets ecology. *Evolution*, 60 (8): 1731–1742.
Simpson G.G. 1940. Mammals and land bridges. *Journal of the Washington Academy of Sciences*, 30: 137–163.
Sondaar P.Y. 1977. Insularity and its effect on mammal evolution. In: Hecht M.N., Goody P.L., Hecht B.M. (eds), *Major Patterns in Vertebrate Evolution*. Plenum, New York, pp. 671–707.
van der Geer A.A.E., Lomolino M., Lyras G.A. 2017. Islands before man: The species-area relationship during the late Pleistocene. *Journal of Biogeography*, 44: 995–1006.
van der Geer A.A.E., Lyras G.A., Lomolino M., Palombo M.R., Sax D. 2013. Body size evolution of palaeo-insular mammals: Temporal variations and interspecific interactions. *Journal of Biogeography*, 40: 1440–1450.
van der Geer A.A.E., van den Bergh G., Lyras G.A., Prasetyo U.W., Due R.A., Setiyabudi E., Drinia H. 2016. The effect of area and isolation on insular dwarf proboscideans. *Journal of Biogeography*, 43: 1656–1666.
Van Valen L. 1970. Evolution of communities and late Pleistocene extinctions. *Proceedings North American Paleontological Convention*, 1 (E): 469–485.
Van Valen L.M. 1973. Body size and numbers of plants and animals. *Evolution*, 27 (1): 27–35.
Wallace A.R. 1855. On the law which has regulated the introduction of new species. *Annals and Magazine of Natural History*, 16 (Series 2): 184–196.
Wallace A.R. 1858. On the tendency of varieties to depart indefinitely from the original type. *Journal of the Proceedings of the Linnean Society (Zoology)*, 3 (9): 53–62.
Wallace A.R. 1869. *The Malay Archipelago: The Land of the Orang-Utan and the Bird of Paradise: A Narrative of Travel with Studies of Man and Nature*. Tyron Press, Scotland.
Wallace A.R. 1876. *The Geographical Distribution of Animals; With A Study of the Relations of Living and Extinct Faunas as Elucidating the Past Changes of the Earth's Surface*, 2 vols. Macmillan & Co., London and Harper & Brothers, New York.
Wallace A.R. 1880. *Island Life, or the Phenomena and Causes of Insular Faunas and Floras, Including a Revision and Attempted Solution of the Problem of Geological Climates*. Macmillan, London. Reprinted 1911, AMS Press.
Whittaker R.J., Fernández-Palacios J.M. 2007. *Island Biogeography: Ecology, Evolution, and Conservation*, 2nd edn. Oxford University Press, Oxford.

CHAPTER THREE

Island Faunas: Types and Origins

Evolution of Island Mammals: Adaptation and Extinction of Placental Mammals on Islands, Second Edition.
Alexandra van der Geer, George Lyras and John de Vos.

© 2021 John Wiley & Sons Ltd. Published 2021 by John Wiley & Sons Ltd.

'I do not deny that there are many and grave difficulties in understanding how several of the inhabitants of the more remote islands, whether still retaining the same specific form or modified since their arrival, could have reached their present homes'. (Charles Darwin 1859: 396.)

The composition of island faunas regarding type and number of species and their ecological specialisation depends on many variables. These are, amongst others, the geology and age of the island, the dispersal abilities of the colonising species, the distance to the mainland and island area, the faunal composition as a whole and characteristics of its elements, and the physiography and climate of the island. Naturally, these variables are interwoven and subject to changes over time, influencing each other constantly. For example, the physical geography of the island together with local and global climate puts a heavy stamp on island size in terms of suitable surface area.

Types of Islands

Early studies on fossil island faunas often presented a rather artificial but convenient categorisation of island types, based on palaeo-zoogeographic evidence and urged by the wish to compare fossil faunas mutually and with extant insular faunas. Two types of islands were recognised by Philip Darlington (1957) – continental versus oceanic islands – to which Josep Alcover and colleagues (1998) added a third type – oceanic-like islands. The usefulness of such a classification is limited, partly because of the interchangeability of the terms *continental and oceanic-like islands* (see the following text). In addition, the palaeogeography of many islands is complex, and the same island may belong to different classes through time. The only clear distinction between oceanic islands and the other islands is perhaps that saltwater intolerant taxa – such as salamanders – are unlikely to occur on oceanic islands.

The geological history of an area is in principle reflected in the faunal evolution of that area. For example, when a highly disharmonic fossil fauna consisting of few endemic taxa is found at a site on the mainland, then this area might very well have been an island in the past. Fossil faunas are thus useful as a palaeogeographic tool.

Continental Islands

Continental islands are part of a continental shelf. Generally, they become isolated from the mainland through subsidence of the isthmus of a peninsula. Often, they are separated from the mainland by relatively shallow water. Continental islands are subject to being reconnected with the mainland by a relatively low lowering of the sea level through a land bridge or land span. Many islands with fossil endemic faunas belong to this category, such as Sardinia, Sumatra, and the Californian Channel Islands. Continental islands sometimes drift away due to tectonics and are never reconnected again, such as the Balearics.

Oceanic and Oceanic-like Islands

Oceanic islands arise beneath the sea and are surrounded by water since their origin. Oceanic-like islands, on the other hand, are continental islands that were connected to the mainland in a very remote past and have since remained isolated (Box 3.1). Both types of island – oceanic and oceanic-like – are colonised overseas. Only very few taxa will be successful enough to found a new population because terrestrial mammals are poor dispersers across wide water barriers. Because of the observed similarity in faunal composition, Alcover and colleagues refer to these two types of island as true islands. Examples of oceanic islands are Cyprus, Flores, and the Galápagos. An example of an oceanic-like island is Madagascar, which got separated from Africa already early in the Cretaceous. Subsequent tectonic and volcanic processes may minimise the distance to continent or in rare cases even lead to a connection with the latter. This is the case with Java, which consisted of one or several small oceanic islands in the Early Pleistocene but was large and connected to the mainland in the Late Pleistocene. After this point in geological time, such islands behave like continental islands and are separated again with rising sea levels, such as today (see below).

BOX 3.1

Oceanic-like Islands

The term *oceanic-like* is somewhat unfortunate without a clear indication of the duration of the isolation to distinguish a continental island from an oceanic-like island. Mallorca, one of the Balearic Islands in the Mediterranean, for example, began its existence as part of the Iberian Peninsula from which it broke off in the Early Oligocene as a continental island, in parallel with the Sardinia–Corsica block. During the Messinian salinity crisis of the terminal Miocene, the island could be reached overland from the mainland. Since then, a long-term isolation started, lasting up to the present day. Its lineages of highly endemic bovids (*Myotragus*) and giant dormice (*Hypnomys*) stem from the period of the Late Miocene land connection and thus represent a vicariance effect, not overseas dispersal. Apparently, despite its long-term isolation, Mallorca cannot be considered an oceanic-like island. Crete, on the other hand, with its much shorter isolation, is often classified as an oceanic-like island. Crete was still part of the mainland during the Miocene. During the Pliocene, it submerged for the greater part. During the Pleistocene, the emerged island is considered an oceanic-like island, based on the composition of its faunas which suggest that the colonising populations came overseas and not by land connection.

From Oceanic to Continental

The opposite – the change from oceanic to continental island – is known as well. For example, Gargano, now a peninsula of Italy, and Las Murchas, now part of the mainland of Spain, were islands in the Late Miocene. Tectonics led to collision of their microplates with continental areas, resulting in the extinction of the islands. The process can be less dramatic or can be intermittent. Java, for example, started its geological history in the Pliocene as an oceanic island through volcanic activity. Gradually, the distance to the continental shelf – the Sunda Shelf – decreased through tectonic movements, while the island grew through continuing volcanic eruptions and tectonic uplift. Eventually, the distance was sufficiently small in the early Middle Pleistocene to form a corridor connection with the mainland during colder periods when the sea level was lower. During subsequent warmer periods, sea levels rose again, and Java was again isolated. The last complete connection was during the Late Pleistocene, when a tropical rainforest fauna – shared with Borneo, Sumatra, and mainland Southeast Asia – entered Java. Today, Java is again an (continental) island.

Dispersals to Islands

There are several ways for vertebrates, including mammals, to reach an island and maintain a viable population. Naturally, the different species may disperse differently, and not all dispersal types are available to all species.

Roughly speaking, three main types of dispersal routes to the islands can be distinguished, respectively, over land, over water, and through the air. The first type has in principal no restrictions – except for obligate aquatic taxa – depending on the habitats available on the connecting land corridor (Box 3.2). The second type is available for taxa that can swim, float, or raft on a floating mass across wide water barriers, while the last type is restricted to bats and birds. For invertebrate taxa, hitchhiking on a host during the journey has been recorded, for example, freshwater snails sticking to the feathers of a bird, but this kind of dispersal seems irrelevant for mammals. The total distance may be broken up into smaller units, for which the popular term *island hopping* is sometimes used.

Corridor and Filter Dispersal

The first type of dispersal – over land – is on its turn artificially divided into two routes, corridor dispersal and filter dispersal. The corridor route applies to cases in which faunal interchange between two areas is possible; this is probable for some animals but improbable for others in the case of a filter route or filter bridge. The former route is by definition over land (land bridge), whereas the latter may include (very) short distances over water ('stepping stones', not to be confused with island hopping). Land bridges are supposed to provide the only possible way of dispersal for groups that are intolerant to saltwater, such as freshwater fishes and amphibians.

Continental islands are typically colonised by a normal mainland fauna through corridors (at low sea level) or filters (at higher sea levels). During further disconnection and isolation, the filter becomes stronger and eventually no new direct influx takes place anymore. At this point, the taxa are confined to the island and from here on undergo so-called vicariance, the effect of being separated from the rest of the original group by a geographic barrier, in this case, a body of water. This separation often results in a differentiation into new varieties or species (anagenesis). In case an array of new species arises (cladogenesis), the term *adaptive radiation* is applied. An example of such a vicariance phenomenon is provided by the Pleistocene fauna of the Balearics and perhaps by the ground sloths of the West Indies. Isolation not necessarily leads to speciation. In some rare cases, hardly or no change takes place at all, and some sort of living fossil, or relic taxon, is preserved on the island, often amidst species that did evolve and radiate. This seems the case with the monotremes of Australia but also with the primitive Amami rabbit (*Pentalagus furnessi*) of the central Ryukyu Islands. Paradoxically, islands often function as laboratories for speciation, as we will see, sometimes to the absurd, and at the same time as sanctuaries for the preservation of rare and primitive taxa.

The opposite situation takes place when an island gradually becomes connected to the mainland, and a mainland fauna arrives, first through filter dispersal and later through corridor dispersal. This is seen in Japan where during the Late Pleistocene, a mainland fauna with a large Chinese deer (*Sinomegaceros*) met the endemic dwarf elephant (*Palaeoloxodon naumanni*) and its accompanying fauna, resulting in the latter's extinction.

BOX 3.2

Land Bridges and Land Spans

The theory of vanished intercontinental land bridges and island arcs across ocean basins in order to explain the distribution of species, held in vogue in the early days by, amongst others, Scharff (1912) and Schuchert (1935), became discredited since the acceptance of plate tectonics in the latter part of the twentieth century. Land bridges as explanation behind the presence of organisms on islands are entirely dismissed by some authors, while others hold on to it. The reality lies, as often, most likely somewhere in the middle. A dry land connection is a very reasonable explanation for cases like Sicily during the Late Pleistocene, while it is a totally impossible option for cases like the Galápagos. For Madagascar, though, Robert McCall (1997) suggested the existence of an at least partially exposed land bridge in the Mozambique Channel during the mid-Eocene to the Early Miocene, which consisted of uplifted crustal blocks along the activated fault of the Davie Ridge. In this way, large areas of dry land in the Channel could have functioned as step stones to Madagascar's north-east margin. These areas subsided in the Early Miocene, separating the Malagasy mammals from their African relatives. Extensive ocean floor research, however, could not detect terrestrial deposits along the ridge, indicating that the uplifted blocks were never exposed above the water. In addition, such a bridge would have allowed the dispersal of a far richer fauna than actually took place. Also, this Miocene bridge fails to explain the arrival of Late Pleistocene hippos.

A more likely land bridge, or land span, is found in the Caribbean region. The Aves Ridge, now almost entirely submerged, may have formed a land bridge at the Eocene–Oligocene transition. This theory, the land span theory, is promoted by Ross MacPhee and Manuel Iturralde-Vinent since 1994. Holcombe and Edgar (1990) discussed the Aves Ridge in great detail and showed what the ridge would look like if it were 600 m and 1000 m higher, exposing many islands, or a land bridge if subsidence had been even greater. The ridge would have provided a filter in the first case, as followed by Pindell (1994) and Droxler and colleagues (1998), and a corridor in the latter case, as followed by MacPhee and Iturralde-Vinent.

'Pendel' Dispersal

Closely related to corridor and filter dispersals – and often not distinguishable from these – is the 'pendel' route, named thus by Michael Dermitzakis and Paul Sondaar (1978) after the Dutch word for pendulum, as it swings back and forth. The distance is now easily crossed by some mammals, and subsequent invasions of the island by the same species take place. Genetic exchange with mainland populations takes place on a regular basis (rescue effect). As in the case of filter dispersal, the term *pendel route* actually applies to the taxon itself, because the same distance during the same period may form an insurmountable barrier for other taxa. The main difference between pendel route on one side and corridor and filter dispersal on the other side is that, in the former case, overseas dispersal over a very short distance is assumed.

Sweepstake Dispersal

Long-distance overseas dispersal is generally referred to as *sweepstake dispersal* – the chance crossing of large bodies of water – which proves to be a highly efficient filter. The idea derives from the work done by William Diller Matthew (1918),

> 'These [= the mammalian faunas] I conceive to have arrived at various times during the Tertiary [....], by accidents of transportation, of which the most probable for the mammals would perhaps be the so-called "natural rafts" or masses of vegetation dislodged from the banks of great rivers during floods and drifted out to sea'. (Matthew 1918, 665)

Matthew's idea was reformulated by Gaylord Simpson (1940, 1965) (Figure 3.1) and adopted by Paul Sondaar (1977) and subsequent palaeontologists. Simpson added, 'spread is impossible for the most and very improbable for some, but does occur accidentally'. A sweepstake dispersal is thus on itself not limited to overseas dispersal over large distances – filter dispersal can also provide a case – but the opposite is always true – overseas dispersal over large distances is always a sweepstake dispersal. That is the reason why terrestrial mammals have colonised so

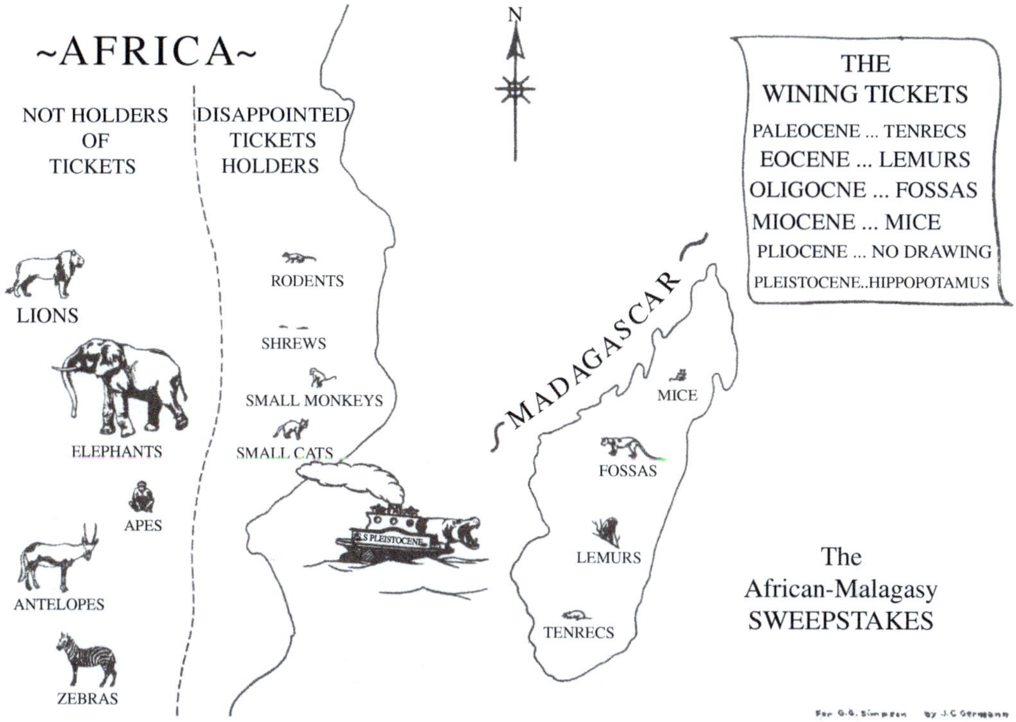

Figure 3.1 Conceptional sketch by Gaylord Simpson (1940) explaining the principles of sweepstake dispersal. Source: From Simpson G.G. 1940, Figure 6. Mammals and Land Bridges. Journal of the Washington Academy of Sciences 30 (1940): 137-163. © Washington Academy of Sciences. Reprinted with permissions of Washington Academy of Sciences.

few isolated islands, as concluded by Lawlor (1986). Natural phenomena such as hurricanes may increase the chance, as Matthew (1918) noted in a footnote 'Tropical storms, as Wallace pointed out years ago, probably play a principal part in transportation of very small animals or their eggs'. (Matthew 1918, 665). Today, this is rarely observed given the extremely short window we live in compared to geological periods spanning thousands or millions of years. A case of cows accidentally dispersing in this way has been reported from Cedar Island in North Carolina. Here, three feral cows were swept into the ocean by Hurricane Dorian in 2019. They were considered lost, until they were rediscovered, happily grazing in good health, a month later on one of the Outer Banks islands, about 6.4 km away. The only way they could have reached their new home was by swimming, being pushed along in the right direction by the storm.

Regarding the water mass to be crossed, the greater the distance and the more challenging the crossing is – unfavourable currents, wave types, temperatures, etc. – the lower the chance is to make it safely to the other site (Box 3.3). That is why, the only large herbivorous animals are giant tortoises instead of ruminants and elephants on remote oceanic islands like the

Galapágos, Mauritius, and the Seychelles. On nearby islands, on the other hand, several herbivores of different size classes may be found together, proboscideans as well as ruminants. This is the case with the central Ryukyu Islands and Honshu, Japan, during the Pleistocene. Where the return voyage is practically impossible, adaptive radiations may evolve from the single coloniser – as seen in the two to four genera of Late Pleistocene giant rats of Flores, all descendants from the earlier middle-sized *Hooijeromys nusatenggara* – or, more commonly, a single lineage undergoing progressive change – as seen in many species of dwarf elephant, for example, the four-tusked stegodon (*Stegolophodon pseudolatidens*) of the late Early and early Middle Pleistocene of Honshu, Japan.

BOX 3.3

Transatlantic Boats?

Rafting was presumably the means of migration of the ancestors of the New World monkeys (Platyrrhini) and the caviomorph rodents from Africa to South America across the Atlantic Ocean some 40 million years ago (Middle Eocene), as suggested by Alain Houle (1999). At that time, the ocean was much less wide, but still between 1000 (50 million years ago) and almost 2000 km (30 million years ago) wide. Recently, Erik Seiffert and colleagues (2020) discovered a third taxon in South America that undertook this almost improbable transatlantic journey: *Ucayalipithecus perdita*, a very small monkey, closely related to Oligocene parapithecid primates of Africa. They estimate the migration might have occurred at the Eocene–Oligocene boundary around 34 million years ago when the sea level lowered.

The raft could have consisted of a vast piece of floating mangrove forests that storms occasionally break off from the tropical African coast. Houle (1998), in his review of oceanic rafting by small vertebrates in the tropical Atlantic, mentioned a floating island measuring 60 m by 23 m, with trees as high as 15 m. Equatorial currents can transport larger floating objects with wind-exposed surfaces in less than two weeks from the river deltas to the South American coast, as summarised by Susanne Renner (2004). In the Early Tertiary, when the Atlantic was narrower than today, such transport probably took less time, increasing survival changes of the passengers.

Evidence of long-distance over-water rafting in the past is limited to small-bodied mammals. Not only do they need less resources and water, but several of these species are able to go into torpor or hibernation, thus drastically reducing their metabolic needs.

Two-way Tickets

No matter how low the changes may be for successful colonisation of an island, and the more so for a subsequent adaptive radiation into a number of new species, even more unlikely is recolonisation of the mainland of one of these new lineages and subsequent radiation into, again, new species (Box 3.4).

Successful Returns

BOX 3.4

Recolonisation of the mainland has been proposed to explain the present distribution of *Anolis*, a group of Caribbean lizards. One group of these lizards has successfully colonised the Americas, as explained by Glor and colleagues (2005) and Nicholson and colleagues (2005). The former concluded that one *Anolis* species of Florida can be traced back to the pre-Pleistocene of Cuba. The latter showed that several dozen *Anolis* species in the continental Neotropics descend from a single Caribbean coloniser. Two-way invasions were also used to explain the phylogeny of a bird lineage in Polynesia and New Guinea by Filardi and Moyle (2005). Dávalos (2007) followed the same line of thinking regarding the phylogeny of extant short-faced fruit bats of the Caribbean and of Central and South America. The latter group of species shares a more recent common ancestor than does the former, which indicates evidence for colonisation from the West Indies and not vice versa. The bats invaded the island prior to the Pleistocene and radiated into several lineages of which one returned to the continent. The way back to the continent has even been brought to the fore in the case of Madagascar and Africa by Raxworthy and colleagues (2002) in respect to chameleons, assuming the common ancestor to crown chameleons had dispersed from Africa to Madagascar in the Late Cretaceous, evolved there, and dispersed back between 90 and 47 million years ago. This study was based on mitochondrial markers of about 40% of chameleon species, using molecular clocks to estimate the time of divergence. A broader analysis of over 90% named chameleon species using dated phylogenies led by Krystal Tolley and colleagues (2013), however, concluded the opposite, namely that chameleons originated in Africa and rafted twice to Madagascar, once some 65 million years ago in the Early Palaeocene, and once in the Oligocene, and not vice versa. During the late Cretaceous and the Oligocene, the oceanic currents flowed from Africa towards Madagascar, so opposite of the situation today.

Reverse colonisation is propagated by Eva Bellemain and Robert Ricklefs (2008). Their basic assumption is that adaptation to island life does not necessarily result in reduced dispersal ability, and that island inhabitants are thus capable to undertake the journey back. However, the examples provided throughout this book demonstrate that most fossil insular taxa clearly had lost their dispersal ability – e.g. flightlessness in birds and short-leggedness in large mammals. The hypothesis of reverse colonisation will not be considered further, as the fossil record lacks any substantial evidence.

Time of the Dispersal

Apart from the dispersal itself – the way in which the founder population reaches the island – there is another interesting aspect of dispersal: time, or when did the dispersal take place. One way to approach the time range, or dispersal window, is by dating the fossiliferous deposit in which fossils of the most primitive form of the endemic lineage are found. This is in most cases impossible, either because no primitive form – that is, closest to the founder species – has been recovered or because no reliable dating can be done. Another method is by using the so-called molecular clock on the basis of DNA or RNA in the living endemic species of the island. The idea was first proposed by Linus Pauling and Emile Zuckerkandl (1962). The working principle, the mechanics of the clock as it were, is based on stochastic events or randomness. Molecular clocks track minor mutations in the cell's DNA or RNA, which not only occur at a constant rate but also randomly. The pattern of randomness can be quantified and used as a predictor. The genetic mutations occurring over time are compared against the fossil record, and the time of split of a species along the evolutionary tree can thus be 'calculated'. At present, molecular clocks are amongst the standard tools for evolutionary biologists, especially when calibrated with dated fossils or deposits.

A possible error though lies in the choice of protein used for the clock. A discrepancy in the clocks based on RNA and DNA, respectively, led to a difference of at least 11 million years in dispersal dates for the Malagasy rodents (Nesomyinae). Jean-Yves Du Bois and colleagues (1996) calculated the arrival of these mice at around 12 million years ago, based on DNA hybridisation. Céline Poux and colleagues (2005), however, calculated a much earlier colonisation (20–25 million years ago) based on RNA. This early date is unlikely because advanced cricetids and murids do not appear in the fossil record before 14 million years ago. Why then their clock gives such an early timing? Du Bois showed that there is a huge difference between the clocks. The 12S rRNA clock indeed suggests an age of 20 million years ago for the dispersal, but the DNA hybridisation clock tells that this happened much later, at ca 8–9 million years ago. The 12S rRNA gene appears to evolve faster in the Nesomyinae than in other tested murids.

The Candidate Species

Few taxa appear to be successful islanders in the long term. The first filter consists of the way of dispersal to the island. For continental islands, this filter is not particularly strong, and, practically, all taxa may pass it. For oceanic islands, the filter is very strong. Practically, only some large herbivores (deer, hippos, and proboscideans), rodents, small primates, bats, birds, and some reptiles (tortoises and crocodiles) are potentially successful over-water dispersers, as we will see. They swim, float, fly, or travel by floating masses. In addition, they have a relatively high potential to survive for prolonged time in open sea.

The second filter that the aspirant colonisers have to pass is simply to survive long enough within a restricted area, often in an entirely different habitat than the one which they were used to. Eventually, only very few taxa survive long enough to become successful colonisers, gradually adapt to their new environment, and evolve into endemic forms.

The Holdovers

In principle, the original founding population of a gradually disconnecting area (corridor and filter dispersals) is identical to the mainland population. Soon after the disruptive event, however, taxa gradually go extinct, either because they are outcompeted by fitter species or simply because of lack of resources and suitable habitat. Resource limitation applies, first of all, to the top predators; they are the first to disappear from the record. Only few taxa, especially small herbivores, are able to survive for a longer time. Other taxa typically found on continental islands that were populated through corridor or filter dispersal are pigs, moles, and lagomorphs. The Middle Pliocene to Early Pleistocene fauna of Sardinia, for example, is mainly characterised by a pig, a mole, a rabbit, and a profusion of bovids, indicators of dispersal over land, and subsequent vicariance. Other typical overland dispersers of this fauna are a hyena and a macaque.

The Swimmers

Fossil faunas of oceanic and oceanic-like islands show that especially elephants, deer, water buffaloes, and less so, hippos were successful in reaching these islands. This is best explained by their shared characters: they swim very well in open sea, and their digestive systems produce gasses which add to their buoyancy. In addition, the trunk of an elephant serves as a snorkel, further enhancing their potential to survive for longer periods in the water. Finally, these animals live and swim in herds. Carnivores, too, can swim, but they do not have the above described floating capacity. Furthermore, if they reach the island, it is often not in a group, and, therefore, the founder population may be too small to establish a viable population. Exceptions are otters, which are frequently found as part of fossil insular faunas. Other taxa are limited to one or two islands only: e.g. wolf-like canids (*Xenocyon*) on Sardinia and Java and tigers on Java.

Especially proboscideans are notoriously good swimmers, as documented by Donald Lee Johnson (1980). This is further confirmed by the presence of their fossils on several islands that were not connected to the mainland by the time of dispersal. For example, extinct stegodons are reported from the Philippines, Sulawesi, Flores, Timor, and Sumba; mammoths from the Californian Channel Islands, Japan, and Crete; and elephants from many Mediterranean islands, the Philippines, and Java. Practically, all islands with a fossil record of the Northern Hemisphere contain a proboscidean species. Deer colonised amongst others Crete, Karpathos, Sardinia, the central Ryukyu Islands, Japan, Java, and the Philippines, while hippopotami managed to reach Cyprus, Crete, Sicily, Malta, Madagascar, and Java (for details, see the relevant chapters). On the latter island, they may even have arrived twice, once overseas during the Early Pleistocene when Java was an oceanic island and once via filter dispersal during the Middle Pleistocene when Java was intermittently connected to the mainland.

Hippos are semi-aquatic but do not swim. Instead, they 'walk' over the bottom of the river or lake in which they live. Only calves aged up to about one year swim actively. In the sea, however, adults would likely float, and swimming hippos have indeed been reported by Frädrich (1968) (see for discussion on hippo dispersal, van der Geer, Anastasakis and Lyras, 2015).

The situation with the pigs is not entirely clear. Generally, pigs were considered evidence of a land connection and, thus, indicators of a corridor or filter dispersal, as in the case of Sardinia. The extant babirusa, however, is an excellent swimmer. This endemic pig is known to often swim in the sea to reach small islands. Roland Melisch (1994) reported that they swam across the 10-km-wide Lake Poso. Presumably, the endemic Pleistocene pigs (*Celebochoerus*) of Sulawesi and the Philippines – which were never connected to the

Asian mainland – had thus reached these islands by swimming. This might theoretically also have been the case for the Antillean sloths, because some sloths seem to have been adapted to marine habitats (*Thalassocnus*), as demonstrated by Muizon and McDonald (1995). The two extant sloth species are excellent swimmers and regularly cross rivers.

Excellent swimmers amongst the rodents are the capybaras (*Hydrochoerus hydrochaeris*), which readily take to water and lead an aquatic life. Their fossils have been recovered from Curacao in the West Indies. Most rodents, however, are unlikely long-distance swimmers, though the extant common or black-bellied hamster (*Cricetus cricetus*) occasionally swims and crosses large rivers in the event of mass population movements forced by food shortages. Before taking to the water, it inflates its cheek pouches with air for greater buoyancy. In addition, the common hamster is large, with a body mass of almost a kilo and a head and body length of up to 340 mm. It can be imagined that at least part of such a migrating group is swept away by the river and ends up in the open sea, eventually perhaps landing on an island, though the changes are undoubtedly extremely limited. The survival changes increase somewhat by the ability of hamsters to enter torpor for up to seven to ten days. Fossil giant island hamsters have been reported only from the Late Miocene of Gargano (*Hattomys* and *Mystemys*) and Tuscany (*Kowalskia*) and the early Early and late Early Pliocene of Mallorca (*Apocricetus darderi* and *Tragomys macpheei*). The latter, however, likely arrived over land during the Messinian Salinity Crisis, and not overseas.

The Rafters

Most proficient amongst mammals in crossing long stretches of open seas are without doubt rodents, found on most oceanic islands, including even the Galapágos Archipelago, 972 km off the shore of continental Ecuador, as shown by Robert Dowler and colleagues (2000). They did not arrive there by swimming, but likely travelled on a floating mass or vegetation mats, known as flotsam. Flotsams, consisting of tree logs, bushes, grasses, and floral detritus, are regularly seen in river mouths where they drift into the open sea, as reported by Guppy (1917), King (1962), and Heatwole and Levins (1972). Currents may push these floral 'boats' further sea inward and eventually lead them to an island where they wash ashore. This is supposed to be the way of transport of insects and small vertebrates such as lizards, mice, and shrews. The distance covered may be huge, for example, the endemic Cuban *Tarentola* lizards are supposed to have rafted all the way from Africa, according to Blair Hedges

(1996). Actual rafting has only been observed in the case of green lizards in the West Indies. A flotsam, which had landed on a beach in Anguilla in October 1995, contained green iguanas, as reported by Censky and colleagues (1998). The journey was supposed to have started in Guadeloupe a month earlier as a result of hurricanes. Hurricanes in the Caribbean are by no means rare, and hundreds of thousands of hurricanes likely occurred since the onset of the Miocene. If only a small fraction, say 0.01%, of these hurricanes caused the transportation of mammals between the islands, even then, tenths of possibilities for dispersal occurred. Within these, only a fraction may eventually have led to successful colonisation.

The past record of endemic island mammals shows that such long-distance over-water rafting was an extremely rare event indeed and has been confirmed only for Madagascar (Box 3.5), Galápagos, Canaries, São Tomé and Príncipe, Christmas Island, and Enggano near Sumatra, as summarised by Jason Ali and Miguel Vences (2019) in a letter to the editor in defence of the possibility of such dispersal. They excluded arrivals in complex areas as the Mediterranean, the West Indies, insular Southeast Asia, and the western Pacific, as these likely involve island hopping, thus reducing the covered distance.

Perhaps, the most extreme evidence of long-distance rafting is that of the Galápagos. It appears that two or three independent colonisations by New World rats and mice (Sigmodontinae) took place, being the extinct, Late Pleistocene giant rice rat *Megaoryzomys*, the extant rice rats *Nesoryzomys indefessus* and *Aegialomys galapagoensis*, and the recently extinct *Nesoryzomys darwini*. Molecular data seems to resolve these latter two genera as sister taxa, in which case, there were only two successful colonisations.

BOX 3.5

Lemurs on a Boat

The theory of rafting on flotsams has been brought to the fore to explain the presence of lemurs on Madagascar, first mentioned by Simpson (1940; Figure 3.1). The resolved lemur phylogeny of the team of Julie Horvath (2008), based on nuclear and mitochondrial DNA, clearly indicates that the time of lemuriform divergence postdates the separation of Madagascar from Africa and India by many millions of years. They calculated the time of dispersal to Madagascar to between 50 and 80 million years ago, which confirm earlier calculations by Anne Yoder and Yang (2004) and Céline Poux and colleagues (2005). This means that lemurs must have arrived by overseas sweepstake dispersal. The same applies to the Malagasy latecomers – the tenrecs, the rodents, and the fossas, – which arrived during the Middle Miocene, as calculated by Poux and colleagues (2005), based on nuclear genes. The prevailing ocean currents ran from Africa to Madagascar in these periods, contrary to today, greatly increasing the chance of a successful trip.

Hibernation may have made the long journey more endurable to the ancestral lemurs. Today, the only hibernation primate worldwide is the fat-tailed dwarf lemur (*Cheirogaleus medius*) of Madagascar. When the season becomes cool and dry, they huddle together in tree holes and go into a winter sleep for up to seven months. Their metabolic activity drops to about 2%, and their body temperature assumes the ambient temperature, as low as about 5°C. This unique feature may have been inherited from an ancestral form, and if so, it may have enabled the overseas journey.

A very peculiar sweepstake dispersal – in the sense of a chance dispersal – by long-distance rafting is provided by the Falkland wolf or fox (*Dusicyon australis*). Its ancestors likely did neither swim or float to the islands nor crossed a persistent land bridge between South America's southern tip and the island group. Darwin held a different explanation for the enigmatic presence of a wolf-like animal on these remote islands:

> 'The Falkland Islands, which are inhabited by a wolf-like fox, come nearest to an exception; but this group cannot be considered as oceanic, as it lies on a bank connected with the mainland; moreover, icebergs formerly brought boulders to its western shores, and they may have formerly transported foxes, as so frequently now happens in the arctic regions' (Charles Darwin 1859, 394).

Instead of drifting on an iceberg on the move, 'land' bridges made of ice could have shortened the distance considerably, or even have connected the Falkland Islands with South America's southern tip. Dispersal over such an icy bridge seems to have been the case elsewhere as well, such as for reindeer to Svalbard (= Spitzbergen) (*Rangifer tarandus platyrhynchus*), which got isolated and evolved dwarf size when the ice bridge disappeared at the end of the last ice age.

Composition of Island Faunas

The depauperate or disharmonic taxonomic composition of the fauna – with poor diversity at higher levels and impressive adaptive radiations at lower levels – of several islands or island groups was already noted by Wallace as early as 1881. Similar observations were done by Matthew (1918), Simpson (1956), and Darlington (1957). In the case of the West Indies, this observation was termed the *central problem* in Caribbean biogeography by Ernest Williams (1989), because he could not explain this taxonomic composition by vicariance alone. The great radiation and morphological diversity seen in, for example, the hystricognath rodents – now filling niches normally occupied by other orders of mammals – is best explained by assuming that these other orders were absent. In the case of vicariance alone, the founder fauna would have been balanced, including all orders that were present at the mainland at that particular time.

Several types of fossil insular faunas are mentioned in the palaeontological literature. These are generally referred to as *balanced mainland faunas, balanced depauperate* (*or impoverished*) *faunas, unbalanced depauperate faunas*, and *unbalanced endemic faunas*. Such classification is artificial though practical, as any classification.

A typical feature of depauperate faunas is that taxonomic diversity is poor, with many major groups absent. At the same time, though, some of the insular genera or species may have undergone radiations to fill in the ecological gaps. The lack of diversity on higher taxonomic levels is balanced as it were by higher than normal diversity on lower taxonomic levels, as compared to the mainland. A further typical feature of a depauperate fauna is the occupation of some endotherm niches by ectotherms, for example, giant tortoises as megaherbivores instead of ruminants. The terms *balanced* and *unbalanced* are applied in relation to the number of carnivore species in relation to herbivore species. A typical mainland fauna is considered balanced, whereas typical insular faunas are devoid of (larger) terrestrial mammalian carnivores and are thus named unbalanced. This lack of mammalian carnivores (see, however, Box 3.6) leaves the niche of top predator open, which is subsequently filled by birds of prey, such as the giant eagles (*Garganoaetus freudenthali*) in Gargano and Haast's eagle (*Harpagornis moorei*) in New Zealand, or reptiles, such as the Komodo dragon (*Varanus komodensis*) in Flores, Rinca, and Komodo. There is a difference in the understanding of the term *balanced* between palaeontology and ecology. The way the term is used by most palaeontologists only means that large carnivores are present and that thus major trophic niches are occupied. In ecology, a balanced fauna means it is ecologically stable, for instance, the trophic niches are not only occupied but also occupied in ratios which do not encourage transitions.

BOX 3.6

Insular Carnivores

In some cases, a large or medium-sized canid species – foxes, dogs, and wolves – forms part of an otherwise typical (unbalanced) Pleistocene Island fauna. These exceptions are just a few, including two large fossil species (*Chasmaporthetes melei* from Sardinia and *Xenocyon merriami* from Java), two medium-sized fossil species (*Cynotherium sardous* from Sardinia and *Xenocyon trinilensis* from Java), one still living species (*Urocyon littoralis* from the Californian Channel Islands), one undescribed and possibly already extinct species (*Urocyon* sp. n. from Cozumel Island of Mexico), and one recently extinct species (*D. australis* from the Falkland Islands). The two Sardinian species – a hyena and a dog from subsequent geological periods – likely arrived through filter dispersal over land, whereas the other species presumably arrived overseas. The dispersal of the Falkland species – a large South-American fox – is still unresolved, and hypotheses so far are a land bridge to Argentina during the last Ice Age, or, more likely, transport on a floating ice berg during the same period as mentioned by Darwin (1859), or, most unlikely, introduction as a pet by the first visitors to the islands. In any case, the extreme rarity of carnivores on islands, except for otters, underlines the stochastic, though taxon-dependent, nature of sweepstake dispersals and vicariance.

Once an insular fauna is established, it has the tendency to remain stable in composition through time. A striking example is the five-million-year long faunal history of the Balearic Islands during which the faunal composition was not affected by intermittent global climatologic changes. On the mainland, on the contrary, dramatic changes took place – extinctions, immigrations, and evolution of new taxa – during the very same period. The arrival of humans on the Balearic Islands approximately 8000 years ago finally caused dramatic changes to the fauna and flora, as on most Mediterranean islands. Generally, insular faunas are only disturbed through extinctions and new arrivals in relation to changes in distance to the mainland, area of the island, human arrival, and, less so, climate (Figure 3.2).

Figure 3.2 Simplified evolutionary history of a hypothetical island. In Stage I, the island is still part of the mainland. The fauna is balanced and of mainland character. In Stage II, the island is formed. The fauna of the newly formed island is at the beginning a subset of the continental fauna of Stage I. However, soon, the fauna melts down and only a few (if any) mammalian species survive. In Stage III, new mammalian species of mainland origin colonise the island. In Stage IV, the island is inhabited by endemic species. Meanwhile, more colonisers arrive from the mainland. In Stage V, the successful establishment of new colonisers to the island, as well as other factors, such as the reduction of the island size or its habitats, leads to extinction of many insular species. In Stage VI, the arrival of humans and the subsequent habitat alternation and introduction of alien species eventually leads to the extinction of the majority (if not all) of the native mammalian species. The sequence of these stages does not need to be a straightforward line. For example, the extinction of insular species of Stage V increases the chance of successful colonisation by mainland species and thus Stages II and III may be repeated. Also, the island may, at a particular time, reconnect to the mainland and thus start again from Stage I. (Drawing George Lyras.)

References

Alcover J.A., Campillo X., Macias M., Sans A. 1998. Mammal species of the world: Additional data on insular mammals. *American Museum Novitates*, 3248: 1–29.

Ali J., Vences M. 2019. Mammals and long-distance over-water colonization: The case for rafting dispersal; the case against phantom causeways. *Journal of Biogeography*, 46: 2632–2636.

Bellemain E., Ricklefs R.E. 2008. Are islands the end of the colonization road? *Trends in Ecology and Evolution*, 23 (8): 461–468.

Censky E.J., Hodge K., Dudley J. 1998. Over-water dispersal of lizards due to hurricanes. *Nature*, 395: 556.

Darlington P.J., Jr. 1957. *Zoogeography: The Geographic Distribution of Animals*. Wiley, New York.

Darwin Ch. 1859. *On the Origin of Species by Means of Natural Selection, or the Preservation of Favoured Races in the Struggle for Life.* John Murray, London.

Dávalos L.M. 2007. Short-faced bats (Phyllostomidae: Stenodermatina): A Caribbean radiation of strict frugivores. *Journal of Biogeography*, 34 (2): 364–375.

Dermitzakis M.D., Sondaar P.Y. 1978. The importance of fossil mammals in reconstruction paleogeography with special reference to the Pleistocene Aegean Archipelago. *Annales Géologiques des Pays Helléniques*, 46: 808–840.

Dowler R.C., Carroll D.S., Edwards C.W. 2000. Rediscovery of rodents (Genus *Nesoryzomys*) considered extinct in the Galápagos islands. *Oryx*, 34 (2): 109–118.

Droxler A.W., Burke K.C., Cunningham A.D., Hine A.C., Rosencrantz E., Duncan D.S., Hallock P. Robinson E. 1998. Caribbean constraints on circulation between Atlantic and Pacific Oceans over the past 40 million years. In: Crowley T.J., Burke K.C. (eds), *Tectonic Boundary Conditions for Climate Reconstructions*. Oxford University Press, New York, pp. 160–191.

Du Bois J.-Y., Rakotondravony D., Hänni C., Sourrouille P., Catzeflis F.F. 1996. Molecular evolutionary relationships of three genera of Nesomyinae, endemic rodent taxa from Madagascar. *Journal of Mammalian Evolution*, 3: 239–260.

Filardi C.E., Moyle R.G. 2005. Single origin of a pan-Pacific bird group and upstream colonization of Australasia. *Nature*, 438: 216–219.

Frädrich H. 1968. Das Flusspferd. In: Grzimek B (ed.), *Grzimeks Tierleben: Enzyklopädie des Tierreiches*, vol. 13: 120–141. Kindler Verlag, Zürich.

Glor R.E., Losos J.B., Larson A., 2005. Out of Cuba: Overwater dispersal and speciation among lizards in the *Anolis carolinensis* subgroup. *Molecular Ecology* 14: 2419–2432.

Guppy H.B. 1917. *Plants, Seeds, and Currents in the West Indies and Azores*. Williams and Northgate, London.

Heatwole H., Levins R. 1972. Biogeography of the Puerto Rican Bank: Flotsam transport of terrestrial animals. *Ecology*, 53: 112–117.

Hedges S.B. 1996. The origin of West Indian amphibians and reptiles. In: Powell R., Henderson R. (eds), *Contributions to West Indian Herpetology: A Tribute to Albert Schwartz.* Society for the Study of Amphibians and Reptiles, Ithaca, New York, pp. 95–127.

Holcombe T.L., Edgar N.T. 1990. Late Cretaceous and Cenozoic evolution of Caribbean ridges and rises with special reference to paleogeography. *International Symposium on Biogeographical Aspects of Insularity, Rome, 18–22 May 1987. Accademia Nazionale dei Lincei, Atti Dei Convegni Lincei*, 85: 611–626.

Horvath J.E., Weisrock D.W., Embry S.L. Fiorentino I., Balhoff J.P., Kappeler P., Wray G.A., Willard H.F., Yoder A.D. 2008. Development and application of a phylogenomic toolkit: Resolving the evolutionary history of Madagascar's lemurs. *Genome Research*, 18: 489–499.

Houle A. 1998. Floating Islands: A mode of long-distance dispersal for small and medium-sized terrestrial vertebrates. *Divers Distrib*, 4: 201–216.

Houle A. 1999. The origin of platyrrhines: an evaluation of the Antarctic scenario and the floating island model. *American Journal of Physical Anthropology*, 109: 541–559.

Johnson D.L. 1980. Problems in the land vertebrate zoogeography of certain islands and the swimming powers of elephants. *Journal of Biogeography*, 7: 383–398.

King F.W. 1962. The occurrence of rafts for dispersal of land animals into the West Indies. *Quarterly Journal of the Florida Academy of Sciences*, 25: 45–52.

Lawlor T.E. 1986. Comparative biogeography of mammals on islands. *Biological Journal of the Linnean Society*, 28: 99–125.

MacPhee R.D.E., Iturralde-Vinent M. 1994. First Tertiary land mammal from Greater Antilles: An early Miocene sloth (Xenartha, Megalonychidae) from Cuba. *American Museum Novitates*, 3094: 1–13.

Matthew W.D. 1918. Affinities and origin of the Antillean mammals. *Bulletin of the Geological Society of America*, 29: 657–666.

McCall R.A. 1997. Implications of recent geological investigations of the Mozambique Channel for the mammalian colonization of Madagascar. *Proceedings of the Royal Society of London (Series B)*, 264: 663–665.

Melisch R. 1994. Observation of swimming babirusa in Lake Poso, central Sulawesi, *Indonesia. Malayan Nature Journal*, 47: 431–432.

Muizon C.D., McDonald, H.G., 1995. An aquatic sloth from the Pliocene of Peru. *Nature*, 375: 224–227.

Nicholson K.E., Glor R.E., Kolbe J.J., Larson A., Hedges S.B., Losos J.B. 2005. Mainland colonization by island lizards. *Journal of Biogeography*, 32: 929–938.

Pindell J.L. 1994. Evolution of the Gulf of Mexico and the Caribbean. In: Donovan S.K., Jackson T.A. (eds), *Caribbean Geology: An Introduction.* The University of the West Indies Publishers' Association, Kingston, Jamaica, pp. 13–39.

Poux C., Madsen O., Marquard E., Vieites D. R., de Jong W.W., Vences M. 2005. Asynchronous colonization of Madagascar by four endemic clades of primates, insectivores, carnivores and rodents as inferred from nuclear genes. *Systematic Biology*, 54: 719–730.

Raxworthy C.J., Forstner M.R.J., Nussbaum R.A. 2002. Chameleon radiation by oceanic dispersal. *Nature*, 415: 784–787.
Renner S., 2004. Plant dispersal across the tropical Atlantic by wind and sea currents. *International Journal of Plant Science*, 165 (4 Suppl.): S23–S33.
Scharff R.S. 1912. *Distribution and Origin of Life in America*. New York: Macmillian.
Schuchert C. 1935. *Historical Geology of the Antillean: Caribbean Region or the Lands Bordering the Gulf of Mexico and the Caribbean Sea*. J. Wiley and Sons Inc., Chapman and Hall, New York, London.
Seiffert E.R., Tejedor M.F., Fleagle J.G., Novo N.M., Cornejo F.M., Bond M., de Vries D., Campbell K.E. Jr. 2020. A parapithecid stem anthropoid of African origin in the Paleogene of South America. *Science*, 368: 194–197.
Simpson G.G. 1940. Mammals and land bridges. *Journal of the Washington Academy of Sciences*, 30: 137–163.
Simpson G.G. 1956. Zoogeography of West Indian land mammals. *American Museum Novitates*, 1759: 1–28.
Simpson G.G. 1965. *The Geography of Evolution*. Capricorn Books, New York.
Sondaar P.Y. 1977. Insularity and its effect on mammal evolution. In: Hecht M.N., Goody P.L., Hecht B.M. (eds), *Major Patterns in Vertebrate Evolution*. Plenum, New York, pp. 671–707.
Tolley K.A., Townsend T.M., Vences M. 2013. Large-scale phylogeny of chameleons suggests African origins and Eocene diversification. *Proceedings of the Royal Society B*, 280: 20130184.
van der Geer A.A.E., Anastasakis G., Lyras G.A. 2015. If hippopotamuses cannot swim, how did they colonize islands: A reply to Mazza. *Leithaia*, 48: 147–150.
Wallace A.R. 1881. *Island Life*. Harper, New York.
Williams E.E. 1989. Old problems and new opportunities in West Indian biogeography. *In*: Woods C.A. (ed.), *Biogeography of the West Indies: Past, Present, and Future*. Sandhill Crane Press, Gainesville, Florida, pp. 1–46.
Yoder A., Yang Z. 2004. Divergence dates for Malagasy lemurs estimated from multiple gene loci: Geological and evolutionary context. *Molecular Ecology*, 13: 757–773.

PART II

The Islands and Their Faunas

In this part, island faunas are presented per island or island group. Each chapter starts with a short introduction on geology and palaeogeography, followed by an overview of the island's history of palaeontology, a geochronological treatment of the insular faunas – biozones and faunal units – and ending with a section on the peculiarities and evolution of the most important endemic species. Madagascar and the West Indies are treated somewhat different, due to the lack of a fossil record with sufficient depth. Their faunas are treated taxonomically. Only those islands are discussed in detail that either yielded several faunas of different ages or a single fauna with a number of elements. Islands from which only one taxon is known are briefly mentioned elsewhere. The discussed islands are, in this order, Cyprus, Crete, Gargano, Sicily, Malta, Sardinia, the Balearics, Madagascar, Java, Flores, Sulawesi, the Philippines, Japan, the Ryukyu Islands, the Californian Channel Islands, and the Greater and the Lesser Antilles.

The chronology of the Cenozoic era. The listed numerical ages follows the chart drafted for the International Commission on Stratigraphy by Kim Mikkel Cohen and colleagues in January 2020 (Cohen and colleagues, 2013; updated). In June 2009, the International Commission on Stratigraphy agreed upon lowering the base of the Pleistocene Epoch such that it includes the Gelasian Age. This book follows the new scheme. However, the older cited literature (prior to 2009) follows the old definition of the Plio-Pleistocene boundary and thus often cite as Late Pliocene what we now define as part of Early Pleistocene.

CHART before June 30, 2009			CHART January, 2020		
Age	**Epoch**			**Age**	**Time in millions of years**
	Late	Pleistocene	Late		0.0117
	Middle	Pleistocene	Middle	Chibanian	0.129
Calabrian	Early	Pleistocene	Early	Calabrian	0.774
Gelasian	Late	Pliocene (before 2009) / Pleistocene (2020)	Early	Gelasian	1.80
Piacenzian	Middle	Pliocene	Late	Piacenzian	2.58
Zanclean	Early	Pliocene	Early	Zanclean	3.600
					5.333

Epoch		**Age**	**Time in millions of years**
Miocene	Late	Messinian	5.333
Miocene	Late	Tortonian	7.246
Miocene	Middle	Serravallian	11.63
Miocene	Middle	Langhian	13.82
Miocene	Early	Burdigalian	15.97
Miocene	Early	Aquitanian	20.44
Oligocene	Late	Chattian	23.03
Oligocene	Early	Rupelian	27.82
Eocene	Late	Priambonian	33.9
Eocene	Middle	Bartonian	37.8
Eocene	Middle	Lutetian	41.2
Eocene	Early	Ypresian	47.8
Palaeocene	Late	Thanetian	56.0
Palaeocene	Middle	Selandian	59.2
Palaeocene	Early	Danian	61.6
			66.0

CHAPTER FOUR

Cyprus

Evolution of Island Mammals: Adaptation and Extinction of Placental Mammals on Islands, Second Edition.
Alexandra van der Geer, George Lyras and John de Vos.
© 2021 John Wiley & Sons Ltd. Published 2021 by John Wiley & Sons Ltd.

Cyprus is the largest oceanic island of the Mediterranean and had a unique fossil fauna. This fauna, Pleistocene in age, was extremely depauperate, consisting only of the world's smallest hippopotamus, a dwarf elephant, and likely bats. The time of arrival of the genet and mouse is unknown. The faunal stasis was dramatically interrupted in the Early Holocene with the extinction of all Pleistocene species.

Geology and Palaeogeography

Cyprus is the third largest Mediterranean island, situated opposite the coasts of Turkey and Syria, and one of its few oceanic islands. The island owes its existence entirely to volcanism, which explains the many copper mines from which the name of the island is said to have been derived. During the Cretaceous – about 120 million years ago, Cyprus was a mountain range on the bottom of the Tethys Sea. Much later, during the Miocene, this part of the oceanic crust was gradually uplifted when the African and the Eurasian plates collided. This resulted amongst others in the emergence of the island's main mountainous area, the Tróodos Massif, and later, near the end of the Miocene, the Pentadaktylos Range. Eventually, around the beginning of the Pleistocene, Cyprus emerged above the water and formed the present-day island. Cyprus was thus never connected to the mainland and has a long history of isolation since the Miocene, as evidenced by geological data.

Historical Palaeontology

The first recorded discovery of vertebrate fossils was by the fifteenth-century Cypriot historian Leontios Machairas. His *Chronicle of Cyprus* is known from two manuscripts only: one from the Oxford Library, copied in 1555 at Paphos, Cyprus, and one from Venice, Italy, printed in 1873 by Sathas. In this extensive work, Machairas (1555) mentioned petrified bones, visible at the surface at Casa Epifani (=Kazaphani) on Mount Pentadactylos, south-east of Kyrenia (Figure 4.1). The bones were said to be locally known as the remains of 300 Maronites – members of one of the Syrian Eastern Catholic Churches – who had fled from Syria to escape persecution, but shipwrecked at Cyprus' inhospitable coast. The villagers referred to them as the *Manifested Saints* (Agioi Fanentes).

The next report that survived is the one from the astronomer and cartographer Benedetto Bordone from Padua, Italy (see also Chapter 2). In his *Isolario* (1528), he described a hill at Kyrenia at the foot of Mount Pentadactylos, which entirely consisted of

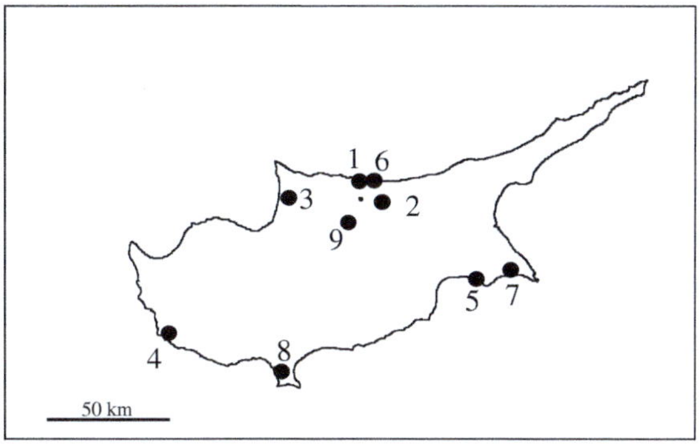

Figure 4.1 The most important Pleistocene localities of Cyprus: (1) Kyrenia, (2) Kazaphani Agios, Chrystostomos and Bellapais (3) Agia Irini, (4) Paphos, (5) Xylophagos, (6) Agios Georgios, (7) Agia Napa and Cape Pyla, and (8) Akrotiri Aetokremnos.

the bones of animals and humans. The book was an illustrated guide about the then known islands for seafarers. Bordone wrote that the local population powdered the bones from the hill to make a potion against all sorts of diseases. Likely, Makhairas and Bordone had seen similar deposits.

The French traveller André Thevet (1554) and the English traveller Fynes Moryson (1617) both mentioned a cave dedicated to the Seven Sleepers at Paphos at the south-western coast. The latter wrote:

> 'On Sunday the nineteenth of May, we came to the first Promontory of the Iland Cyprus, towards the West, and after eight houres sayling, we came to the old City Paphos (or Paphia), now called Baffo, and the wind failing us, and gently breathing upon this Castle of Venus, we hovered here all the next night, gaining little or nothing on our way. This place is most pleasant, with fruitful hills, and was of old consecrated to the Goddesse Venus, Qgeene of this Iland [......]. A mile from this place is the Cave, wherein they faigne the seven sleepers to have slept, I know not how many hundred yeeres'. (F. Moryson, 1617, Part 1, Book 3, Chapter 1)

The ancient legend about the Seven Sleepers stems from the sixth century or perhaps earlier. According to one variety of the story, once, seven young Christians were locked up in a cave at Ephesus by the non-Christian emperor in order to die. By miraculous intervention, however, they did not die but slept for two or three hundred years. One day, a shepherd passed nearby and discovered the cave. He opened the cave and saved the young men who had not even aged in the meantime. The legend is known in Asia Minor, and also on Cyprus.

The Cypriot monk and philosopher Neophytos Rhodinos, though, wrote in his book on the heroes, warriors, and saints of Cyprus, published posthumous in 1659, that the petrified remains in the cave at Paphos did not belong to the Seven Sleepers but to the Seven New Martyrs instead, local Cypriot saints who have nothing to do with those of Ephesus.

The Dutch traveller Cornelius de Bruyn (1698) also mentioned a bone deposit in the Pentadactylos range with petrified remains of animals and humans. One of his plates figures a human upper arm, which in reality is a fossil of the extinct Pleistocene dwarf hippopotamus. De Bruyn considered these remains as evidence of the biblical Flood, a common explanation for fossils in that time. This view was shared by some Cypriots as well, because an exposed fossiliferous layer – in reality, a collapsed rock shelter and perhaps the same as de Bruyn's deposit – at Agia Irini in the Pentadactylos range is locally known as Dragondovounari, literally the Dragon Hill. According to local myth, dragons had drowned here during the Flood and got petrified thereafter. De Bruyn noticed similar petrified bone remains at Agios Chrysostomos near Koutsovendi, not far from Kazaphani and possibly the same as noticed earlier by Makhairas.

A variation on the theme of 300 Marionites who had fled from Syria was recorded by the British consul Alexander Drummond (1754). The coastal bone breccia, locally known as Agios Fanentis (Manifested Saints), not far from the ruins of the Bellapais monastery at the foot of the Pentadactylos, was believed to consist entirely of the petrified remains of pirates or barbarians, who had come to the island for plunder but shipwrecked. As punishment for their planned crime, their bones turned into stone. Some good-hearted fellows though were converted to the Christian faith and lived happily thereafter on the island and eventually became saints.

Luigi Palma di Cesnola (1877), who visited Cape Pyla between 1872 and 1875, published a description of two caves with bones at the village Xylophagos, one of which he said was dedicated to Forty Saints (Agioi Saranda).

In 1901, Dorothea Bate went on her own, without formal training and with minimal support, to explore the island for fossils and zoological specimens. Her attention was drawn to several regions, such as the Pentadactylos range, by the earlier descriptions of others. She was successful because she discovered 12 new fossiliferous caves, for the greater part near Kyrenia. She also relocated earlier mentioned sites, such as the cave of the Forty Saints at Cape Pyla, already mentioned by Cesnola.

A year later, Bate returned to Cyprus with funding from the Royal Academy of Sciences. This time, she found remains of a dwarf elephant, which she (Bate, 1903a) described as *Elephas cypriotes*. In the same year (1903b), she described subfossil remains of a genet as *Genetta plesictoides*. Part of the material she collected was purchased by Dr F. Krantz Rheinisches

Figure 4.2 Rock-cut chapel dedicated to St Fanourios at a locality with abundant remains of the Cypriot pygmy hippo (*Hippopotamus minor*). (Photograph archive: John de Vos.)

Mineralien-Kontor in Bonn (Germany) – the first and still running geological warehouse worldwide, from where the material went to various collections.

Much later, in the mid-1965s, Paul Sondaar and Bert Boekschoten went to Cyprus to relocate Bate's sites and collect material. At Agios Georgios, west of Kyrenia, they found a collapsed rock shelter halfway the steep sea cliff with remains of the Manifested Saint (Agios Fanourios) (Figure 4.2), just below the Byzantine rock-cut grotto. Perhaps, this is the place mentioned by Bordone, Cesnola, and, maybe later, Drummond. The site contains hundreds and hundreds of bones, firmly stuck in the breccia. The villagers go there to collect petrified bones since centuries to make powerful potions as a remedy against most diseases and discomforts. Based on the geomyth, Boekschoten and Sondaar (1972) renamed the small hippo of Desmarest as *Phanourios minor* (now again *Hippopotamus minor*) in honour of the saint to whom the fossil bones were attributed by the local villagers.

Several smaller expeditions to collect hippo fossils were undertaken, including by William Dreghorn in the late 1960s or early 1970s; Malatesta and Girotti of the University of Rome ('La Sapienza') in 1970–1971; J. Nienhuis, Richard Reyment, Horace Richards, and Richard White, all in 1971 but independently; and David Reese in 1973–1974.

Biozones and Faunal Units

Despite the incredible amount of mammalian fossils, the species diversity amongst this rich material is extremely low and constant throughout its history. Cyprus provides a classic example of stasis in an endemic island fauna. The single faunal unit consisted of only two herbivores – a dwarf elephant and a pygmy hippopotamus – which hardly underwent any further change once established on the island. Small mammals – a mouse and a genet – are only known from the latest Pleistocene or the beginning of the Holocene. The long-term stasis and extreme impoverishment are best explained by the large distance to the nearest mainland in combination with the oceanic origin of the island.

Pleistocene

The single fossil fauna of Cyprus consists of a pygmy hippopotamus (*H. minor*; Figure 4.3; Plate 2) and a dwarf elephant (*Palaeoloxodon cypriotes*; Figure 4.4) during the Late Pleistocene. A second, slightly larger elephant species (*Palaeoloxodon xylophagou*) is likely an older chronospecies of the same lineage. More than forty Pleistocene localities are known, the majority of which is found at the foot of the Pentadactylos range and at the coast near Kyrenia roughly between Koutsovendis and Ayia Irini. Another important cluster is located at Cape Pyla, especially near Xylophagos.

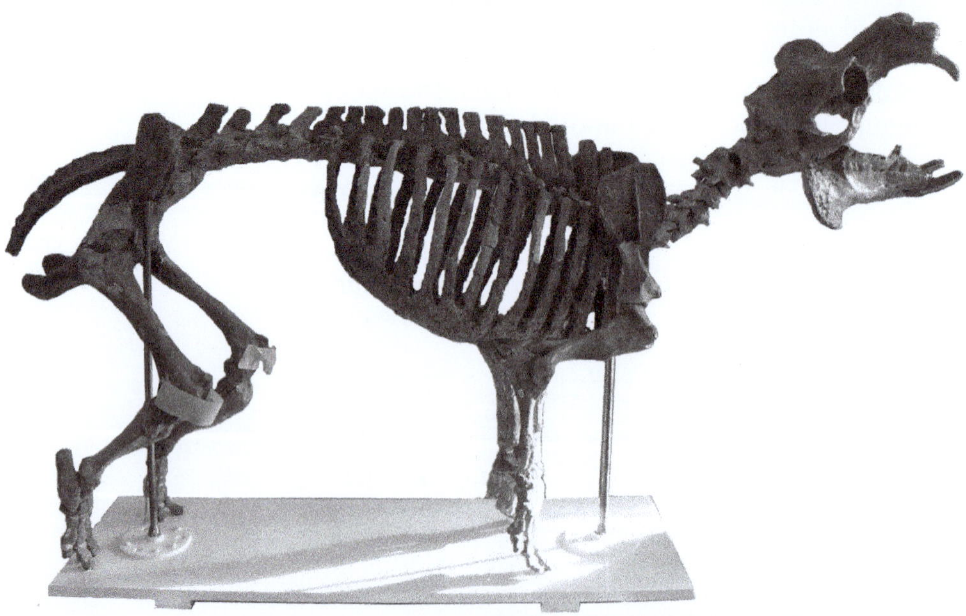

Figure 4.3 A composite mounted skeleton of the Cypriot pygmy hippo (*Hippopotamus minor*). Museum of Geology and Palaeontology, University of Athens, Greece. (Photograph: George Lyras.)

Figure 4.4 Proximal part of a femur of the Cypriot dwarf elephant (*Palaeoloxodon cypriotes*) compared with that of a woolly mammoth in antero-lateral view. National Museum of Natural History, Leiden. (Photograph: Eelco Kruidenier.)

Datings of hippo fossils by David Reese (1995) yielded ages of about 22,000 years ago for the site Agios Phanentis and about 767,400 years ago for Asproyi. However, these datings are now considered unreliable, as they were based on bone collagen. Geological evidence indicates a Late Pleistocene age for most of the sites, and a late Middle Pleistocene age for the site with the larger elephant, as discussed by Athanassiou and colleagues (2015). The extremely small size of the hippo (see below) is also in favour of an earlier, Middle Pleistocene colonisation, which would on its turn be in line with the geological age of other Mediterranean insular hippos (Malta, Sicily, and Crete).

Towards the end of the Pleistocene or at the beginning of the Holocene, the elephants and hippos were accompanied by a genet (*Genetta plesictoides*; Figure 4.5), mice (*Mus cypriotes*), a

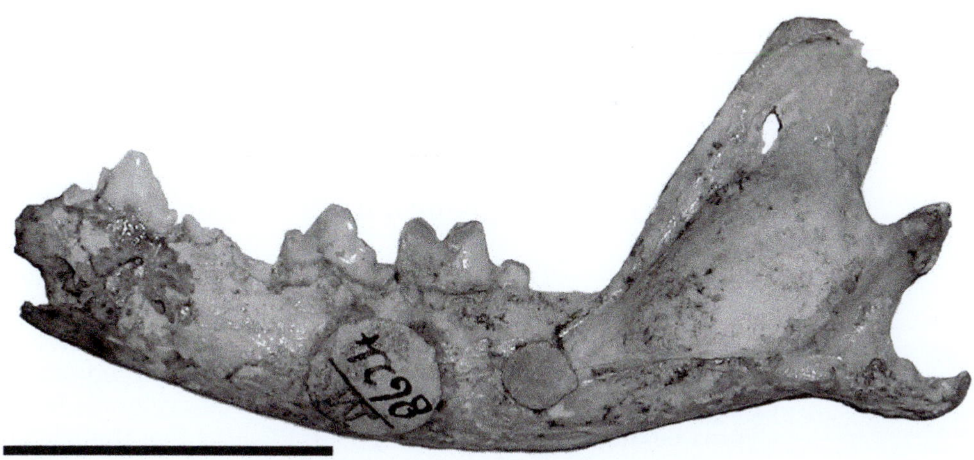

Figure 4.5 Mandible of the Late Pleistocene Cypriot genet (*Genetta plesictoides*). Scale bar: 2 cm. Natural History Museum, London. (Photograph: George Lyras.)

fruit bat (*Rousettus aegyptiacus*), and some undetermined bats. The rare murid remains are not found in association neither with hippos nor with elephants. Molecular clock analysis by Thomas Cucchi and colleagues in 2006, though, indicates an arrival during the Middle Pleistocene. Fossils of the fruit bat (*Rousettus aegyptiacus*) are known only from the locality Vokolosspilios. The Cypriot genet is known only from the localities Vokolosspilios, Aetokremnos, and Aghia Napa. It is not certain whether the genet fossils originate from the Late Pleistocene layer with the pygmy hippo or from the Holocene layer. In any case, the Cypriot genet is larger in comparison with the common genet (*Genetta genetta*). In addition, its dentition was more apt for cutting, possibly indicating a larger percentage of meat in its diet, as proposed by Theodorou and colleagues (2007). If so, it is likely to assume that it co-occurred with the mice on the island. The causes for the end of the island's dwarfed megafauna remain unclear (Box 4.1).

Holocene

The Holocene fauna is in principle a mainland fauna, though rather impoverished, unbalanced, and endemic on at most subspecies level. The only native carnivore today is the red fox (*Vulpes vulpes indutus*) and the only large herbivore is the Cypriot mouflon (*Ovis orientalis ohion*), a feral descendent of an ancient introduction. Small mammals, like the commensal rats and mice (*Rattus rattus*, *Rattus norvegicus*, and *Mus musculus*), the lesser white-toothed shrew (*Crocidura suaveolens cypria*), and 15 bat species are abundantly present. Presumably, these mammals, too, except for the bats, were introduced through human agency.

The ancient Cypriot shrew (*Crocidura suaveolens praecyprioa*) is likely a Holocene intrusion as well. Its subfossil remains were recovered from a Bronze Age settlement, as reported by Jelle Reumer and Urs Oberli (1988).

Peculiarities and Evolution of Endemic Mammals

Dwarf Hippopotamus

The pygmy hippo of Cyprus is the smallest hippo of all known insular hippos (Figure 4.3, Plate 2) with a body mass reduction of about 96%. More than 90% of the mammalian fossils from Cyprus belong to this pygmy hippo, indicating that the island offered quite favourable conditions to this herbivore. The pygmy hippo could afford a very small body size because neither terrestrial predators nor food competitors were around.

Its lophodont dentition shows adaptations towards a higher degree of browsing instead of grazing contrary to what is seen in the dentition of its ancestor, *Hippopotamus amphibius* or *H. antiquus*. The development of a more lophodont dentition is in fact a return to a more primitive condition, as it is a typical feature of stratigraphically older hippopotamids. This is why, the American palaeontologist Henry Osborn (1921) assumed that Cyprus was the earliest isolated island of the Mediterranean Sea and thus harboured the most primitive species through vicariance. However, the primitive dentition should not be seen as retention of an ancient character, but as an adaptation to a new environment and thus secondarily derived. This is confirmed by another feature – the loss of the fourth premolar in many specimens – which was interpreted by Sondaar (1977) as due to a change in jaw movement with a larger transversal component.

The Cypriot pygmy hippo could climb and walk very well on rugged terrain, but likely was no good runner, as indicated by the anatomy of its limb bones. The toes are those of an unguligrade walker as bovids and deer, not those of a plantigrade walker as its ancestor. Its small size, together with its browsing food habits and unguligrade locomotion, is suggestive of a shift towards the deer niche. Notwithstanding the hippo's extreme adaptations, it is a direct descendant of mainland hippopotamus – *H. amphibius* or *Hippopotamus antiquus* depending on the time of arrival, and thus its generic name is *Hippopotamus*, not *Phanourios*.

The nomenclatorial confusion around the Cyprus dwarf hippo starts in the early nineteenth century. The famous French anatomist and palaeontologist Georges Cuvier (1804, 1824) described several fossil hippopotamuses of various sizes, which he named accordingly. One of his hippos was of a small size, 'petit hippopotame fossile', but, unfortunately, he did not know where it had come from. In his view, the fossilisation and the surrounding matrix indicated that it originated from between Dax and Tartas in southern France. The small hippo remained further unnoticed in the collection of the natural history museum at Paris till Charles Forsyth Major's attention was drawn to them again after seeing Dorothea Bate's materials, excavated by her in 1901. She had sent him some fossils with a note saying that they possibly belonged to some sort of pig. Forsyth Major realised that these 'pig' fossils were extremely similar to those depicted and described by Cuvier (1804) as a small hippo ('hippopotame petit') and given the name *H. minor* by Desmarest (1822). Forsyth Major (1902) concluded that the Paris' material obviously originated from Cyprus, not from France.

BOX 4.1

The Last Hippos

Towards the end of the Pleistocene or at the beginning of the Holocene, humans arrived on the island, but whether or not they initiated the extinction of the megafauna is still a matter of debate since Charles Ducloz's (1968) suggestion, followed by David Reese (1974, 1989), and Alan Simmons (1988). Our book, however, is not the place to present all arguments and data, not only because this is not within the scope of this book but also because such an extensive overview already exists, written by Simmons and associates (2000). In their book, all the then available data concerning the faunal extinction in Cyprus are presented. From these data, they conclude that humans gave the final blow to the 'mini'-megafauna, as is evident from the subtitle of the book: *Pygmy Hippopotamus Hunters of Cyprus*. The arguments entirely focus on the findings at the collapsed rock shelter Aetokremnos, literally the Vulture's Cliff, on the Akrotiri peninsula. These findings are surely spectacular, though controversial. The site was discovered in 1961 by the schoolboy David J. Nixon. Here, evidence for one of the oldest human occupations of the Mediterranean, dated to the tenth millennium before present, was found in the 1980s. The cultural horizon is known as the Aetokremnos Phase. In this oldest cultural layer, blackish hippo bones are present, while in the overlying younger layers, attributed to the Neolithic, no hippo bones are found but burnt shells and bird and hare bones instead. On the platform in front of the former cave, many hippo fossils were found, plus a few elephant fossils, shells, and primitive tools. The idea is that the slow-moving hippos and dwarf elephants formed an easy and nutritious prey for the pre-Neolithic colonisers. The pygmy hippos could not cope with extensive hunting because of their slow rate of reproduction. By the time Neolithic people arrived on the island, the endemic fauna was already driven to extinction by the earlier inhabitants.

However, some counterarguments were raised by others. For example, Sandra Olsen (2000) found neither cut marks nor any evidence for marrow extraction in the more than 16,000 bone fragments from Aetokremnos I she studied. The observed damage and fragmentation are alternatively explained by trampling. Furthermore, the small blades and abundant thumbnail scrapers found at the side seem better suited to processing molluscs than to butchering hippos. Regarding the burnt hippo bones, she demonstrated that also bones buried beneath a fire are cracked, split, and blackened. This is further proven by a more extensive experiment done by Bennet and Klippel (1995). The presence of internally burnt bones confirms the hypothesis that these bones were already cracked and split naturally before the fire above them was lit. Furthermore, the black colour of many hippo bones is not caused by fire but by a chemical staining involving manganese oxides.

Simmons and colleagues conclude that the Aetokremnos Phase is unique to Cyprus and not found on the mainland. However, Arie Boomert and Alistair Bright (2007) demonstrated that the premise of island archaeology that insular human societies show intrinsic characteristics essentially dissimilar from those on mainlands is incorrect. Generally, island people have intense and frequent contact with people on other islands and/or on mainland coasts, though not necessarily with other people on the same island. Islanders are certainly not by definition isolated. The absence from Akrotiri of larger choppers and blades can thus not be explained simply by an autochthonous cultural development under isolation.

Dwarf Elephants

The Cypriot dwarf elephant (Figure 4.4; *Palaeoloxodon cypriotes*) stood about 1.4 m at the shoulder. It had evolved a body mass reduction of about 98%, relative to its giant straight-tusked mainland ancestor, *Palaeoloxodon antiquus*, similar to the smallest Sicilian dwarf elephant. Fossils of the Cypriot dwarf elephant are very rare, especially compared to the abundance of hippo fossils. At the rock shelter of Agia Napa, George Theodorou excavated fossils of dwarf elephants and dwarf hippos at the same site from the same layer, proving their co-existence.

Excavations at the beginning of 2000 at a site near Xylophago yielded evidence for the presence of a larger elephant as well, as reported by Athanassiou and colleagues (2014). A year later, they named it *Palaeoloxodon xylophagou* after the site. The relation of this larger species to the smaller dwarf elephant – ecomorph, large size variation, or chronomorph – is as yet unresolved.

References

Athanassiou A., Reese D., Iliopoulos G., Herridge V., Roussiakis S., Mitsopoulou V., Tsiolakis E., Theodorou G. 2014. The endemic elephants of Cyprus: A reconsideration of their variation and taxonomy based on new fossil finds. In: Kostopoulos D.S., Vlachos E., Tsoukala E. (eds), *VIth International Conference on Mammoths and their Relatives, Grevena – Siatista, Greece 2014. Scientific Annals, School of Geology, Aristotle University of Thessaloniki, Greece*, 102: 24.

Athanassiou A., Herridge V., Reese D.S., Iliopoulos G., Roussiakis S., Mitsopoulou V., Tsiolakis E., Theodorou G., 2015. Cranial evidence for the presence of a second endemic elephant species on Cyprus. *Quaternary International*, 379: 47–57.

Bate D.M.A. 1903 (a). Preliminary note on the discovery of a Pigmy Elephant in the Pleistocene of Cyprus. *Proceedings of the Royal Society of London*, 71: 498–500.

Bate D.M.A. 1903 (b). On an extinct species of genet (*Genetta plesictoides*, sp. n.) from the Pleistocene of Cyprus. *Proceedings of the Zoological Society of London*, 1903 (2), 121–124.

Bennet J., Klippel V. 1995. Thermal alteration of subsurface faunal remains. *Society for American Archaeology, Abstracts of the 60th Annual Meeting: 36.*

Boekschoten G.J., Sondaar P.Y. 1972. On the fossil mamalia of Cyprus, I & II. *Proceedings of the Koninklijke Nederlandse Akademie van Wetnschappen (Series B)*, 75 (4): 306–338.

Boomert A., Bright A.J. 2007. Island archaeology: In search of a new horizon. *Island Studies Journal*, 2 (1): 3–26.

Bordone B. 1528. *Libro di Benedetto Bordone. Nel qual si ragiona di tutte l'Isole del mondo, con li lor nomi antichi & moderni, historie, favole, & modi del loro vivere.* Niccolo Zoppino, Venice. In *facsimile, Edizioni Aldine, Modena, 1982.*

Cucchi T., Orth A., Auffray J.-C., Renaud S., Fabre L., Catalan J., Hadjisterkotis E., Bonhomme F., Vigne J.-D. 2006. A new endemic species of the subgenus *Mus* (Rodentia, Mammalia) on the Island of Cyprus. *Zootaxa*, 1242: 1–36.

Cuvier G. 1804. Sur les ossement fossiles d'hippopotame. *Annales du Muséum National d'Histoire Naturelle, Paris*, 5: 99–122.

Cuvier G. 1824. *Recherches sur les Ossements Fossiles de Quadrupeds où l'on rétablit les Caractères de Plusieurs Animaux dont les Révolutions du Globe ont détruit les Espèces*, 4 vols, 3rd edn, Paris.

de Bruyn C. 1698. *Reisen door de vermaardste delen van Klein Asia: de Cylanden Scio, Rhodus, Cyperus, Metelino, Stanchio.* Delft.

Desmarest A.G. 1822. *Mammalogie ou description des espèces de mammifères.* Supplèment. Agasse, Paris.

Di Cesnola L.P. 1877. *Cyprus: Its Ancient Cities, Tombs and Temples. A Narrative of Researches and Excavations during Ten Years' Residence in that Island.* Harper and Brothers, New York.

Drummond A. 1754. *Travels through Different Cities of Germany, Italy, Greece, and Several Parts of Asia.* London.

Ducloz C. 1968. Les formation Quaternaires de la region de Klepini (Chypre) et leur plac dans la chronologie du Quaternaire Mediterraneen. *Archives des Sciences Genève*, 20: 123–198.

Forsyth Major C.I. 1901. Exhibition of, and remarks upon, the skull of a new fossil mammal (*Enhydrictis galictoides*). *Proceedings of the Zoological Society of London*, 1901 (2): 625–628.

Forsyth Major C.I. 1902. On the Pigmy Hippopotamus from the Pleistocene of Cyprus. *Proceedings of the Zoological Society of London*, 2 (1): 238–239, 2 (2): 107–112.

Machairas L. 1555. *Chronicle of Cyprus*. Manuscript, Paphos. [In Greek.]

Moryson F. 1617. *An Itinerary written by Fynes Moryson gent. first int he Latine tongue, and then translated by him into English: Containing His Ten Yeeres Travell through the Twelve Dominions of Germany, Bohmerland, Sweitzerland, Netherland, Denmarke, Poland, Italy, Turky, France, England, Scotland, and Ireland.* Diuided into III parts. Glasgow: James MacLehose 1907. Reprint of London: John Beale, 1617.

Olsen S. 2000. Investigation of Phanourios bones for evidence of cultural modification. In: Simmons A.H. (ed.), *Faunal Extinction in An Island Society: Pygmy Hippopotamus Hunters of Cyprus.* Interdisciplinary Contributions to Archaeology, Kluwer Academic, Dordrecht, pp. 230–237.

Osborn H.F. 1921. *The Age of Mammals in Europe, Asia and North America.* New York: The MacMillan Company.

Reese D.S. 1974. *New Sites and Finds from the Pleistocene in Cyprus.* Ms. on file, Cyprus American Archaeological Research Institute, Nicosia.

Reese D.S. 1989. Man meets Hippo. *American School of Oriental Research Newsletter* (Fall), 5.

Reese D.S. 1995. The Pleistocene vertebrate sites and fauna of Cyprus. Ministry of Agriculture, natural resources and environment. *Geological Survey Department Bulletin*, 9: 1–203.

Reumer J.W.F., Oberli U. 1988. Shrews (Mammalia: Soricidae) from a Bronze Age deposit in Cyprus, with the description of a new subspecies. *Bonner zoologische Beiträge*, 39 (4): 305–314.

Rhodinos N. 1659. *On Heroes, Generals, Philosophers, Saints and other important figures who came from the Island of Cyprus.* Rome. [in Greek.]

Simmons A.H. 1988. Extinct pygmy hippopotamus and early man in Cyprus. *Nature*, 333: 554–557.

Simmons A.H. and associates 2000. *Faunal Extinction in an Island Society. Pygmy Hippopotamus Hunters of Cyprus.* Interdisciplinary contributions to Archaeology. Kluwer Academic/Plenum, New York.

Sondaar P.Y. 1977. Insularity and its effect on mammal evolution. In: Hecht M.N., Goody P.L., Hecht B.M. (eds), *Major Patterns in Vertebrate Evolution.* Plenum, New York, pp. 671–707.

Theodorou G.E., Roussiakis S.I., Athanassiou A., Giaourtsakis I., Panayides I. 2007. A Late Pleistocene endemic genet (Carnivora, Viverridae) from Aghia Napa, Cyprus. *Proceedings of the 11th International Congress, Athens, May 2007. Bulletin of the Geological Society of Greece*, 2007: 201–208.

Thevet A. 1554. *Cosmographie de Levant. Ian de Tournes et Guillaume.* Gazeau, Lyon.

CHAPTER FIVE

Crete

Evolution of Island Mammals: Adaptation and Extinction of Placental Mammals on Islands, Second Edition.
Alexandra van der Geer, George Lyras and John de Vos.
© 2021 John Wiley & Sons Ltd. Published 2021 by John Wiley & Sons Ltd.

The fossil fauna of Crete, the largest island of Greece, is mainly known for its Pleistocene endemic taxa, amongst which are dwarf deer, dwarf elephants and mammoths, dwarf hippopotamuses, and giant mice. Two different main biozones can be discerned, separated by an almost complete faunal turnover towards the end of the Middle Pleistocene. The pre-Pleistocene mammalian fossils are of a Late Miocene age and are part of a mainland fauna.

Geology and Palaeogeography

Crete was not always an island. Up to the Late Miocene, Crete was connected to the mainland of Asia Minor. This is, amongst others, evidenced by fossil remains of mainland faunas (for an overview, see Jan van der Made, 1996). Gradually, Crete broke up into smaller pieces, starting during the late Late Miocene. As a result, marine deposits and late Tortonian (8–7 million years ago) foraminifera beds overlay the remains of the Miocene terrestrial faunas. The region became largely submerged during the warmer Pliocene – known as the mid-Pliocene transgression – and consisted of at least three islands; the remainder of the former land area was transformed into a shallow sea with shoals, as shown by Dermitzakis and Sondaar (1978) and Sondaar and Dermitzakis (1982). No fossils of terrestrial mammals have been reported from this period. Later, the region emerged again, and finally, in the Early Pleistocene, Crete got its present configuration and could be colonised only by overseas sweepstake dispersal.

Historical Palaeontology

The first known report on fossils from Crete is that of the British traveller Richard Pococke (1745), who described a fossiliferous cave on the peninsula of Chania. The cave consists of two chambers, one of which houses a chapel of St George, possibly because of the presumed bone relics of the saint that are found here. Pococke's early exploration of this cave was followed a century later by the French geologist Félix Victor Raulin (1869). In his extensive work on the physical geography of Crete, he reported unusual, large mammal bones in the cave floor. In 1904, Dorothea Bate returned to the cave in the hope to find fossils of dwarf hippos and elephants, but found only ruminant fossils at the base of the cave walls, now known as *Candiacervus*, the Cretan deer. The larger fossils apparently had disappeared by then.

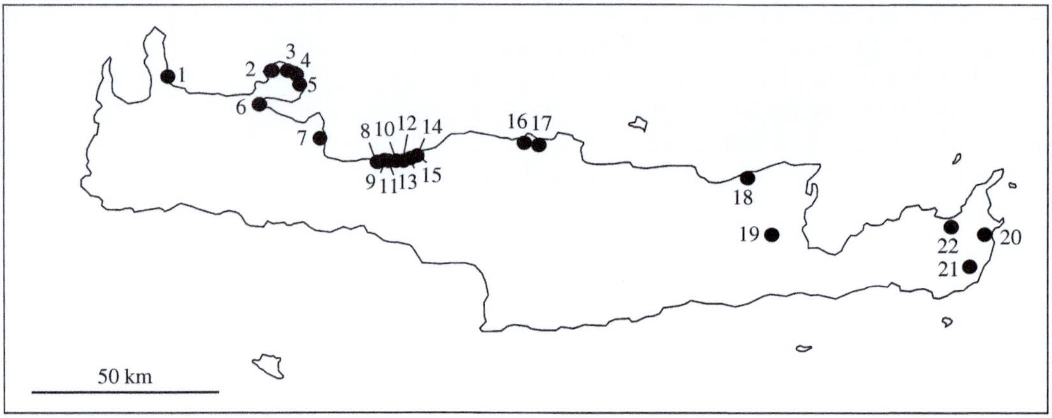

Figure 5.1 The most important Pleistocene localities of Crete. (1) Tripiti cave, (2) Stavros cave, (3)Agios Georgios cave, (4) Avlaki, (5) Cape Meleka, (6) Chania/Souda cave deposit, (7) Liko caves, (8) Gerani caves, (9) Bate cave, (10) Zourida cave, (11) Mavromouri caves, (12) Simonelli cave, (13) Koumpes cave, (14) Panayia caves, (15) Rethymnon fissure, (16) Mpali caves, (17) Kalo Choraphi cave, (18) Milatos caves, (19) Katharo basin, (20) Karoumpes, (21) Kato Zakro, and (22) Siteia I. For a complete overview, see Lax, 1996.

During his stay in Crete, Raulin received some hippopotamus fossils from the people of Kritsa village at the foot of the Dikti Mountains. They had been found at the Katharo (Figure 5.1), an upland basin high in the mountains above the village. He forwarded the fossils to the French anatomist and zoologist Henri Marie Ducrotay de Blainville, who ascribed them in 1847 to *Hippopotamus amphibius*, the extant hippo. Two years before, Richard Owen had mentioned fossil hippo teeth from the same Katharo basin, which had been forwarded to him by Captain Graves from Mr Ittas from Heraklion. Owen (1845) applied the name *Hippopotamus medius*, based on their intermediate size. The species name *medius* had been applied by the French naturalist Anselme Gaétan Desmarest (1822) to remains of what he and his teacher Georges Cuvier had considered a fossil medium-sized hippo ('moyen Hippopotame fossile') from Saint-Michel-en-Chasine, western France. In reality, the remains belonged to the Miocene sea cow *Metaxytherium medium*. Owen simply applied the same name to the Cretan species, based upon the illustration of a left lower jaw by Cuvier (1804) and the size.

In 1865, the British navy officer Thomas Abel Spratt explored Crete, after his successful excavations at Malta a few years before. One of his motifs was to investigate the Katharo basin, the locality of dwarf hippopotamuses. He applied the name *Hippopotamus minor* to the Katharo fossils, using the name Desmarest (1822) had given to the dwarf hippo which Cuvier (1804) had illustrated as 'hippopotame petit'; neither Cuvier,

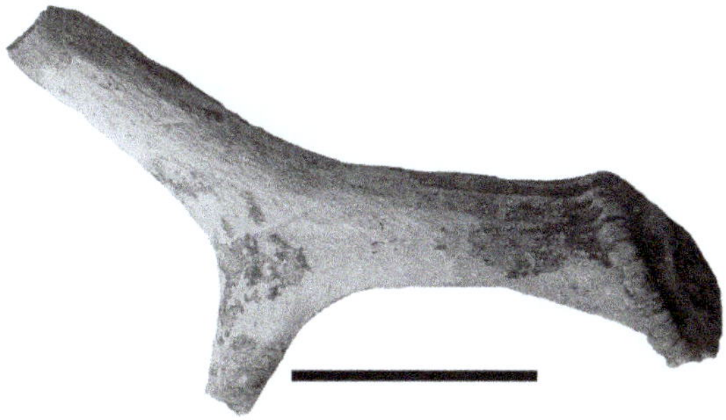

Figure 5.2 Example of an antler of Cretan deer on the basis of which Kuss named it *Anoglochis cretensis*. Right antler, posterior view; brow tine (T1) pointing downward. Scale bar: 5 cm. Museum of Geology and Palaeontology, University of Athens, Greece. (Photograph: George Lyras.)

Desmarest, nor Spratt knew that the fossils on which the drawing was based originated from Cyprus (see Chapter 4). Spratt discovered several more fossiliferous deposits, amongst others, a cave deposit of 36 m long between Chania and Souda and Tripiti caves on the Rodopou peninsula. Spratt sent some fossils from the latter cave to Hugh Falconer of the British Natural History Museum in London for identification. Falconer considered them the remains of a goat, a roebuck, and a small dormouse (*Myoxus*) but did not publish his findings. Unfortunately, his identifications cannot be verified because at present the cave is largely devoid of its former bone bed, and in addition, the remaining bone bed is very hard, which makes it difficult to obtain good specimens.

The Italian geologist Vittorio Simonelli was the first who identified elephant remains on Crete. In 1893, he found fossils of a quite large elephant in three caves near Rethymnon (Agios Antonios cave, Koulouridi cave, and Bali cave), which, a year later, he described as *Elephas priscus*, a species which is now considered synonymous with *Elephas* (= *Palaeoloxodon*) *antiquus* since Vaufrey's publication (1929). Simonelli (1908) mentioned additional *E. priscus* remains from Koumpes cave 2 at the small gorge Grida Avlaki near Koumpes (= Gumbes).

Simonelli also found some deer remains, which he described as *Anoglochis cretensis* (Simonelli, 1907). Pleistocene deer of France had been classified into two groups by Croizet and Jobert (1826–1828) on the basis of the position of the brow-tine. Deer with a sub-basilar brow-tine constituted the genus *Anoglochis* and deer with a basilar brow-tine *Cataglochis*. Simonelli followed the French classification when he described his Cretan deer (Figure 5.2). He noted similarities with the Sardinian megacerine deer, which was then attributed to *Eucladoceros*, a subgenus of *Anoglochis* in the system of Croizet and Jobert. In his view, the Cretan deer was a Miocene relict.

Bate went in the spring of 1904 to Crete to relocate the deposits described by Spratt, Raulin, and Pococke in order to find fossils of dwarf elephants and dwarf hippopotamuses as she had found at Cyprus. In total, she (Bate, 1905) described no less than thirteen caves and karst holes, but in only a very few of them she found elephant remains. The trip was not easy, travelling by foot and on horseback in cold and windy weather and sleeping in flea-infested rooms. Against official regulations, she managed to ship a lot of material back to London. In the years thereafter, she devoted part of her time to the study of this material. She described the dwarf mammoth as *Elephas creticus*, now known as *Mammuthus creticus*, based on a molar from the collapsed coastal cave Cape Melekas site 1 (Bate, 1907). Although she recognised the similarity with molars of the southern mammoth (*Mammuthus meridionalis*), she decided in favour of an ancestry from *E.* (=*Palaeoloxodon*) *antiquus* because she had found remains of the latter elsewhere on the island at Karoumpes 3 (= Charoumbes) near the sea on the mountain Traostalos, Siteia. A few years thereafter, she described a lower murid jaw from Sphinari cave as type for *Mus catreus* (Bate, 1912).

In the same paper, she described an upper jaw from Cape Melekas site 1 with all three molars preserved as belonging to a large rat (*Epimys*, now *Rattus*) and another one from Spratt's cave between Chania and Souda as belonging to a spiny mouse (*Acomys*). Much later, she (Bate, 1942) revised her original list of 1912 and assigned the latter two species to two new, endemic species, respectively, *Rattus kiridus* and *Mus minotaurus*. Kuss and Misonne (1968) transferred the rat species as well as Bate's *Mus catreus* to the new genus *Kritimys*.

A second endemic mouse species from Stavros cave was described by David Mayhew (1977) as *Mus bateae*, in honour of Dorothea Bate. Mayhew suggested that this species was ancestral to Bate's large mouse *Mus minotaurus*.

Remains of an elephant were described as *Loxodonta creutzburgi* by Sigfried Kuss (1965), in honour of the German geologist Creutzburg, based on fossils from Kalo Chorafi cave near the village of Sisses, Rethymnon. He noted that the size and molar wear pattern were similar to that of the African elephant, *Loxodonta africana* and hence was his choice of the generic name. He attributed Simonelli's (1894) *E. priscus* to his new species *creutzburgi* as well. In the view of Kuss, all Cretan elephants are loxodontine, including Bate's pygmy mammoth. Kuss considered five proboscidean species for Crete, including *E.* (= *Palaeoloxodon*) *antiquus* of which he had found large quantities from the Panagia cave I, near the village of Koumpes. The cave is named after Mary, Mother of Christ, but whether or not the large bones were connected with Her is unclear. Later, Kuss (1973) moved his five Cretan species to the genus *Elephas* as was already proposed by Raymond Vaufrey (1929) in his revision of the Mediterranean dwarf elephants. This designation – or, alternatively, *Palaeoloxodon* – was maintained till Dick Mol and colleagues (1996) proposed a mammoth ancestry for Bate's pygmy species *creticus*.

In the early 1970s, figurative rock art at the small Asphendou cave, high in Sphakia mountains of western Crete, came to the attention of archaeologists. The small cave is located on the side of the Asphendou gorge that forms a north–south pass through the mountains. The figures represent, amongst others, a herd of animals. Thomas Strasser and colleagues (2018) analysed the petroglyphs with photogrammetry and argued that the basal images date to the Palaeolithic period and represent a herd of the endemic but extinct Cretan dwarf deer, *Candiacervus*.

Biozones and Faunal Units

Excavations in the many cave localities – of which Michael Dermitzakis (1977), Elliot Lax (1996), and George Iliopoulos and colleagues (2010) gave overviews – revealed a clear zonation in the Pleistocene fossil record. In total, two main biozones, consisting of, respectively, three and two subzones, are recognised by Mayhew (1977, 1996), based on the phylogeny and the evolutionary stage of the murid species (Figure 5.3). Sondaar and colleagues (1986), de Vos and Dermitzakis (1986), and Dermitzakis and de Vos (1987) elaborate on this and confirm the presence of a single major faunal turnover during the Middle Pleistocene. The disappearance in the faunal record of the Cretan mouse (*Kritimys*) and the appearance of the common mouse (*Mus*) characterise this turnover. At the same time, the Cretan mammoth and Cretan dwarf hippo are replaced by the Cretan dwarf elephant and the Cretan deer. The two murid lineages – *Kritimys* and *Mus* – follow each other in time, as is evidenced by the cave Milatos 3, where remains of the former were found in older, cemented cave sediments, while remains of the latter were found in uncemented clayish deposits overlying and laterally adjacent to the cemented deposits. Based on the murids, two biozones are recognised. There is a minimal temporal overlap between the two biozones – *Kritimys* zone and the *Mus* zone – because fossils from the oldest species of the latter (*Mus bateae*) are found together with those of the youngest species of the former (*Kritimys catreus*) in the same block of cemented sediment near Stavros cave, according to Mahyew.

Early–early Middle Pleistocene

Typical fauna elements of this period – the *Kritimys* zone – are the giant Cretan mouse (*Kritimys*) with two or three

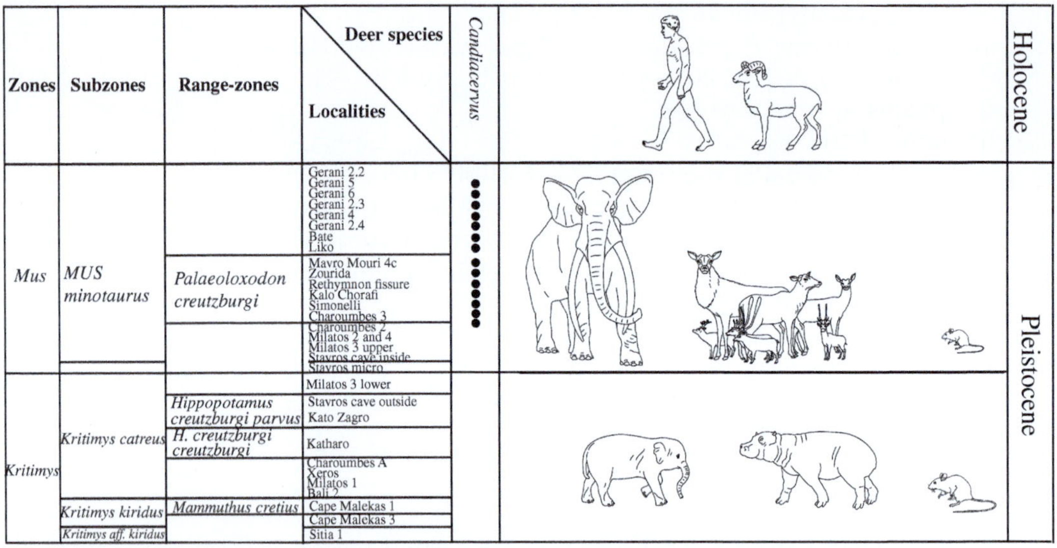

Figure 5.3 Stratigraphic scheme showing the land vertebrate faunal succession of Crete. Characteristic elements of the successive faunal units are shown to the right.

chronospecies, the Cretan dwarf mammoth (*Mammuthus creticus*) and Creutzburg's dwarf hippo (*Hippopotamus creutzburgi*), apart from the Caspian turtle (*Mauremys* cf. *caspica*) and the marsh frog (*Pelophylax* cf. *ridibundus*). Absolute dates (amino acid racemisation [AAR] and electron spin resonance [ESR]) were provided for hippopotamus molar fragments from Katharo, and range between 375 and 850 thousand years ago. The radiocarbon dating (^{14}C) on hippo bone from the Katharo indicates a much younger age of only about 12,500 years ago, but these are considered unreliable by Reese and colleagues (1996) and in conflict with the tooth enamel dating and geology of the site.

The most famous dwarf hippo locality is the Katharo basin (Figure 5.4). This basin is at a height of 1100 m above the sea, which is not exactly a place where one would expect to find hippos. The other sites with hippo bones are all collapsed coastal caves in which the fossils are found in situ in cemented breccias. The amount of fossils from these caves is very small, and the material is often badly preserved, with the exception of Kato Zakros at the east coast. The earliest finding of the dwarf hippo is at Siteia in eastern Crete, the oldest Pleistocene locality. Here, also, the earliest form of the giant Cretan mouse is found. In younger localities, the dwarf hippo is found together with the intermediate form of the Cretan mouse, but never with the youngest, largest form, and neither with *Mus* – the mouse of the following biozone – as observed by Andries Spaan (1996).

Figure 5.4 Mandible of the Cretan dwarf hippo (*Hippopotamus creutzburgi*), occlusal view. Katharo, Middle Pleistocene. Scale bar: 5 cm. Natural History Museum, London. (Photograph: George Lyras.)

Sites with *Kritimys*, all coastal, are characterised by solid, cemented red-coloured breccias. The only site with a dwarf mammoth (*Mammuthus creticus*) is Akrotiri Melekas 1, or Cape Melekas (= Cape Kiamou), already discovered by Bate (1905). Sondaar (1973) found here a lower molar with an attached *Kritimys kiridus* jaw (Figure 5.5), as reported in Mol and colleagues (1996), which indicates that the dwarf mammoth was already present at the beginning of the *Kritimys* biozone. The deposits here are somewhat younger in age than those of Siteia, based upon the evolutionary stage of *Kritimys*.

The Cretan mouse is represented by two or three species, together forming a lineage from the geologically oldest *Kritimys* cf. *kiridus*, found at Sitia, to *Kritimys kiridus* and the youngest species *K. catreus*. The different *Kritimys* species are never found together.

Late Middle–Late Pleistocene

The typical fauna elements of this biozone – the *Mus* zone – are two chronospecies of mice, *Mus bateae* and *Mus minotaurus* – unrelated to the Cretan mice of the previous biozone – Creutzburg's dwarf elephant (*Palaeoloxodon creutzburgi*; Figure 5.6), Cretan deer (*Candiacervus*, with the

Figure 5.5 Lower molar of the Cretan dwarf mammoth (*Mammuthus creticus*), Cape Meleka, occlusal view (posterior end to the left). Scale bar: 5 cm. Attached to it is a jaw of the smaller form of the Cretan rat (*Kritimys kiridus*) (not visible here). Museum of Geology and Palaeontology, University of Athens, Greece. (Photograph: George Lyras.)

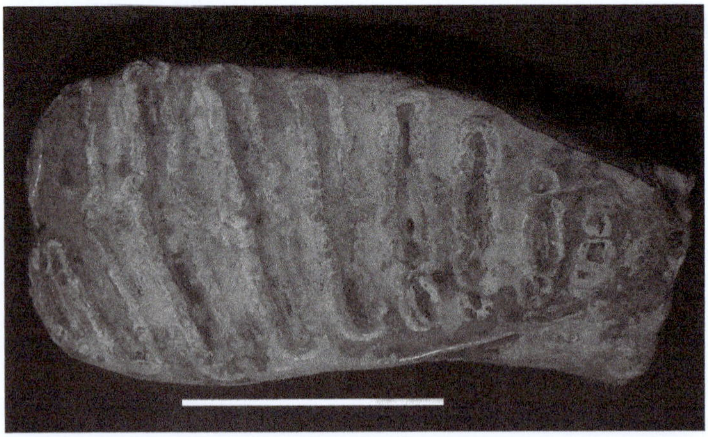

Figure 5.6 Lower molar of the Cretan elephant (*Palaeoloxodon creutzburgi*), Katharo, occlusal view (posterior end to the right). Scale bar: 5 cm. Museum of Natural History, University of Crete. (Photograph: George Lyras.)

eight species, namely *ropalophorus*, *devosi*, *listeri*, *reumeri*, *cretensis*, *rethymnensis*, *dorothensis*, and *major*; Figure 5.7), the Cretan otter (*Lutrogale cretensis*) (Plate 3), and the Cretan shrew (*Crocidura zimmermanni*). The herpetofauna contains amongst others the endemic Cretan marginated tortoise

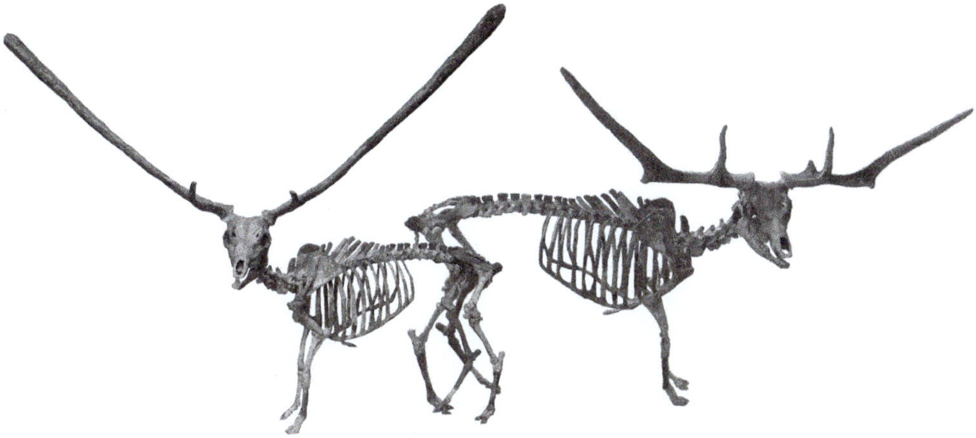

Figure 5.7 Two composite mounts of dwarf deer species of Late Pleistocene Crete. Left: *Candiacervus ropalophorus* (size 1), right: *Candiacervus devosi* (size 2) (virtual mount). (Photograph: George Lyras.)

(*Testudo marginata cretensis*). The avifauna contains two endemic species, a long-legged walking owl (*Athene cretensis*), and a yet undetermined species of bird – originally classified as a golden eagle, *Aquila chrysaetos simurgh*, but rejected by Antoine Louchart and colleagues in 2005. Other birds do not differ from mainland birds.

The faunal turnover between this period and the previous may be incomplete, because close to the hippopotamus remains at the Katharo, a molar and a tusk fragment of the elephant and an antler fragment of the deer were found as briefly reported by Dermitzakis and colleagues (2007). Furthermore, remains of the Cretan shrew were found together with fossils of the last *Kritimys* species at Xeros, as reported by Lax (1996). This implies that there was an overlap between the biozones to an unknown extent.

The youngest and oldest absolute date for this biozone are, respectively, 33,600 and 127,500 years ago, based on electron spin resonance (ESR) dating on tooth enamel of the deer from Koumpes cave and Gerani cave 5 by Reese and colleagues (1996; we here dismiss the results obtained from bone).

The most important localities yielding fossils of this biozone are the cave Liko, situated below the village Likotinara at the northeast coast of the Chania Province, and a series of coastal caves near the village Gerani in the Rethymnon area. Thousands of fossils were retrieved from these caves by German-Greek and Dutch-Greek teams (see Historical Palaeontology). In Liko cave, Gerani 4, and Simonelli cave, the fossiliferous layer is a red-brown, sandy clay. In all localities, the elements are disarticulated.

The fauna of the second Pleistocene period went extinct just before or after the arrival of the first humans on the island. Problems of dating and the lack of Palaeolithic artefacts and human remains associated with the fauna obscure this point. An indication of temporal overlap of humans and endemic fauna is the presence of Palaeolithic artefacts from the Plakias region (dated to about 94 to 114 thousand years ago) and rock art representing a herd of what seem Cretan dwarf deer at the Asphendou rock shelter, reported by Thomas Strasser and colleagues (2010, 2018), respectively.

In any case, the fauna of the second biozone was already completely extinct by the time of the Aceramic Neolithic level of Knossos and replaced by newcomers who came together or along with the humans. The new fauna included ancestors of the extant native species, such as the kri-kri, agrimi or Cretan ibex (*Capra aegagrus cretica*), the Cretan stone marten (*Marten foina bunites*), and the Cretan badger (*Meles meles arcalus*). The Holocene fossils of these two mustelids are intermediate in morphology between the living forms and their Near Eastern relatives, as pointed out by Karel Steensma and David Reese (1996). This indicates an eastern origin of the human settlers and the accompanying fauna, as already suggested by the archaeologist John Evans (1968).

Peculiarities and Evolution of Endemic Mammals

Dwarf Mammoth

The dwarf mammoth is extraordinarily small compared to mainland mammoths with its shoulder height of only 1.5 m, and it is the world's smallest mammoth as announced by Victoria Herridge and Adrian Lister (2012). Most molars of this mammoth have a low and wide crown with broad cement areas and thick and simple enamel ridges. Its limbs are very short in relation to body length, when compared with living elephants. It has been suggested that its sturdy and solid limbs, together with a possible higher degree of bone fusions (synostosis), indicate an adaptation to a more uneven and rugged terrain. Mol and colleagues (1996) suggested the Early Pleistocene southern mammoth (*Mammuthus meridionalis*) as its most likely mainland ancestor, which was confirmed by Herridge and Lister (2012). If so, the Cretan dwarf mammoth evolved a body size reduction of about 96%, as calculated by Mark Lomolino and colleagues (2013). A mammoth ancestry for the Cretan pygmy species was also confirmed earlier by Nikos Poulakakis and colleagues (2006), based on ancient DNA dated back to about 800,000 years ago. However, a year later, Binladen and colleagues questioned the validity of their conclusions because the warm climate of that period would have hampered the preservation of DNA. Also, Orlando and colleagues (2007) questioned the transfer of the species *creticus* to *Mammuthus*, because Poulakakis put mammoths as sister taxon to African elephants and not to Asian elephants, contrary to what complete mitochondrial genomes, nuclear genes, and genomic screenings have suggested. Notwithstanding, morphological characters of the molar all point to a mammoth, not an elephant. Already, Bate (1907) admitted that the molars of the pygmy form most closely resembled those of *Mammuthus meridionalis*, but she concluded that it must be related to the straight-tusked elephant (*P. antiquus*), because a similar evolution from a large *antiquus* to a pygmy form had occurred on Sicily as well. This is a nice example of how theory and interpretation of the data may influence taxonomy. Now we know that also at Sicily, the smallest form predated the larger forms and not vice versa.

Dwarf Elephants

The Cretan dwarf elephant (*Palaeoloxodon creutzburgi*) (for dwarf elephants on the other Greek islands, see Box 5.1) was a bit over half as tall as the mainland species. Its average body mass has been estimated at 3200 kg by Gary Burness and colleagues (2001). Lomolino and colleagues (2013) calculated a body mass reduction of more than 60% – using a body mass of 3063 kg for the Cretan species. The species *creutzburgi* was reduced to at most subspecies level by Nikos Poulakakis and colleagues (2002), based on the ground that measurements of molars ascribed to *creutzburgi* fall within the size range of *P. antiquus* when female specimens are included. On the other hand, they did not take morphology of the molars into account and did not consider that *creutzburgi* is on average about 60% less heavy than *antiquus*, which is a considerable difference for intra-specific variation.

Most literature of the previous century dealing with the elephants of the late Middle and Late Pleistocene of Crete mentions either two separate species – the large *P. antiquus* of mainland proportions and the medium-sized *P. creutzburgi* – or just one, *Palaeoloxodon* cf. *antiquus*.

A second, slightly larger dwarf species, *Elephas* (=*Palaeoloxodon*) *chaniensis*, was described by Nikos Symeonidis and colleagues (2000, 2001) for fossils from a submerged cave near Chania, dated to about 18,000 years. Poulakakis and colleagues (2002) and Athanassiou and colleagues (2019) show, however, that the material overlaps in size with *creutzburgi* and represents at most the larger individuals of the latter species. In addition, no differential diagnosis with *creutzburgi* was given in the original diagnosis. That said, only one elephant species inhabited Crete in the Late Pleistocene, *Palaeoloxodon creutzburgi*.

Why the Cretan elephant did not shrink further to extreme pygmy proportions as did the mammoth of the Early to Middle Pleistocene is unclear. Two explanations have been brought to the fore. First, there was a periodic contact between the Cretan and the mainland population. Herds of *P. antiquus* may have swam to and from the island in search of forage in a similar way as herds of Asian elephants do. Unexplained in this scenario is the dwarfing of the deer, which indicates a considerable isolation. Second, the presence of the deer blocked the dwarfism of this large competitor, as suggested by Palombo (2007, 2009). The ecological niche of a dwarf elephant was already occupied by one or more of the deer species. This scenario may be the most likely and is confirmed by a similar observation on Sardinia, where a rather large mammoth is found together with an abundance of deer as well.

Dwarf Hippos

The Cretan dwarf hippo (*Hippopotamus creutzburgi*) is intermediate in size between those from Malta and Cyprus with a body size reduction of over 85%. The dwarf hippo was not only very small in size but also walked more on its hooves instead of on its footpad, compared to mainland hippos. This is explained as an adaptation to a less aquatic life in a rockier environment, which seems to fit with the environment of the Katharo. The body proportions of the Cretan dwarf hippo differed slightly from those of mainland *Hippopotamus antiquus*. The humerus was relatively long, whereas the radius was relatively short. Also, the feet – pes and manus – were relatively shorter. The femur and tibia seem to have been less vertically arranged than in the extant hippo, regarding the morphology of the knee as described by Spaan (1996).

The most likely ancestor of the Cretan dwarf hippo is *H. antiquus*. This species – described as *H. major* by Cuvier because of its large size – lived in Europe during the Early and early Middle Pleistocene and was replaced by the smaller *H. amphibius* (= *H. incognitus*) in the late Middle and Late Pleistocene, perhaps with a small overlap in time.

Differences in size between materials from various localities led Kuss (1975a) to the opinion that there were two subspecies, the larger *creutzburgi* from Katharo (Plate 4) and the smaller *parvus* from Kato Zakros and Stavros. He attributed the smaller form to a younger geological age, which implies an evolutionary trend towards smaller size. The measurements of the molars of the small form constitute the lower range of *H. creutzburgi* with minimal overlap with the molars of the larger form. This difference is indeed best explained as a temporal difference between the sites.

The Cretan dwarf hippopotamus had to wait for its own specific name until Bert Boekschoten and Paul Sondaar (1966) named it *Hippopotamus creutzburgi*, based on material from the Katharo collected by Bate 60 years earlier together with some new materials. Before 1966, the hippo had been variously given the names of the Sicilian, Maltese, and Cypriot dwarf hippopotamus, respectively, *Hippopotamus pentlandi*, *H. melitensis*, and *H. minor* or *minutus*, a situation in parallel to Mediterranean dwarf elephants, for which the names of the Sicilian dwarf elephants were applied based on size alone.

Dwarf and Giant Deer, An Adaptive Radiation

The Cretan deer is represented by no less than eight different species, ranging from dwarf size with withers height of about 0.40 m for the smallest species to very large with withers height of about 1.65 m for the largest species (Figure 5.8). De Vos (1996) and de Vos and van der Geer (2002) explain this phenomenon as sympatric speciation followed by an adaptive radiation to occupy all possible empty niches ranging from dense forest to prickly rocks. The coexistence of various environments has been confirmed by studies on the rich fossil avifauna by Peter Weesie (1988).

The remarkably large variability in size and morphology within the Cretan deer species '*Cervus*' *cretensis* was already noted by Kuss (1965). Ten years later, Kuss (1975b) erected a new genus to accommodate the Cretan deer, *Candiacervus*, after the Latin name for the island. At that occasion, he also described a new species, *Candiacervus rethymnensis,* for a larger form, in size comparable to red deer (*Cervus elaphus*). He explained the size differences as a trend in time towards smaller size. A year later, an Italian team led by Tassos Kotsakis (1976) announced the existence of two even larger species, recovered from Bate cave in the cliffs below the Rethymnon-Chania National Road. They provisionally referred to the middle-sized one as 'Cervo taglia media' and to the large-sized species as 'Cervo taglia grande'. The former species was formally named *Cervus dorothensis* by Lucia Capasso Barbato (1992) in honour of Dorothea Bate, while the latter species was named *Cervus major*, based on its exceptionally large size by Capasso Barbato and Carmelo Petronio (1986). Both are now recognised as species of *Candiacervus*. Meanwhile, de Vos (1984) had described the smallest variety – his size 1 – of the Cretan deer

Sizes	Names used by de Vos in 1979	Names used by Caloi and Palombo in 1996
I	*Candiacervus ropalophorus*	*Megaceroides (Candiacervus) ropalophorus*
II	*Candiacervus* sp.IIa, sp.IIb, sp.IIc	*Megaceroides (Candiacervus) ropalophorus*
III	*Candiacervus cretensis*	*Megaceroides (Candiacervus) cretensis*
IV	*Candiacervus rethymnensis*	*?Pseudodama rethymnensis*
V	*Candiacervus* sp.V	*?Pseudodama (Leptocervus) dorothensis*
VI	*Candiacervus* sp.VI	*?Pseudodama (Leptocervus) major*

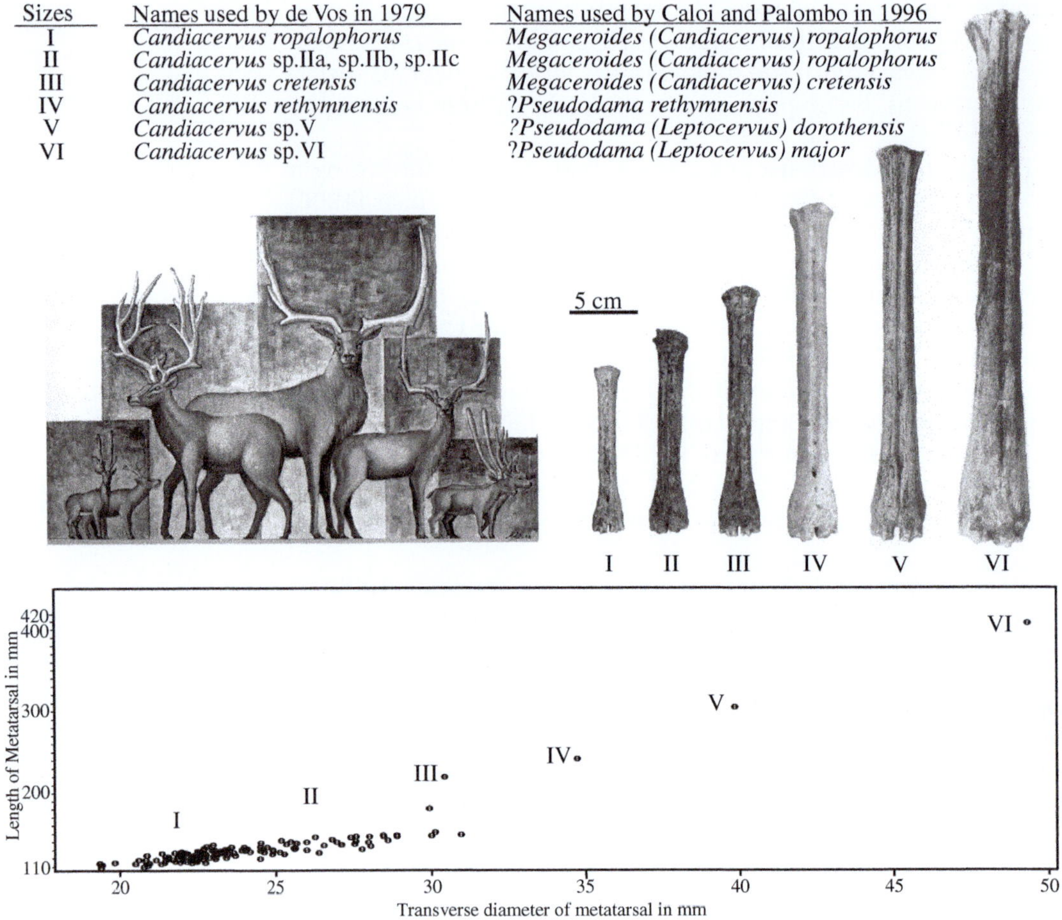

Figure 5.8 Metatarsals of the six different size groups of the Cretan deer (*Candiacervus*), dorsal view. Museum of Geology and Palaeontology, University of Athens (sizes 1–4: *ropalophorus, listeri—devosi—reumeri, cretensis, rethymnensis*) and Museum of Palaeontology, University of Rome (sizes 5 and 6: *dorothensis, major*). (Reconstruction of the deer by Alexis Vlachos.)

as *Candiacervus ropalophorus* (Figure 5.9; Plate 5) based on a few thousand fossils from the cave Gerani 4 situated west of Rethymnon. The very peculiar antler ending in a knob-like extension (Figure 5.7, left) inspires the name. De Vos (1979) also distinguished three more dwarf species – his size 2 – larger than *ropalophorus* but each with a different type of skull and antler. Van der Geer (2018) formally named these three species as *Candiacervus devosi*, *Candiacervus reumeri*, and *Candiacervus listeri* (Plate 6). The main beam of the antler in these latter four species is either relatively straight or gently curved (*ropalophorus*-type), with (*ropalophorus*) or without (*listeri*) a bludgeon-shaped distal end, with an additional back tine (*devosi*) or with three tines (*reumeri*). The antler is unusually

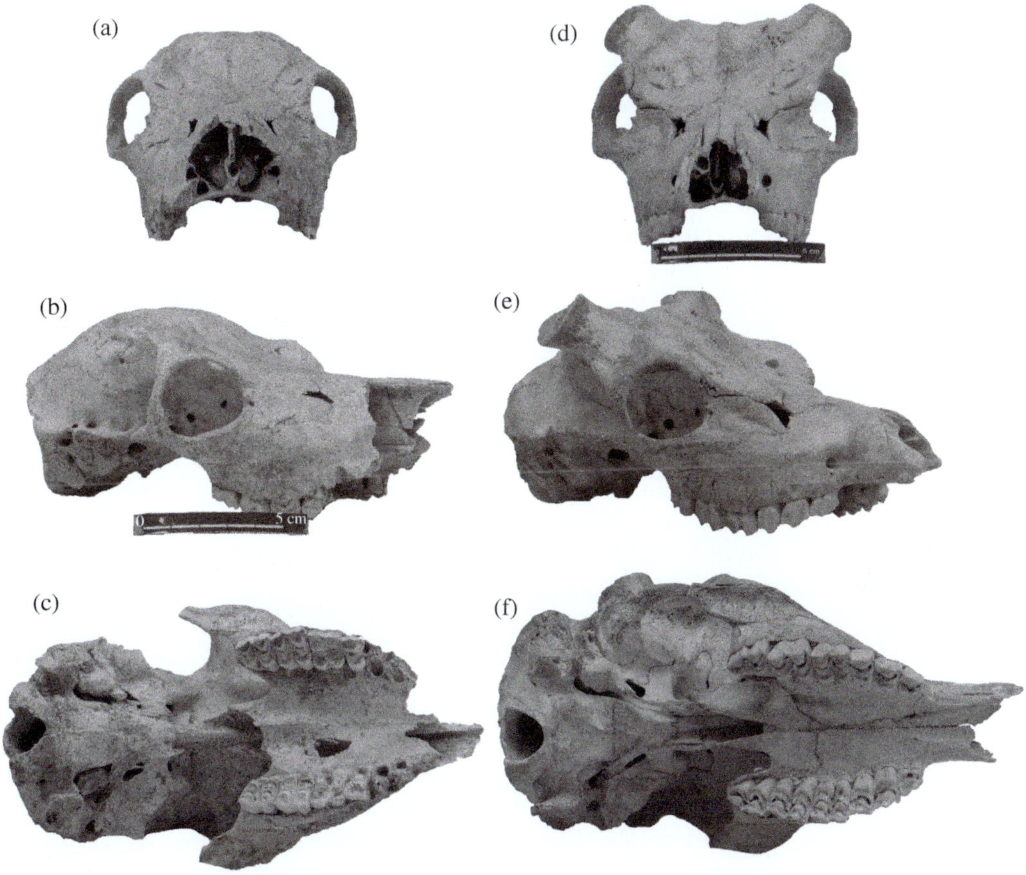

Figure 5.9 Female (a–c) and male (d–f) skull of the smallest Cretan deer (*Candiacervus ropalophorus*). (a, d) anterior, (b, e) anterolateral, and (c, f) ventral views. Scale bar: 5 cm. National Museum of Natural History, Leiden. (Photograph Eelco Kruidenier.)

long in the smallest species (*ropalophorus*) and could almost match the length of their bodies.

The most typical Cretan deer are the smallest species (Figure 5.7). They have not only relatively and absolutely short limbs (van der Geer and colleagues, 2006), but also long and simplified antlers (Plate 5). These species occupied a niche close to that of the wild goat of Crete today: barren rocks with thorny bushes, as shown by features of their osteology and goat-like body proportions. Fossils of males and females of the smallest sizes are found in the ratio 1 : 1, indicating a gregarious species with mixed herds. A further taphonomic peculiarity is that all ontogenetic ages are represented from very young to very old (van der Geer and colleagues, 2014).

The massivity of the limbs of the smaller sizes of the Cretan deer led Falconer and Bate to assume that the fossils belonged to goats or antelopes. Goats, however, made their first appearance on Crete with the colonisation by human settlers and are unknown from the fossil record. The roebuck which Falconer identified from Tripiti cave is likely based on the small-sized deer teeth or antler fragments. He could not combine the deer cranial remains with the 'bovid' postcranial material because the existence of insular deer with robust limbs was unknown to science at the time.

The largest species are equally bizarre. They developed extremely thin and elongated metacarpals and metatarsals and do not resemble any known mainland form. Their limbs were even longer than those of the giant deer *Megaloceros giganteus.* Probably the larger forms were typically browsers, based on the molars which show more ridges. Both the small and the large forms deviated so much from mainland deer that it is impossible to indicate with certainty their ancestors, other than that it is most likely a megacerine deer (*Praemegaceros, Dama*).

The Cretan Otter

The Cretan otter is the only carnivore known from the Pleistocene of Crete; the other fossil carnivores (beech marten and badger) are of a Holocene age. Its only remains consist of a nearly complete skeleton (Plate 3) found in the deer layer of Liko cave, dated to the late Middle Pleistocene. The skull and mandible (Plate 3) were originally described as *Isolalutra cretensis* by Symeonidis and Sondaar (1975). Five years later, Gerard Willemsen (1980) described the skeleton, and later (Willemsen, 1992), he transferred the species to the genus *Lutrogale*, based on similarities with the extant smooth-coated otter of Asia (*Lutrogale perspicillata*), though acknowledging several differences with *Lutrogale*. Later, a left mandible, two left first molars and a left femur, also from Liko cave, were added to this species, as reported by Willemsen (1996).

The Cretan otter was less aquatic than both the common otter (*Lutra lutra*) and the smooth-coated otter (*Lutrogale perspicillata*) and comparable in this aspect to the African clawless otter (*Aonyx capensis*). This is explained as a secondary development as an adaptation to the different conditions on Crete. Its robust teeth with a large talon of the upper carnassial indicates that its diet consisted mainly of fish with crustaceans as an important addition and possibly also small land vertebrates such as mice. The most likely ancestor is, according to Willemsen (1992), *Lutrogale perspicillata*. At present, this species is restricted to Asia, but it is thought to have had a much wider distribution – including Asia Minor or even Greece – in the past. Fossils, however, of *Lutrogale* are very rare and the only fossils species known so far are two Javanese species and the Cretan species, both insular species. No fossils are known from the extant species, so conclusions about its Pleistocene distribution are based on weak evidence. The main reason to attribute the Cretan otter to the genus *Lutrogale* is found in peculiarities of the dentition and the thoracic vertebra. The body mass is estimated between 9 and 14 kg as calculated by Lyras and colleagues (2010).

Giant Mice

The lineage of the Early-early Middle Pleistocene Cretan mouse (*Kritimys*) contains two nominal and one referred species, being the geologically oldest *Kritimys* cf. *kiridus, Kritimys kiridus*, and the youngest species *K. catreus*. This latter species was almost seven times the size of its suggested ancestor (mainland *Praomys*), as calculated by Van den Hoek Ostende and colleagues (2017). The earliest species, restricted to Siteia, is also the smallest but already almost twice the size as mainland *Praomys*. This oldest form had relatively low-crowned teeth with relatively simple cusp morphology. It has been suggested that the earliest *Kritimys* is an immigrant from Rhodes, via Kassos and Karpathos, based on a similar finding at the Late Pliocene (MN16) Damatria formation on Rhodes, reported as *Kritimys* aff. *kiridus* by Mayhew (1977).

The common mouse is represented by two species, of which the earlier is *Mus bateae*, which is smaller than the later *Mus minotaurus*, but already larger than the common house mouse (*Mus musculus*). Freudenthal and Martín-Suárez (2013) calculated an over three times body mass increase for the largest species. The two species belong to a single lineage. Like in the case of the Cretan mouse of the previous biozone, a long-term trend for increasing size is attested.

Cretan Shrew

The Cretan white-toothed shrew is a remarkable example of long-term resilience, as it managed to survive on Crete till the present day. The fossil and the living *Crocidura zimmermanni* are the same species as shown by Jelle Reumer (1986). Apart from this species, two more shrews live today on Crete. The white-toothed shrew *Crocidura suaveolens caneae* was introduced around 3700–3550 years ago, and the musk shrew *Suncus etruscus* is found in sites dated to as early as 3370–3200 years ago. There is no genetic relation between the Cretan shrew and the later introduced white-toothed shrew, according to Vogel (1980), nor between the Cretan shrew and the other living European species. Most likely, the Cretan shrew is related to an Early or Middle Pleistocene species, for example, *Crocidura kornfeldi*. Unfortunately, the Cretan shrew lives today only in two elevated regions of Crete. It is a unique relic species, being the only survivor of a once existing completely different world of endemic species, and deserves, therefore, active conservation.

BOX 5.1

Insular Elephants from Other Greek Islands

Crete is not the only Greek island with dwarf elephants. During the Late Pleistocene, dwarf elephants, all descendants of *P. antiquus*, lived on Tilos, Rhodes, Naxos, Delos, Kasos and a few other Aegean islands too (Figure Box 5.1.1). The most important findings come from Charkadio cave on Tilos, discovered by Nikos Symeonidis in the early 1970s. He and his team started to excavate here and retrieved about 2000 elephant fossils. The variation in size was large, and Symeonidis and colleagues (1973) ascribed the material to two different subspecies (*Palaeoloxodon antiquus falconeri* and *P.a. mnaidriensis*; both names of Sicilian dwarf elephants). The excavations were continued in the years thereafter and up to 11 thousand elephant fossils were unearthed, besides long bones of the marginated tortoise (*Testudo marginata*), some bird fossils and, in a much older layer than the elephant-bearing layer, some limb bones and antlers of a middle-sized deer. George Theodorou (1983), who continued the excavations of Symeonidis, concluded in his doctoral dissertation that there was only one species and that the variation was entirely due to sexual dimorphism.

Theodorou and colleagues (2007) described a new species (*Elephas* (=*Palaeoloxodon*) *tiliensis*) for the Tilos material (Plate 7), based on no less than 15 thousand fossils. The Tilos elephant stood less than 1.5 m at the shoulder and had a body mass of slightly over 700 kg. Compared to its giant ancestor of almost 8 ton, this means a body mass reduction of about 90%. The limbs were less pillar-like than those of mainland elephants, a result of changes in tarsus and carpus and an inward turn of the metapodials. In addition, the morphology of the limb bones indicates a higher degree of mobility than seen in mainland elephants. The absolute age of the Tilos elephant ranges from about 45,000 to 3500 years before present. This means that it possibly survived into the Bronze Age and could have been observed by prehistoric humans. At present, however, unequivocal evidence for this is lacking.

The species from the Cycladic island of Naxos (*Palaeoloxodon lomolinoi*; Plate 7) was similar, with a body mass reduction of 92%. The maxilla from Naxos was found in the Trypiti River, as mentioned by Maximus Mitzopoulos (1961) but was recognized as a distinct species by van der Geer and her colleagues in 2014. A number of Cycladic Islands probably were connected to each other during the Late Pleistocene, forming one large island. The material from Paros may belong to the same species, but the material from Delos – a

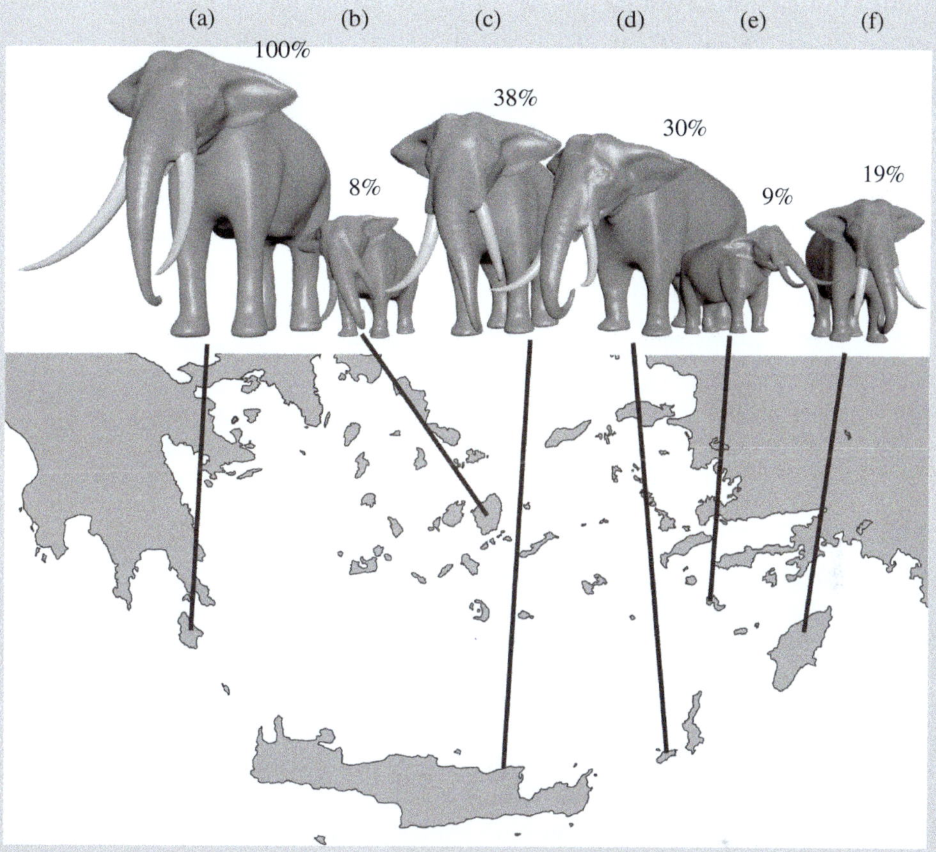

Figure Box 5.1.1 Map of the Eastern Mediterranean showing islands with Pleistocene elephant fossils belonging to the genus *Palaeoloxodon*. The dwarf forms have variable degrees of body size reduction, given here as percentage of *Palaeoloxodon antiquus* of mainland Greece, according to the particular ecological conditions of each palaeo-island. (a) Kýthera, (b) Naxos, (c) Crete, (d) Karpathos, (e) Tilos, and (f) Rhodos. Body mass reduction based on Athanassiou and colleagues, 2019. (Map: George Lyras.)

left upper molar – is considerably larger (body mass reduction about 83%) and may belong to an earlier form of the Naxos dwarf elephant. The molar from Delos was found near the temple of Apollo in a deposit of the Inopos River. Michel Cayeux (1908) doubted if it was *E.* (=*Palaeoloxodon*) *antiquus* on the ground of larger distances between the plates and showed the molar to Marcellin Boule in Paris. Sevket Sen and colleagues (2014) re-evaluated the material based on a cast (the original being lost) and confirmed the original ascription to *Palaeoloxodon*. At the end of the Pleistocene, the Cycladic landmass fragmented due to a rising sea level (Figure Box 5.1.2).

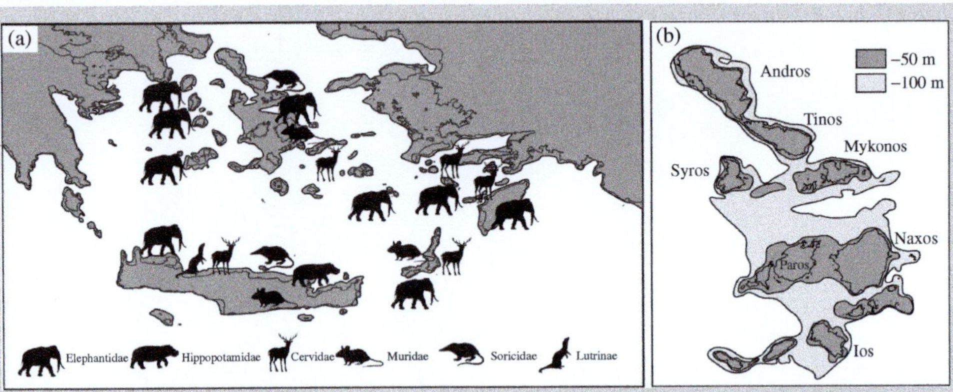

Figure Box 5.1.2 Map of the southern Aegean Sea indicating the presence of fossil insular mammals. The grey area represents the approximate extent of the exposed land during a sea-level drop of 100 m. Note that not all taxa lived simultaneously on each island and thus their presence does not necessarily represent natural faunal compositions. (Map: George Lyras.)

Other dwarf elephant remains have been reported from Ladiko cave on Rhodes and described by Symeonidis and colleagues (1974). According to Theodorou and colleagues (2007), the size of the Rhodes elephant falls within the upper size limit of the Tilos dwarf elephant.

In the literature, there are further confirmed reports of dwarf elephant remains from Astypálaia, Kasos, and Karpathos. Remains from Kythnos, Seriphos, and Milos, however, cannot be determined further, whereas the remains from Kephallenia, Kalymnos, and Kythera are indistinguishable on the species level from mainland *P. antiquus*. For an overview of the literature on fossil proboscideans from the Greek islands, see the work of Athanassios Athanassiou and colleagues (2019).

References

Athanassiou A., van der Geer A.A.E., Lyras G.A. 2019. Pleistocene insular Proboscidea of the Eastern Mediterranean: A review and update. *Quaternary Science Reviews*, 218: 306–321.

Bate D.M.A. 1905. Four and a half months in Crete in search of Pleistocene Mammalian remains. *Geological Magazine (Decade 5)*, 2 (5): 193–202.

Bate D.M.A. 1907. On elephant remains from Crete, with description of *Elephas creticus* n. sp. *Proceedings of the Zoological Society of London*, 1907: 238–250.

Bate D.M.A. 1912. On a new species of mouse and other rodent remains from Crete. *Geological Magazine (Decade 5)*, 9: 4–6.

Bate D.M.A. 1942. New Pleistocene Murinae from Crete. *Annals and Magazine of Natural History (Series 11)*, 9: 41–49.

Binladen J., Thomas M., Gilbert P., Willerslev E. 2007. 800,000 year old mammoth DNA, modern elephant DNA or PCR artefact? *Biology Letters*, 2007 (3): 55–56.

Boekschoten G. J., Sondaar P.Y. 1966. The Pleistocene of the Katharo Basin (Crete) and its *Hippopotamus. Bijdragen tot de Dierkunde*, 36 (8): 17–44.

Burness G.P., Diamond J., Flannery T. 2001. Dinosaurs, dragons, and dwarfs: the evolution of maximal body size. *Proceedings of the National Academy of Sciences of the USA*, 98: 14518–14523.

Capasso Barbato L. 1992. Nuova specie di cervide del pleistocene di Creta. *Atti della Accademia Nazionale dei Lincei, Memorie Lincee, Scienze Fisische e Naturali (Serie 9)*, 1 (7): 183–220.

Capasso Barbato L., Petronio C. 1986. *Cervus major* n. sp. of Bate Cave (Rethymnon, Crete). *Atti della Accademia Nazionale dei Lincei, Memorie, classe di Scienze fisischi, matematische e naturali (Serie 8, 2a)*, 18 (2): 59–100.

Cayeux L. 1908. Découverte de l'*Elephas antiquus* à l'île de Délos (Cyclades). *Comptes Rendus hebdomadaires des Séances de l'Académie des Sciences*, 147: 1089–1090.

Croizet J.B., Jobert A.C.G. 1826–1828. *Recerces sur les Ossemens Fossiles du Département du Puy-de-Dôme*. Paris.

Cuvier G. 1804. Sur les ossement fossiles d'hippopotame. *Annales du Muséum National d'Histoire Naturelle, Paris*, 5: 99–122.

De Blainville H.M.D. 1847. Sur les Hippotames (Buffon) (*Hippopotamus* L.) et les cochons (Buffon) (*Sus* L.). In: *Ostéographie ou description iconographique comparée du squelette et du sistème dentaire des mammifères récents et fossiles pour servir de base á la zoologie et á la géologie*, A (22). J.B. Baillière, Paris, pp. 1–248.

de Vos J. 1979. The endemic Pleistocene deer of Crete. *Proceedings of the Koninklijke Akademie van Wetenschappen (Series B)*, 82 (1): 59–90.

de Vos J. 1984. The endemic Pleistocene deer of Crete. *Proceedings of the Koninklijke Nederlandse Akademie van Wetenschappen, Afdeling Natuurkunde, Eerste reeks, Deel*, 31. North-Holland Publishing Company, Amsterdam, Oxford, New York, pp. 1–100.

de Vos J. 1996. Taxonomy, ancestry and speciation of the endemic Pleistocene deer of Crete compared with the taxonomy, ancestry and speciation of Darwin's finches. In: Reese D. (ed.), *Pleistocene and Holocene Fauna of Crete and its First Settlers. Monographs in World Archaeology*, 28. Prehistory Press, Madison, Wisconsin, pp. 111–124.

de Vos J., Dermitzakis M.D. 1986. Models of the development of Pleistocene deer on Crete. *Modern Geology*, 10 (2–3): 243–248.

De Vos J., van der Geer A.A.E. 2002. Major patterns and processes in biodiversity: taxonomic diversity on islands explained in terms of sympatric speciation. In: Waldren W.H., Ensenyat J.A. (eds), *World Islands in Prehistory, International Insular Investigations, V Deia International Conference of Prehistory. British Archaeological Reports International Series* 1095: 395–405.

Dermitzakis M., Iliopoulos G., van der Geer A., Lyras G. 2007. The rise and fall of the Cretan Deer. XVII INQUA Congress, July 28–August 3, 2007, Cairns, Australia. *Quaternary International*, 167–168 (suppl. 1): 97.

Dermitzakis M.D. 1977. The quaternary fossil mammals in Caves and Karstic Holes of Crete Island and their significance. *Bulletin de la Société Spéléologique de Grèce*, 14: 152–190.

Dermitzakis M.D., de Vos J. 1987. Faunal succession and the evolution of mammals in Crete during the Pleistocene. *Neues Jahrbuch Geologische und Paläontologische Abhandlungen*, 173 (3): 377–408.

Dermitzakis M.D., Sondaar P.Y. 1978. The importance of fossil mammals in reconstruction paleogeography with special reference to the Pleistocene Aegean Archipelago. *Annales Géologiques des Pays Helléniques*, 46: 808–840.

Desmarest A.G. 1822. *Mammalogie ou description des espèces de mammifères. Supplèment.* Agasse, Paris.

Evans J.D. 1968. Knossos Neolithic, Part II: Summary and conclusion. *The Annual of the British School of Archaeology at Athens*, 63: 267–276.

Freudenthal M., Martín-Suárez E., 2013. Estimating body mass of fossil rodents. *Scripta Geologica*, 145: 1–130.

Herridge V.L., Lister A.M. 2012. Extreme insular dwarfism evolved in a mammoth. *Proceedings of the Biological Sciences*, 279 (1741): 3193–3200.

Iliopoulos G., Eikamp H., Fassoulas C. 2010. A new Late Pleistocene mammal locality from western Crete. *Bulletin of the Geological Society of Greece*, 43: 918–925.

Kotsakis T., Melentis J., Petronio C., Sirna G. 1976. Seconda spedizione paleontologica Lincea nell'isola di Creta (1975). *Academia Nationale dei Lincei, Quaderno,* 223: 3–10.

Kuss S.E. 1965. Eine pleistozäne Säugetierfauna der Insel Kreta. *Berichte der Naturforschenden Gesellschaft zu Freiburg im Breisgau*, 55 (2): 271–348.

Kuss S.E. 1973. Die pleistozänen Säugetierfaunen der ostmediterranen Inseln. *Berichte der Naturforschenden Gesellschaft zu Freiburg im Breisgau*, 93 (1–2): 49–71.

Kuss S.E. 1975 (a). *Hippopotamus creutzburgi parvus* n. spp., ein pleistozänes Zwergflusspferd von der Insel Kreta. *Berichte der Naturforschenden Gesellschaft zu Freiburg im Breisgau*, 65 (1–2): 5–23.

Kuss S.E. 1975 (b). Die pleistozänen Hirsche der ostmediterranen Inseln Kreta, Kasos, Karpatos und Rhodos (Griechenland). *Berichte der Naturforschenden Gesellschaft zu Freiburg im Breisgau*, 65 (1–2): 25–79.

Kuss S.E., Misonne X. 1968. Pleistozäne Muriden der Insel Kreta. *Neues Jahrbuch für Mineralogie, Geologie, und Paläontologie Abhandlungen*, 132 (1): 55–69.

Lax E.M. 1996. A Gazetteer of Cretan Paleontological localities. In: Reese D.S. (ed.), *Pleistocene and Holocene fauna of Crete and Its First Settlers. Monographs in World Archaeology* 28: 1–32. Prehistory Press, 1996.

Lomolino M.V., van der Geer A.A.E., Lyras G.A., Palombo M.R., Sax D., Rozzi R. 2013. Of mice and mammoths: Generality and antiquity of the island rule. *Journal of Biogeography*, 40: 1427–1439.

Louchart A., Bedetti C., Pavia M. 2005. A new species of eagle (Aves: Accipitridae) close to the Steppe Eagle, from the Pleistocene of

Corsica and Sardinia, France and Italy. *Palaeontographica A*, 272: 121–148.

Lyras G.A., van der Geer A., Rook L. 2010. Body size of insular carnivores: Evidence from the fossil record. *Journal of Biogeography*, 37: 1007–1021.

Mayhew D.F. 1977. The endemic Pleistocene murids of Crete I-II. *Proceedings of the Koninklijke Akademie van Wetenschappen (Series B)*, 80 (3): 182–214.

Mayhew D.F. 1996. The extinct murids of Crete. In: Reese D.S. (ed.), *Pleistocene and Holocene fauna of Crete and Its First Settlers. Monographs in World Archaeology*, 28. Prehistory Press, Madison, pp. 167–171.

Mitzopoulos M.K. 1961. Über einen pleistozänen Zwergelefanten von der Insel Naxos (Kykladen). *Praktika tes Akademias Athenon*, 36: 332–340.

Mol D., de Vos J., van den Bergh G.D., Sondaar P.Y. 1996. The taxonomy and ancestry of the fossil elephants of Crete. Faunal Turnover and Comparison with Proboscidean faunas of Indonesian Islands. In: Reese D.S. (ed.), *Pleistocene and Holocene fauna of Crete and Its First Settlers. Monographs in World Archaeology*, 28. Prehistory Press, Madison, pp. 1–32.

Orlando L., Pages M., Calvignac S., Hughes S., Hanni C. 2007. Does the 43BP sequence from an 800,000 year old Cretan dwarf elephantid really rewrite the textbook on mammoths? *Biology Letters*, 2007 (3): 57–59.

Owen R. 1845. *Descriptive and Illustrated Catalogue of the Fossil Organic Remains of Mammalia and Aves Contained in the Museum of the Royal College of Surgeons of England*. London: R. and J. Tayler.

Palombo M.R. 2007. How can endemic proboscideans help us understand the "island rule"? A case study of Mediterranean Islands. *Quaternary International*, 169–170: 105–124.

Palombo M.R. 2009. Body size structure of Pleistocene mammalian communities: What support is there for the "island rule"? *Integrative Zoology*, 4: 341–356.

Pococke R. 1745. *A Description of the East and Some Other Countries*. Vol. 2, Part 1. London: W. Bowyer.

Poulakakis N., Mylonas M., Lymberakis P., Fassoulas C. 2002. Origin and taxonomy of the fossil elephants of the island of Crete (Greece): Problems and perspectives. *Palaeogeography Palaeoclimatology Palaeoecology*, 186 (1–2): 163–183.

Poulakakis N., Parmakelis A., Lymberakis P., Mylonas M., Zouros E., Reese D.S., Glaberman S., Caccone A. 2006. Ancient DNA forces reconsideration of evolutionary history of Mediterranean pygmy elephantids. *Biology Letters*, 2006 (2): 451–454

Raulin F.V. 1869. *Description Physique de l'île de Crète*. 2 volumes and atlas. Paris.

Reese D.S., Belluomini G., Ikeya M. 1996. Absolute dates for the Pleistocene fauna of Crete. In: Reese D.S. (ed.), *Pleistocene and Holocene fauna of Crete and Its First Settlers. Monographs in World Archaeology*, 28. Prehistory Press, Madison, pp. 47–52.

Reumer J.W.F. 1986. Notes on the Soricidae (Insectivora, Mammalia) from Crete, part I. The Pleistocene species *Crocidura zimmermanni*. *Bonner zoologische Beiträge*, 37: 3.

Sen S., Barrier E., Crété X. 2014. Late Pleistocene dwarf elephants from the Aegean islands of Kassos and Dilos, Greece. *Annales Zoologici Fennici*, 51: 27–42.

Simonelli V. 1894. Appunti sopra i terreni neogenici e quaternari dell'Isola di Candia. *Rendiconti della R. Academia dei Lincei (Classe di scienze, Matematichee naturali 3, serie 5)*, 2: 265–268.

Simonelli V. 1907. Mammiferi quaternari dell'isola di Candia I. *Memorie, Accademia delle Scienze. Istituto di Bologna (Classe di Scienze Fisiche, serie 6)*, 4: 455–471.

Simonelli V. 1908. Mammiferi quaternari dell'isola di Candia II. *Memorie, Accademia delle Scienze (Instituto di Bologna Classe di Scienze Fisische serie 6)*, 5: 103–111.

Sondaar P.Y., de Vos J., Dermitzakis M.D. 1986. Late Cenozoic faunal evolution and palaeogeography of the South Aegean island arc. *Modern Geology*, 10: 249–259.

Sondaar P.Y., Dermitzakis M.D. 1982. Relation migration land vertebrates, paleogeography and tectonics. In: Pichon X.L., Augustidis S.S., Mascle J. (eds), *International Symposium on the Hellenic Arc and Trench (H.E.A.T.), April 8–10 1981*, 2: 283–308. Athens.

Spaan A. 1996. *Hippopotamus creutzburgi*: The case of the Cretan *Hippopotamus*. In: Reese D.S. (ed.), *Pleistocene and Holocene fauna of Crete and Its First Settlers. Monographs in World Archaeology* 28: 99–111. Prehistory Press, Madison.

Spratt T.A.B. 1865. *Travels and Researches in Crete*, 2 Vols. John van Voorst, London.

Steensma K.J., Reese D.S. 1996. The mustelins of Crete. In: Reese D. (ed.), *Pleistocene and Holocene Fauna of Crete and Its First Settlers. Monographs in World Archaeology*, 28: 159–166. Prehistory Press, Madison, Wisconsin.

Strasser T., Murray S., van der Geer A.A.E., Kolb C., Ruprecht L., Jr. 2018. Palaeolithic cave art from Crete, Greece. *Journal of Archaeological Science Reports*, 18: 100–108.

Strasser T.F., Panagopoulou E., Runnels C.N., Murray P.M., Thompson N., Karkanas P., McCoy F.W., Wegmann K.W. 2010. Stone Age seafaring in the Mediterranean: Evidence from the Plakias Region for Lower Palaeolithic and Mesolithic habitation of Crete. *Hesperia*, 79: 145–190.

Symeonidis N., Bachmeyer F., Zapfe H. 1973. Grabungen in der Zwergelefanten-Höhle "Charkadio" auf der Insel Tilos (Dodekanes, Griechenland). *Annalen des Naturhistorischen Museums in Wien*, 77: 133–139.

Symeonidis N., Bachmeyer F., Zapfe H. 1974. Entdeckung von Zwergelefanten auf der Insel Rhodos (Ausgrabungen 1973). *Annalen des Naturhistorischen Museums in Wien*, 78: 193–202.

Symeonidis N.K., Sondaar P.Y. 1975. A new otter from the Pleistocene of Crete. *Annales Géologiques des Pays Helléniques*, 27: 11–24.

Symeonidis N.K., Theodorou G., Gianopoulos B. 2000. The new species *Elephas chaniensis* from the submerged Pleistocene deposits of Vamos Cave at Chania, Crete. *Bulletin, Society for Speleology of Greece*, 22: 95–108. [in Greek with English abstract.]

Symeonidis N.K., Theodorou G.E., Giannopoulos V.I. 2001. New data on *Elephas chaniensis* (Vamos cave, Chania, Crete). In: Cavaretta G., Gioia P., Mussi M., Palombo M.R. (eds), *The World of Elephants. Proceedings of the 1st International Congress, Roma, 16–20 October, 2001*, pp. 510–513.

Theodorou G., Symeonidis N., Stathopoulou E. 2007. *Elephas tiliensis* n. sp. from Tilos island (Dodecanese, Greece). *Hellenic Journal of Geosciences*, 42: 19–32.

Theodorou G.E. 1983. The Dwarf Elephants of the Charkadio Cave on the Island of Tilos (Dodekanese, Greece). PhD thesis, University of Athens. [In Greek.]

van den Hoek Ostende L.W., van der Geer A.A.E., Wijngaarde, C. 2017. Why are there no giants at the dwarf's feet? Insular micromammals in the eastern Mediterranean. In: Mol D. et al. (eds), *Proceedings of the VIth International Conference on Mammoths and their Relatives, May 2014, Grevena-Siatista, Greece. Quaternary International*, 445: 269–279.

van der Geer A.A.E. 2018. Uniformity in variety: Antler morphology and evolution in a predator-free environment. *Palaeontologia Electronica*, 21.1: 9A.

van der Geer A.A.E., de Vos J., Lyras G., Dermitzakis M. 2006. New data on the Pleistocene Cretan deer *Candiacervus* sp. II (Mammalia, Cervinae). *Courier Forschungsinstitut Senckenberg*, 256: 131–137.

van der Geer A.A.E., Lyras G.A., MacPhee R.D.E., Lomolino M., Drinia H. 2014. Mortality in a predator-free insular environment: The extinct dwarf deer from Crete. *American Museum Novitates*, 3807, 1–29.

van der Made J. 1996. Pre-Pleistocene Land mammals from Crete. In: Reese D.S. (ed.), *Pleistocene and Holocene fauna of Crete and Its First Settlers. Monographs in world Archaeology*, 28: 69–79. Prehistory Press, Madison.

Vaufrey R. 1929. Les éléphants nains des îles méditerranéennes et la question des isthmes pléistocènes. *Archives de l'Instityte de Paléontologie Humains, Memoire*, 6: 1–220.

Vogel P. 1980. Metabolic levels and biological strategies in shrews. In: Schmidt-Nielsen K., Bolis L., Tayler C.R. (eds), *Comparative Physiology: Primitive Mammals*. Cambridge University Press, Cambridge, pp. 170–180.

Weesie P.D.M. 1988. The Quaternary avifauna of Crete, Greece. *Palaeovertebrata*, 18 (1): 1–94.

Willemsen G.F. 1980. Comparative study of the functional morphology of some Lutrinae, especially *Lutra lutra, Lutrogale perspicillata* and the Pleistocene *Isolalutra cretensis. Verhandelingen der Koninklijke Akademie van Wetenschappen (Series B)*, 83: 289–326.

Willemsen G.F. 1992. A revision of the Pliocene and Quaternary Lutrinae from Europe. *Scripta Geologica*, 101: 1–115.

Willemsen G.F. 1996. The Cretan otter *Lutrogale cretensis*. In: Reese D.S. (ed.), *Pleistocene and Holocene fauna of Crete and Its First Settlers. Monographs in World Archaeology* 28: 153–157. Prehistory Press, Madison.

CHAPTER SIX

Gargano

Evolution of Island Mammals: Adaptation and Extinction of Placental Mammals on Islands, Second Edition.
Alexandra van der Geer, George Lyras and John de Vos.

© 2021 John Wiley & Sons Ltd. Published 2021 by John Wiley & Sons Ltd.

The palaeo-island Gargano is now part of mainland Italy. During the Late Miocene and Early Pliocene, it was an island, harbouring a highly endemic fauna, including five-horned deer, giant gymnures, giant pikas, and several rodent lineages. The mid-Pliocene flooding of the area led to the extinction of this peculiar fauna. After the area emerged again and became connected to Italy in the Early Pleistocene, it was colonised by a balanced mainland fauna.

Geology and Palaeogeography

The Gargano region (Figure 6.1) forms a promontory at the south-east coast of Italy, province of Apulia (Puglia), consisting of a block of uplifted Jurassic and Cretaceous limestone. This carbonate block descends into the Foggia Graben in the west, which separates Gargano from the main Italian mountain ridges. During the Late Miocene, and possibly already since the Early Miocene, the Gargano area formed an island, referred to as Gargano.

Late Miocene land mammal faunas of Italy indicate the presence of three separate palaeo-bioprovinces, as suggested by Lorenzo Rook and colleagues (2000). Between 11 and 9 million years ago, Gargano may have belonged to the Apulia portion of the Apulia-Abruzzi palaeo-bioprovince, which was located at the Adriatic side of the Apennines. This province either formed one large island or, alternatively, a series of islands or archipelago. Between about 7.2 and 5.3 million years ago, Gargano was isolated from the Abruzzi portion. The Early Pliocene flooding of the Mediterranean around 5.3 million years ago – ending the Messinian salinity crisis – greatly reduced the emerged areas and eventually led to the extinction of the endemic fauna of Gargano.

The Late Miocene faunas of the Apulia portion of the palaeo-bioprovince are found in fissure fillings – *terre rosse* – in numerous limestone quarries situated between Apricena, San Nazario, and Poggio Imperiale, in the province of Foggia. The more than hundred fissures are named differently by Matthijs Freudenthal since the 1970s and Claudio De Giuli and colleagues since the mid-1980s, respectively, which obscures direct comparison. This is partly due to the constantly changing aspect of the region as a direct result of quarrying activities. The majority of the fissures of the 1970s had vanished entirely by the mid-1980s. The faunas of the Abruzzi portion of the palaeo-bioprovince on the other hand are known from marine calcarenite deposits at Scontrone at the southern border of the National Park of Abruzzo, province of L'Aquila.

Gargano is sometimes considered part of a larger archipelago (Apulo-Dalmatic Realm) to explain the variation and trends

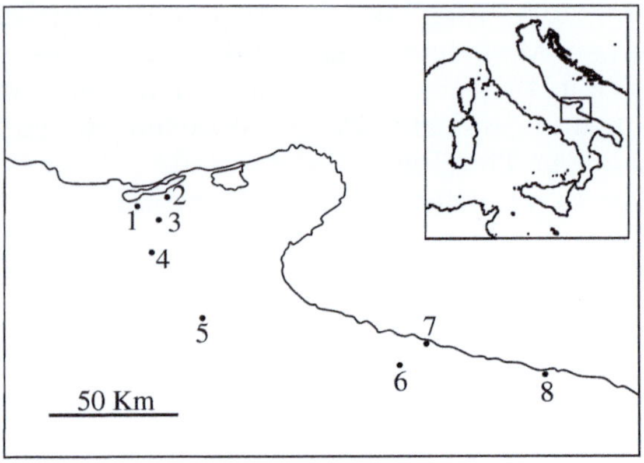

Figure 6.1 Map showing nearest villages to quarries with Late Miocene–Early Pliocene sediments, Gargano Promontory (1–3) and the rest of northern Puglia (4–8). (1) Poggio Imperiale, (2) San Nazario, (3) Apricena, (4) San Severo, (5) Foggia, (6) Andria, (7) Trani, and (8) Bari. Inset: Present location of the Gargano promontory at Italy's south-eastern coast. During the Miocene, Gargano was an island.

observed in the small mammals, as first published by Freudenthal (1976) and later worked out by De Giuli and Danilo Torre (1984). The endemic taxa are in this model seen as products of local evolution, which regularly came into contact with each other when the separate islands became connected and got isolated when the regions disconnected again. The present Gargano and Le Murge are assumed to have been two of these islands. If so, the frequency of contacts must have been high because the fauna of the entire archipelago forms a homogeneous unit.

The endemic ruminants and galericine insectivores were considered leftovers of a mainland stock by Mazza and Rustioni (2008). Their isolation was due to the submerging of a land bridge – vicariance effect – across the Adriatic Sea between the Gargano region and the Dalmatian coast via the current Sušac, Lastovo, Vela and Mala Palagruža, and the Tremiti islands. The supposed primitive character of the ruminants places this land bridge in the Early Oligocene (29–30 million years ago). Seismostratigraphic and stratigraphic data are interpreted by Etta Patacca and colleagues (2008) as evidence for the existence of a series of structural highs, stretching from the Balkan mainland to the Gargano area. This, however, does not necessarily mean that these highs were emerged above the water during the Early Oligocene global sea level drop. Furthermore, morphological and phylogenetic features of the various endemic taxa favour Late Miocene colonisation instead of colonisation during the Early Oligocene. The composition of the mammalian fauna, consisting of only rodents, gymnures (moonrats), and deer-like ruminants, is typical for overseas sweepstake dispersal. The pikas of Gargano on the other hand are more difficult to explain by sweepstake dispersal. A much later land connection to the Balkans at the end of the Miocene has thus been supposed to have facilitated their dispersal to Gargano. Geological evidence for this is, however, lacking;

furthermore, such a late land connection would have led to the invasion of a mainland fauna, which is not the case.

From the Early Pleistocene on, the region was interested by a general uplifting, resulting in the regression and continentalisation of the area. This enabled a typical Early Pleistocene (late Villafranchian) mainland fauna to invade the area, known as the Pirro Nord faunal unit (1.3–1.6 million years ago). A second karst cycle (neokarst) started in the late Early Pleistocene, which removed part of the palaeokarst fill, as shown by Laura Abbazzi and colleagues (1996).

Historical Palaeontology

In 1969, Cornelis Beets, Hendrik Schalke, and Matthijs Freudenthal from the Dutch Rijksmuseum van Geologie en Mineralogie (now Naturalis) discovered a number of fossiliferous fissure fillings in marble quarries on the Gargano Peninsula between the villages Apricena and Poggio Imperiale (Plate 8). The red clay deposits of these fissures appeared to contain fossils of several vertebrate groups, especially mice. Apart from the quarries, similar fossiliferous deposits were discovered along the road. The extraordinary character of the fossil taxa – especially the gigantism of the small mammals, including a giant gymnure – was recognised immediately. The same year, the group returned with funding and further assistance and excavated for a month, joined by a team of the Geological Institute at Bari, Italy. The next two summers, the excavations were continued, assisted by Jacques Michaux of the University of Montpellier, France. The preliminary results of these two campaigns were published by Freudenthal (1971). Afterwards, several monographs appeared in the 1970s and early 1980s on the various taxa by various specialists. The first of this series is that of Freudenthal (1972) where he described an almost complete skeleton of the giant gymnure as *Deinogalerix koenigswaldi*, in honour of his teacher. Eight years later, Percy Milton Butler added four additional species, being *Deinogalerix freudenthali*, *Deinogalerix brevirostris*, *Deinogalerix minor*, and *Deinogalerix intermedius*.

The next papers were by Peter Ballmann (1973, 1976) on the avifauna. The planned collaboration of Freudenthal with Michaux regarding the murids did not proceed, and Freudenthal (1976) published on the endemic murids. On this occasion, he established a new genus, *Microtia* – which he laterrenamed as *Mikrotia* (Freudenthal, 2006)– to which he ascribed three species (*Mikrotia magna*, *Mikrotia maiuscola*, and *Mikrotia parva*) of decreasing size (Figure 6.2). The next monograph is that of Gerard Willemsen (1983) where he

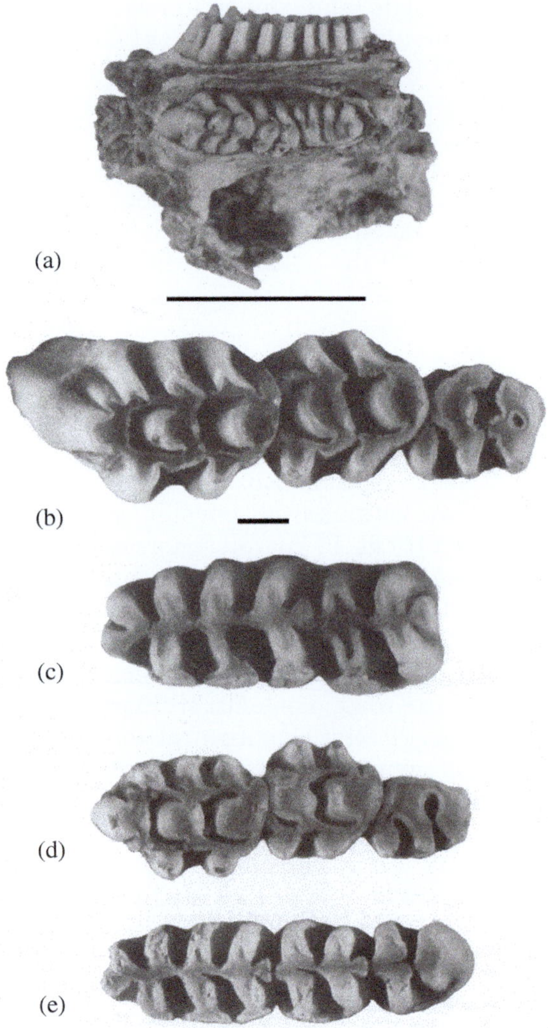

Figure 6.2 Maxilla fragment (a) and molar series (b–e) of *Mikrotia*, occlusal view. (a) *Mikrotia Magna*; scale bar 2 cm, (b) *Mikrotia maiuscula* upper M1–M3, (c) *Mikrotia maiuscula* upper M1–M2, (d) *Mikrotia parva* upper M1–M3 and (e) *Mikrotia parva* upper M1–M3. National Museum of Natural History, Leiden. Scale bar: 1 mm. Source: From Freudenthal M. 1976. Rodent stratigraphy of some Miocene fissure fillings in Gargano (prov. Foggia, Italy). Scripta Geologica, 37: 1–23.

describes a new otter species as *Paralutra garganensis*, based on a calcaneus and a maxillary fragment with the carnassial and the first molar preserved (Figure 6.3).

A year later, Joseph Leinders (1984) erected a new ruminant family (Hoplitomerycidae) to accommodate the bizarre ruminant of Gargano. As type species, he named *Hoplitomeryx matthei* in honour of the discoverer of the Gargano fauna. This enigmatic ruminant family displays a unique combination of diagnostic features which makes it impossible to group it with the other deer. These are, for example, the presence of orbital and nasal horns, sabre-toothed canines (Plate 9), two lacrimal orifices, and a closed metatarsal gully. None of these features is unique by itself; it is the combination which makes the family unique.

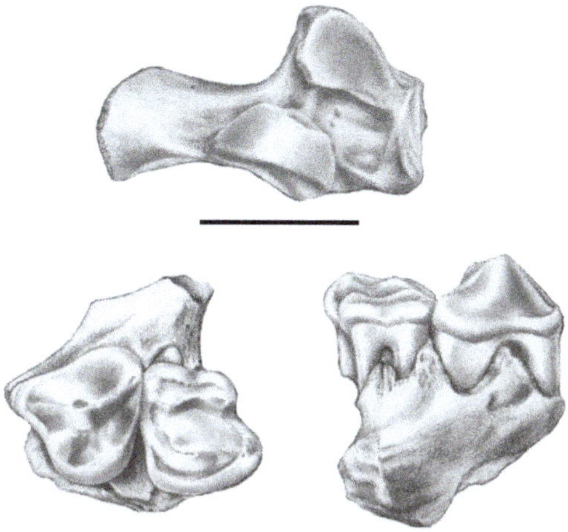

Figure 6.3 Right calcaneum (top; medial view, proximal tuber to the left) and left maxillary fragment (bottom; occlusal and lingual views) with P4 and M1 of the Gargano otter (*Paralutra garganensis*). Scale bar: 3 cm. National Museum of Natural History, Leiden. (Adapted from Willemsen, 1983, https://repository.naturalis.nl/pub/317451.)

Freudenthal (1985) described a new giant hamster genus to accommodate his three new species (*Hattomys beetsi*, *Hattomys nazarii*, and *Hattomys gargantua*), with increasing size, and one intermediate or transitional form (*nazarii-gargantua*). The smallest form is found in the oldest locality and the largest form in the youngest, though not all specimens can be properly assigned to one of the three species. In the same year, he also described a new dormouse, together with Remmert Daams, as *Stertomys laticrestatus*, based on teeth morphology. This dormouse is a giant form with broad-crested molars, as its name already suggests.

Two new ochotonid species were described by Paul Mazza (1987): the small-sized species from practically all fissures as *Prolagus apricenicus* and the large-sized species from San Giovannino as *Prolagus imperialis*, named, respectively, after the villages Apricena and Poggio Imperiale, between which the localities are situated.

In the early 1990s, a rich Late Miocene vertebrate locality was discovered at Scontrone, a small village bordering the National Park of Abruzzo, as announced by Marco Rustioni and colleagues (1992). The vertebrate fauna resembles that of Gargano and consists mainly of crocodiles, turtles, and *Hoplitomeryx* (=*Scontromeryx*) apart from a few otter bones, as reported by Paul Mazza and Rustioni (1996). The vertebrate remains, though, are all very fragmentary, hampering a thorough comparison. Small mammal remains were entirely missing, so no conclusions could be drawn considering the relative age of Scontrone in relation to that of Gargano. In 1999, however, a fragment of a right maxilla with associated third molar and partial second molar of the smallest species of a giant gymnure *Deinogalerix* were found at Scontrone, originally assigned to the species *freudenthali* of Gargano, but named *samniticus* by Savorelli and colleagues (2017), acknowledging an independent evolutionary lineage (inter alia, this would then call for a separate genus, as in the case of the ruminant *Scontromeryx*).

During the 1980s, excavations at the Gargano palaeokarst fissure fillings of the Apricena-Poggio Imperiale limestone quarries were continued by the University of Florence, led by Claudio De Giuli and Danilo Torre. They retrieved a very rich and diversified fauna, the bulk of which consists of small mammals, described in detail in several papers on individual taxa, biochronology, or palaeogeography of the region (see the following text).

Later, during the first decade of the 2000s, excavations were continued by the University of Turin, led by Giulio Pavia. In the Dell'Erba quarry, they discovered the oldest locality, M013. Here, they recovered the ancestral form of *Mikrotia* and the earliest representatives of the other endemic lineages. Other noteworthy finds of the team are that of a second skeleton of the giant gymnure *Deinogalerix*.

Biozones and Faunal Units

Two fossil faunal units are described from Gargano, being a Late Miocene insular fauna and an Early Pleistocene (late Villafranchian) continental fauna. The latter assemblage is referred to in the literature as the Pirro Nord unit or, alternatively, Dell'Erba, and famous for perhaps the earliest Palaeolithic tools of Europe.

The fossils belonging to the first biozone, characterised by *Mikrotia*, are retrieved from buried palaeokarst fissures – often referred to as *Cava* in Italian literature – in the Mesozoic limestone substrate, which are overlain by Pliocene–Early Pleistocene sediments of a subsequently marine, shallow water and terrigenous origin, as shown by Abbazzi and colleagues (1996).

A biostratigraphy of Gargano was first proposed by Freudenthal (1976). He based his sequence primarily on the stage of evolution in the various lineages of the murid genus *Mikrotia*, added with data on the cricetids (hamsters) and glirids (dormice). The oldest fissure fillings in this scheme are Biancone 1, Rinascita 1, and Trefossi 1, followed by, amongst others, Cantatore 3A, Fina D, San Nazario 4, and Fina H, whereas the youngest fissures are Gervasio 1, Chiro 4, Gervasio 2, and San Giovannino. The oldest phase is characterised by the presence of continental hamsters (e.g. *Cricetus* and *Cricetulodon*), and the end of the middle phase (Fina H) by the last occurrence of the endemic hamsters (*Hattomys*).

The last *Hattomys* is contemporaneous with the first giant form of the endemic mouse (*M. magna*). Rinascita is not only the species richest fissure regarding dormice but also regarding hamsters. These giant hamsters, however, were the first endemic mammals that went extinct, as they are not reported from fissures younger than Fina H, unlike the dormice, which continue into the youngest localities. Mice fossils are abundantly present in all fissures, equalled or outnumbered only by ochotonid fossils, a situation comparable to the Late Pleistocene deposits in Sardinia.

The largest *Mikrotia* species is restricted to the younger localities, more or less coinciding with the extinction of the endemic giant hamsters (*Hattomys*), though there is some temporal overlap between Chiro 27 and Fina H. In all younger localities, the three species are present together.

De Giuli and Torre (1984) and De Giuli and colleagues (1985, 1986, 1987a) apply a different naming system for the fissures, consisting of capital F followed by a number, except for San Giovannino which name they retain. In the model of Federico Masini and colleagues (2008) – a refinement of De Giuli's scheme of 1986 – the oldest fissure filling is F15, followed by F21b and F21c (Phase 1). Phase 2 is represented by fissure F1, Phase 3 by fissures F8, F9 and San Giovanni, and Phase 4 by fissure F32. This model is based on the evolutionary stages of both *Mikrotia* and *Prolagus*.

Direct comparison between the two schemes is hampered by the different names and the different fissures that were sampled, in part due to the constantly changing geography of active quarries. A few comparisons are, however, possible. F15 is younger than Freudenthal's three oldest fissures – Biancone, Rinascita 1, and Trefossi 1. The three fissures F21a, b, and c coincide with the three fissures Trefossi 1, 2, and 3, respectively. Fissure F9 is the same as Cava Fina and possibly comparable to Freudenthal's Fina E, contemporaneous with Gervasio 1 of the youngest phase. F32 is younger than all Freudenthal's fissures. Whereas both schemes are based on – and thus perfectly fit – the evolutionary stages of the rodents, the pikas, gymnures, and ruminant species are more difficult to fit in these schemes, likely as a result of a chronological discrepancy in the speed of evolution or of some taphonomic peculiarity. Finally, the oldest locality, M013, was named by the team of the University of Turin in early 2000, which uses the prefix M for their Late Miocene localities (M = *Mikrotia*). Here, the most primitive members of the endemic lineages are present.

Vertebrate remains found at Scontrone, part of the Abruzzi portion, were ascribed by Rustioni and colleagues (1992) and papers since then to the same endemic fossil assemblage as those from Gargano, be it geologically older, so probably ancestral.

The fauna indicates a Late Miocene age of colonisation of Scontrone. This is also inferred from the detailed stratigraphic and palaeo-environmental analysis by Etta Patacca and colleagues (2013). They show that the vertebrate remains are embedded in marginal-marine deposits, documenting a complex wave-dominated river-mouth deposition setting that developed over a large, flat, and semi-arid carbonate ramp. This is in line with the earlier suggestion by Mazza and Rustioni (2008) of lag deposits formed by the gradual removal of elements by tides in a coastal lagoon. A regional stratigraphic correlation with northern Majella narrows the time window to between 7.2 and 11.6 million years ago (Tortonian).

Age constraints of the *Mikrotia* biozone are poor and have been subject to fierce discussions, as summarised by van den Hoek Ostende and colleagues (2009) and Freudenthal and colleagues (2013). Diagnostic primitive species are *Parasorex ibericus*–like gymnures, *Dryomys*, and *Cricetulodon*, as well as the single *Megacricetodon* molar from Biancone 1, reported by Freudenthal (1985). This places the Gargano fauna in the Late Miocene (late Tortonian or Messinian) rather than in the Early Pliocene, and thus in the same time frame as other Italian insular faunas such as the Baccinello-Cinigiano fauna of Tuscany and the Fiume Santo fauna of Sardinia. From the detailed studies of the individual lineages and species (see below), however, it appears that the colonisation may not have been a single event, but, as in the case of many islands, in waves, as is the case with extinctions. Island demographics is a fluctuating process, where arrivals, extinctions, and speciation contribute to the composition of the fauna. Gargano, with its high number of endemic lineages and multitude of adaptive radiations each starting at different points in time constitutes, thus one of the most complex insular regions.

Late Miocene

The endemic fauna of the *Mikrotia* biozone of Gargano consisted of several endemic mammal lineages, all represented by at least two species. These lineages are those of a giant gymnure (*Deinogalerix*), a burrowing murine (*Mikrotia*), a giant dormouse (*Stertomys*), a giant hamster (*Hattomys*), a giant ochotonid (*Prolagus*), and an enigmatic deer-like ruminant (*Hoplitomeryx*; Box 6.1). Represented by a single species are an otter (*Paralutra garganensis*), a field mouse (*Apatodemus degiulii*, earlier attributed to *Apodemus*), a second genus of giant hamster (*Mystemys giganteus*), and a second genus of gymnure (*Apulogalerix pusillus*). In addition, some non-endemic species have been reported as well, all restricted to the oldest localities. These are three hamster species (*Cricetulodon*, *Megacricetodon*, and *Cricetus*), a shrew (*Lartetium* sp.), and a dormouse (*Dryomys apulus*, earlier attributed to *Eliomys*). The oldest localities also contain the more primitive members of the later lineages. Most mammals of Gargano show extraordinary morphological signs of insularity and are endemic on a genus level.

The avifauna, on the other hand, described by amongst others Peter Ballmann (1973, 1976) and Marco Pavia (2013), is mainly represented by taxa with a wider distribution, such as a pigeon (*Columba omnisanctorum*), a swift (*Apus wetmorei*), and a duck (*Anas velox*). Some genera, however, evolved into endemic forms, such as the giant eagle (*Garganoaetus freudenthali*) and the giant barn owl (*Tyto gigantea*). The latter may have evolved locally from the endemic large barn owl (*Tyto robusta*), which in turn likely derived from the widespread Eurasian owl *Tyto sanctialbani*. The giant eagle evolved from the endemic eagle *Garganoaetus murivorus*, which is likely closely related to true eagle *Aquila delphinensis* from La Grive-Saint-Alban, France, according to Ballmann (1973). Its closest living relatives are the small true eagles (*Hieraaetus*, *Spizaetus* and *Lophaetus*). A giant waterfowl, *Garganornis ballmanni*, described by Hanneke Meijer (2014), was flightless and seems to have been an apt fighter, as shown by Marco Pavia and colleagues (2017).

Apart from mammal and bird fossils, remains of *Crocodylus*, a genus of African origin, have been found in Gargano and Scontrone, as reported by Tassos Kotsakis and colleagues (2004) and Delfino and Rossi (2013), respectively, but whether this species shows endemic characters or not is unknown. During the Late Miocene, crocodylids occurred only in the Tusco-Sardinian and the Apulo-Abruzzi palaeo-bioprovinces, both insular domains. Their limited distribution and the fact that they are amongst the last crocodiles of Europe indicate a relict effect, the prolonged survival of a taxon on an island.

From Scontrone, less vertebrates are known, including several species of *Scontromeryx* (an early form, probably ancestral to *Hoplitomeryx*), *Deinogalerix samniticus* (earlier attributed to *freudenthali*), and an otter, as far as mammalian remains are concerned, plus reptiles and giant waterfowl. The giant waterfowl is assigned to the same species as the one from Gargano (*G. ballmanni*) by Pavia and colleagues (2017), but the only element retrieved so far is a right leg bone (tarsometatarsus) with the diagnostic proximal and distal end heavily fragmented and eroded, hampering a diagnosis on the species level. From a locality between Trani and Andria, former province of Bari, only fragmentary *Mikrotia* material has been reported by Freudenthal (1976), but no further excavations were carried out here. A fourth locality representing the Abruzzo-Apulia palaeo-bioprovince might be Capo di Fiume near Palena in Abruzzo, from where an articulated skeleton of a pika was reported by Mazza and colleagues (1995) and tentatively ascribed to *Prolagus apricenicus*.

BOX 6.1

The Phylogeny of the Hoplitomerycidae

The phylogeny of Hoplitomerycidae is unresolved, because some diagnostic characters are parallelisms as a result of adaptation to the island environment. In addition, the phylogeny of many other early cervoid taxa is equally poorly understood, due to the rarity of complete skeletons and lack of diagnostic postcranial material. Furthermore, a clear consensus regarding the interfamilial relationships of the living Pecora has not yet been reached. *Hoplitomeryx* might be a holdover of a primitive deer-like stock. Phylogenetically primitive forms, or 'relicts', tend to be preserved in island biota, and *Hoplitomeryx* might be such an example. In that light, it is likely to assume that it descended from a mainland ancestor decorated with horns and long canines – presumably, the palaeomerycids in the broadest sense – which persisted longer on the island than on the mainland. Palaeomerycids, including the dromomerycines, had supra-orbital bony outgrowths (except for *Amphitragulus*), though these were probably skin covered instead of keratin covered. The orbital position of cranial appendages is considered a primitive configuration and is typical for the earliest cervids. The palaeomerycids survived into the Late Miocene in Eurasia and the Pliocene in Asia. There is no support for considering the hornless extinct moschids *Micromeryx* and *Amphimoschus* as sister taxa of *Hoplitomeryx*. The former was suggested as its closest relative by Leinders (1984), but *Hoplitomeryx* differs from this genus by the presence of cranial appendages, loss of the lower second premolar, a non-bifurcated protocone, weakly developed entostyle, and ectostylid. *Amphimoschus* was suggested as its close relative by Moyà-Solà and colleagues (1999), but *Hoplitomeryx* differs from this genus by the presence of cranial appendages, a non-molarised lower fourth premolar, and double lacrimal orifices. Apart from the palaeomerycids, also the antilocaprids, including the merycodontines, developed supra-orbital horns like *Hoplitomeryx*, but these are keratin covered. In addition, antilocaprids are restricted to North America.

Late Early Pleistocene

After the uplift of the foreland, Gargano became again populated, but this time by a balanced mainland fauna, recognised by typical members of the late Early Pleistocene Eurasian fauna (late Villafranchian, 1.3–1.6 million years ago). A second major phase of superficial karst development led to the deposition of cave sediments in which fossils belonging to this biozone are found.

The megafauna included the giant hyena (*Pachycrocuta brevirostris*), sabre-toothed cats (*Homotherium latidens* and *Megantereon megantereon*), the giant cheetah (*Acinonyx pardinensis*), the Mosbach wolf (*Canis mosbachensis*), a stenoid horse (*Equus altidens*), a bison (*Bison degiulii*), and a large-sized gelada (*Theropithecus* sp.). The fauna is further characterised by several small mammals, amongst others, a vole (*Allophaiomys ruffoi*), a shrew (*Episoriculus gibberodon*), and a large porcupine (*Hystrix refossa*). A small number of flint lithic artefacts, ascribed to *Homo erectus*, were found in association with this fauna, as reported by Marta Arzarello and colleagues (2006). They suggest that this is the first evidence for humans in Europe, predating the evidence from Atapuerca in Spain and roughly contemporaneous with Dmanisi in Georgia.

The fauna was first described by De Giuli and colleagues (1987b) from the site Pirro Nord – or Cava Dell'Erba – close to the village of Apricena. The assemblage of large mammals of Pirro Nord is recognised as the Pirro Nord Faunal Unit for Western European biochronology, as defined by Elsa Gliozzi and colleagues (1997). The total absence of any endemic form excludes an insular character for Pirro Nord since the Early Pleistocene.

Evolution and Peculiarities of Endemic Mammals

Five-horned Deer: Hoplitomerycidae

The doubtlessly most striking characteristic of the Gargano ruminants is the presence of five horns, of which one projects between the eyes on the caudal part of the nasals. The other four horns arise in pairs above the orbit and can be considered pronged horns. Leinders (1984) described the cranial and dental material of these enigmatic ruminants as *H. matthei* and established a new cervoid family Hoplitomerycidae (Plate 9). Apart from the five 'horns', Hoplitomerycidae are further characterised by the presence of large, flaring, and sabre-like canines in the upper jaw (Figure 6.4) and a short, massive snout. They are considered cervoids, based on amongst others

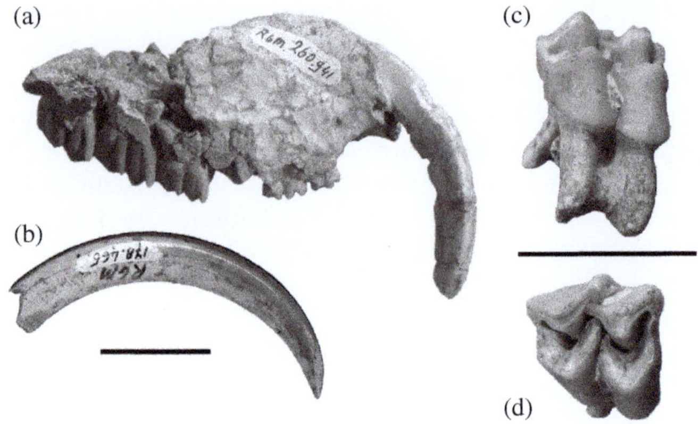

Figure 6.4 Dental and cranial elements of *Hoplitomeryx*. (a) part of a maxilla with upper canine, (b) isolated upper canine, and (c–d) upper molar. External view (a, b), lingual view (c), and occlusal view (d). Scale bar: 2 cm. National Museum of Natural History, Leiden. (Photographs: Alexandra van der Geer.)

Figure 6.5 Metatarsus of *Hoplitomeryx* showing a distally closed gully, as typical for deer, and a fused navicocuboid tarsal bone that is fused with the metatarsal, as typical for several insular deer and bovids. Dorsal view, proximal end to the right. Scale bar 5 cm. National Museum of Natural History, Leiden. (Photograph: Alexandra van der Geer.)

a deer morphology of the molars (Figure 6.4), the distally closed metatarsal gully (Figure 6.5), and the presence of a double lacrimal orifice on the rim of the orbit.

A number of diagnostic characters of *Hoplitomeryx* likely represent secondary derived parallel characters without phylogenetic value, as summarised by van der Geer (2005, 2014a, b). These are the fused metatarso-navicocuboid (Figure 6.5), the non-parallel-sided astragal, the elongated patella, and the absence of the *Palaeomeryx* fold in the lower molars. The fused metatarso-navicocuboid is a condition shared with other island ruminants, deer – e.g. *Candiacervus* from Crete – as well as bovids – e.g. *Myotragus* from the Balearics. The non-parallel-sided astragal is seen in living tragulids, in extinct hypertragulids, leptomeryids, protoceratids but also in pigs, strongly indicative of a reversal to the primitive condition, due to a relatively heavy body, typical for an insular ruminant (van der Geer, 1999). This is confirmed by *Myotragus balearicus*, in which species the astragal is slightly non-parallel sided, although not to the extent as seen in *Hoplitomeryx*. Also, the elongated patella, shared with tragulids, is best explained as due to endemic insular adaptations, and not necessarily of phylogenetic importance. The absence of the *Palaeomeryx* fold, finally, is likely linked to the higher molar crowns, since this fold is typically expressed on low crowned molars only. An increase in crown height (hypsodonty) is common for endemic insular ruminants, and in that case, the lack of the *Palaeomeryx* fold is inherent and of no phylogenetic value. The same is valid for the loss of premolars; this is seen in other insular species as well (for details, see Chapter 24). These primitive characters may confound cladistic analyses when not recognised as such, and the false impression of a primitive, Oligocene basal ruminant surviving in isolation up to the latest Miocene is easily, but understandably, created.

Leinders (1984) recognised five morphotypes in the skull material, based on the morphology of the horn cores and the ear regions. The dental elements also show great size differences, and no less than six morphotypes might be present, as described by Mazza and Rustioni (2011). The *Hoplitomeryx* limb bones are not homogeneous in size either, and range from pygmy size to an extremely large size (van der Geer, 2005, 2008) (Figure 6.6). Based on the differences in postcranial, dental, and cranial material (van der Geer, 2014a, b) two distinct, but likely phylogenetically closely related genera, can be distinguished, being *Scontromeryx* for the Scontrone (Early Tortonian) and *Hoplitomeryx* for the Gargano region (Late Tortonian or Messinian). Both genera are characterised by orbital appendages. Main differences are that *Scontromeryx* is characterised by the presence of second upper and lower premolars and no evidence for the presence of a nasal (median) horn, while *Hoplitomeryx* is characterised by the loss of the second premolar and presence of a nasal horn. This difference is best explained as an evolutionary trend towards loss of premolars, as noted by Mazza and Rustioni (2011), based on the geological time difference between the localities. *Scontromeryx* is represented by six species (*minutus*, *falcidens*, *apulicus*, *apruthiensis*, *magnus*, and *mazzai*), mainly based on lower teeth, for which the holotypes were selected by Mazza and Rustioni (2011) from Scontrone material, except for *mazzai*, for which van der Geer (2014a) assigned a holotype. *Hoplitomeryx* is represented by four species (*matthei*, *devosi*, *macpheei*, and *kriegsmani*), mainly based on limb bones, with holotypes from Gargano material. *Hoplitomeryx* species were contemporaneous, as two or more occur in each fissure and biozone, except in the youngest fissure. The two multispecies assemblages of Gargano and Scontrone represent independent adaptive radiations with the two genera as sister taxa.

Interestingly, the tallest species of Gargano – *H. kriegsmani* – is represented by juvenile material only, and the largest cranial fragment – ear region type 3, juvenile –can perhaps be

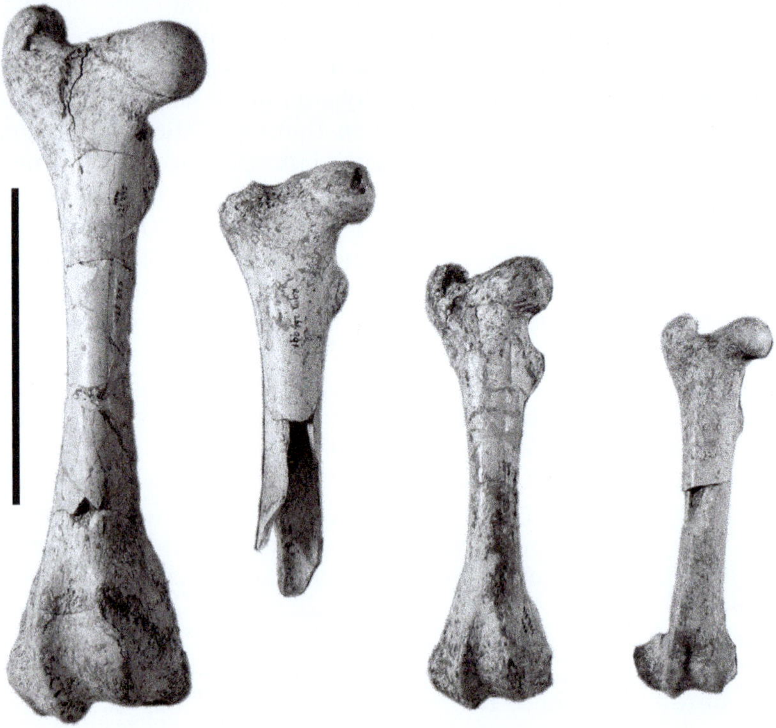

Figure 6.6 Femurs of the five-horned Late Miocene deer *Hoplitomeryx* of Gargano. The variation in body size within this insular genus is large and reflects the existence of at least four species in the postcranial material. Scale bar: 10 cm. National Museum of Natural History, Leiden. (Photograph: Alexandra van der Geer.)

associated with it. This gigantic juvenile provides an interesting parallel between *Hoplitomeryx* and the Cretan *Candiacervus major*. The latter had extremely elongated limbs, relatively and absolutely much longer than known in any mainland deer, including *Megaloceros giganteus*. *Hoplitomeryx* presents a second example of such 'giraffid / camelid' legs. From Scontrone, a possibly equally large specimen has been recovered, though the distal and proximal parts are largely missing, reported by Mazza and colleagues (2015). This specimen is, like *H. kriegsmani* and *C. major*, not only tall but also has a slender build, unlike mainland ruminants of that height such as giraffes and moose (*Alces*).

Daniel DeMiguel (2016) reconstructed the diet and ecology of *Hoplitomeryx*. He found that *Hoplitomeryx* was primarily a browser of forbs and dicots in closed areas, as reflected by the low degree of mesowear, the presence of sharp molar cusps, and low or moderately low crowns. The species differ slightly in the average crown heights, with the smallest forms having slightly higher crowns than the larger ones, but still, none has high crowns typical for grazers and graze-dominated mixed feeders. DeMiguel proposes that (slight) differences in diet and body size, and thus niche segregation, may have allowed *Hoplitomeryx* species to coexist. If so, dietary differences promoted the speciation as reflected in dental, cranial, and limb bone morphotypes. When ordering the mesowear data stratigraphically, DeMiguel indeed noticed a fluctuation in diet composition, with a trend towards more abrasive food in all species. This indicates an increasing aridification of the region, supported by dental data from the burrowing mice (*Mikrotia*) and pikas (*Prolagus*), as well as the disappearance of the hamsters. Decreasing humidity also agrees with a late Tortonian (MN11–early MN12) or Messinian (late MN12–MN13) geological age, which eventually culminated in the Messinian salinity crisis (mid-MN13) around 5.96 million years ago.

Giant Gymnures: Deinogalerix

The genus of gymnures, or moonrats, of Gargano (*Deinogalerix*) is probably the largest member of its family (Erinaceidae: hedgehogs and moonrats) that ever lived, being about 1.5 to twice the size of extant moonrat *Echinosorex*. They belong to the subfamily Galericinae and are best known by the giant form (*D. koenigswaldi*) with a skull length of approximately 20 cm (Figure 6.7). This species was described based on a

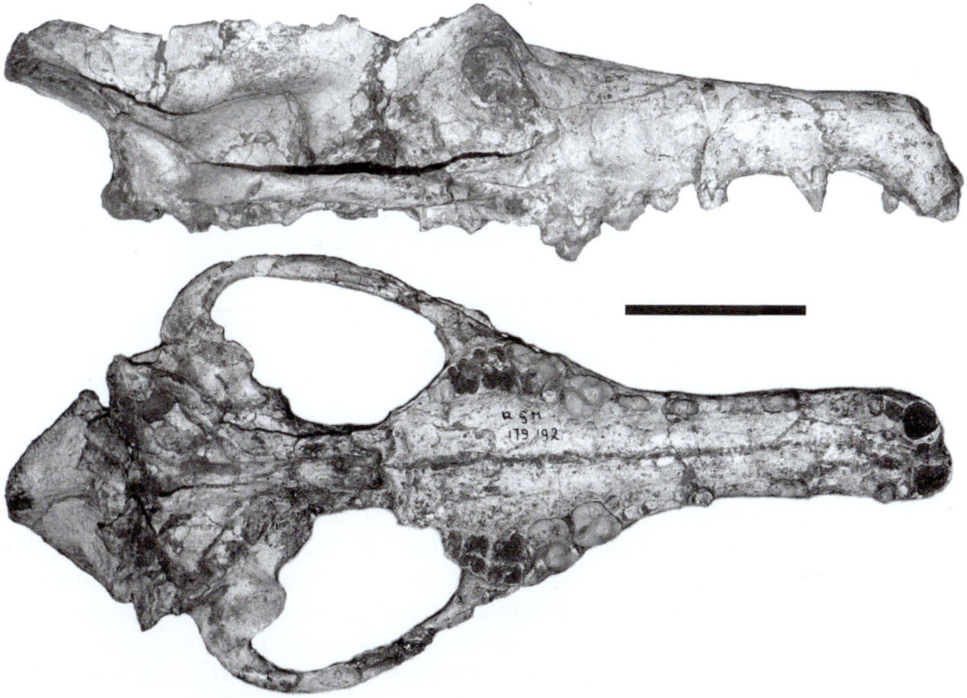

Figure 6.7 Skull of the Gargano giant gymnure (*Deinogalerix koenigswaldi*). Late Miocene. Scale bar: 5 cm. National Museum of Natural History, Leiden. (Photograph: George Lyras.)

unique, almost complete skeleton of a subadult individual by Freudenthal (1972) from San Giovannino, one of the geologically youngest fissure fillings. A second almost complete skeleton was recovered from fissure M 010, a fissure without associated fauna, hampering its chronological placement. The skeleton, described by Boris Villier (2010), belongs to a juvenile with permanent molar teeth but first upper incisor still in eruption with open root. Villier assigns the specimen tentatively to the middle-sized species *Deinogalerix brevirostris* based mainly on teeth characteristics, acknowledging the juvenile features which hamper a diagnosis.

Van den Hoek Ostende (2001) proposed the Late Miocene *Parasorex* as the ancestor of *Deinogalerix*. Main typical features of the dentition of *Deinogalerix* are a much larger upper first incisor than second and third incisors, lack of lower third incisor, lack of premolariform upper canine, and spacing between the canine and the third premolar in both upper and lower jaws in the larger species. Regarding mastication, Villier (2010) confirmed the hypothesis by Butler (1980) that *Deinogalerix* could close its jaws very fast at the expense of strength, based on the anatomy of the ascending part of the lower jaw. This would be suggestive of a diet primarily consisting of fast-moving prey, such as fish.

Butler (1980) observed a large size variation between specimens from the same fissure. He distinguished five species in total, mainly based on size, within two lineages of overlapping size and common ancestry. One lineage is, from oldest to youngest, represented by *Deinogalerix freudenthali–D. minor–D. brevirostris*, and the other by *Deinogalerix freudenthali–D. intermedius–D. koenigswaldi* in accordance with Freudenthal's biozones. Much later, Boris Villier and colleagues (2013) added a sixth species, *Deinogalerix masinii*, from the oldest fissure, M013. The largest form, *Deinogalerix koenigswaldi*, is present only in the youngest fissure (San Giovannino) in association with the most advanced form of the other lineage, *Deinogalerix brevirostris*. The one but most primitive species, *Deinogalerix freudenthali*, is present from the older fissures (Biancone 1, Rinascita 1) onwards, whereas the intermediate forms of both lineages are sporadically found in all but the oldest fissures and in association with each other at Fina H, Gervasio 1, and Chiro 20E. Also, in Fina H, the smallest species occurs. The situation is thus rather complicated and not easy to explain. To complicate matters even further, *Deinogalerix minor* is known mainly from jaws and lower teeth, whereas its supposedly direct ancestor is based on upper teeth, hampering direct comparison. Butler tentatively suggested that they might be the same species in fact. The species *freudenthali* was originally reported also from

Scontrone, but was described as a new species *Deinogalerix samniticus*, representing an independent phyletic lineage, by Savorelli and colleagues (2017) to distinguish the Scontrone gymnures from those of Gargano.

Boris Villier (2008) illustrated *Deinogalerix brevirostris* in much detail and pointed out the differences with the other species. The various species overlap in size and morphology, and Butler's species are not so clear-cut as previously assumed. Likely, the diversity within *Deinogalerix* was already present early in the sequence.

A second gymnure, *A. pusillus*, occurring throughout the stratigraphy and already noted by Freudenthal (1971, 1972), is finally named and described by Masini and Fanfani (2013), based on remains from F32, the youngest site. *Apulogalerix* is smaller than *Deinogalerix*, and characterised by, amongst others, very large and stout premolars—larger than in all other gymnures except for *Deinogalerix*—with conical, blunt and stout main cusps, and by a long premolar row relative to the molar row, indicative of a specialist snail eater. Like *Deinogalerix*, *Apulogalerix* evolved from a *Parasorex* ancestor, as shown by Lars van den Hoek Ostende (2001) in his revision of gymnures. Throughout the stratigraphy, Masini and Fanfani observed a trend in *Apulogalerix* of a gradual loss of the posterior cusp of the two lingual cusps of the upper third premolar. They attribute the loss of the third lower incisor as inherited from its *Parasorex ibericus*-like ancestor, and not an in situ acquired trait. Based on the primitive, double-rooted second premolar, they point to either a very primitive *Parasorex ibericus* or an advanced *Parasorex socialis* between 8.7 and 9.7 million years ago (MN 10, or late Vallesian).

Giant Dormice: Stertomys

The Gargano dormice, members of the subfamily of mouse-tailed dormice (Myomiminae), were first known by a giant form only, named *S. laticrestatus* by Daams and Freudenthal (1985). Its fossils were found in San Giovannino, one of the youngest localities. In size, it approaches the giant dormice of Pleistocene Malta (*Leithia*). Much later, four additional species were named by Elvira Martín-Suárez and Freudenthal (2006). These are the small *Stertomys daamsi* and the medium-sized *Stertomys daunicus* from the oldest fissure (Biancone) and the small *Stertomys simplex* and the medium-sized *Stertomys lyrifer* from the second oldest fissure (Rinascita). They further mention a *Stertomys* aff. *daamsi* from Rinascita, which is clearly smaller than the species to which it shows close affinities. In addition, Freudenthal and Martín-Suárez in the same year mentioned three unnamed

Stertomys species from Rinascita. Paolo Rinaldi and Federico Masini (2009) described yet another species, *Stertomys degiulii*, from fissure F9.

The molars of *Stertomys* are characterised by additional crests, and this complexity progressively increases from the smallest to the largest species. The molar pattern is complicated, with low and very wide crests, especially on the upper molars. In most fissures, two or more species or morphotypes are found together, one of a smaller size with simpler molar pattern, and a medium-sized or giant one with complicated molar pattern. The continental mouse-tailed dormouse *Myomimus dehmi* or another Messinian *Myomimus* is considered ancestral to *Stertomys* according to Freudenthal and Martín-Suárez (2006). A second dormouse genus, represented by *Dryomys apulus* in the oldest fissures, Biancone and Rinascita, closely resembles the continental species *Dryomys nitedula*.

The evolution of *Stertomys* is complex and suggestive of an early adaptive radiation into five or more lineages, represented by two main branches with different body size and teeth morphology (either complex or primitive), as concluded by Rinaldi and Masini (2009). This mosaic evolution is suggestive of the existence of a palaeo-archipelago or a very large island with high elevation, similar to Luzon, Philippines.

Giant Mice: Mikrotia

The Gargano mice genus *Mikrotia* is represented by several lineages and species (Figure 6.2), of which the largest, *Mikrotia magna* (the type species) has a skull length of about 10 cm (Plate 10 (a–c)). Blanca Moncunill-Solé and colleagues (2018) reconstructed its body mass to between around 1.3 kg in older localities and 1.9 kg in San Giovannino, based on the upper limb bones (humerus and femur). The heaviest individual weighted even over 3.2 kg. This estimated body mass is much higher than the about 400 g calculated by Millien and Jaeger (2001), based on incisor size. The heavier mass estimate is more in line with the very large skull, as indicated by metrics of the unrelated Southern Luzon giant cloud rat (*Phloeomys cumingi*), which has a skull length of 59.2 mm and a body mass range of 1.45–2.1 kg. The smallest species, *Mikrotia parva*, had about the size of a large *Stephanomys* of the mainland. *Mikrotia maiuscola* is of intermediate size.

The genus appears to be the only burrowing murine genus known till now, according to Virginie (Millien) Parra and colleagues (1999). Its incisors were perfectly adapted to burrowing, in parallel with those of voles. Its diet was abrasive, likely including rhizomes and roots. The shape of the lower jaw, the elongation of the lower first molar and of the upper third molar, and the hypsodonty of all teeth reminds of that of the extant mainland vole *Microtus*, hence the original choice of the name *Microtia* for this genus – *Mikrotia* is *nomen novum* for *Microtia*, see Freudenthal's correction in 2006.

The largest species – *Mikrotia magna* – lived long, based on the high number of seasonal growth lines – 15 in total – as briefly reported by Moncunill-Solé and colleagues (2013) at a conference. This is not unexpected, considering its huge size and the parallel provided by the extant Luzon giant cloud rats, who live over ten years in captivity. Longevity of *Mikrotia* was confirmed by Christian Kolb and colleagues (2015a, b), who observed a high amount of remodelling bone in a sectioned femur, typical for very old individuals.

The Gargano burrowing mice follow a tendency towards more complex molar patterns. The lower molars of *Mikrotia parva* from the early locality Biancone resemble those of *Parapodemus*, yet differ in some features. Two smaller forms are restricted to Rinascita and Trefossi, in time just following Biancone. Their molars are morphologically much closer to those of *Stephanomys* from the mainland. Initially, either *Stephanomys* or *Apodemus* were mentioned as possible ancestor. However, the ancestral form from the oldest fissure (M013) – provisionally referred to as Murinae n. gen., sp. n. – is considered unrelated to *Stephanomys* or *Apodemus*, while a relation to *Parapodemus* cannot be ruled out, according to Masini and colleagues (2013). In a conference, Masini and colleagues (2014) mentioned *Hansdebruijnia* as the most plausible ancestor.

A second endemic mouse, *Apatodemus*, has been described by Andrea Savorelli and colleagues (2016). The genus occurs in all biozones but is absent from the oldest fissure, M013. They distinguish two species, *A. degiulii* (type species) and a yet unnamed and larger species. A moderate increase in size is observed between the two species. *Apatodemus* shows a combination of primitive and derived features, obscuring its ancestry, including, for example, the derived feature of the presence of a seventh tubercle on the first and second upper molars, and the primitive feature of lower crowned molars. In all likelihood, such enigmatic mosaic is due to insularity, where a number of derived features are lost – a secondary return to the primitive condition – while others are retained. Savorelli and colleagues list the Eastern European genus *Hansdebruijnia* as a probable ancestor, in parallel with that of *Mikrotia*. This would indicate a dispersal route from the Balkan Peninsula, where the genus occurs at Maramena in Greece (MN13; 4.9–7.2 million years ago). Already, Freudenthal and Martín-Suárez (2006) remarked the similar taxonomic composition and age of the Gargano and Maramena small mammal faunas.

Giant Pikas: Prolagus

Lagomorphs are represented by two species of pika (Figure 6.8), of which the younger species (*Prolagus imperialis*) is not only much larger than the older species (*Prolagus apricenicus*, Plate 10 (d)) but also larger than any other known *Prolagus* known so far, as calculated by Moncunill-Solé (2021), first observed by Freudenthal (1971). The oldest species shows the same general size as the *Prolagus* in other Miocene localities in Europe. This smaller species underwent a gradual size increase and continues into the youngest localities. In the two youngest localities, it occurs together with the giant species. The pikas further show a progressive abundance: they are very rare in the oldest localities but constitute the dominant element in the

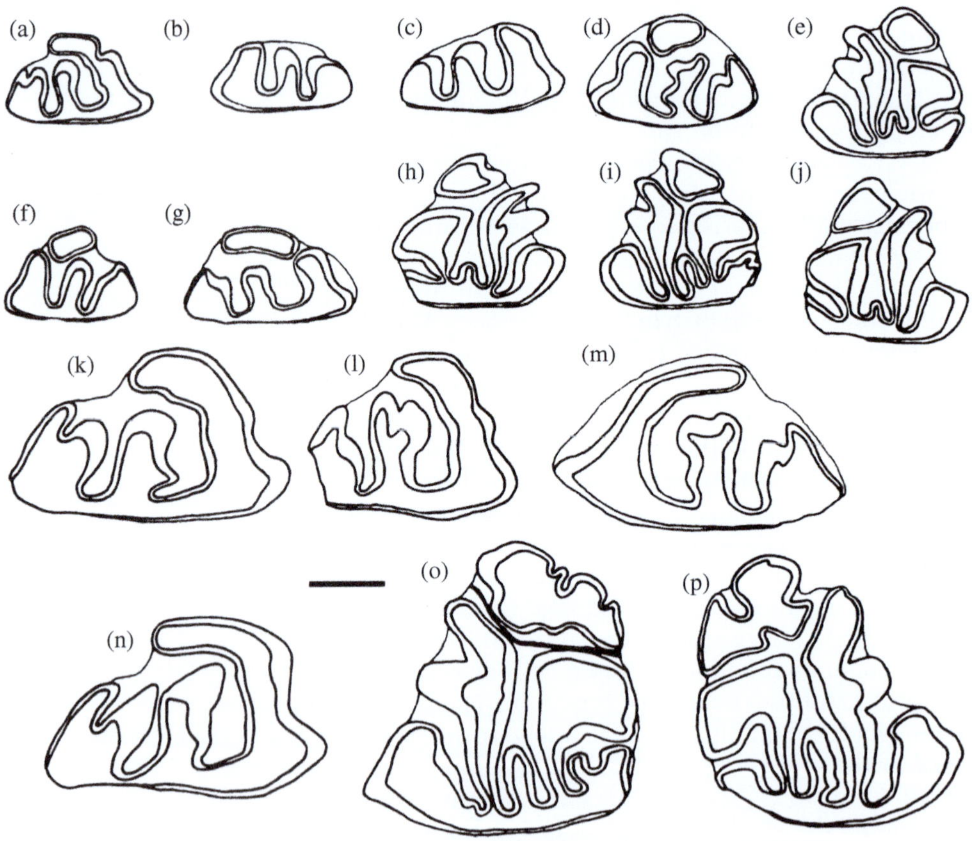

Figure 6.8 *Prolagus* molars from the Gargano fissure fillings, occlusal view. (a–j), *Prolagus apricenicus*, (k–p), *Prolagus imperialis*. (a) holotype, right upper P2; (b) left upper P2; (c) right upper P2; (d) left upper P2; (e) right lower P3; (f) left upper P2; (g) right lower P3; (h) left lower P3; (i) right lower P3; (j) left lower P3; (k) holotype, right upper P2; (l) right upper P2; (m) left upper P2; (n) right upper P2; (o) left lower P3; and (p) right lower P3. Scale bar: 1 mm. (Adapted from Mazza and colleagues, 1995.)

younger localities. The older species most closely resembles the continental *Prolagus oeningensis* in dental morphology, according to Chiara Angelone (2007). Apart from size increase, a progressive lengthening and complication of enamel crests in the lower third premolar took place, possibly in response to an increasingly abrasive diet. The characters present in the smaller species are all present in the giant species but more pronounced, which calls for a common origin. The adaptive success of the giant species seems to have been limited, because the giant is outnumbered by its smaller sister species in the youngest locality (F32). A complete skeleton attributed to the more primitive species (*Prolagus* cf. *Prolagus. apricenicus*) has been found at Capo di Fiume, Campania, also in southern Italy. Comparison by Angelone (2007) with a complete skeleton of *Prolagus crusafonti* that has been preserved in much the same way led her to the conclusion that it resembles most closely *Prolagus oeningensis*. No taxonomic attribution is possible yet, because diagnostic features are not available. Angelone suggested colonisation of Gargano by an *oeningensis*-like species during the earliest Messinian transgression of the Late Miocene.

Giant Hamsters: Hattomys

The lineage of giant hamsters is represented by three species (*H. beetsi*, *H. nazarii*, and *H. gargantua*) with progressively increasing body size, as inferred from molar size. In addition to an evolution towards a larger body mass, Savorelli (2013) noted a marked trend in the molars towards, amongst others, thicker enamel, higher crests, deeper valleys, a larger mesoloph, and equal-sized cusps of the anteroconid. The three species, however, cannot be delimited sharply, according to Freudenthal (1985), hence his transitional or intermediate form (*H. nazarii-gargantua*) instead of a fourth species. The smallest *Hattomys* was about as large as the common hamster (*Cricetus cricetus*, head and body length between 200 and 340 mm), and its fossils occur in the three oldest localities (Biancone, Rinascita and Trefossi). In the very same localities, remains of Late Miocene (Turolian) mainland hamsters have also been found: *Cricetulodon* in all three fissures, *Megacricetodon* in Biancone, and *Cricetus* in Rinascita. It is thus likely to assume that hamsters colonised Gargano in the Late Miocene. The presence of a mainland *Apodemus* in the oldest fissures is in line with this. The mainland hamsters are lacking in all younger fissures. One of these mainland hamsters likely gave rise to the lineage of giant hamsters. The molars of *Hattomys* resemble those of *Cricetulodon* most, but are larger. In addition, the anterior part of the first molar is more complex, the third molars are less reduced, and the mesolophids and anterior hypolophulids are different.

A second endemic hamster, *Mystemys giganteus*, is described by Savorelli and Masini (2016). This giant hamster has exceptionally large first upper molars, up to 6.23 mm long, in combination with a primitive molar pattern, moderately high-crowned molars with robust and swollen cusps. The genus can be referred to the subfamily of Cricetinae.

Gargano Otter: Paralutra

The Gargano otter (*P. garganensis*) was perhaps larger than its most likely mainland ancestor *Paralutra jaegeri*. The first molar of the latter species is very characteristic and strongly resembles that of the Gargano otter. The fourth premolar (Figure 6.3) of the Gargano species has, however, a larger talon than the mainland species, which justifies a separate species for the Gargano otter. Its body size seems to have been increased somewhat, at least based on the dental measurements. The relatively larger talon of the fourth premolar indicates a larger proportion of shellfish in its diet. A shift in diet is not uncommon for island species. The calcaneus is very unlike that of sea otters (*Enhydra*) but resembles more that of the semi-terrestrial common otter (*Lutra*). It can thus be inferred that the Gargano otter was either semi-terrestrial or somewhat more terrestrial like the smooth-coated otter (*Lutrogale*), while a strictly aquatic lifestyle can be excluded.

The retrieved material is extremely poor, limited to a left upper jaw fragment (the type specimen) with fourth premolar and first molar and a calcaneus (the paratype) from Fina H, described by Willemsen (1983), and a left upper jaw fragment with fourth premolar from Fina 9, an epistropheus (axis vertebra) from Gervasio 1, and a left upper canine from Fina H, described by Villier and colleagues (2011). Apart from these five otter elements, a juvenile humerus of a further unidentified otter species was recovered from fissure F4/A at Cava Dell'Erba. Villier and colleagues (2011) notice similarities with the common otter (*Lutra lutra*) rather than with *Paralutra jaegeri*.

References

Abbazzi L., Benvenuti M., Boschian G., Dominici S., Masini F., Mezzabotta C., Piccini L., Rook L., Valleri G., Torre D. 1996. Revision of the Neogene and Pleistocene of the Gargano region (Apulia, Italy). The marine and continental successions and the mammal faunal assemblages in an area between Apricena and Poggio Imperiale (Foggia). *Memorie della Società Geologica Italiana*, 51: 383–402.

Angelone C. 2007. Messinian *Prolagus* (Ochotonidae, Lagomorpha) of Italy. *Geobios*, 40: 407–421.
Arzarello M., Marcolini F., Pavia G., Pavia M., Petronio C., Petrucci M., Rook L., Sardella, R. 2006. Evidence of earliest human occurrence in Europe: The site of Pirro Nord (Southern Italy). *Naturwissenschaften*, 94 (2): 107–112.
Ballmann P. 1973. Fossile Vögel aus dem Neogen der Halbinsel Gargano (Italien). *Scripta Geologica*, 17: 1–75.
Ballmann P. 1976. Fossile Vögel aus dem Neogen der Halbinsel Gargano (Italien) zweiter Teil. *Scripta Geologica*, 38: 1–59.
Butler M. 1980. The giant erinaceid insectivore, *Deinogalerix* Freudenthal, from the upper Miocene of Gargano, Italy. *Scripta Geologica*, 57: 1–72.
Daams R., Freudenthal M. 1985. *Stertomys laticrestatus*, a new glirid (dormice, Rodentia) from the insular fauna of Gargano (Prov. of Foggia, Italy). *Scripta Geologica*, 77: 21–27.
De Giuli C., Masini F., Torre D. 1985. Effetto arcipelago: Un esempio nelle faune fossili del Gargano. *Bollettino della Società Paleontologica Italiana*, 24 (2–3): 191–193.
De Giuli C., Masini F., Torre D., Boddi V. 1987 (a). Endemism and bio-chronological reconstructions: The Gargano case history. *Bolletino della Società Paleontologica Italiana*, 25: 267–276.
De Giuli C., Masini F., Torre D. 1987 (b). The latest Villafranchian faunas of Italy. The Pirro Nord local fauna (Gargano). *Palaeontografia Italiana*, 74: 52–62.
De Giuli C., Masini F., Torre D., Boddi V. 1986. Endemism and bio-chronological reconstructions: The Gargano case history. *Bollettino della Società Paleontologica Italiana* 25 (3): 267–276.
De Giuli C., Torre D. 1984. Species interrelationships and evolution in the Pliocene endemic faunas of Apricena (Gargano Peninsula – Italy). *Géobios (Mémoire Spécial)*, 8: 379–383.
Delfino M., Rossi M.A. 2013. Fossil crocodylid remains from Scontrone (Tortonian, Southern Italy) and the late Neogene Mediterranean biogeography of crocodylians. *Geobios*, 46: 25–31.
DeMiguel D. 2016. Disentangling adaptive evolutionary radiations and the role of diet in promoting diversification on islands. *Scientific Reports*, 6: 29803
Freudenthal M. 1971. Neogene vertebrates from the Gargano Peninsula, Italy. *Scripta Geologica*, 3: 1–10.
Freudenthal M. 1972. *Deinogalerix koenigswaldi* nov. gen., nov. spec., a giant insectivore from the Neogene of Italy. *Scripta Geologica*, 14: 1–19.
Freudenthal M. 1976. Rodent stratigraphy of some Miocene fissure fillings in Gargano (prov. Foggia, Italy). *Scripta Geologica*, 37: 1–23.
Freudenthal M. 1985. Cricetidae (Rodentia) from the Neogene of Gargano (Prov. of Foggia, Italy). *Scripta Geologica*, 77: 29–74.
Freudenthal M. 2006. *Mikrotia* nomen novum for *Microtia* Freudenthal 1976 (Mammalia, Rodentia). *Journal of Vertebrate Paleontology*, 26 (3): 784.
Freudenthal M., Martín-Suárez E. 2006. Gliridae (Rodentia, Mammalia) from the Late Miocene fissure filling Biancone 1

(Gargano, Province of Foggia, Italy). *Palaeontologia Electronica*, 9 (2): 6A, 1–23.

Freudenthal M., van den Hoek Ostende L.W., Martín-Suárez E. 2013. When and how did the *Mikrotia* fauna reach Gargano (Apulia, Italy)? *Geobios*, 46: 105–109.

Gliozzi E., Abbazzi L., Azzaroli A., Caloi L., Capasso Barbato L., Di Stefano G., Esu D., Ficcarelli G., Girotti O., Kotsakis T., Masini F., Mazza P., Mezzabotta C., Palombo M.R., Petronio C., Rook L., Sala B., Sardella R., Zanalda E., Torre D. 1997. Biochronology of selected mammals, molluscs and ostracods from the Middle Pliocene to the Late Pleistocene in Italy. The state of the art. *Rivista Italiana Paleontologia e Stratigrafia*, 90: 369–388.

Kolb C., Scheyer T.M., Lister A.M., Azorit C., de Vos J., Schlingemann M.A.J., Rössner G.E., Monaghan N.T., Sánchez-Villagra M.R. 2015a. Growth in fossil and extant deer and implications for body size and life history evolution. *BMC Evolutionary Biology*, 15: 1–15.

Kolb C., Scheyer T.M., Veitschegger K., Forasiei A.M., Amson E., van der Geer A.A.E., van den Hoek Ostende L.W., Hayashi S., Sánchez-Villagra M.R. 2015b. Mammalian bone paleohistology: A survey and new data with emphasis on island forms. *PeerJ*, 3: e1358.

Kotsakis T., Delfino M., Piras P. 2004. Italian Cenozoic crocodilians: Taxa, timing and palaeobiogeographic implications. *Palaeogeography, Palaeoclimatology, Palaeoecology*, 210: 67–87.

Leinders J.J.M. 1984. Hoplitomerycidae fam. nov. (Ruminantia, Mammalia) from Neogene fissure fillings in Gargano (Italy); part 1: The cranial osteology of *Hoplitomeryx* gen. nov. and a discussion on the classification of pecoran families. *Scripta Geologica*, 70: 1–51.

Martín-Suárez E., Freudenthal F. 2006. Gliridae (Rodentia, Mammalia) from the late Miocene fissure filling Rinascita 1 (Gargano, prov. Foggia, Italy). *Treballs del Museu de Geologia de Barcelona*, 14: 37–59.

Masini F., Fanfani F. 2013. *Apulogalerix pusillus* nov. gen., nov. sp., the small-sized Galericinae (Erinaceidae, Mammalia) from the "Terre Rosse" fissure filling of the Gargano (Foggia, South-Eastern Italy). *Geobios*, 46: 89–104.

Masini F., Petruso D., Bonfiglio L., Mangano G. 2008. Origin and extinction patterns of mammals in three central Western Mediterranean Islands from the Late Miocene to Quaternary. *Quaternary International*, 182: 63–79.

Masini F., Rinaldi P.M., Savorelli A., Pavia M. 2013. A new small mammal assemblage from the M013 Terre Rosse fissure filling (Gargano, South-Eastern Italy). *Geobios*, 46: 49–61.

Masini F., Savorelli A., Maul L.C. 2014. New data on the evolution of *Mikrotia* (Muridae, Rodentia) in the Late Miocene of Gargano (Apulia, Italy). In: Dela Pierre F., Lozar F., Natalicchio M. (eds), *Abstract Book, RCMNS Interim Colloquium, 25–28 September 2014, Torino (Italy): 38*. Museo Regionale di Scienze Naturali, Turin, Italy.

Mazza P. 1987. *Prolagus apricenicus* and *Prolagus imperialis*: Two new Ochotonids (Lagomorpha, Mammalia) of the Gargano (Southern Italy). *Bollettino della Società Paleontologica Italiana*, 26 (3): 233–243.

Mazza P., Rustioni M., 1996. The Turolian fossil artiodactyls from Scontrone (Abruzzi, Central Italy) and their paleoecological and paleogeographical implications. *Bollettino della Società Paleontologica Italiana*, 35: 93–106.

Mazza P., Rustioni M., 2008. Processes of island colonization by Oligo-Miocene land mammals in the central Mediterranean: New data from Scontrone (Abruzzi, Central Italy) and Gargano (Apulia, Southern Italy). *Palaeogeography Palaeoclimatology Palaeoecology*, 267: 208–215.

Mazza P., Rustioni M., Aruta G., Di Carlo E. 1995. A Messinian *Prolagus* from Capo di Fiume Quarry (Palena, Abruzzi, Central Italy). *Bollettino della Società Paleontologica Italiana*, 34: 55–66.

Mazza P.P.A., Rossi M.A., Agostini S. 2015. Hoplitomerycidae (Late Miocene, Italy), an example of giantism in insular ruminants. *Journal of Mammal Evolution*, 22: 271–277.

Mazza P.P.A., Rustioni M. 2011. Five new species of *Hoplitomeryx* from the Neogene of Abruzzo and Apulia (central and southern Italy) with revision of the genus and of *Hoplitomeryx matthei* Leinders, 1983. *Zoological Journal of the Linnean Society*, 163: 1304–1333.

Meijer H.J.M. 2014. A peculiar anseriform (Aves: Anseriformes) from the Miocene of Gargano (Italy). *Comptes Rendus Palevol*, 13: 19–26.

Millien V., Jaeger, J.-J. 2001. Size evolution of the lower incisor of *Microtia*, a genus of endemic murine rodents from the late Neogene of Gargano, southern Italy. *Paleobiology*, 27 (2): 379–391.

Moncunill-Solé B. 2021. Eco-evolutionary adaptations of ochotonids (Mammalia: Lagomorpha) to islands: new insights into Late Miocene pikas from the Gargano palaeo-archipelago (Italy). *Biological Journal of the Linnean Society*, 132: 400–413.

Moncunill-Solé B., Jordana X., Köhler M. 2018. Where did *Mikrotia magna* originate? Drawing ecogeographical inferences from body mass reconstructions. *Geobios*, 51: 359–366.

Moncunill-Solé B., Marín-Moratalla N., Jordana X., Casanovas-Vilar I., Rook L., Köhler M. 2013. Preliminary results on life history traits of the insular fossil rodent *Mikrotia* (Muridae, Rodentia) *from Gargano archipelago* (Apulia, Italy). In: Lamm E.T., Bailleul A., Flynn K. (eds), *Book of Abstracts of the Second International Symposium on Paleohistology. Bozeman, Montana, July 18–20, 2013*: 68. Museum of the Rockies, Bozeman.

Moyà-Solà S., Quintana J., Alcover J.A., Köhler M. 1999. Endemic island faunas of the Mediterranean Miocene. In: Rössner G.E., Heissig K. (eds), *The Miocene Land Mammals of Europe*. Pfeil, Munchen, pp. 435–442.

Parra V., Jaeger J.-J., Bocherens H. 1999. The skull of *Microtia* (Freudenthal, 1976), an extinct burrowing murine rodent genus of the late Neogene of Gargano, South Italy. *Lethaia*, 32 (2): 89–100.

Patacca E., Scandone P., Carnevale G. 2013. The Miocene vertebrate-bearing deposits of Scontrone (Abruzzo, Central Italy): Stratigraphic and paleoenvironmental analysis. *Geobios*, 46: 5–23.

Patacca E., Scandone P., Mazza P., 2008. The Miocene land-vertebrate fossil site of Scontrone (Central Apennines). *Bollettino della Società Geologica Italiana*, 127 (1): 51–73.

Pavia M. 2013. The Anatidae and Scolopacidae (Aves: Anseriformes, Charadriiformes) from the late Neogene of Gargano, Italy. *Geobios*, 46: 43–48.

Pavia M., Meijer H.J.M., Rossi M.A., Göhlich U.B. 2017 The extreme insular adaptation of *Garganornis ballmanni* Meijer, 2014: A giant Anseriformes of the Neogene of the Mediterranean Basin. *Royal Society Open Science*, 4: 160722.

Rinaldi P.M., Masini F. 2009. New data on the taxonomy of the endemic Myomiminae (Gliridae, Rodentia) from the Late Miocene-Early Pliocene of Gargano (southern Italy) with the description of the new species *Stertomys degiulii*. *Bollettino della Società Paleontologica Italiana*, 48 (3): 189–233.

Rook L., Mazza P., Rustioni M., Torre D. 2000. Lands and endemic mammals in the Late Miocene of Italy: Paleogeographic outlines of Tyrrhenian and Adriatic areas between 11–9 and 7–4 Ma. *'State of the Art' Workshop, Lyon, France, 16–18 November 2000. Environments and Ecosystem Dynamics of the Eurasian Neogene (EEDEN).* http://www.esf.org.

Rustioni M., Mazza P., Azzaroli A., Boscagli G., Cozzini F., Di Vito E., Masseti M., Pisanò A. 1992. Miocene Vertebrate remains from Scontrone, National Park of Abruzzi, Central Italy. *Rendiconti della Classe Scienze Fisiche, Matematiche e Naturali dell'Accademia dei Lincei*, 9: 227–237.

Savorelli A. 2013. New data on the Cricetidae from the Miocene "Terre Rosse" of Gargano (Apulia, Italy). *Geobios*, 46: 77–88.

Savorelli A, Masini F. 2016. *Mystemys giganteus* n. gen. et sp.: An enigmatic and rare cricetid from the Terre Rosse M013 fissure filling (Gargano, Southeastern Italy). *Palaeontographica, A: Palaeozoology-Stratigraphy*, 306: 1–23.

Savorelli A., Colombero S., Masini F. 2016. *Apatodemus degiulii* n. gen. et sp. (Rodentia, Muridae), a hitherto undescribed endemite from the Terre Rosse of Gargano (Late Miocene, Southeastern Italy). *Palaeontographica, A: Palaeozoology-Stratigraphy*, 306: 25–49.

Savorelli A., Masini F., Mazza P.P.A., Rossi M.A., Agostini S. 2017. New species of *Deinogalerix* (Mammalia, Eulipotyphla) from the late Miocene of Scontrone (Abruzzo, central Italy). *Palaeontologia Electronica*, 20.1: 16A.

van den Hoek Ostende L.W., 2001. A revised generic classification of the Galericini (Insectivora, Mammalia) with some remarks on their palaeobiogeography and phylogeny. *Géobios*, 34 (6): 681–695.

van den Hoek Ostende L.W., Meijer H.J.M., van der Geer A.A.E., 2009. A bridge too far, reply to "Processes of island colonization by Oligo–Miocene land mammals in the central Mediterranean: New data from Scontrone (Abruzzo, Central Italy) and Gargano (Apulia, Southern Italy)" by Mazza P. P.A. and Rustioni M. 2008, *Palaeogeography, Palaeoclimatology, Palaeoecology*, 267: 208–215. *Palaeogeography, Palaeoclimatology, Palaeoecology*, 279: 128–130.

van der Geer A.A.E. 1999. On the astragalus of the Miocene endemic deer *Hoplitomeryx* from the Gargano (Italy). *Deinsea*, 7: 325–336.

van der Geer A.A.E. 2005. The postcranial of the deer Hoplitomeryx (Mio-Pliocene; Italy): Another example of adaptive radiation on Eastern Mediterranean Islands. In: Alcover J.A., Bover P. (eds),

Proceedings of the International Symposium Insular Vertebrate Evolution: The Palaeontological Approach. Monografies de la Societat d'Història Natural de les Balears, 12: 325–336.

van der Geer A.A.E. 2008. The effect of insularity on the Eastern Mediterranean early cervoid *Hoplitomeryx*: The study of the forelimb. *Quaternary International*, 182 (1): 145–159.

van der Geer A.A.E. 2014 (a). Systematic revision of the family Hoplitomerycidae Leinders, 1984 (Artiodactyla: Cervoidea), with the description of a new genus and four new species. *Zootaxa*, 3847: 1–32.

van der Geer A.A.E. 2014 (b). Parallel patterns and trends in functional structures in island mammals. *Integrative Zoology*, 9: 167–182.

Villier B. 2008. *Les faunes endemiques Miocènes de Gargano (Italie). Nouvelles decouvertes attribues au genre Deinogalerix*. Unpublished MSc thesis, Muséum National d'Histoire Naturelle, Paris.

Villier B. 2010. *Deinogalerix*: A giant hedgehog from the Miocene. *Annali dell'Università di Ferrara*, 6: 93–102.

Villier B., Pavia M., Rook L. 2011. New remains of *Paralutra garganensis* Willemsen, 1983 (Mustelidae, Lutrinae) from the Late Miocene "Terre Rosse" of Gargano (Apulia, Italy). *Bollettino della Società Paleontologica Italiana*, 50 (2): 135–143.

Villier B., van den Hoek Ostende L.W., de Vos J., Pavia M. 2013. New discoveries on the giant hedgehog *Deinogalerix* from the Miocene of Gargano (Apulia, Italy). *Geobios*, 46: 63–75.

Willemsen G.F. 1983. *Paralutra garganensis* sp. nov. (Mustelidae, Lutrinae), a new otter from the Miocene of Gargano, Italy. *Scripta Geologica*, 72: 1–8.

CHAPTER SEVEN

Sicily

Evolution of Island Mammals: Adaptation and Extinction of Placental Mammals on Islands, Second Edition.
Alexandra van der Geer, George Lyras and John de Vos.

© 2021 John Wiley & Sons Ltd. Published 2021 by John Wiley & Sons Ltd.

Sicily is a large island in the central Mediterranean, close to the south-western tip of the Italian peninsula. During the Pleistocene, Sicily was inhabited by successive endemic faunas characterised by first only small mammals and a marten then by dwarf elephants, dwarf hippos, and giant dormice. Gradually, Sicily lost its isolation, and an increasing number of mainland taxa populated the island. In the early Middle Pleistocene, Sicily was connected to Malta, sharing faunal elements. This connection got lost before the end of the Late Pleistocene, and both islands harboured their own endemic faunas.

Geology and Geography

During the Pliocene and the Early Pleistocene, Sicily consisted of two islands, separated by deep marine basins, one coinciding with the northern part of the present-day island and the other one with the south-eastern part, as described by Laura Bonfiglio and Piperno (1996). Southern Calabria too consisted of several smaller islands – the present-day Cape Vaticano, Aspromonte, and Serre – so Sicily could be reached only through a series of islands. The end of the Early Pleistocene is characterised by volcanic activity and a strong uplift of Sicily and Calabria. During colder stages of the Middle and the Late Pleistocene, the sea level was again lower, and Sicily had acquired its present-day shape though was connected to Malta. The marine cycles in interaction with uplift caused the development of marine terraces and caves in ancient shorelines. The distance between Sicily and Calabria, the Strait of Messina, varied in accordance with changing sea levels during the Ice Ages. Apart from this strait, the isthmus of Catanzaro between the north and south part of Calabria also played an important filter role in the colonisation events of Sicily and Malta.

Historical Palaeontology

Literally, tonnes of fossil vertebrate bones had been excavated at the Sicilian caves to serve as carbon for the refinement of sugar. For example, by the late 1820s, the San Ciro cave near Palermo (Figure 7.1) alone had provided at least 20 tonnes of fossil bones, for the greater part belonging to extinct hippos. Unfortunately, these fossil bones were subsequently all discarded as useless by the refinery due to the heavy fossilisation. Eventually, the Bourbon government intervened and put the abbot Scinà in charge of the excavation of this cave and the guarding of the findings. The aim was to settle a dispute between Antonio Bivona Bernardi, who had identified the

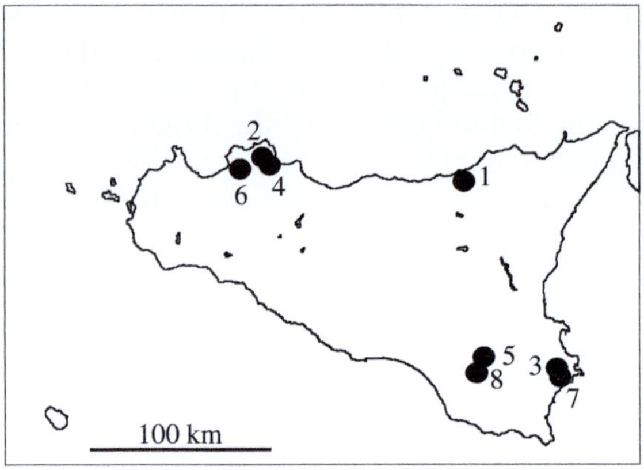

Figure 7.1 Pleistocene localities of Sicily. (1) San Teodoro cave, (2) Monte Pellegrino, (3) Spinallo caves, (4) Luparello cave, (5) Iblei (=Hyblean) Plateau, (6) Puntali cave, (7) Acquedolci, and (8) Ragusa.

fossil remains as such, and the supporters of the existence of giants in the past (for details, see Bruno Accordi, 1955). This would lead to the very start of palaeontological interest in Sicily in early 1830. The excavated fossils were sent to Georges Cuvier in Paris, who confirmed the true nature of the remains. Scinà mentioned in his report of 1831 small elephant molars and two types of hippos of different size.

The first scientific description of endemic mammals from Sicily was published a year after Scinà's report. In 1832, the Irish naturalist Joseph Pentland briefly reported some hippopotamus fossils from the Ben Fratelli cave near Palermo. He remarked that they belong to a new species, without however suggesting a name for it. Christian Hermann von Meyer seized this opportunity the same year and named the new species *Hippopotamus pentlandi* in honour of its discoverer, adding that it was 'hardly larger than our large domestic ox' (Plate 11).

In 1859, Francesco Anca from Palermo searched for fossiliferous caves, advised to do so by his friend Hugh Falconer. In Grotta di San Teodoro, or San Teodoro cave, near Acquedolci in north-eastern Sicily, he discovered evidence for human occupation dating back to the Palaeolithic. In an older layer, he found bones of a rather large 'cave elephant' together with remains of the spotted hyena or 'cave hyena'. He sent two elephant molars and some twenty jaws of the hyena to Falconer in England for identification. Falconer considered Anca's elephant identical to the living African elephant and the hyena identical to the living African spotted hyena. In a letter written on 9 July 1860 to Charles Darwin, Falconer wrote that this was the proof that Sicily and Africa had been connected during a recent geological period, because remains of living species had been found in association with those of extinct species (see, also, Chapter 2). [Decades later, Vaufrey (1929) showed that the Sicilian small elephant bears no direct relation with the African elephants (*Loxodonta*). This put an end to the hypothesis of a dry land connection between Sicily and Africa as postulated by Falconer.]

Between 1868 and 1870, Gaetano Giorgio Gemmellaro, after whom the current natural history museum of Palermo is named, extensively excavated Puntali cave, near Carini. He retrieved hundreds of mainly elephant remains, including complete skulls, which belonged to a middle-sized dwarf species (Plate 12). The cave was known before to contain remains of 'giant men', as reported by the Dominican friar Tommaso Fazello (1558) in his encyclopaedic history of Sicily.

Leith Adams (1874) described the elephant fossils from Mnaidra on Malta as *Elephas* (= *Palaeoloxodon*) *mnaidriensis*. Based on the similar size, the Sicilian remains from San Teodoro cave and Puntali cave were considered conspecific with the former. Since then, the scientific names of the three Maltese elephants – as well as those of the dormice – were applied to Sicilian elephants, leading to taxonomic confusion and biogeographic errors.

In 1886, Antonio De Gregorio reported sediments with vertebrate fossils on Mount Pellegrino not far from Palermo. He immediately recognised the importance of these fossils, so different from the ones that were known from the rest of Sicily. De Gregorio (1886) described three new species: the gundi *Pellegrinia panormensis*, the mustelid *Mustela arzilla*, and the mouse *Mus piletus*. During his further excavations, he found more material and renamed his earlier mentioned *Lepus cuniculus* into *Lepus* sp. n. (De Gregorio, 1925), without further naming it. At present, it is known as the endemic hare *Hypolagus peregrinus*. Furthermore, De Gregorio concluded that his earlier described mouse must be related to the fat dormouse, based on its aberrant large size. He thus synonymised his murid with the large Maltese dormouse *Myoxus* (=*Leithia*) *melitensis*, described by Leith Adams (1863), but with the addition var. *piletus*. De Gregorio erected a new genus for his mustelid, *Mustelercta* (Figure 7.2) Almost fifty years later, the French palaeontologist Louis Thaler (1972) discovered a new site on the western slope of Mount Pellegrino and described two new rodent species from here, a large dormouse resembling the one from Malta (*Leithia* sp.), and the giant field mouse, *Apodemus maximus*.

Over the years, thousands of fossil remains have been recovered from various deposits of different ages by various teams, apart from the many illegal excavations for trade. Noteworthy are those by Vaufrey (Luparello cave, late 1920s), G. Bonafede and R. Fabiani (e.g. Cannita cave), Graziosi (San Teodoro cave, 1943–1947), Bruno Accordi (e.g. Spinagallo cave, 1958 to 1960), Laura Bonfiglio (Acquedolci, since 1982), and Fabio Martini (San Vito area, since 2004).

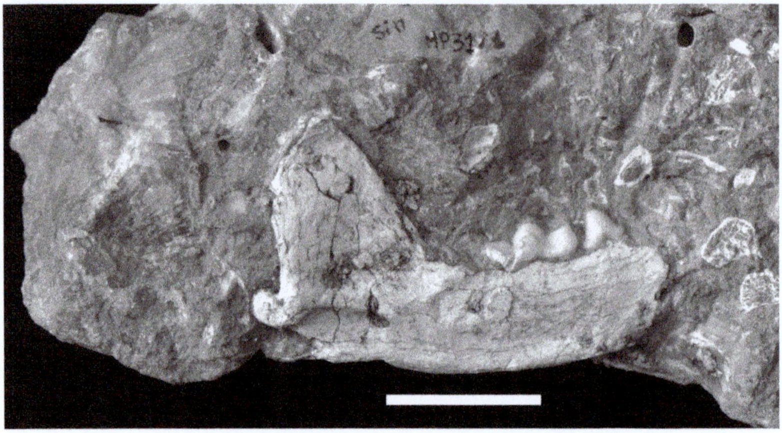

Figure 7.2 Right mandible of the marten (*Mustelercta arzilla*) of the Early Pleistocene Pelegrino fauna. Scale bar: 2 cm. Museo Geologico G.G. Gemmelaro, Palermo, Sicily. (Photograph: Bartholomeus van der Geer.)

Biozones and Faunal Units

The Pleistocene vertebrate assemblages of Sicily are generally known as *Faunal Complexes*, as first proposed by Bonfiglio and colleagues (2000) and updated by Masini and colleagues (2008). The biozones of Sicily are partly based upon the occurrence of elephant species (*Palaeoloxodon falconeri*, *Palaeoloxodon* aff. *mnaidriensis*) (Figure 7.3), species that were originally defined for Malta (see also Chapter 8 and below).

Figure 7.3 Two Sicilian dwarf elephants (*Palaeoloxodon falconeri*, *Palaeoloxodon* aff. *mnaidriensis*) from subsequent biozones compared to their mainland relative (*Palaeoloxodon antiquus*) from Italy. Forschungsinstitut Senckenberg, Frankfurt am Main, Germany; Museo Geologico G.G. Gemmelaro, Palermo, Sicily; Palaeontological Museum, University La Sapienza, Rome. (Photographs: George Lyras.)

Early Pleistocene

The oldest endemic fauna of Sicily has been found only on Monte Pellegrino and consists of a marten (*Mustelercta arzilla*), a large red-toothed shrew (*Asoriculus burgioi*), a large field mouse (*Apodemus maximus*), two large dormice (*Leithia* sp. and *Maltamys* cf. *gollcheri*), a gundi (*Pellegrinia panormensis*; Figure 7.4), and a hare (*Hypolagus peregrinus*); large mammals were not retrieved so far. The fauna is generally known as the *Monte Pellegrino Faunal Complex* after the single known locality.

The dormouse *Leithia* is likely a descendant of a common dormouse of the genus *Eliomys*. The Monte Pellegrino species has not yet been studied in detail, but there is a certain consensus in the literature to consider it the ancestor of the better described *Leithia cartei* and *Leithia melitensis* from the Middle Pleistocene of Sicily and Malta. It is considered to be ancestral to *Maltamys* as well, but this is less likely because a form similar to the Middle Pleistocene *Maltamys gollcheri* is already present in the Monte Pellegrino fauna. Both *Leithia* and *Maltamys* are regarded as relics of an older fauna of Messinian age (Late Miocene) by Remmert Daams and Hans de Bruijn (1995). This shrew and field mouse were about twice as large as their representatives from the mainland. The hare, described by Florian Fladerer and Matilde Fiore (2003), was less cursorial than most other hares. The marten (Figure 7.2) seems to be a descendant of the mainland *Pannonictis*. If so, then this poses a nomenclatorial problem, because the genus name *Mustelercta* was erected by De Gregorio (1925), six years before Kormos gave the name *Pannonictis* to the mainland form.

Figure 7.4 Skull of the gundi (*Pellegrinia panormensis*) from the Early Pleistocene of Sicily. Ventral view. Scale bar: 1 cm. Museo Geologico G.G. Gemmelaro, Palermo, Sicily. (Photograph: Bartholomeus van der Geer.)

Early Middle Pleistocene

The early Middle Pleistocene fauna of Sicily is highly endemic and impoverished. Its only known elements are the pygmy elephant (*Palaeoloxodon falconeri*), a shrew (*Crocidura esuae*), a giant dormice (*L. melitensis*), two large dormice (*L. cartei*, *Maltamys gollcheri*), the Sicilian otter (*Lutraeximia trinacriae*; Figure 7.5), and bats. Apart from these mammals, five endemic bird species were part of this fauna, as reported by Marco Pavia (2000, 2004) and Pavia and Mourer-Chauviré (2002). These are a large barn owl (*Tyto mourerchauvireae*), a small swan (*Cygnus equitum*), the Maltese crane (*Grus melitensis*), a long-legged little owl (*Athene trinacriae*), similar to the species on Crete (*Athene cretensis*), and a new, yet undescribed corvid species (Corvidae sp. n.). The fauna is generally known as the *Palaeoloxodon falconeri* Faunal Complex, after its most famous member.

The most important sites are Spinagallo cave, Luparello cave, Marasà cave, and Poggio Schinaldo cave. The dwarf elephant remains from Spinagallo (Figure 7.6) were dated to 455,000 ± 90,000 years ago by amino acid racemisation by Bada and colleagues (1991). However, results from a more recent geochronology fieldwork, reported at a conference by David Richards, Victoria Herridge, and colleagues (2019), indicate a somewhat younger age. They noticed that dwarf elephant (*Palaeoloxodon falconeri*) bones from Spinagallo, now in museum collections, originated from reddish clay deposits, which they dated to minimally between 230,000 and 350,000 years ago, based on uranium-thorium ages of intercalated calcite flowstones. Optically stimulated luminescence dating (OSL) of this earth yielded ages of over 200,000 years, while calcite veneers on two dwarf elephant bones from the cave were dated to at least 220,000 and 300,000 years ago, respectively. If this is the case, then the following fauna with *Palaeoloxodon* aff. *mnaidriensis* evolved in a shorter time span than previously thought.

Figure 7.5 Almost complete skeleton of the Sicilian otter (*Lutraeximia trinacriae*), Poggio Schinaldo cave near Palermo. Museo Geologico G.G. Gemmelaro, Palermo, Sicily. (Photograph: George Lyras.)

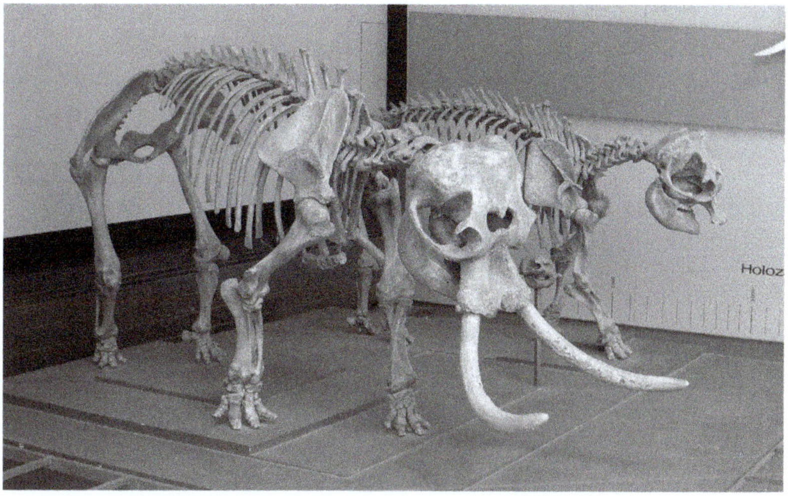

Figure 7.6 Two skeletons (casts; male and female) of the pygmy elephant (*Palaeoloxodon falconeri*) from Spinagallo caves. Forschungsinstitut Senckenberg, Frankfurt am Main, Germany. (Photograph: John de Vos.)

Late Middle Pleistocene–early Late Pleistocene

In this period, a new invasion of mainland species took place. The fauna of this period is now balanced with a few endemic species. New endemic species of this fauna are the dwarf hippopotamus *Hippopotamus pentlandi* and the middle-sized dwarf elephant *Palaeoloxodon* aff. *mnaidriensis* (Figure 7.7) – species reminding of the Maltese species but likely representing a new species. The hippo dispersed to Malta as well, where it evolved to a smaller size (*Hippopotamus melitensis*); the same may be valid for the elephant.

Other new arrivals are fallow deer (*Dama carburangelensis*), aurochs (*Bos primigenius*), red deer (*Cervus elaphus*), wild boar (*Sus scrofa*), brown bear (*Ursus arctos*), European bison (*Bison priscus*), grey wolf (*Canis lupus*), lion (*Panthera leo*), spotted hyena (*Crocuta crocuta*), European hedgehog (*Erinaceus europaeus*), all typical for a warm period. This is confirmed by the presence of the European pond turtle (*Emys*

Figure 7.7 Partial mandible of the Sicilian dwarf elephant (*Palaeoloxodon* aff. *mnaidriensis*); occlusal view. Scale bar: 5 cm. National Museum of Natural History, Leiden. (Photograph: Eelco Kruidenier.)

orbicularis). Remains of over fifty bird species were recovered as well, including an endemic swan (*Cygnus falconeri*) and crane (*G. melitensis*), the latter being a survivor from the previous period, as reported by Pavia (2001).

The shrew, the giant dormice, and the otter of the previous period survived the impact. A new dormouse species (*Maltamys wiedincitensis*) likely is a descendant of the *Maltamys* of the previous faunal complex. Remarkable is the absence of new colonising small mammals and otters, indicating that the dispersal was through a filtered connection, or via stepping stones, as already suggested by Cinzia Marra (2009). The survival of giant dormice apparently could not safeguard the large endemic barn owl from extinction. The position of the owl likely was now occupied by mainland *Bubo bubo*, as suggested by Pavia (2004).

Electron spin resonance dating on teeth enamel of the elephant and hippopotamus from Contrada Fusco by Rhodes (1996) provided an age ranging between 146,800 ± 28,700 and 88,200 ± 19,500 years ago. Deposits with the elephant at Puntali and Zà Minica have a somewhat older age and have been dated 200,000 ± 40 by amino acid racemisation (AAR) by Bada and colleagues (1991). The fauna is sometimes referred to as *Maccagnone Faunal Complex* or the *Palaeoloxodon mnaidriensis* Faunal Complex after its most typical member. Major sites are Maccagnone, Puntali, Zà Minica, and sediments at the base of cliff where San Teodoro cave is situated.

The ungulates as well as the dwarf hippopotamus of this period are generally some 20% less in shoulder height than their mainland conspecifics, on the ground of which they are assigned subspecific status, e.g. *Bos primigenius siciliac*, *Cervus elaphus siciliae*, and *Bison priscus siciliae*. The fallow deer (*D. carburangelensis*) evolved about 35% body mass reduction. These body size reductions confirm the existence of a filter connection with the mainland. Free influx was somehow hampered either by the Strait of Messina or the fragmented geography of Calabria, or both. This may also explain the absence of landfowl (Galliformes) and songbirds. Similar taxa, potentially ancestral to the Sicilian species, were found in southernmost Calabria at Bovetto in sediments of a late Middle or early Late Pleistocene age. These include remains of fallow deer (*Dama dama tiberina*), a hippopotamus (*Hippopotamus amphibius*), and an elephant (*Palaeoloxodon antiquus*).

Late Pleistocene

At the end of this period, during the Late Pleistocene, more mainland taxa arrived on the island, including the European ass (*Equus hydruntinus*). The new small mammals of this period are the common field mouse (*Apodemus sylvaticus*), the pine vole (*Microtus savii*), and the Sicilian shrew (*Crocidura sicula*). The fauna of this period is found in association with most large mammal species previously considered restricted to the previous faunal complex, such as *Palaeoloxodon* aff. *mnaidriensis*. The small mammals of the previous periods are now extinct. This stage is sometimes referred to as the *Grotta San Teodoro-Contrada Pianetti Faunal Complex*, after the two main localities.

The minimal isolation of Sicily since the Late Pleistocene resulted in Sicilian varieties of otherwise typical mainland species (see also Box 7.1). During this process, some species went extinct, such as the elephant. Its youngest remains date to about 32,000 years ago ($^{230}Th/^{234}U$ dating) from the San Teodoro cave as reported by Bonfiglio and colleagues (2008). The environment included glacial landscapes with low percentage of temperate (mesophilic) plant species, as indicated by pollen analysis of hyena coprolites by Yll and colleagues (2006).

The end of this period, known as the *Castello Faunal Complex*, is marked by the arrival of the horse (*Equus caballus*) and other mainland species. This stage is characterised by findings of lithic artefacts and other cultural evidences of a Late Palaeolithic culture.

BOX 7.1

The Islet of Pianosa

Progressive endemisation of terminal Pleistocene–Holocene faunas is also seen on the islet Pianosa, near the larger island Elba, on which the same mainland fauna as on Sicily arrived during the Late Pleistocene. When Pianosa became smaller, due to eustatic sea level rising, the aurochs (*Bos primigenius bubaloides*) became smaller as well and the deer (*Cervus elaphus*) developed somewhat shortened limbs before they both went extinct after the Neolithic invasion by humans and their domestic stock. Small mammals, due to their fast life and high reproduction, evolve easier into endemic species or subspecies, such as the Late Pleistocene mouse *Mus lopadusae* on the island of Lampedusa between Sicily and Tunisia's east coast, and the extant *Crocidura sicula aegatensis* of the Aegadian islets near Sicily's north-western coast and *Crocidura sicula calypso* of the Maltese island Gozo.

Peculiarities and Evolution of Endemic Mammals

Dwarf Elephants

Two different species of endemic dwarf elephants can be recognised on Sicily: *Palaeoloxodon falconeri*, a remarkably small dwarf form of the early Middle Pleistocene, and *Palaeoloxodon* aff. *mnaidriensis*, a medium-sized dwarf form of the late Middle–Late Pleistocene (Box 7.2). The straight-tusked elephant *Palaeoloxodon antiquus* is regarded as the ancestor of the Sicilian dwarf elephants. Bate (1907) saw similarities in morphology of the molars between the smallest species, *falconeri*, and mammoths. However, the microstructure of its tusks is like that of elephants, not mammoths, according to Palombo and Villa (2001). Furthermore, its external brain morphology is unlike that of a mammoth (own observation, G.L.).

Palaeoloxodon falconeri (Figure 7.6) is the smallest elephant species known, with a shoulder height of about 0.9 m in females and 1.3 m in males and an estimated body weight of around 100 kg, which means a body size reduction of about 98% relative to its mainland ancestor, the straight-tusked elephant (*Palaeoloxodon antiquus*), the largest elephant that ever lived, reaching a height of over four metres and a weight of almost 15 tons. Its habitat consisted of an open environment with sparse tree cover, dominated by grasses, and was much like the habitat of the extant African elephant as indicated by pollen data, analysed by Bertoldi and colleagues (1989) and Suc and colleagues (1995). In addition, the environment was highly seasonal with severe summer drought resulting in annually inconstant though not necessarily scarce resources. Falconer's dwarf elephant is mainly known from the thousands of fossils excavated from the Spinagallo caves near Syracuse.

Palaeoloxodon aff. *mnaidriensis* (Figure 7.3) of the next biozone had a shoulder height of almost 2 m, thus slightly less than half the size of its ancestor and with a body mass reduction of about 83%. Its molars have less plates but proportionally thicker enamel (Figure 7.7). The addition aff. (= affinity with) is needed, pending a new binominal name, because the species *mnaidriensis* is based on material from Malta; see, for discussion, Marco Ferretti (2008).

A third dwarf species may be represented by the material from Luparello cave. When Vaufrey excavated this cave, he noted the somewhat larger size of the elephant remains, relative to those of *Palaeoloxodon falconeri* from Spinallo. Also, Victoria Herridge (2010) in her doctoral thesis noted the morphological and size differences between elephant molars from the two sites. This means that independent processes of isolation took place on each part of Sicily, corresponding to the existing two separate palaeo-islands during the early part of the Pleistocene.

A fourth, much larger form, *Palaeoloxodon antiquus leonardi,* was suggested by Emiliano Aguirre (1969) based on material from Via Liberta, Palermo, from late Middle Pleistocene deposits. This subspecies is only slightly smaller than the mainland form. These large forms occur together with remains of *Palaeoloxodon* aff. *mnaidriensis* in Contrada Fusco at Syracuse, which indicates that *leonardi* might be nothing else than very large males of aff. *mnaidriensis*. Elephants are highly sexually dimorphic and bulls, may be some 20–40% larger than cows. The subspecies *leonardi* is not well-defined – holotypes and measurements are missing – which hampers correct identification.

Related middle-sized elephant remains were also found outside Sicily. The remains from the Aegadian island Favignana, only some 15 km off the north-western coast of Sicily, are not yet described, but they likely belong to *Palaeoloxodon* aff. *mnaidriensis*. The remains found at San Francesco di Archi in southern Calabria are somewhat smaller than those of continental straight-tusked elephants and may be intermediate between those and the Sicilian middle-sized dwarf species.

BOX 7.2

From Large to Small?

Originally, the smallest elephants of Sicily and Malta were considered the evolutionary end result of a progressive dwarfing, starting from the mainland straight-tusked elephant *Palaeoloxodon antiquus*, first brought to the fore by Vaufrey (1929). In that view, the first step in this dwarfing was represented by the subspecies *leonardi*, just slightly smaller than its ancestor and occurring only in Sicily. The next steps were the species *mnaidriensis*, *melitensis*, and *falconeri* standing, respectively, 1.9 m, 1.4 m, and 0.9 m at the shoulder, all three living in Sicily as well as in Malta, as was the general opinion.

Stratigraphical and geochemical data, however, spoiled this comprehensive picture: *falconeri* turned out to be older than all other forms. Datings by Belluomini and Bada (1985) on the Spinagallo and Luparello cave materials suggested a much older age for *Palaeoloxodon falconeri* in regard to that for *Palaeoloxodon* aff. *mnaidriensis*. The first stratigraphic evidence to confirm this observation came from the travertine quarries at Alcamo on the slope of Monte Bonifato, excavated by Burgio and Cani (1988). The fossils from the travertine formation include *Palaeoloxodon falconeri*, whereas those from the red soil infillings of the karst fissures in the travertine contain elements typical of the *Palaeoloxodon* aff. *mnaidriensis* fauna. The former fauna, therefore, preceded the latter fauna. The Sicilian elephant species thus represent subsequent invasions, not an evolution in situ from one species into the other, in parallel to what we see in Crete and Flores.

Dwarf Hippos

The Sicilian dwarf hippopotamus (Plate 11), *Hippopotamus pentlandi*, is the largest hippo from the Mediterranean islands. It only differed from mainland *Hippopotamus amphibius* in its smaller size with a body mass reduction of about 25% only. Regarding the entire population, there is another difference: the Sicilian hippo shows large size variability, a feature seen more often in insular species. *Hippopotamus pentlandi* was reduced to the level of subspecies of the extant hippo *Hippopotamus amphibius* by Vaufrey (1929), followed by Capasso Barbato and Petronio (1983). The majority of scholars, however, maintain its specific status.

The best record of *Hippopotamus pentlandi* consists of composite skeleton of rich material found in Grotta della Cannita near Palermo and described by Bruno Accordi (1955). Furthermore, thousands of remains of this hippo have been recovered at Acquedolci in north-eastern Sicily during systematic excavations in the 1980s.

Dirk Hooijer (1946) erected a second species, *Hippopotamus siculus*, in an attempt to correct the erroneous description of new hippo findings from Gravitelli as the Asian *Hippopotamus* (=*Hexaprotodon*) *sivalensis* by Seguenza (1902, 1907), who assumed a Late Miocene age for the fossils he described. Hooijer based his new species on Seguenza's plates and measurements, obviously without comparing the plates with the actual *pentlandi* material. Hooijer's short-lived species was justly ignored by later scientists.

The Otter

An almost complete skeleton of an otter (Figure 7.5) was described by Enzo Burgio and Matilde Fiore (1988) as *Nesolutra trinacriae*. The skeleton was found in Middle or Late Pleistocene deposits at Poggio Schinaldo cave near Palermo. The genus name had been established fifty years earlier by Bate for the Maltese fossil species *euxena* and applied by

Malatesta to the fossil Sardinian otter species *ichnusae* as well. The genus was reviewed by Willemsen (1992) who considered *Nesolutra* a junior synonym of *Lutra*, the common otter, and suggested that the Sicilian and Maltese otters probably evolved from the mainland species *Lutra simplicidens*. Later, Marco Cherin and colleagues (2016) erected a new genus, *Lutraeximia*, for a short-muzzled Italian mainland species and moved the Sicilian otter to this new genus. They further observed a close phylogenetic relation between the two *Lutraeximia* species, the Cretan otter (*Lutrogale cretensis*) and one of the Sardinian otters (*Sardolutra ichnusae*). According to them, the Sicilian otter had a 'fish specialist' dentition, feeding mainly on fast-moving fishes, parallel to that observed in the Eurasian otter (*Lutra lutra*) and the Early and Middle Pleistocene *L. simplicidens*, which has been suggested as ancestral to the Sicilian otter by Willemsen (2006). The teeth of these otters are relatively slender with sharp cutting blades.

Red-toothed and White-toothed Shrews

The Monte Pellegrino shrew of the Early Pleistocene is thought to have evolved in parallel with the other Mediterranean insular red-toothed shrews – *Nesiotites* from the Balearics and *Asoriculus* from Sardinia/Corsica – all originating from a clade of common ancestors. Federico Masini and Maurizio Sara (1998) named this large shrew *Asoriculus burgioi*. They consider it an insular form of the continental genus *Asoriculus*, which occurred in mainland Eurasia as well in the Maghreb (North Africa) between the end of the Miocene and the beginning of the Middle Pleistocene.

The red-toothed shrew is replaced by a white-toothed shrew in the early Middle Pleistocene. Remains from Spinagallo cave were described by Tassos Kotsakis (1986) as *Crocidura esu* (now *esuae*). Rainer Hutterer (1991) suggested that the fossil and the extant shrew (*Crocidura sicula*) of Sicily might be in actual fact one and the same species, while Kotsakis regarded the fossil species as the direct ancestor of the living species. The latter was assigned to the mainland species *Crocidura russula* by Corbett (1978) and thought to have been imported by humans in the Holocene. Molecular evidence, however, pointed out that the extant Sicilian shrew is identical to the extant shrew of Gozo Island near Malta, which was named *Crocidura sicula* by Gerrit Miller (1900). The Sicilian and Maltese extant shrew might on their turn have descended from the Saharan shrew *Crocidura tarfayaensis*, as recently found out by Sylvain Dubey and colleagues (2007), based on DNA. They calculated the split between the Saharan and the Siculo-Maltese clades to the Messinian (Late Miocene), the period during which the salinity crisis resulted in a large 'Lago Mare'. The connection between Africa and the Siculo-Maltese mountain chains acted as a filter, enabling the passage of only a few taxa. If so, *Crocidura esuae* must already have been present during the Monte Pellegrino fauna or even before. The split between the extant shrew populations of Sicily and Gozo has been calculated by the same authors to the Middle Pleistocene. This confirms that Sicily and Malta each had their own faunas since the late Middle Pleistocene on.

Giant Dormice

Two different, but likely related, genera of giant dormice were endemic to Sicily and Malta, the larger *Leithia* and the smaller *Maltamys*. The Middle Pleistocene species *L. melitensis* was a true giant, twice the size of the extant edible dormouse (*Glis glis*), and about 13 times the size of its ancestor, a form close to the garden dormouse (*Eliomys quercinus*). *Leithia cartei* was 'only' three times the ancestral size. Both *Maltamys* species – the early Middle Pleistocene *gollcheri* and the late Middle Pleistocene *wiedincitensis* – were much smaller but for the rest similar. The dormice are thought to have evolved from the dormice (*Leithia* sp.) of the Early Pleistocene Pellegrino fauna. The genus *Leithia* was conceived by Richard Lydekker (1895) as a squirrel, based upon the dentition and features of the skull. He considered the restoration of the lower jaw as presented by Leith Adams probably incorrect.

Perhaps in relation to the large size of these dormice– their main prey – also the owls had increased considerably in size. The Sicilian owl (*Tyto mourerchauvireae*) had about the size of *Tyto robusta*, according to Marco Pavia (2004). *Tyto robusta* is a Late Miocene barn owl, endemic to the Gargano (Chapter 6), which has no direct phylogenetic relation with the Sicilian large owl. Both owls, however, likely derived from a much smaller form like *Tyto balearica*.

References

Accordi B. 1955. *Hippopotamus pentlandi* von Meyer del Pleistocene della Sicilia. *Palaeontografia Italiana*, 50: 1–52.

Aguirre E. 1969. Revisión sistemática de los Elephantidae por su morfología y morfometría dentaria. *Estudios Geológicos*, 25: 317–367.

Bada J.L., Belluomini G., Bonfiglio L., Branca M., Burgio E., Delitala L. 1991. Isoleucine epimerization ages of Quaternary Mammals of Sicily. *Il Quaternario*, 4 (1A): 5–11.

Bate D.M.A. 1907. On elephant remains from Crete, with description of *Elephas creticus* n. sp. *Proceedings of the Zoological Society of London*, 1907: 238–250.

Belluomini G., Bada J.L. 1985. Isoleucine epimerization ages of the dwarf elephants of Sicily. *Geology*, 13: 451–452.

Bertoldi R., Rio D., Thunell R. 1989. Pliocene-Pleistocene vegetational and climatic evolution of the South-Central Mediterranean. *Palaeogeography, Palaeoclimatology, Palaeoecology*, 72: 263–275.

Bonfiglio L., Esu D., Mangano G., Masini F., Petruso D., Soligo M., Tuccimei P. 2008. Late Pleistocene vertebrate-bearing deposits at San Teodoro Cave (North-Eastern Sicily): Preliminary data on faunal diversification and chronology. *Quaternary International*, 190 (1): 26–37.

Bonfiglio L., Marra A.C., Masini F. 2000. The contribution of Quaternary vertebrates to the paleoenvironmental and paleoclimatic reconstructions in Sicily. In: Hart M.B. (ed.), *Climates: Past and Present. Geological Society, Special Publications*, 181. London, pp. 169–182.

Bonfiglio L., Piperno, M. 1996. Early faunal and human populations. In: Leighton R. (ed.), *Early Societies in Sicily: New Developments in Archaeological Research. Accordia Specialist Studies on Italy*, Vol. 5. Accordia Research Centre, London, pp. 21–29.

Burgio E., Cani M. 1988. Sul ritrovamento di elefanti fossili ad Alcamo (Trapani, Sicilia). *Il Naturalista Siciliano*, 12 (3–4): 87–97.

Burgio E., Fiore M. 1988. *Nesolutra trinacriae*, nuova specie di lontra quaternaria della Sicilia. *Bolletino della Società Paleontologica Italiana*, 27: 259–275.

Capasso Barbato L., Petronio C. 1983. Considerazioni sistematiche e filogenetiche su "*Hippopotamus pentlandi*" von Meyer, 1832 (Mammalia). *Atti Società Italiana delle Scienze Naturali, Museo Civico della Storia Naturale, Milano*, 124 (3–4): 229–248.

Cherin M., Iurino D.A., Willemsen G., Carnevale G. 2016. A new otter from the Early Pleistocene of Pantalla (Italy), with remarks on the evolutionary history of Mediterranean Quaternary Lutrinae (Carnivora, Mustelidae). *Quaternary Science Reviews*, 135: 92–102.

Corbett G.B. 1978. *The Mammals of the Palaearctic Region: A Taxonomic Review*. British Museum of Natural History, London.

Daams R., de Bruijn H. 1995. A classification of the Gliridae (Rodentia) on the basis of dental morphology. *Hystrix*, 6 (1–2): 3–50.

De Gregorio A. 1886. Intorno a un deposito di roditori e di carnivori sulla vetta di Monte Pellegrino. *Atti Società Toscana Scienze Naturali Pisa*, 8 (1): 3–39.

De Gregorio A. 1925. Mammiferi quaternari di Sicilia. *Annales de Geologie et de Paleontologie*, 43: 3–18.

Dubey S., Koyasu K., Parapanov R., Ribi M., Hutterer R., Vogel P. 2007. Molecular phylogenetics reveals Messinian, Pliocene, and Pleistocene colonizations of islands by North African shrews. *Molecular Phylogenetics and Evolution*, 47 (2): 877–882.

Fazello T. 1558. *De Rebus Siculis Decades Duae*. Palermo.

Ferretti M. 2008. The dwarf elephant *Palaeoloxodon mnaidriensis* from Puntali Cave, Carini (Sicily; Late Middle Pleistocene): Anatomy, systematics and phylogenetic relationships. *Quaternary International*, 182 (1): 90–108.

Fladerer F.A., Fiore M. 2003. The Early Pleistocene insular hare *Hypolagus peregrinus* sp. nov. from Northern Sicily. *Palaeontografia Italiana*, 89: 37–63.

Herridge V.L. 2010. Dwarf elephants on Mediterranean islands: a natural experiment in parallel evolution. Ph.D. thesis. University College London, UK.

Hooijer D.A. 1946. Notes on some pontian mammals from Sicily, figured by Seguenza. *Archives Neerlandaises de Zoologie*, 7: 301–333.

Hutterer R. 1991. Variation and evolution of the Sicilian shrew: Taxonomic conclusions and description of a possibly related species from the Pleistocene of Morocco (Mammalia: Soricidae). *Bonner Zoolol-gische Beiträge*, 42: 241–251.

Kormos T. 1931. Pannonictis pliocaenica n.g.n.sp., a new giant mustelid from the Late Pliocene of Hungary. *Annalibus Instituti Regii Hungarici Geologici*, 29: 1–16.

Kotsakis T. 1986. *Crocidura esui* n. sp. (Soricidae, Insectivora) du Pléistocène supérieur de Spinagallo (Sicilia orientale, Italie). *Geologica Romana*, 23: 51–64.

Leith Adams A. 1863. Observations on the Fossiliferous caves of Malta. *Journal of the Royal Society*, 4 (2): 11–19.

Leith Adams A. 1874. On the dentition and osteology of the Maltese fossil Elephants, being a description of remains discovered by the Author in Malta between the years 1860 and 1866. *Transactions of the Zoological Society of London*, 9 (1): 1–124.

Lydekker R. 1895. On the affinities of the so-called extinct giant dormouse of Malta. *Proceedings of the Zoological Society of London*, 1895: 860–863.

Marra A.C. 2009. Pleistocene mammal faunas of Calabria. *Bollettino della Società Paleontologica Italiana*, 48: 113–122.

Masini F., Petruso D., Bonfiglio L., Mangano G. 2008. Origin and extinction patterns of mammals in three central Western Mediterranean Islands from the Late Miocene to Quaternary. *Quaternary International*, 182: 63–79.

Masini F., Sara M. 1998. *Asoriculus burgioi* sp. nov. (Soricidae, Mammalia) from the Monte Pellegrino faunal complex (Sicily). *Acta Zoologica Cracoviensia*, 41: 111–124.

Miller G.S. 1900. Five new shrews from Europe. *Proceedings of the Biological Society of Washington*, 14: 41–45.

Palombo M.R., Villa P. 2001. Schreger lines as support in the Elephantinae identification. In: Cavaretta G., Gioia P., Mussi M., Palombo M.R. (eds), *The World of Elephants. Proceedings of the 1st International Congress, Roma, 16–20 October, 2001*, pp. 656–660.

Pavia M. 2000. *Le avifauna pleistoceniche dell'Italia meridionale*. PhD Thesis, University of Torino.

Pavia M. 2001. The Middle Pleistocene fossil avifauna from the "*Elephas mnaidriensis* Faunal Complex" of Sicily (Italy): Preliminary results. In: Cavaretta G., Gioia P., Mussi M., Palombo M.R. (eds), *The World of Elephants, Proceedings of the 1st International Congress, Rome, 16–20 October 2001*. Consiglio Nazionale delle Ricerche, Rome, pp. 497–501.

Pavia M. 2004. A new large barn owl (Aves, Strigiformes, Tytonidae) from the Middle Pleistocene of Sicily, Italy, and its taphonomical significance. *Geobios*, 37: 631–641.
Pavia M., Mourer-Chauviré C. 2002. An overview of the genus *Athene* in the Pleistocene of the Mediterranean Islands, with the description of *Athene trinacriae* n. sp. (Aves: Strigidae). In: Zhou Z., Zhang F. (eds), *Proceedings of the 5th Symposium of the Society of Avian Paleontology and Evolution*. Science Press, Beijing, pp. 13–27.
Pentland M.W. 1832. Note contenant la détermination des Ossemens fossiles des cavernes voisines de Palerme, par M. W. Pentland. *Annales des Sciences Naturelles*, 15: 208–210.
Rhodes E.J. 1996. ESR dating of tooth enamel. In: Basile, B., Chilardi, S. (eds), *Le ossa dei Giganti. Lo scavo paleontologico di Contrada Fusco*. Arnaldo Lombardi, Siracusa, pp. 39–44.
Richards D., Herridge V., Nita D., Schwenninger J.-L., Lister A., Mangano G., Bonfiglio L. 2019. Constraining the chronology of dwarf elephant evolution using Quaternary deposits in coastal caves of the Mediterranean. In: *20th Congress of the International Union for Quaternary Research (INQUA), Dublin, Ireland, 25th–31th July 2019*, Abstracts: P2810.
Scinà D. 1831. *Rapporto sulle ossa fossili di Mardolce e degli altri contorni di Palermo*. Reale Tipografia di Guerra, Palermo.
Seguenza L. 1902. I Vertebrati fossili della Provincia di Messina. Parte seconda: Mammiferi e Geologia del Piano Pontico. *Bolletino della Societa geologica Italiana*, 21: 115–175.
Seguenza L. 1907. Nuovi resti di mammiferi Pontici di Gravitelli presso Messina. *Bollettino della Societa Geologica Italiana*, 26: 89–122.
Suc J.P., Bertini A., Combourieau-Nebout N., Diniz F., Leroy S., Russo-Ermolli E., Zheng Z., Bessais E., Ferrier J. 1995. Structure of West Mediterranean vegetation and climate since 5.3 ma. *Acta Zoologica Cracoviensia*, 38: 3–16.
Thaler L. 1972. Les rongeurs (Rodentia et Lagomorpha) du Monte Pellegrino et la question des anciens isthmes de la Sicilie. *Comptes Rendus de l'Académie des Sciences de Paris (D)*, 274: 188–190.
Vaufrey R. 1929. Les éléphants nains des îles méditerranéennes et la question des isthmes pléistocènes. *Archives de l'Instityte de Paléontologie Humains, Memoire*, 6: 1–220.
Von Meyer H. 1832. *Palaeologica zur Geschichte der Erde und ihrer Geschöpfe*. Frankfurt.
Willemsen G.F. 1992. A revision of the Pliocene and Quaternary Lutrinae from Europe. *Scripta Geologica*, 101: 1–115.
Willemsen G.F. 2006. *Megalenhydris* and its relationship to *Lutra* reconsidered. *Hellenic Journal of Geosciences*, 41: 83–87.
Yll R., Carrión J.S., Marra A.C., Bonfiglio L. 2006. Vegetation reconstruction on the basis of pollen in Late Pleistocene hyena coprolites from San Teodoro Cave (Sicily, Italy). *Palaeogeography, Palaeoclimatology, Palaeoecology*, 237: 32–39.

CHAPTER EIGHT

Malta

Evolution of Island Mammals: Adaptation and Extinction of Placental Mammals on Islands, Second Edition.
Alexandra van der Geer, George Lyras and John de Vos.
© 2021 John Wiley & Sons Ltd. Published 2021 by John Wiley & Sons Ltd.

Malta is a small island south of Sicily, to which it was connected in the Early and early Middle Pleistocene. Since that period, Malta harboured its own endemic fauna, mainly consisting of dwarf proboscideans, dwarf hippopotamuses, dwarf deer, and giant dormice. The Maltese endemics likely evolved from Sicilian taxa through vicariance effects. Due to the greater isolation, smaller surface area, and even more limited number of founder species, the composition of the Maltese fauna differed essentially from that of the Sicilian fauna at any time.

Geology and Palaeogeography

Malta, including the smaller islands Gozo and Comino and the inhabited islets around it, emerged during the Late Miocene. During the salinity crisis of the Messinian period of the Late Miocene, the islands were likely connected to each other and with Sicily. At the onset of the Pliocene, when the Mediterranean was refilled with water during the Zanclean flood, Malta became isolated. Nevertheless, a submerged ridge connected it with the southern tip of Sicily. This ridge is supposed to have been partly exposed at sea level lowering of more than hundred and fifty metres. Since Malta's isolation, such a drop in sea level occurred several times, likely enabling colonisations by Sicilian terrestrial mammal species.

The first significant sea level drop took place at the beginning of the early Middle Pleistocene. The sea level rose again and isolated Malta from Sicily, resulting in endemic taxa through vicariance effects. The history repeated itself twice, once during the late Middle Pleistocene and once during the Late Pleistocene.

Historical Palaeontology

The very first fossil findings on Malta were done in the first half of the seventeenth century by Jesuit father Giovanni Francesco Abela – Malta's first historian and archaeologist – who actively collected archaeological remains from the islands. In his monumental work on the description of Malta (1647), he describes bones of giants, presumably the builders of the Maltese megalithic temples, for example, the temple at Haqar Qim. He also mentions a molar belonging to a giant skull, found at Imriehel (= Mriehel) near Birkirkara. His collection of antiquities was large and would form the basis of the National Museum. It included a tusk of a small elephant, as noted down in a list of his collection at the time of his death on 4 May 1655, made by notary Michael Ralli and kept in the National Library.

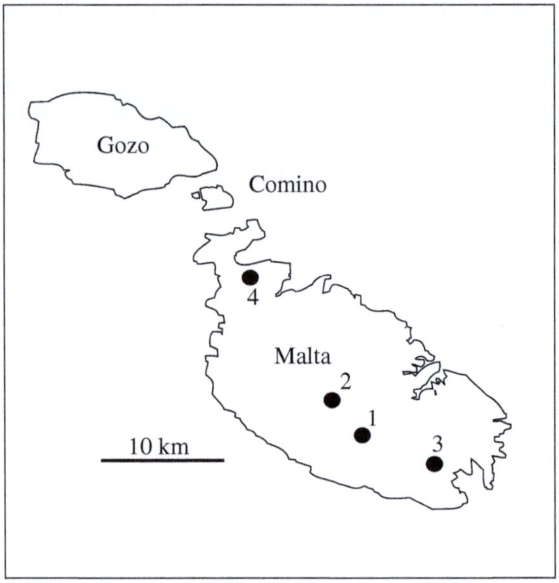

Figure 8.1 The most important Pleistocene localities of Malta. (1) Mqabba, Halq is-Sigar Maghlaq, Mnaidra, and Crendi; (2) Żebbuġ; (3) Għar Dalam; and (4) Mellieħa.

Two centuries past till in 1857, abbot Francesco Spiteri-Agius found a small, incomplete elephant molar in a fissure filling at a quarry at Ta' Gandat near Mqabba (Figure 8.1). The piece was given to the National Natural History Museum at Mdina, but not before the second half of the twentieth century, it was given a proper catalogue number by George Zammit Maempel. The original label – in the meantime largely consumed by moths and bookworms – is hardly readable: 'Animal tooth which led to the discovery of the Maltese Elephant in 1857'. Captain Thomas Abel Brimage Spratt of the British navy, an enthusiastic naturalist in his spare time, had a cast sent to Hugh Falconer in London, an expert on fossil elephants at the British Museum. According to Falconer, the molar resembled that of an African elephant most, be it in miniature form.

This diagnosis prompted Spratt to start excavations in search of more material of this enigmatic dwarf elephant. He simply called his localities (Spratt, 1867) after the nearest village – 'Crendi Cave' (1858), 'Żebbuġ Cave', (1859; a subterraneous fissure or cavern), and 'Mellieħa Caves' (1863), respectively. The Crendi (= Qrendi) cave is locally known as Halq is-Sigar and is one of the Maghlaq caves of Andrew Leith Adams. Here, he found remains of a dwarf hippopotamus, which was regarded as identical to the earlier described Sicilian form, *Hippopotamus pentlandi*. In the Żebbuġ cave, he finally found the elephant remains he had hoped for, but, in the Mellieħa caves, he found only hippopotamus remains. The difference in sedimentation – firm breccias with hippopotamus fossils in the caves against soft, reddish clay with elephant fossils in the fissures – was interpreted by him as due to different geological ages.

His search resulted in a large number of fossils, which he dispatched to several experts for further study. The elephant remains were studied by Falconer, who presented his findings during a meeting of the Royal Academy of Sciences. At that occasion, he proposed the name *Elephas melitensis* for the Maltese dwarf elephant. His untimely death, however, delayed the scientific publication of the new name. His findings and notes were posthumously edited and published by his colleague Charles Murchinson (as Falconer, 1868). A year before, George Busk had divided Falconer's material from Malta in two size groups and assigned the smallest molars to a new species, *Elephas falconeri*, in honour of his colleague and friend. Together with the description, he depicted material from the Żebbuġ cavern excavated by Spratt.

In 1863, Leith Adams discovered and described another fossiliferous deposit. It was a large opening in the limestone, which he named Mnajdra Gap, a few metres away from and above the Maghlaq middle cave. Leith Adams was a Scottish military physician, posted on Malta for seven years, who spent his spare time studying the natural history of the countries where he was working. Later, after his retirement from the army, he became a professor in natural history, for which he is better known. At Mnajdra Gap, Leith Adams found large amounts of elephant fossils, including partial skeletons, in association with enormous quantities of dormouse fossils. The large bones were locally believed to be the remains of giants who had built the famous megalithic temples at nearby Mnajdra, as written down by Giovanni Abela (1647). Leith Adams (1868) described the very large form of dormouse from the Maqhlaq middle cave as *Myoxus melitensis* and the small form as *Myoxus cartei*. Both species were assigned to a new genus *Leithia* in his honour by Richard Lydekker (1895). Leith Adams (1870) gave the remains of the small elephant the scientific name *Elephas mnaidrae*, which he changed four years later into *Elephas mnaidriensis*. In that same year, Leith Adams published an article figuring the first reconstruction of the Maltese fauna, including a dwarf hippopotamus, a giant dormouse, a giant swan, a giant tortoise, and three large-eared elephants with different sizes, closely resembling the African elephants (Figure 8.2). His country fellow and famous illustrator of his time Jemima Wedderburn made the drawing.

Forsyth Major (1902) based his new species of dwarf hippopotamus, *Hippopotamus melitensis*, on material from Għar Dalam. The cave – *għar* means 'cave' in Maltese – had been excavated for its Neolithic remains since 1865, starting with the

Figure 8.2 Reconstruction of the Maltese fauna with a dwarf hippopotamus, a giant dormouse, a giant swan, a giant tortoise, and three large-eared elephants of different sizes. From Leith Adams, 1870. (Drawing by Jemima Wedderburn.)

Italian geologist Arturo Issel. The material studied by Forsyth Major originated from the 1892 excavation of John H. Cooke. The bulk of his material remained in Malta, but a comparative collection was sent to the British Museum. Għar Dalam is a 144-m-long phreatic tube, or '*cul de sac*', in which the disarticulated hippo bones form a bone bed. These remains appeared to be clearly smaller than those of the already earlier described Sicilian form, thus warranting a separate species name. Already, earlier, it had been noted that the Maltese hippo was smaller than the Sicilian hippo, hence the occasionally used name *Hippopotamus minor*, or its junior synonym *Hippopotamus minutus*. Both names, however, were restricted to the Cypriot dwarf hippo, as pointed out by Forsyth Major and, therefore, cannot be applied to other island forms.

Excavations of the Għar Dalam cave were continued by, amongst others, the Maltese naturalist Giuseppe Despott with various colleagues between 1912 and 1920 and the Maltese archaeologist Joseph Baldacchino in the 1930s.

In 1920, Bate described a new vole from the Għar Dalam cave as a water vole, *Arvicola melitensis*, based on a right mandible. Later, using the new extensive collection of the same cave, she published in 1935 again on this vole, moved it to the subterranean genus *Pitymys* – now a subgenus of *Microtus*, restricted to America – and identified a second species, *Pitymys pauli*.

Hans de Bruijn (1966) described a new dormouse, *Eliomys gollcheri*, from Pleistocene deposits of Mnajdra Gap, which he later transferred to the new genus *Maltamys* (Zammit Maempel and de Bruijn, 1982). In the same publication, they described a second Maltese dormouse, *Maltamys wiedincitensis* from Late Pleistocene deposits at Wied Incita. It has been suggested that *gollcheri* is actually synonymous with *Leithia cartei* from the same locality as described by Leith Adams (1870).

Excavations of the Għar Dalam cave have since continued, amongst others, by a Maltese-German team led by Gerhard Storch in 1970. David Richards and Chris Standish undertook geochronological sampling of Għar Dalam cave in the early 2010s, but results still await publishing (Standish and colleagues, in prep.).

Biozones and Faunal Units

The biostratigraphy of Malta can be reconstructed, thanks to the relatively well-preserved stratigraphy of Għar Dalam. The excavated section of the cave shows from bottom to top the following layers: a bone free clay layer, a bone bed consisting of mainly hippopotamus bones, a pebble layer, a layer with deer bones, a calcareous sheet, and, at the top, a layer with remains of domestic animals and the now lost skull of a Neolithic child. The latter layer is the cultural layer, which holds evidence of the first humans on the island. Apart from the buried archaeological remains, there are also – for the greater part recently destroyed – rock paintings and bruising on the cave walls. The thickness of the separate layers is, by approximation, 1.25 m, 1.20 m, 0.35 m, 1.75 m, 0.60 m, and 0.75 m, respectively. The deer layer is dated to about 18,000 years ago and the cultural layer to about 7400 years ago.

Għar Dalam yielded mainly megafauna, and as a consequence, the biostratigraphic and phylogenetic situation of the Maltese dormice remained unclear. In addition, the oldest period was not preserved in Għar Dalam. There are practically speaking as many small-mammal faunal lists as there are authors. In any case, Malta harboured giants as well as species of a more modest size, apparently living simultaneously. The species that were described for Malta are the dormice *Leithia melitensis* from Mnajdra, Tal Gnien fissure, Wied Incita, and

Benghisa Gap; *L. cartei* from Mnajdra; *Maltamys gollcheri* from Mnajdra; and *M. wiedincitensis* from Wied Incita. *Maltamys gollcheri* might be a junior synonym of *L. cartei*. These four species names are also applied to the Sicilian material, adding to the confusion. Most likely, a difference between the dormice of Sicily and Malta existed since the late Middle Pleistocene with independent lineages but originating from the same founder, close to the *Leithia* sp. from the Late Pliocene Sicilian Monte Pellegrino fauna. A Messinian (Late Miocene) age has been suggested for the ancestor of this *Leithia* sp., likely an *Eliomys* from the mainland of Europe.

Early Middle Pleistocene

The oldest preserved Maltese fossil vertebrate fauna is unbalanced and partly endemic. It is characterised by a dwarf elephant (*Palaeoloxodon falconeri*), a giant dormouse (*L. melitensis*), two large dormice (*M. gollcheri* and *M. wiedincitensis*), the Maltese otter (*Lutra euxena*), and a white-toothed shrew (*Crocidura esuae*). The dwarf elephant was the only 'large' mammal of this period in Malta. The species of elephant, shrew, and dormice are shared with Sicily of the same geological period, indicating the presence of a filter connection, or a 'stepping stones' bridge. A certain degree of isolation from Sicily seems, however, to have been the case, seeing the specific status of the Maltese otter.

Maltamys wiedincitensis is slightly larger than *Glis glis*, the largest extant dormouse. The presence of three species of giant dormice – probably all descendants of a *Leithia* from the Monte Pellegrino fauna of Sicily, and thus representing an adaptive radiation – indicates a long-term isolation of this fauna.

The most important locality of fossils from this period is the older layer of the Żebbuġ fissure.

Late Middle Pleistocene

Towards the end of the Middle Pleistocene, a new invasion of mammals took place. New elements of this period are the Maltese dwarf hippopotamus (*H. melitensis*), a small elephant (*P. mnaidriensis*; Figure 8.3), a red deer (*Cervus* sp.), several bats, and perhaps a vole (*Microtus* (*Pitymys*) *pauli*). Survivors from the previous period are the dormice, the shrew, and the otter. The new species of large dormouse (*L. cartei*) is a descendant of the dormice of the previous period. Bate found remains of *Microtus pauli* at the Għar Dalam cave, but Franz Malec and Gerhard Storch only retrieved *Microtus melitensis* during their 1969 excavation of the Late Pleistocene deposits of the cave (reported in 1970). Hence, they concluded that, likely, *pauli*

Figure 8.3 Second upper molar of the middle-sized dwarf elephant, *Palaeoloxodon mnaidriensis* from Ta Gandia Fissure, Malta. Scale bar: 5 cm. Natural History Museum, London. (Photograph: George Lyras.)

with its more complex molar pattern is an older form from the deeper cave layers, and possibly the same as the vole found at Mnajdra Gap by Adams in the 1870s. Apart from the mammals, a giant swan (*Cygnus falconeri*), the giant Malta crane (*Grus melitensis*), the common crane (*Grus grus*), and a large tortoise (*Geochelone robusta*) were part of this biozone.

The most important sites are the caves Crendi (= Maghlaq), Mnajdra (=Mnaidra) Gap, Tal Gnien, Mellieħa, the younger layer of the Żebbuġ fissure, and the older layers of the Għar Dalam cave. The Żebbuġ fissure, or cavern, differs considerably from the Mellieħa and Crendi caves, which are located at about 100 m above the sea level, overlooking the open sea, and containing water-worn pebbles as well as *Hippopotamus* remains in a hard breccia but no trace of elephantine bones. The Żebbuġ fissure lies some 3 m below the ground level deep inland, containing abundant elephant material in clayish sediments without any water-worn pebble or hippo bone. The bone breccia with hippo and elephant fossils at Għar Dalam cave was originally considered to date between about 115 and 180 thousand years ago, as discussed by Hunt and Schembri (1999). More likely, the breccia dates to before 163 thousand years ago, as summarised by Leila D'souza (2019), based on preliminary findings by Chris Standing using Uranium/Thorium on speleothem (publication forthcoming).

This fauna is again impoverished and unbalanced and mainly characterised by dwarf herbivores and large dormice. The composition of this new fauna, together with the local evolution of the pre-existing fauna, strongly indicates an overseas dispersal from Sicily to Malta. The main difference between the Sicilian fauna and the Maltese fauna of this period is the smaller size of the Maltese hippopotamus and elephant in combination with a larger degree of endemism of the Maltese fauna.

Late Pleistocene

The Late Pleistocene fauna from Malta consists mainly of three or more endemic species of short-limbed dwarf deer (*Cervus* sp.), a burrowing vole (*Microtus* (*Pitimys*) *melitensis*), a white-toothed shrew (*Crocidura esuae*), and bats, among which the lesser horseshoe bat (*Rhinolophus hipposideros*). The most important site is Għar Dalam, where the upper layers are of a Late Pleistocene age.

The dormice of the previous periods persisted into this period, while the vole likely evolved from the earlier *Microtus pauli*. The shrew likely is a survivor as well, as the split between the shrew populations of Sicily and Gozo has been calculated to the Middle Pleistocene by Dubey and colleagues (2007), based on DNA. The deer may have evolved from the deer of the previous period. The majority of deer fossils are from a red clay layer dated to between 80,000 and 116,000 years ago by Chris Standing using Uranium/Thorium on flowstone (publication forthcoming; details from the doctoral thesis of D'souza, 2019). The lowest layer of the red clay deposit contains fossils of a wolf (*Canis lupus*), which may thus belong to either the beginning of this faunal zone or the end of the previous. The red clays are separated from the bone breccia with hippos by a pebble layer, likely indicating a moment when the river broke through the cave roof.

The Late Pleistocene fauna of Malta differs from that of Sicily by a higher degree of endemism. The fauna as a whole is much poorer than the one from Sicily, indicating that the distance between the islands was much larger or the road to Malta much more difficult. Furthermore, an adaptive radiation of deer is unique to Malta. The extinction of this fauna may have been caused by a volcanic event, because at Mriehel, a thick volcanic ash layer was found by George Zammit Maempel, overlying the deer layer.

Holocene

At the end of this period, new small mammals arrive from the mainland, for example, the Etruscan shrew (*Suncus etruscus*) and the field mouse (*Apodemus sylvaticus*). Their subfossils were retrieved from the top layer of Għar Dalam. Also, subfossil remains of the bicoloured white-toothed shrew (*Crocidura* cf. *leucodon*) were reported from the same cave, while the common shrew (*Sorex araneus*) is entirely lacking, indicating a filtered connection. The extant shrew of Gozo (*Crocidura sicula*) is likely a descendant of the Pleistocene shrew of Malta and Sicily (*Crocidura esuae*).

In the upper Holocene layers, remains of introduced red deer (*Cervus elaphus*) were found. This deer differs in its larger size, slender limbs, and lack of bone fusions from the dwarfed deer of the previous period. Also, horse, goats, and domestic dogs arrived on Malta, imported by Phoenician settlers. Remains were found in several deposits, for example, the Valletta fissure.

Peculiarities and Evolution of Endemic Mammals

Dwarf Elephants

Three species of dwarf elephants have been described for Malta. These were, in chronological order of discovery and with increasing body size, *Palaeoloxodon falconeri* by Busk (1867), *Palaeoloxodon melitensis* by Falconer (1868, though already mentioned by Falconer, 1862), and *P. mnaidriensis* by Leith Adams (1870). Soon after, these names were applied to elephant finds of approximately similar sizes from Sicily, and later to those from other islands as well. Small species were named *falconeri*; medium-sized *melitensis* and larger forms that were only somewhat smaller than mainland elephants were named *mnaidriensis*. This was the start of a long history of nomenclatorial confusion.

To add to the confusion, the species *melitensis* and *falconeri* are not clearly defined, and the molars described by Busk as belonging to two different species cannot be sufficiently separated based on size. Pierluigi Ambrosetti (1968) thus proposed to consider the two species as synonymous, back to the original situation, and to retain only *falconeri* as a valid species. At that occasion, he ascribed the material from the Spinagallo cave of Sicily to *Elephas* (=*Palaeoloxodon*) *falconeri*. Since then, *falconeri* is regarded as a typical Sicilian dwarf elephant, while the type specimens are from Malta.

Although the elephant from Mnajdra Gap (late Middle Pleistocene; *P. mnaidriensis*) is a little larger than the largest forms from the Żebbuġ fissure (early Middle Pleistocene; *P. falconeri*), it did not derive from the latter. Its most likely ancestor is the contemporaneous middle-sized dwarf elephant from Sicily (*Palaeoloxodon* aff. *mnaidriensis*).

Dwarf Hippopotamus

The Maltese dwarf hippo (*H. melitensis*) is somewhat smaller than the Sicilian species (*H. pentlandi*) and evolved about 40% body mass reduction relative to their mainland ancestor, *Hippopotamus amphibius*. The size variation within the Maltese hippopotamus exceeds that of a mainland population, but this is not uncommon for island species. The large variation gives the false impression of two different species on Malta, a smaller *H. melitensis* and a larger *H. pentlandi*, but only the first species is valid for the Maltese material. Its direct ancestor probably is the latter species, which dispersed to Malta.

Falconer was of the opinion that the dwarf hippopotamuses from Crete, Malta, and Sicily all belonged to the same species. The name *pentlandi* was since then regularly used for middle-sized dwarf hippos.

Dwarf Deer

We mentioned and figured in our 2009 popular book the remarkable variation in size of the Maltese Pleistocene deer from Għar Dalam, in combination with typical insular ruminant characters like a relative shortening of the lower limb bones and a fusion of the navicocuboid – two fused tarsal bones – with the metatarsus. We recognised six sizes and compared this speciation with the adaptive radiation observed in Cretan Pleistocene deer (Figure 8.4). Leila D'souza (2019)

Figure 8.4 The variety in length of the metatarsals of the Maltese deer (*Cervus* sp.) is remarkable. To the left, the Holocene population of red deer (*Cervus elaphus*). In the middle and to the right, endemic Middle or Late Pleistocene forms of *Cervus*, likely representing two different species. All three size classes are dimorphic, resulting in six size groups in total. Scale bar: 10 cm. Għar Dalam Natural History Museum, Malta. (Photograph: Alexandra van der Geer.)

studied the Maltese deer from the Pleistocene levels of Għar Dalam in much detail. She confirmed the presence of six sizes, and explained the six sizes as three sexually dimorphic size groups. All size groups are smaller than the Sicilian red deer. Based on skeletal morphology, she indicates *Cervus elaphus* as most likely ancestor, presumably originating from Sicily during the late Middle Pleistocene at a low sea level stand. Indeed, the earliest deer fossil in the cave dates back to before 160,000 years ago, towards the end of the late Middle Pleistocene.

The Holocene deer (*Cervus elaphus*) are entirely different, with a slender build and lacking bone fusions. These were brought to the island by Neolithic people, along with domesticates.

The Otter

Bate (1935) described some postcranial elements of an otter from Pleistocene deposits of the Tal Gnien fissure near Mqabba as *Nesolutra euxena*. Much later, Willemsen (1992) reviewed the genus and concluded that the genus name was a junior synonym of *Lutra*, the common Eurasian otter. Cherin and colleagues (2016) moved the Sicilian otter to the new genus of short-muzzled otters *Lutraeximia*. If the Maltese otter evolved from the Sicilian otter, which is likely – or vice versa – the Maltese otter should be named *Lutraeximia euxena*. No molar teeth have been recovered, so any inference about its diet remains speculative. Willemsen (1992) notes that it shows stronger aquatic adaption than the common otter.

A Brown Bear?

A single fossil of a bear (Figure 8.5) was retrieved in 1882 by Cooke from the red clay layers at Għar Dalam cave that also yielded abundant fossils of the endemic small red deer (*Cervus* cf. *elaphus*), a small fox (*Vulpes* sp.), and a vole (*Microtus melitensis*). Cooke (1892), who reported the find, mentioned

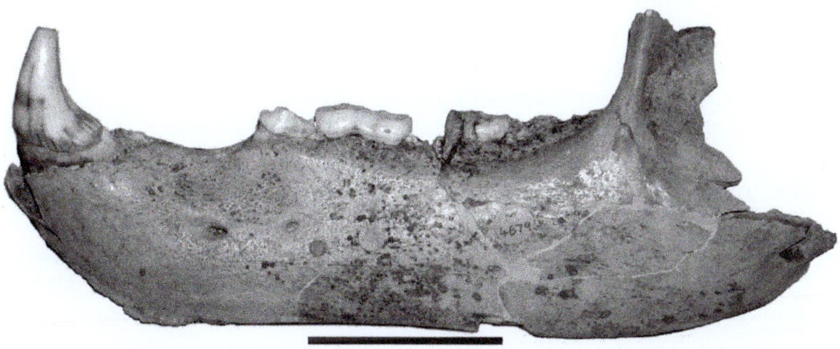

Figure 8.5 Fossil mandible of a brown bear (*Ursus arctos*) from Għar Dalam. Scale bar: 5 cm. Natural History Museum, London. (Photograph: George Lyras.)

the presence of human artefacts and remains in the very same layer, which he considered of Neolithic age. The bear fossil was described as the left lower mandible of a brown bear (*Ursus arctos*), by Woodward (1893), and considered smaller than a continental brown bear by Zammit Maempel (1989). Unfortunately, its context is unclear because Għar Dalam cave has a complex stratigraphy including reworked sediments, and faunas were likely mixed in the early excavations. If genuine, the fossil is one of the very rare examples of a bear on an island. Alternatively, the jaw may be a cultural artefact, such as a talisman or a trophy.

References

Abela G.F. 1647. *Della descrittione di Malta Isole nel Mare Siciliano con le sue antichità, ed altre notitie*. Malta.

Ambrosetti P. 1968. The Pleistocene dwarf elephants of Spinagallo (Siracusa South-Eastern Sicily). *Geologica Romana*, 1: 217–229 and 7: 277–398.

Bate D.M.A. 1920. Note on a new vole and other remains from the Ghar Dalam cavern, Malta. *Geological Magazine*, 57: 208–211.

Bate D.M.A. 1935. Two new mammals from the Pleistocene of Malta, with notes on the associated fauna. *Proceedings of the Zoological Society of London*, 112: 247–264.

Busk G. 1867. Description of the remains of three extinct species of elephant, collected by Capt. Spratt, C.B.R.N., in the ossiferous cavern of Zebbug, in the island of Malta. *Transactions of the Zoological Society of London*, 6: 227–306.

Cherin M., Iurino D.A., Willemsen G., Carnevale G. 2016. A new otter from the Early Pleistocene of Pantalla (Italy), with remarks on the evolutionary history of Mediterranean Quaternary Lutrinae (Carnivora, Mustelidae). *Quaternary Science Reviews*, 135: 92–102.

Cooke J.H. 1892. On the occurrence of *Ursus arctos* in the Maltese Pleistocene. *The Mediterranean Naturalist*, 2 (18): 267–269.

D'Souza L.C. 2019. *An Endemic Radiation of Deer in the Late Pleistocene of Malta*. PhD thesis, Bristol University.

de Bruijn H. 1966. On the Pleistocene Gliridae (Mammalia, Rodentia) from Malta and Mallorca. *Proceedings of the Koninklijke Akademie van Wetenschappen (Series B)*, 69 (4): 480–496.

Dubey S., Koyasu K., Parapanov R., Ribi M., Hutterer R., Vogel P. 2007. Molecular phylogenetics reveals Messinian, Pliocene, and Pleistocene colonizations of islands by North African shrews. *Molecular Phylogenetics and Evolution*, 47 (2): 877–882.

Falconer H. 1862. On ossiferous caves in Malta, explored by Captain Spratt, R.N., C.B., with an account of *Elephas melitensis*, a pigmy species of fossil elephant, and other remains found in them. Parthenon (London) I (25) 780.

Falconer H. 1868. On the fossil remains of *E. melitensis*. An extinct pygmy species of elephant and other mammalia from the ossiferous caves of Malta. In: Murchinson C. (ed.), *Palaeontological Memoirs and Notes of the Late Hugh Falconer*, Vol. 2. Robert Hardwicke, Picadilly, UK, pp. 292–308.

Forsyth Major C.I. 1902. On the Pigmy Hippopotamus from the Pleistocene of Cyprus. *Proceedings of the Zoological Society of London*, 2 (1): 238–239, 2 (2): 107–112.

Hunt C.O., Schembri P.J. 1999. Quaternary environments and biogeography of the Maltese Islands. In: Mifsud A., Savona Ventura C. (eds), *Facets of Maltese Prehistory*. The Prehistoric Society of Malta, Malta, pp. 41–75.

Leith Adams A. 1863. Observations on the Fossiliferous caves of Malta. *Journal of the Royal Society*, 4 (2): 11–19.

Leith Adams A. 1868. On a species of dormouse (*Myoxus*) occurring in the fossil state in Malta. *Transactions of the Zoological Society of London*, 6 (5): 307–308.

Leith Adams A. 1870. *Notes of a Naturalist in the Nile Valley and Malta*. Edmonton and Douglas, Edinburgh.

Lydekker R. 1895. On the affinities of the so-called extinct giant dormouse of Malta. *Proceedings of the Zoological Society of London*, 1895: 860–863.

Malec F., Storch G. 1970. Zur Kenntnis der jungpleistozänen Wühlmaus *Pitymys melitensis* (Mammalia, Rodentia). *Zeitschrift für Säugetierkunde*, 35 (2): 75–80.

Spratt T.A.B. 1867. On the Bone-Caves near Crendi, Zebbug, and Melliha, in the Island of Malta. *Quarterly Journal of the Geological Society*, 23 (1–2): 283–297.

van der Geer A., de Vos J., Dermitzakis M., Lyras G. 2009. *Hoe dieren op eilanden evolueren. Majorca, Ibiza, Kreta, Sardinië, Sicilië, Japan, Madagaskar, Malta*. Wetenschappelijke Bibliotheek 98. Utrecht, Veen.

Willemsen G.F. 1992. A revision of the Pliocene and Quaternary Lutrinae from Europe. *Scripta Geologica*, 101: 1–115.

Woodward A.S. 1893. Report on the organic remains. *Proceedings of the Royal Society London*, 54: 274–283.

Zammit Maempel G. 1989. *Ghar Dalam Cave and Deposits*. Mid-Med Bank, Malta.

Zammit Maempel G., de Bruijn H., 1982. The Plio-Pleistocene Gliridae from the Mediterranean Islands reconsidered. *Proceedings of the Koninklijke Akademie van Wetenschappen (Series B)*, 85 (1): 113–128.

CHAPTER NINE

Sardinia and Corsica

Evolution of Island Mammals: Adaptation and Extinction of Placental Mammals on Islands, Second Edition.
Alexandra van der Geer, George Lyras and John de Vos.

© 2021 John Wiley & Sons Ltd. Published 2021 by John Wiley & Sons Ltd.

Today, Sardinia and Corsica are two separate islands. Before the Holocene, however, they formed a unit with a shared geological history. Their fossil faunas are therefore practically identical. With the start of the Miocene, the Sardo-Corsican microplate separated from southern Europe and was inhabited by successive endemic insular faunas since then. During the Late Miocene, a temporary land connection with the island Tuscany resulted in a shared, insular endemic fauna. The post-Miocene faunas of Sardinia and Corsica are relatively rich, likely due to the large surface area combined with a relatively small distance to the Apennine Peninsula. The most common elements are goral-like bovids in the Miocene and Pliocene and dwarf megacerine deer and giant pikas in the Pleistocene.

Geology and Palaeogeography

Until the Early Oligocene, the Sardinia-Corsica block was an integral part of the southern margin of the European plate and bordered the region that constitutes today the Iberian Peninsula and southern France. The separation process, due to crustal rifting, took place from the Eocene to the Late Oligocene. This gradual process was superimposed by changes in eustatic sea level, which gave rise to intermittent land connections before the microplate with Sardinia and Corsica became completely isolated from continental Europe.

Starting from the Early Miocene, an anticlockwise rotation of the microplate took place. In the Middle Miocene, major parts of the Mediterranean lands were flooded during a high sea level stand when the seaway to the Indian Ocean opened again. Exposed areas of Sardinia were limited in area. Then, during the Late Miocene, the Sardinia-Corsica block formed a unit with lands that today are part of Tuscany on the Tyrrhenian side of the Apennine Peninsula. This large land area constituted one of the three major palaeo-bioprovinces of Italy's Late Miocene, the other two being Apulia-Abruzzi at the Adriatic side and Sicily-Calabria in the South. It might be that the region consisted of several large islands, forming an archipelago, as indicated by small differences between the mutual faunas. At the onset of the Late Miocene Messinian salinity crisis, the Tusco-Sardinian palaeo-bioprovince disrupted. The Tuscany part became connected to the continent when the Apennine mountain chain was uplifted due to tectonic activity, resulting in a broad land connection over which a continental Eurasian fauna reached southern Tuscany. Approximately, at the same time, the Sardinia-Corsica part of the bioprovince got separated from Tuscany by the opening of the Tyrrhenian Sea resulting from post-collision phases of the Northern Apennines. While

Sardinia and Corsica became again isolated from the mainland, Tuscany remained part of the Italian Peninsula.

Historical Palaeontology

The very first scientific report of a fossil from Sardinia aimed at describing a pika, a small-sized member of the order of hares and rabbits (Lagomorpha). The new species was named *Lagomys sardus fossilis*, literally, the fossil Sardinian hare mouse, by the German anatomist and physiologist Rudolf Wagner (1829). At this occasion, Wagner also described in much detail the fossiliferous deposit in which he had found the fossils a year before, named Monreale (= Montereale) at Bonaria near Cagliari (Figure 9.1). At present, the species is known as *Prolagus sardus* and is the symbol of the endemic Pleistocene faunas of Sardinia and Corsica (Plate 13) because of the abundance of its fossils. After Wagner's graduation in 1826, he had spent a year or more in Paris with the French comparative anatomist Georges Cuvier. Cuvier had shown him a toothless jaw of this species, originating from a small piece of bone breccia he had found near Cagliari in 1825, and which he had described and figured in the same year, however, without formally naming it. In the same work, he also reported a similar lagomorph from Corsica. Cuvier considered the two species resembling most closely the extant pikas (*Lagomys*, now *Ochotona*). Wagner (1829) assigned the Corsican species thus to *Lagomys corsicanus*, now a junior synonym of the Sardinian species. More than a century later, Heinz Tobien (1935) studied the Sardinian pika extensively, while Mary Dawson (1969) reviewed the diagnosis of the species.

The same piece of bone breccia that was studied by Cuvier contained the remains of a vole and a shrew as well. Cuvier noticed similarities between this shrew and the European water shrew *Sorex fodiens*, now *Neomys fodiens*. Wagner shared his teacher's view in his description of the Monreale deposit. The German naturalist Reinhold Hensel though had different ideas. He (1855) renamed the Monreale shrew to *Sorex similis*, now known as *Asoriculus similis*. The next year (Hensel, 1856), he described *Mus orthodon*, now known as *Rhagamys orthodon*, from an unknown locality on Sardinia or Corsica, but possibly from the same Monreale site.

A year later, the Italian physiologist Cesare Studiati (1857) described a new canid species, the Sardinian dhole *Cuon sardous*, now *Cynotherium sardous*, a species unrelated to the dholes. The fossils had been recovered in 1832 during an excavation of the fossiliferous breccias of Monreale, organised by general Giuseppe Alberto Ferrero della Marmora, governor

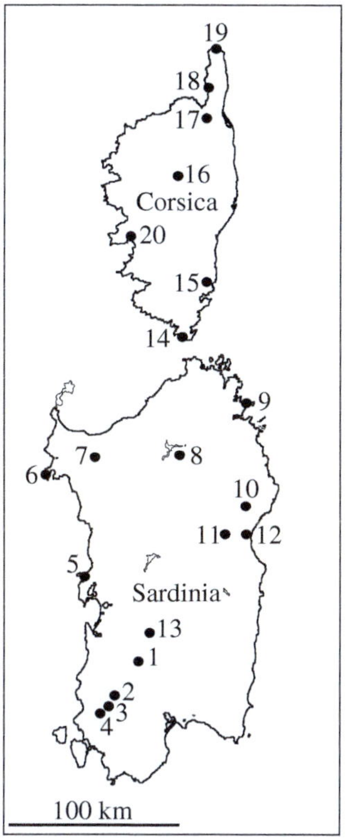

Figure 9.1 The most important fossil localities of Sardinia and Corsica. (1) Monreale or Montereale, Bonaria; (2) Monte San Giovanni, Iglesias; (3) Funtana Morimenta, Gonnesa; (4) Terras de Collu and Bacu Abis mines, Gonnesa; (5) Capo Mannu; (6) Neptuno cave and Dragonara Cava, Capo Caccia; (7) Fiume Santo, Sassari; (8) Oschiri; (9) Capo Figari, Olbia; (10) Monte Tuttavista, Orosei; (11) Corbeddu cave, Oliena; (12) Ispiginoli cave, Dorgali; (13) Sardara; (14) Araguina-Sennola abri; (15) Punta di Calcina; (16) Corte; (17) Castiglione cave; (18) Grotta del Margina, Nonza; (19) Grotta de la Coscia; and (20) Vaccio, Ajaccio.

of Sardinia, a keen naturalist in his spare time, and author of 'Voyage en Sardaigne' (1826). Studiati's description formed part of the appendix of this monumental monograph on Sardinia. Probably, the fossils of Hensel's shrew originated from the very same excavation.

Fifteen years later, the French malacologist Arnould Locard and palaeontologist Charles Depéret studied deer fossils from Nonza, Corsica, but its formal description had to wait till Depéret (1897) finally named it *Cervus cazioti* (now *Praemegaceros cazioti*) in honour of Eugène Caziot for his extensive work on the molluscs of Corsica. Depéret considered his species related to but different from the *Eucladoceros* from England, based on antler morphology.

The most important early scholar on Sardinian and Corsican extinct faunas was the British palaeontologist Charles Forsyth Major. Forsyth Major (1882) erected a new species of large vole, *Tyrrhenicola* (=*Microtus*) *henseli*, in honour of Reinhold Hensel. The fossils originated from Monte San Giovanni near Iglesias in south-western Sardinia. Hereafter, he published several papers on the biogeography and palaeontology of Sardinia, based on the huge number of fossils he and others had collected on Sardinia.

Amongst them were proboscidean foot elements, which he (Forsyth Major, 1883) described as *Elephas* (=*Mammuthus*) *lamarmorae*. The fossils were reported by L. Acconci (1881) and were the first proboscidean fossils from Sardinia. They had been found during railway constructions in petrified sand dunes (aeolian deposits) at Fontana Morimenta (Funtana Murimenta) near the village of Gonnesa in south-western Sardinia and consisted of tarsal and carpal elements, partly in anatomical position.

In 1883, mine engineer Emilio Ferraris had signalled jaw fragments of Eocene age in the brown coal layers (lignites) of Terras de Collu near Gonnesa and sent them to the geological museum of the University of Pisa. Eight years later, Forsyth Major studied them and referred to them as the tapir-like species *Lophiodon isselensis*, without describing or illustrating the fossils. This was done some years later by Camillo Bosco (1902), who renamed it to *Lophiodon sardus*, because of its smaller dimensions and different jaw shape.

Forsyth Major (1901) described a new mustelid from Sardinia, *Enhydrictis* (= *Pannonictis*) *galictoides* from Monte San Giovanni and later (1905) he erected a new field mouse genus *Rhagamys* to accommodate Hensel's species *orthodon*, described fifty years before.

Since 1910, Forsyth Major systematically excavated the fossiliferous breccias at Capo Figari together with the French palaeontologist Emile-George Déhaut who had started there a few years earlier. Déhaut (1911) described a new primate genus and species with very large eyes from Capo Figari as *Ophtalmomegas lamarmorae*, in honour of della Marmora. Later studies revealed, however, that the cranial fragments belonged to a bird instead. In the same publication, he erected a new species for the insular goral-like bovid *Antelope* (= *Nesogoral*) *melonii*. Charles Andrews suggested a few years later that this antelope might be a *Myotragus* – the mouse goat of Majorca – in his description of the skull and skeleton of the latter genus. This notion never received much attention, partly due to the lack of diagnostic fossils. Jan van der Made (2005) reconsiders the link between the two island genera in his description of a new *Nesogoral* species from the Campidano, *Nesogoral cenisae*.

Apart from excavating, Forsyth Major also bought fossils, for example, two upper jaw fragments, acquired in 1903 and originating from near Oschiri. The pieces remained undescribed in the collection of the British Museum for almost a century before they were promoted to holotype and paratype of a new giraffomorph ruminant, *Sardomeryx oschiriensis*, by van der Made (2008).

A second fossil of an odd-toed ungulate was described by the Italian geologist Giambattista Dal Piaz (1929) as *Atalonodon monterinii*, based on a juvenile lower jaw found in an Eocene layer not far from the mine in which the earlier tapir remains had been found. The species was named after the glaciologist and explorer Umberto Monterin who had illustrated the jaw a year earlier tentatively referring it to *L. sardus*. Monterin thought that the missing talon of the third molar was either due to damage or, more likely, to its juvenile age. Dal Piaz, however, considered this feature of phylogenetic importance and erected a new genus. He pointed out that the feature was also present in the Eurasian *Chasmotherium* – considered synonymous with the North American *Hyrachyus* by Leonard Radinsky (1967) – but this genus differs in other characters from *Atalonodon* and is currently considered an early rhinocerotid. In short, Monterin's tapiroid seems an endemic genus, unique to Sardinia, but whether it is an insular taxon or not is a matter of debate.

The Tyrrhenian mole *Talpa tyrrhenica* was described by Bate (1945) on the base of a left maxilla with some teeth from Monte San Giovanni near Iglesias. She had not excavated the fossils herself, but studied them in the collection of Forsyth Major, who on his turn had discovered them in the collection of Richard Lydekker. Lydekker, in his function of cataloguer for the British Natural History Museum, had listed a number of specimens back in 1887, but these had never been described.

Augusto Azzaroli (1946) described teeth of a small macaque from the breccias of Capo Figari and named it *Macaca majori* after Forsyth Major who had collected there more than half a century before.

After WW II, several endemic small mammals were described by various authors from different localities, especially so in the 1970s, when no less than eleven endemic Sardinian taxa were presented. These are the large field mouse *Apodemus mannu* from Capo Mannu, described by Louis Thaler (1973), the moles *Geotrypus oschieriensis* and *Nuragha schreuderae*, the dormouse *Glis major*, and the gundis *Sardomys dawsonae*, *Sardomys antoniettae*, and *Pireddamys rayi*, all six from Oschiri and described by Hans de Bruijn and Cornelia Rümke (1974), the dormouse *Tyrrhenoglis* (=*Eliomys*) *majori* from Capo Figari, named after Forsyth Major, by Burkart Engesser (1976), and, finally, *Rhagamys* (=*Rhagapodemus*) *minor* from Capo Figari by Louis Brandy (1978).

The cave Corbeddu was systematically excavated between 1982 and the mid-1990 by the Dutch palaeontologist Paul Sondaar. This cave, called after the fugitive bandit Giovanni Salis Corbeddu (Figure Box 9.1) who lived in the cave during the last quarter of the nineteenth century, is situated in the

Lanaittu valley between the villages Oliena and Dorgali. During thousands of years, more than twenty metres of clayish sediments with hundreds of mainly deer and ochotonid bones therein had been accumulated throughout the four halls of the cave. Three of these halls have been excavated by Sondaar and his team in a period of about fifteen years (Plate 14). Three human fossils (Box 9.1) were recovered in total, described by Fred Spoor and Sondaar (1986). The phalanx fragment was dated to about 20,000 years ago, while a maxillary and a temporal bone were radiocarbon dated to between about 10.2 and 9.5 years ago by Gerard Klein Hofmeijer and colleagues (1987).

In the early 1990s, an important Late Miocene site was discovered in north-west Sardinia during the construction of a parking area within a thermo-electric power plant. The site, named Fiume Santo, and the first fossils were preliminary described by Jean-Marie Cordy and Sergio Ginesu, who recognised the similarity with of this fauna with that of Late Miocene sites in the coastal Maremma region in southern Tuscany, based on the occurrence of the endemic ape *Oreopithecus* and the bovids *Maremmia* and *Tyrrhenotragus*. This was the first palaeontological indication for the existence of a Late Miocene palaeo-bioprovince connecting Sardinia and southern Tuscany. Several excavations took place since then, and more than 13,000 fossils have been collected so far, partly described by Laura Abbazzi and colleagues (2008). They described several new taxa the giraffid *Umbriotherium azzarolii* and the bovids *Etruria viallii* and *Turritragus casteanensis*. The former two names had already been used by Johannes Hürzeler and Burkart Engesser (1976) but were later considered *nomina nuda* by the ICZN because the remains were not described.

A few years later, another important locality was discovered. In one of the fissure fillings of the limestone quarries at Monte Tuttavista near the village of Orosei (Plate 15), an almost complete skull of a hunting hyena was found by Giampietro Mele in 1995. The skull was described as *Chasmaporthetes melei* by Lorenzo Rook and colleagues (2004). A hyena is a typical mainland carnivore, and its presence in Sardinia was thus unexpected. The finding immediately prompted further excavations of the fissure fillings, starting with the collection of material by the archaeological survey of Nuoro and Sassari. The study of the prepared macrofossils was led by Sondaar and colleagues and students, including two of us (AG, GL), summarised as an unpublished report (1998) kept at the survey, which was partially reworked a year later by Sondaar (2000) in an article. The small mammals were studied by Jean-Marie Cordy and students. Thereafter, Marisa Arca and Caterinella Tuveri of the survey continued the collection and study, from 2002 onwards, assisted by teams from the universities of Florence and Rome. At present, it is the richest Sardinian locality, ranging in time from the Late Pliocene to the earliest Holocene, producing the amazing quantity of over 80,000 fossil specimens. This extensive Orosei assemblage has been described in several papers, either aiming at giving an overview, such as the study by Abbazzi and colleagues (2004), or focussing on one taxon or lineage, for example, on *Cynotherium* by Abbazzi and colleagues (2005), on *Microtus* (*Tyrrhenicola*) by Marcolini and colleagues (2005), and on *Prolagus sardus* by Angelone and colleagues (2008). New mammalian taxa reported from Orosei, apart from the hunting hyena, include *Microtus sondaari* by Marcolini and colleagues (2006) and *Asoletragus gentry* by Maria Rita Palombo and colleagues (2006).

Van der Made (2008) describes three new endemic mammals from Oschiri: the giraffomorph *Sardomeryx oschiriensis*, based on the two jaw fragments bought by Forsyth Major in 1903; a pig with uncertain generic affinities, *Hyotherium insularis*, and the primitive ruminant *Bachitherium sardus*. Although the fossils had been found near Oschiri, they did not originate from the same deposit as the small mammals, hampering biostratigraphic conclusions.

BOX 9.1

Colonisation by Humans

The earliest human remains have been retrieved from the cave Corbeddu in the Lanaittu valley between the villages Oliena and Dorgali, named after Giovanni Salis Corbeddu (Figure Box 9.1). Three of its four halls were systematically excavated since 1982 by the team of Dutch palaeontologist Paul Sondaar, which retrieved hundreds of fossils of mainly deer and pikas. The most important findings were three human fossils, described by Fred Spoor and Sondaar (1986) and contextualised by Sondaar and colleagues (1995). The phalanx fragment was dated to about 20,000 years ago, while a maxillary and a temporal bone were radiocarbon dated to between about 10.2 and 9.5 years ago by Gerard Klein Hofmeijer and colleagues (1987).

The presence of Palaeolithic humans in Sardinia is also attested by large quantities of stone artefacts, found by Fabio Martini in the fields around Perfugas dell'Anglona. These artefacts resemble the Clactonian tools of the Middle Pleistocene of Italy, as postulated by Fabio Martini (1992). Human fossils other than those from Corbeddu cave consist of a complete Early Holocene skeleton found at Araguina-Sennola on Corsica. The human colonisation seems to coincide with the extinction of the dwarf dog (*Cynotherium sardous*; last occurrence at Corbeddu cave, about 11,000 years ago) and the dwarf deer (*Praemegaceros cazioti*; latest occurrence at Giuntu cave, about 7000 years

Figure Box 9.1 The bandit Salis Giovanni Corbeddu, after whom the fossiliferous cave Corbeddu was named. (Photograph: U. Martelli, 1894.)

ago) and the introduction of rats. Only the small mammals manage to survive the impact for some time: the field mouse was still present during the Neolithic (Grotta del Guano, Oliena), and the vole even during the late Bronze Age (Nuraghe Is Paras de Isili), though the mole and the shrew disappeared already in the Early Holocene.

Biozones and Faunal Units

Until the second half of the previous century, only some isolated sites, each with its own particular fauna, were known, and their mutual relation in time and geography was hardly known. The first attempt to reconstruct the biogeography of the Pleistocene of Sardinia and Corsica was made by Forsyth Major in the late nineteenth century. Roughly a century later, Azzaroli reconstructed the palaeo-biogeography of the Palaeogene and Neogene (formerly: Tertiary). Towards the end of the twentieth century, new localities were discovered, amongst which, especially, the many rich fissure fillings of Monte Tuttavista, Orosei, are important (Plate 15). Their faunas span from the Late Pliocene to the Holocene. These mammal assemblages greatly enhance the biostratigraphic precision of the various Sardinian and Corsican fossil deposits.

Early–Middle Eocene

The earliest Sardinian mammalian fauna is hardly known because its fossils are extremely scanty and restricted to the south-western Sulcis peninsula. The relation with the European Archipelago of the Eocene is unclear, but Sardinia may have constituted one of its many southern islands (see Chapter 1, Box 1.1).

The fossil findings are restricted to tapir-like odd-toed ungulates (helaletids) and a large opossum (*Amphiperatherium*). The tapiroids are *Atalonodon monterinii* from late Early Eocene deposits and '*Lophiodon*' *sardus* from Middle Eocene deposits in the Terras de Collu mine at the village of Gonnesa, near Iglesias. *Atalonodon* is represented by half a lower jaw and '*Lophiodon*' by two jaw pieces. The opossum fossils – several teeth – were found in the mine of Bacu Abis, also near Gonnesa, in a Middle Eocene layer. The opossum was most likely contemporaneous with the second tapiroid.

The dentition of the two tapir-like inhabitants indicates a marsh plant eater and a forest browser, respectively. The climate was tropical and humid during the Early Eocene as indicated by the type of foraminifera found in the same limestone beds in which *Atalonodon* was found. During the Middle Eocene, the climate had changed to sub-humid tropical to subtropical with strong seasonal variations, mainly based on the physical aspects of the coal beds in which '*Lophiodon*' was found. Dense vegetation was likely present along the shores of the lake, mainly consisting of tropical palms, Myricaceae, and ferns.

As far as can be concluded from the scanty remains, '*Lophiodon*' might represent an endemic species and *Atalonodon* an endemic genus. The latter genus closely resembles *Lophiodon*, but lacks a talon at the base of the third lower molar (hypoconulid). This character is shared with *Hyrachyus* – a common North American and Eurasian rhinocerotoid of that time – but *Atalonodon* differs from *Hyrachyus* in having much larger canines, lacking a lower first premolar, and having a diastema between the canine and the premolar series, a diagnostic feature of the helaletids or early tapirs. *Atalonodon* looks in fact more like a derived lophiodontid, but this group seems not to have appeared in Europe before the Middle Eocene.

'*Lophiodon*' *sardus* differs from the other *Lophiodon* species and is sometimes attributed to the Middle Eocene genus *Paralophiodon*. The main difference between *Lophiodon* and *Paralophiodon*, apart from their chronology, is the lack of a diastema after the canine in the latter. The Sardinian remains are, however, sparse and a correct diagnosis is hampered by, amongst others, the lack of the jaw part between the canine and the premolar series. To make things worse, the original fossils are at present lost. The endemism, if present, indicates some kind of barrier between Sardinia and southern Europe, as followed by Tassos Kotsakis and colleagues (2008).

Early Miocene

The Early Miocene fauna of Sardinia is known only from a fossiliferous layer cropping out in the surroundings of Oschiri, northern Sardinia. The fauna from Oschiri is one of the oldest insular faunas worldwide yet resembles in composition the Pleistocene insular faunas with a relative abundance of ungulates and small mammals and the absence of terrestrial mammalian carnivores. The most typical endemic elements are two moles (*Geotrypus oschieriensis* and *Nuragha schreuderae*), three gundis (*Sardomys dawsonae*, *Sardomys antoniettae*, and *Pireddamys rayi*), an insular pig (?*Hyotherium insularis*), a primitive ruminant (*Bachitherium sardus*), and a stem palaeomerycid (*Sardomeryx oschiriensis*), a yet undetermined moschid and possibly a primitive odd-toed ungulate. The presence of an odd-toed ungulate has been assumed based on two molar fragments, which differ from the teeth of known even-toed ungulates and also from tapir teeth. Other Oschiri taxa are close to or identical with continental species, such as the shrew (*Oligosorex antiquus*), the dormice (*Glis major*, *Peridyromys* aff. *murinus*, and *Microdryomys* aff. *Koenigswaldi*), and possibly a palaeomerycid ruminant ('*Amphitragulus*' sp.).

Both the island pig and the bachithere ruminant have shortened phalanges, as shown by van der Made (1999). The stem palaeomerycid *Sardomeryx* shows insular adaptations in its dentition: a shorter premolar row and a higher molar crown (hypsodont) compared with continental Early Miocene ruminants.

The largest of the three Sardinian gundis is the largest species of its family, indicating long-term insular conditions. Gundis – family Ctenodactylidae – from mainland Europe has been reported only from Greece (Early Miocene, Chalkidiki and Middle Miocene, Chios). The family is considered to have originated during the Oligocene in Asia, from where it dispersed into North Africa and from there to the islands Sardinia and Sicily, where it has been recorded in deposits of, respectively, Early Miocene and Early Pleistocene ages.

The degree of endemism of the odd-toed ungulates and the gundis indicate a long period of isolation.

The remains of '*Amphitragulus*', a left lower fourth premolar, are similar in size and morphology to the isolated find from Sardara, western Sardinia, which was assigned to *Amphitragulus boulangeri* by Ida Comaschi Caria (1953). The relation of this specimen with the Oschiri premolar and with mainland specimens is unclear. The taxonomy of the species is thus unresolved, and it is sometimes included in *Pomelomeryx*.

Finally, the shrew *Oligosorex antiquus* is sometimes referred to as *Crocidosorex antiquus*, with the former genus considered a junior synonym of the latter genus. However, as van den Hoek-Ostende (2001) pointed out, the type species of these two genera, *antiquus* and *piveteaui*, respectively, differ essentially from each other, which fact validates the retention of the two genera.

Late Miocene

After a long gap in the fossil record – only the gavial *Tomistoma* and other reptiles are cited from the Middle Miocene – a second Miocene fauna has been reported from Sardinia. This Late Miocene fauna is found at Fiume Santo, north-western Sardinia, but also at the lignite mines of Monte Bamboli and the middle levels (V1, V2, see below) of the Baccinello-Cinigiano basin of the Maremma region in southern Tuscany (now mainland Italy). This shared fauna is evidence for a Tusco-Sardinian palaeo-bioprovince that existed between about 6.7 and 8.3 million years ago. The fauna is generally referred to as the *Oreopithecus* faunal assemblage, named after its best-known member, the ape *Oreopithecus bambolii*. The fauna is unbalanced, endemic, and characterised by a profusion of unique artiodactyl genera, pointing to a long-term isolation.

The most typical members of this tropical fauna as found in Fiume Santo are the bovids, amongst which Lorenz' antelope (*Maremmia* cf. *lorenzi*) is the commonest taxon.

Other bovids are dwarf antelopes (*Tyrrhenotragus gracillimus* and Etruria *viallii*), a small bovid (*Turritragus casteanensis*, earlier ascribed to *Maremmia haupti* of Tuscany; Plate 16), and an undetermined bovid (Neotragini indet.). Other ungulates of this fauna are a giraffid (*Umbriotherium azzarolii*) and the Etruscan pig (*Eumaiochoerus* cf. *etruscus*). Initially, two giraffids were reported from Fiume Santo by Cordy and Ginesu (1994), but morphometric analysis of the material by Abbazzi and colleagues indicated the presence of only one form (*U. azzarolii*). The small size differences between dental elements are explained by them as reflecting sexual dimorphism.

The small mammal faunas of southern Tuscany are retrieved from strata of different geological ages – named Baccinello V0 to V3 – each containing different species. The non-endemic mouse *Huerzelerimys vireti* is restricted to the oldest layer – Baccinello V0 – and sets the lower limit of these faunas to the early Turolian (MN11). The upper limit is indicated by the presence of the hazel dormouse *Muscardinus vireti,* the field mouse *Apodemus etruscus*, and the hamster *Kowalskia nestori* of the early Villafranchian (MN13) of Europe in the youngest layer (Baccinello V3). The oldest endemic level – Baccinello V1 – is known by mice (*Huerzelerimys oreopitheci*, *Parapodemus* sp. I.), a hamster (*Kowalskia* sp.), a dormouse (*Anthracoglis marinoi*), a pika (*Paludotona etruria*), a shrew (cf. *Crocidosorex*), and an undescribed giant dormouse (Gliridae nov. gen. et nov. sp., by Engesser, 1983). The youngest endemic level – Baccinello V2 – is known by other mice (*Anthracomys majori*, *Parapodemus* sp. II), dormice (*Anthracoglis engesseri, Anthracoglis lorenzi*), a pika (*Paludotona* aff. *etruria*), and an undescribed shrew. A radiometric ($^{40}Ar/^{39}Ar$) dating of the volcanic layer between layers V1 and V2 sets the boundary between the subzones to 7.55 million years ago, reported by Lorenzo Rook and colleagues (2000). The small mammal species of the V2 subzone presumably evolved *in situ* from the earlier species, though *Parapodemus* sp. II is considered a new immigrant. Small mammals retrieved from Fiume Santo on the other hand are limited to mice (*H. oreopitheci*, *Anthracomys lorenzi*) and dormice (*Anthracoglis engesseri*, *Anthracoglis* sp. I) and correspond best to the Baccinello V2 level, as reported by Isaac Casanovas-Vilar and colleagues (2011).

The carnivore remains of southern Tuscany are extremely scanty and limited to a mustelid (*Mustela major*) and otters. Two species have been described so far for the upper Baccinello level (V2), *Paludolutra maremmana* and *Paludolutra campanii*. In the lower level (V1), another endemic otter genus occurs (*Tyrrhenolutra helbingi*), which is considered ancestral to *Paludolutra campanii* by Rook and colleagues (1999). The carnivore remains of Fiume Santo are even scantier, and only two small dental fragments of a mustelid were found, which are not identifiable below the family level, as reported by Abbazzi and colleagues (2008). The presence of the omnivorous bear (*Indarctos anthracitis*) as earlier reported by Ginesu and Cordy (1997) has not been confirmed.

The small differences between the composition of the Sardinian and Tuscany faunas might be explained as evidence for the existence of an archipelago rather than a single mega-island, or alternatively, for ecological or temporal differences between the sites. For example, the giraffid *Umbriotherium* is the second most abundant form in Sardinia, whereas, in Tuscany, it is represented only by the type specimen. The differences in the dentition of *Maremmia* and the pig further suggest different ecological conditions. Despite the fact that the abundant fossils from Fiume Santo have been described in great detail by Abbazzi and colleagues (2008), they hamper a direct comparison with the southern Tuscany fauna. The reason is that the Fiume Santo fossils are often badly preserved, with the molar roots often eroded or totally absent and with the outer cortex – or sometimes the complete cortex – of bones dissolved. Both factors hamper a thorough comparison between the faunas.

Pliocene–Early Pleistocene

The opening of the Tyrrhenian Sea starting at the end of the Miocene made an end to the Miocene Tusco-Sardinian palaeo-bioprovince. The subsequent isolation with possibly new colonisations gave rise to a new fauna. The fauna of this period is balanced though depauperate, also known as *Nesogoral* fauna. The fauna is characterised by goral-like caprids (*Nesogoral melonii* (Figure 9.2) and *N. cenisae*), a small pig (*Sus sondaari*), a macaque (*M. majori*; Plate 17), a hunting hyena (*Chasmaporthetes melei*), a small bovid (*Asoletragus gentry*), an undetermined caprid, a mustelid (*Pannonictis* sp.), and several small mammals, including a large field mouse (*Apodemus mannu*), a small field mouse (*Rhagapodemus minor*), a dormouse (*Tyrrhenoglis majori*), a shrew (*Asoriculus* sp. I), a mole (*Talpa tyrrhenica*), a pika (*Prolagus figaro*), and a small hare (*Sardolagus obscurus*). The fauna is generally referred to as *Nesogoral* fauna, after the goral-like bovid.

The composition of the fauna favours moderately long-term isolation and a colonisation by land bridge, possibly during the Messinian salinity crisis of the latest Miocene or the mid-Pliocene transgression. The degree of endemism of the fauna differs essentially between the taxa. The hunting hyena (*C. melei*) is only somewhat smaller than the continental form

Figure 9.2 Skull fragment of *Nesogoral melonii*, Capo Figari, Early Pleistocene. Top: labial view. Below: ventral view. Scale bar: 2.5 cm. Collection Malatesta, Palaeontological Museum, University La Sapienza, Rome. (Photograph: Alexandra van der Geer.)

(*Chasmaporthetes lunensis*) of the same period of the rest of Eurasia (Figure 9.3) as reported by Rook and colleagues (2004). Also, the macaque (Plate 17) is smaller than its proposed ancestor, *Macaca florentina*, from the Upper Valdarno of Italy. It has a shortened snout, shorter than expected for its size. The premolars, especially the upper third premolars, are smaller, too, diastemata are lacking, and the canines are extremely small for a monkey of its size.

The Sardinian rabbit remains were initially considered similar to the continental rabbit (*Oryctolagus lacosti*) of the Pliocene of Europe, but Chiara Angelone and colleagues (2018) erected a new genus and species (*Sardolagus obscurus*), characterised by an enigmatic combination of an advanced third lower premolar but primitive second upper premolar. The Sardinian rabbit (*Sardolagus*) is limited to the youngest faunal sub-complex (MN17), but also occurs in the next biozone (subzone Orosei 2) with *Microtus* (*Tyrrhenicola*). The ancestor of the rabbit is probably an early form of the European rabbit (*Oryctolagus*), arriving during the Pliocene regression of about 3.6 million years ago during the Capo Mannu D1 subzone, as suggested by Angelone and colleagues (2018).

The profusion of goat-like bovids is indicative of a vicariance effect in a predator-poor environment. The size of the island seems to have been large enough to sustain a viable population of hyenas, which preyed upon the bovids.

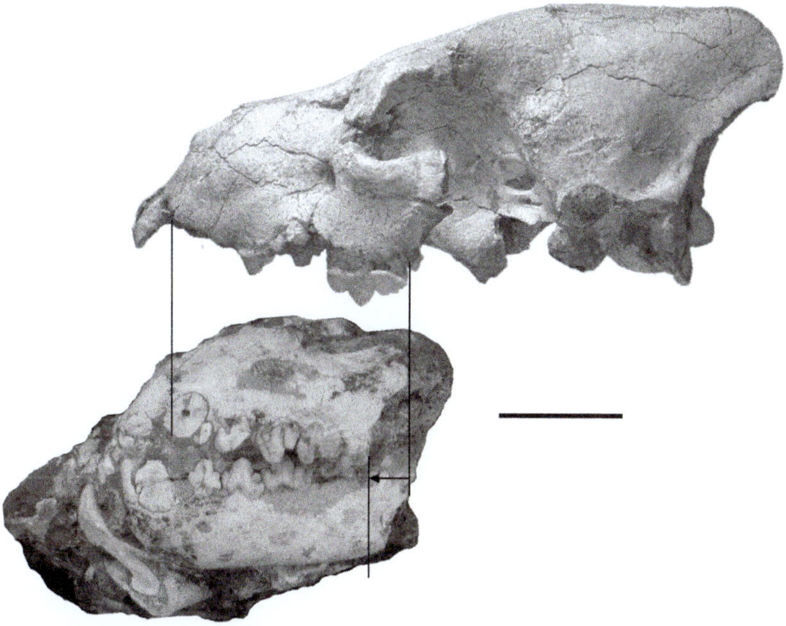

Figure 9.3 Skull of the continental *Chasmaporthetes lunensis* (top; adapted from Antón and colleagues, 2006), compared to the skull of *Chasmaporthetes melei* (below; Civic Archaeological Museum, Nuoro, Sardinia; photograph: Alexandra van der Geer). Scale bar: 5 cm.

The fauna is, however, not constant, and an internal evolution and possibly new invasions can be observed when the oldest locality, Capo Mannu D1 at Mandriola – a weathered coastal deposit of an earliest Late Pliocene age at the west coast of Sardinia – is compared with the younger localities, Capo Figari, north-eastern Sardinia; fissure filling Orosei 1, central-eastern Sardinia; and Capo Mannu D2, western Sardinia.

Capo Mannu D1

The earliest locality – Capo Mannu D1 at Mandriola – is about 3.6 million years old, coinciding with the Early–Late Pliocene transgression. Here, a somewhat smaller and less evolved form (*Rhagapodemus azzarolii*, described by Chiara Angelone and Tassos Kotsakis, 2001) of the small field mouse has been found, closely resembling the continental *Rhagapodemus ballesioi*. The dormouse from Mandriola (*Tyrrhenoglis* aff. *T. figariensis*) is smaller than the dormouse (*T. majori*) present in younger localities.

The large field mouse (*Apodemus mannu*) is already present in Mandriola, suggesting a long-term isolation and accordingly attesting to either an earlier colonisation by *Apodemus*, or a survivor of an earlier fauna. Mention has been made of mole remains which might represent a *Nuragha*, the genus of the Early Miocene of Oschiri. If this is confirmed, then *Apodemus* might also represent a vicariance effect. The pika of Mandriola, earlier listed as having affinities with the continental *Prolagus depereti* from southern France by Angelone and Kotsakis

(2000), is now recognised as a form close to *Prolagus figaro* of the subsequent subzone.

The single, broken tooth of a leporid found at Capo Mannu D1 is probably related or ancestral to the Sardinian rabbit of the subsequent subzone (Orosei 1). The shrew, the hyena, the mustelids, and the macaque are missing in Mandriola and might theoretically represent a later invasion. On the other hand, the deposits of Capo Figari and Orosei are much larger in extent and richer than the single deposit of Mandriola, so the evidence may be biased.

The first occurrence of the shrew was at Nuraghe su Casteddu, Dorgali, which is considered to be only slightly younger than Mandriola. Along with the shrew, the large *Tyrrhenoglis majori* was found as well.

Late Early Pleistocene–Early Holocene

This biozone is without doubt the best known and represented by thousands of fossils, especially of small mammals which are found in large quantities.

Typical elements of this biozone are a lineage of dogs (*Cynotherium*), dwarf deer (*Praemegaceros*), voles (*Microtus* (*Tyrrhenicola*)), red-toothed shrews (*Asoriculus*). Other elements, such as the mammoth and the otters, have a more restricted occurrence. The ancestor of the dwarf mammoth (*Mammuthus lamarmorae*) likely arrived at some time during the late Middle Pleistocene, while the time of arrival of the three Sardinian otters (*Megalenhydris barbaricina*, *Sardolutra ichnusae*, and *Algarolutra majori*) and the Corsican otter (*Lutra castiglionis*) is unclear.

The fauna – generally referred to as *Tyrrhenicola* fauna, after the most abundant rodent – is without doubt unbalanced and entirely endemic, indicating a long-term isolation and sweepstake dispersal for some elements, like the deer and the mammoth but an in situ evolution after vicariance of others, like the pika, the field mouse, and the shrews. During this period, all taxa gradually change over time, evolving into new chronospecies. Apart from the mammals, an endemic slender steppe eagle (*Aquila nipaloides*) was reported by Antoine Louchart and colleagues (2005) from the Late Pleistocene deposits. Based on new arrivals and in situ evolutions, two subzones are recognised – Orosei 2 and Dragonara, in the sense of Palombo (2006).

The limestone quarries at Monte Tuttavista near Orosei yielded the oldest findings of this biozone, known as Orosei 2. This biozone is roughly dated to between 0.8 and 1.1 million years ago (late Early Pleistocene). The transition between this biozone and the previous is gradual, because remains of an archaic form of the vole (*M. sondaari*) – described by Federica Marcolini and colleagues (2006) in honour of Paul Sondaar) – and possibly remains of the canid (?*Cynotherium* sp.) are reported to have been found together with the rabbit (*Sardolagus*), the macaque, the caprid, the dormouse (*Tyrrhenoglis*), and *Pannonictis* of the previous biozone at Orosei 2, the oldest subzone of this biozone. In addition, the small field mouse (*R. minor*) and the shrews (*Asoriculus* sp. I) are continuations of the previous biozone. The canid remains are limited to a juvenile phalanx, which hampers a precise taxonomic identification, other than that it is larger in size than corresponding elements from the younger subzones.

New elements that mark the beginning of the next subzone in the early Middle Pleistocene are a canid (*Cynotherium malatestai*), a large vole (*Microtus (Tyrrhenicola) henseli*), a large form of the dwarf deer (*Praemegaceros sardus*), red-toothed shrews (*Asoriculus similis* and *Asoriculus corsicanus*), and possibly a human (*Homo erectus*). This fauna also occurred in Corsica (Box 9.2). The bovids, the hyena, the monkey, the pig, and the large dormouse of the previous period are now extinct.

The pika had evolved considerably larger size and is assigned to a new species (*P. sardus*; Plate 13). The mole, *Talpa tyrrhenica*, on the other hand seems a continuation of the mole of the previous period. The vole evolved from the older and smaller species *sondaari*. The field mouse also increased in size, and the late Middle Pleistocene forms are known as *R. orthodon*, which was larger than the living broad-toothed field mouse (*Apodemus mystacinus*).

Towards the end of the subzone, perhaps about 300,000 years ago, the deer had evolved with smaller body mass and is known as *Praemegaceros cazioti*. Also, the dog – *Cynotherium sardous* – had evolved dwarf size by this time.

BOX 9.2

Corsica

The fauna of the *Tyrrhenicola* biozone is the oldest faunal complex known for Corsica. In the oldest localities, Punta di Calcine (or Conca) and Corte, the small form of the field mouse (*R. minor*) occurs, and, in the first locality, the archaic small vole (*Microtus (Tyrrhenicola) sondaari*) has been found, according to Pereira and colleagues (2005). Taxa older than the early Middle Pleistocene seem not to have been retrieved from Corsica thus far, except for several teeth of a Late Oligocene mainland moschid (*Pomelomeryx* (or '*Amphitragulus*') *boulangeri*) from Vaccio near Ajaccio, found in 2000. The Corsican sites are known only for the more derived forms of the Middle and Late Pleistocene.

Peculiarities and Evolution of Endemic Mammals

Endemic Primates: Oreopithecus

The large-bodied ape *Oreopithecus* retrieved from several coal mines in the Monte Bamboli area of southern Tuscany – now part of mainland Italy – has puzzled scientists since the discovery of a complete skeleton by Swiss palaeontologist Johannes Hürzeler in the 1950s. This enigmatic primate shows a mixture of cercopithecid (Old World monkeys), hominoid (Apes), and unique features and has thus alternatively been considered to be a member of the cercopithecids, hominoids, or even hominids (Great Apes). Its molars are similar to those of Old World monkeys, whereas some skeletal specialisations seem to show a close phylogenetic relationship with apes (Figure 9.4). *Oreopithecus* is the latest known hominoid from Europe (6.7–8.3 million years ago), close in geological time to the divergence between chimpanzees and humans. The species has been regarded as a model for early hominin anatomy.

Its long arms are considered an indication of its vertical climbing, as described by Terry Harrison (1991). Meike Köhler and Salvador Moyà-Solà (1997), on the other hand, explained some of its anatomical features as evidence for bipedal walking and a hominid-like precision grip in the hand, typical for the Great Apes. They consider its novel way of life an insular adaptation. The mobile and prehensile foot, as described by Harrison (1991), on the other hand, are in conflict with an extremely terrestrial lifestyle. The debate is, not surprisingly, still continuing. Ashley Hammond and colleagues (2020) dismissed the earlier ideas and showed that *Oreopithecus* was neither a tree climber nor walking upright, based on features

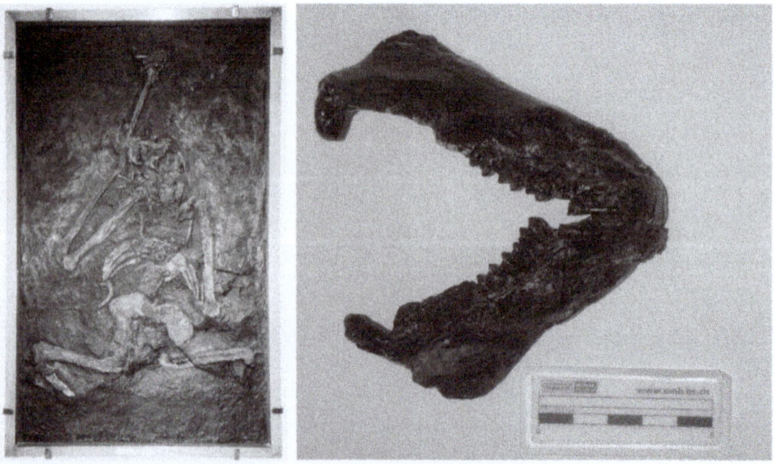

Figure 9.4 Cast of the skeleton of *Oreopithecus bambolii* (Museum of Palaeontology, University of Florence) and the original mandible, occlusal view (Natural History Museum of Basel). (Photographs: George Lyras and Loic Costeur).

of its hips and lower vertebral column. Rather, in anatomical aspects, it was more similar to large arboreal gibbons (*Symphalangus*), though with its body mass of over 30 kg in males, it would be twice that size. Its phylogeny within hominoids, however, remains unresolved as several evolutionary scenarios can explain the observed features of the lower torso. Its diet, finally, consisted mainly of leafs, as indicated by dental microwear patterns and features of the skull that are related to a heavy masticatory activity. The lower third premolar in *Oreopithecus* is extremely variable, as shown by Lorenzo Rook and colleagues (1996). In this light, the differences between the Fiume Santo material and the lower premolars from Tuscany are likely not relevant, despite the suggestion of Cordy and Ginesu (1994).

Dwarf Dogs: Cynotherium

The Pleistocene lineage of Sardinian dogs gradually evolved smaller body size, and the smallest specimens are attributed to the species *Cynotherium sardous* (Figure 9.5), which had an average body mass of about 15 kg. The largest forms come from late Early Pleistocene fissure fillings of Capo Figari and Monte Tuttavista (*Cynotherium* sp.) and in the breccia of Grotta dei

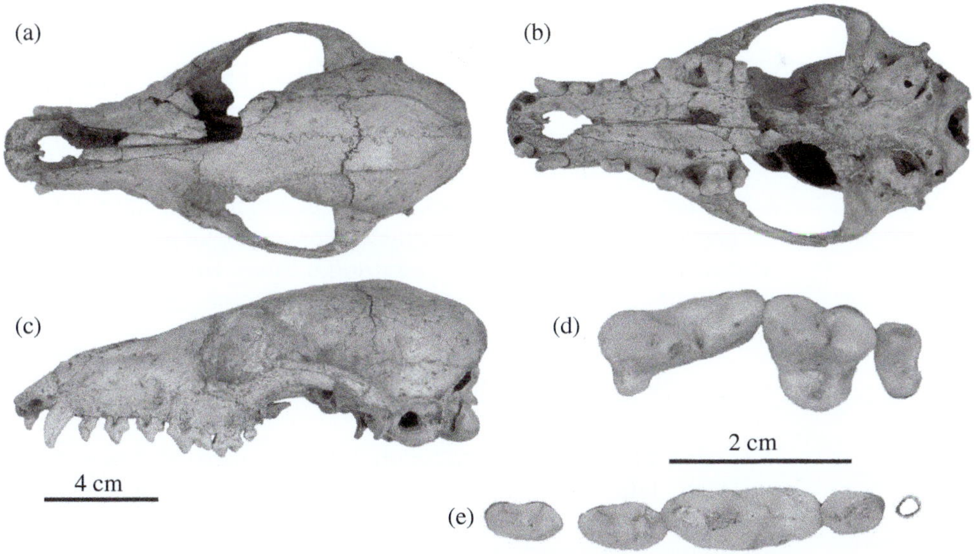

Figure 9.5 Skull and molar series of the Sardinian dog, *Cynotherium sardous*, from Corbeddu cave, terminal Pleistocene. (a–c) Skull in dorsal (a), lateral (b), and ventral (c) view. (d, e) upper P4–M2 and lower P4–M3 (the last molar is represented only by its alveolus), occlusal view. Civic Archaeological Museum, Nuoro, Sardinia. (Photographs: George Lyras.)

Fiori (*C. malatestai*). The smallest individuals were found in latest Pleistocene deposits at Corbeddu cave (Plate 18), having a body mass of only about 10 kg, as calculated by Lyras and van der Geer (2006). Especially at Dragonara cave, *Cynotherium sardous* fossils are abundant: disarticulated bones of almost forty individuals were retrieved.

The lineage *Cynotherium* shares all features that unite the *Canis*-like wolfs and dogs as already noted by Studiati (1857), Malatesta (1970), and Vera Eisenmann and Bartholomeus van der Geer (1999). Within this clade, it is most closely related to mainland *Xenocyon* species, whose genus was widespread in Europe during the Early Pleistocene. The most important similarities between *Cynotherium* and *Xenocyon* are the presence of a lower third molar (absent in *Cuon*), a unicuspid talonid on the lower first molar, and the absence of cuspids on the anterior side of the premolars, as pointed out by Lyras and colleagues (2006). Although *Cynotherium* decreased considerably in size once on the island, it kept its hypercarnivorous dental adaptations. Instead of large prey, it had shifted its diet to small and fast mammals and birds as indicated by the morphology of its skull, which had lost the massive ridges and strong muscular attachments so typical of a hypercarnivorous dog. This feeding strategy requires relatively low hunting costs, but it cannot sustain a large body size, as shown by Carbone and colleagues (1999, 2007). In this way, *Cynotherium* displays a unique combination of a hypercarnivorous 'strong' dentition with a hypocarnivorous 'weak' skull. The postcranial elements indicate that *Cynotherium* was a stalker, able to catch swift, small prey by surprise, such as the abundantly present pikas.

From Castiglione, Corsica, also a *Canis* sp. and an undetermined Cuoninae were reported, but these most likely represent *Cynotherium* as well. The former is somewhat larger than *C. sardous*, much like the specimens from Capo Figari and Monte Tuttavista. The latter retained a lower third molar –seen in *Cynotherium* and *Xenocyon*, its ancestor – thus excluding *Cuon*.

Dwarf Deer: Praemegaceros

The Sardinian deer shows considerable size decrease from a large late Early Pleistocene form (*Praemegaceros* sp.; Figure 9.6), to the intermediate-sized Middle Pleistocene *Praemegaceros sardus*, to the typical *Praemegaceros cazioti* (Figure 9.7) of the Late Pleistocene of Sardinia and Corsica, found at several sites on both islands, including the type locality Nonza, Corsica. The body mass of the smallest species is estimated at about 70 kg by Burness and colleagues (2001). The intermediate form, *Praemegaceros sardus*, dated to the

Figure 9.6 Female skull with associated lower jaw of the largest Sardinian megacerine deer (*Praemegaceros* sp.), Monte Tuttavista, late Early Pleistocene. Civic Archaeological Museum, Nuoro, Sardinia. (Photograph: Alexandra van der Geer.)

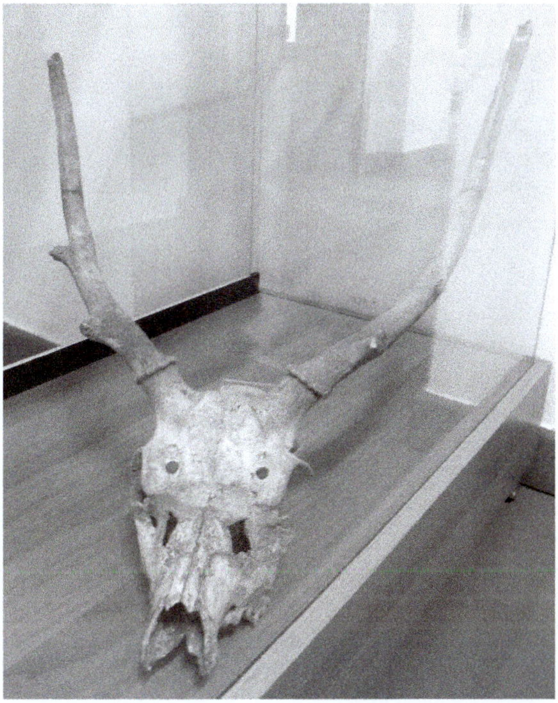

Figure 9.7 Male skull with antler of *Praemegaceros cazioti* from Corbeddu cave. Civic Archaeological Museum, Nuoro, Sardinia. (Photograph: Alexandra van der Geer.)

Middle Pleistocene, was found at Santa Lucia near Iglesias, south-western Sardinia, but is also present in the older layers of Capo Figari, as reported by Jan van der Made and Maria Rita Palombo (2006). Datings from Santa Lucia and Capo Figari 2 yielded dates of about 450,000 and 367,000 years ago, respectively (van der Made and Palombo, 2006). An even larger form was retrieved from the Su Fossu de Cannas cave near Sadali Sardinia, as reported by Palombo and Melis (2005). *Praemegaceros cazioti* fossils are accumulated in huge numbers in caves such as La Coscia in Corsica, and Dragonara and Corbeddu cave in Sardinia. Fossils from the latter cave were extensively described by Gerard Klein Hofmeijer (1997). In addition, numerous fossil tracks and footprints attributed to this deer are present in cemented sands along the south-western coast. Apart from size decrease, the deer also underwent a slight proportional reduction in the length of the metapodials combined with a very slight increase in robusticity of the metapodials, but not to the degree as seen in typical insular ruminants. Microwear analysis of its molars suggests that *Praemegaceros cazioti* consumed a large amount of grasses and may have included ligneous plants and fruits in the autumn and winter. The presence of numerous tracks and trackways found in coastal deposits shows that these endemic deer frequented the beaches, dunes, and lagoons, probably attracted by salt crusts or the grasses and reeds.

The Sardinian Pleistocene deer is generally considered a megacerine deer of the *verticornis* group and thus named *Praemegaceros* – or *Megaloceros*, the genus to which the Irish giant elk *Megaloceros giganteus* belongs, in phylogenies that consider all megacerine deer as members of one genus. Suggested ancestors are *Praemegaceros verticornis*, *Praemegaceros Soleilhacus*, or similar forms, or a form close to *Eucladoceros rhenanus*, *Eucladoceros Giulii*, or *Arvernoceros*. The former have robust metapodials, whereas the latter have more slender metapodials. Ancestry from the former would thus imply a considerable increase in slenderness, while ancestry from the latter, on the other hand, would imply a slight increase in robustness of the metapodials.

In addition, on Corsica, an endemic red deer (*Cervus elaphus rossii*), has been described from Middle Pleistocene localities together with *Praemegaceros* by Pereira (2001, 2005). The Corsican red deer is somewhat smaller than the mainland red deer of the Middle Pleistocene.

Dwarf Bovids and Giraffe-like Ruminants

Pre-Pleistocene Sardinian faunas are characterised by bovids and giraffoids, several of which evolved a (extremely) high-crowned dentition and dwarfism. *Maremmia* of the Late Miocene fauna, for example, is a spiral-horned bovid with very high-crowned (hypsodont) molars, ever-growing incisors, loss of the first two lower premolars, and enlarged lower and upper third molars. The presence of ever-growing incisors (hypselodonty) is not restricted to this genus but is also seen in two otherwise unrelated artiodactyls: the Pleistocene goral-like goat *Myotragus balearicus* of Mallorca and Minorca and the living camelid *Vicugna vicugna* of South America. High-crowned teeth and a reduction of the premolar series are present as well in *Turritragus casteanensis* of the same biozone. *Maremmia* resembles the African Alcelaphini – wildebeests, hartebeests, bonteboks, and allies – as shown by Johannes Hürzeler (1983). However, the earliest occurrence of Alcelaphini is the latest Miocene–Early Pliocene, which poses a chronological problem. The phylogenetic relationships of *Maremmia* are still unresolved.

The typical bovid genus of the Late Pliocene and Early Pleistocene is *Nesogoral* and is considered to be related to the living gorals (*Nemorhaedus*), the Plio-Pleistocene mainland goral (*Gallogoral*), and perhaps (the ancestors of) the mouse goat (*Myotragus*) of Majorca and Minorca. The horn-cores of the type species *melonii* from Capo Figari are almost straight, extending backward on the same plane as the frontals and very little diverging. They are separated at their base by a rather flat surface as width as the transverse width of the horn-cores themselves. The metacarpals and metatarsals are particularly long and slender in *Nesogoral cenisae*, but slender limbs are seen also in the living gorals. The different *Nesogoral* species indicate an anagenetic evolution on the island from *Nesogoral* sp. from Mandriola to *Nesogoral melonii* from Capo Figari, with *Nesogoral cenisae* from an unknown locality of the Campidano area, south-western Sardinia, in an uncertain position.

The second bovid genus (*Asoletragus gentry*) of the Late Pliocene–Early Pleistocene is of a small size, has also straight, almost conical horn-cores, which are only slightly divergent, but these are, in contrast to those of *Nesogoral*, more massive and very closely set at their bases and strongly inclined backwards. The type skull comes from a fissure filling of Monte Tuttavista, associated with a fauna that is reminiscent of that of Mandriola.

The diagnosis of the Late Miocene ruminant *Umbriotherium* as a giraffid is by no means firmly grounded because the main diagnostic features of giraffes – the presence of lower canines consisting of two lobes (bilobate) and ossicones – have not been confirmed neither for the Tuscany nor for the Sardinian material due to insufficient material. The diagnosis is based on the markedly rugose enamel, fusion of enamel folds in much worn teeth and the presence of a highly molarised lower fourth premolar. The lower third premolar on the other hand is not molarised. This is also seen in two other giraffids – *Decennatherium pachecoi* from Spain and *Helladotherium* from Greece – according to Abbazzi and colleagues (2008). The size of *Umbriotherium* is similar to that of the Late Miocene *Palaeotragus*, but as long as its phylogenetic position is unresolved, no conclusion regarding size changes can be drawn. Initially, *Umbriotherium* was considered to have evolved from *Sardomeryx* from the Oschiri fauna of the Early Miocene by van der Made (2008). Based on new material of *Sardomeryx*, Bastien Mennecart and colleagues (2019) showed that it, instead, is likely closely related to *Bedenomeryx* of France, and an early member of the Palaeomerycoidea instead. They explain the derived features of its teeth, such as high-crowned molars, shorter premolar row, as an insular adaptation in arid environments.

Giant Pikas: Prolagus

The giant pika of Sardinia and Corsica (*P. sardus*) is the symbol of the endemic Pleistocene faunas of Sardinia and Corsica (Plate 13) because of the abundance of its fossils. Over time, it evolved doubled body mass relative to that of its ancestor (Moncunill-Solé and colleagues, 2016).

The first *Prolagus* likely arrived during the Pliocene transgression, as shown by Chiara Angelone and colleagues (2015), and evolved from continental *Prolagus sorbinii* from the Italian mainland. The first species, *Prolagus figaro*, is small yet already larger than the mainland form. During the Late Pleistocene, a second chronospecies is recognised with a considerably larger size, *Prolagus sardus*.

The Sardinian pika was intensively hunted by Neolithic humans seeing the large quantities of burnt and damaged bones found at archaeological sites. The amount of pika bones decreased considerably after the Roman occupation of the islands. It seems to have survived till the second century BCE. It is, however, generally believed that it managed into the eighteenth century, based on a report by Francesco Cetti (1774) in his natural history of Sardinia. Cetti wrote about large burrows and heaps of earth as if made by a large mole on Tavolara, an offshore islet of Sardinia, and this was wrongly interpreted as evidence for the large-sized pika. Cetti considered the animals some sort of large mice, but it is not entirely clear which animal he referred to.

Voles and Field Mice: Tyrrhenicola, Rhagamys

The ancestor of the Sardinian vole *Microtus* (*Tyrrhenicola*) is thought to be closely related to the mainland vole *Microtus ruffoi*, as indicated by the close resemblance of the earliest Sardinian species (*sondaari*) to most species of extinct and extant voles of the genus *Microtus*. During its evolution on the island, the molar pattern becomes increasingly more complex as an adaptation to a more abrasive diet. Molars from the geological younger sites – such as La Coscia and Funtaneddu, Corsica, and Corbeddu cave, Sardinia – have more enamel folds than those from older sites, such as Monte Tuttavista fissure VI, Sardinia, and Punta di Calcina, Corsica.

Apart from the size increase observed in the lineage of field mice (*Rhagapodemus-Rhagamys*), the molar pattern became slightly more complex, too, as an adaptation to a more abrasive diet together with an increase in molar crown height (hypsodonty) from the primitive form (*R. azzarolii*) to the derived form (*R. orthodon*). *Rhagamys* is considered to be derived from *Rhagapodemus*. As ancestor, the mainland *R. ballesioi* has been proposed. This, however, poses taxonomical problems, because the name *Rhagamys* has priority over *Rhagapodemus*. The latter generic name is widely used, whereas the former name is restricted to the endemic Sardinian and Corsican species. For reasons of convenience, *Rhagamys* is thus applied only to the species *orthodon*, even though this is in conflict with the rules of taxonomical nomenclature.

Island Pigs

The Early Pleistocene pig of Sardinia (*Sus sondaari*) is small and evolved dental adaptations. The molars have a high crown, thick enamel, and a simple morphology. The premolars are smaller than in similar-sized mainland pigs, and the first premolar is sometimes lost. These adaptations are explained as an adaptation to chew more abrasive food. Sondaar's pig seems to have been more cursorial than its ancestor, *Sus arvernensis*, contrary to what one would expect on an island, based on the peculiarities of its third metacarpal: it is slender, bears a pronounced crest on the distal articulation surface, and has a more flat proximal articulation.

The ancestor is a form of *Sus arvernensis*, which is about 15% larger than Sondaar's pig. Afterwards, an increase in size may have occurred, because the remains from Capo Mannu, western Sardinia, are smaller than those from Capo Figari, a younger site in north-eastern Sardinia. The pig was originally referred to as *Potamochoerus* – the African bushpigs – based on remains from the Capo Figari breccias by Emmanuel Passemard (1925). The formal description of this material was finally done by van der Made (1988), who named it *Sus nanus*. Later, he renamed it *Sus sondaari* in honour of Paul Sondaar, because the species name *nanus* was preoccupied by *Sus scrofa nanus* (Nehring, 1884), which is a prehistoric domestic pig.

The molars of the Late Miocene Fiume Santo pig *Eumaiochoerus* cf. *etruscus* seem to be simpler in morphology than those of the Tuscany pig, in the sense that they lack certain grooves, a feature in common with the pig of the subsequent Sardinian biozone (*Sus sondaari*). Alternate taxonomies of the pig genus are *Hippopotamodon* and *Microstonyx*. This Late

Miocene pig species might have evolved into the pig species of the Late Pliocene-Early Pleistocene fauna, based on the simpler molars with thicker enamel.

The earliest island pig, *Hyotherium insularis* of the Early Miocene, has shortened phalanges, as shown by van der Made (1999). This is explained as an energy-saving adaptation which is typical for insular artiodactyls.

Red-toothed Shrews: Asoriculus

The Corsican and Sardinian red-toothed shrews (Figure 9.8) (*Asoriculus similis* and *Asoriculus corsicanus*) were three times as large as their mainland ancestors and likely had terrestrial habits, as suggested by the relatively deep skull and specialised humerus in contrast to more amphibious living genus *Soriculus*, to which they are thought to be most closely related. They seem to have survived into historical times based upon their estimated times of extinction of, respectively, about 4000 years ago and 2500 years ago. White-toothed shrews of

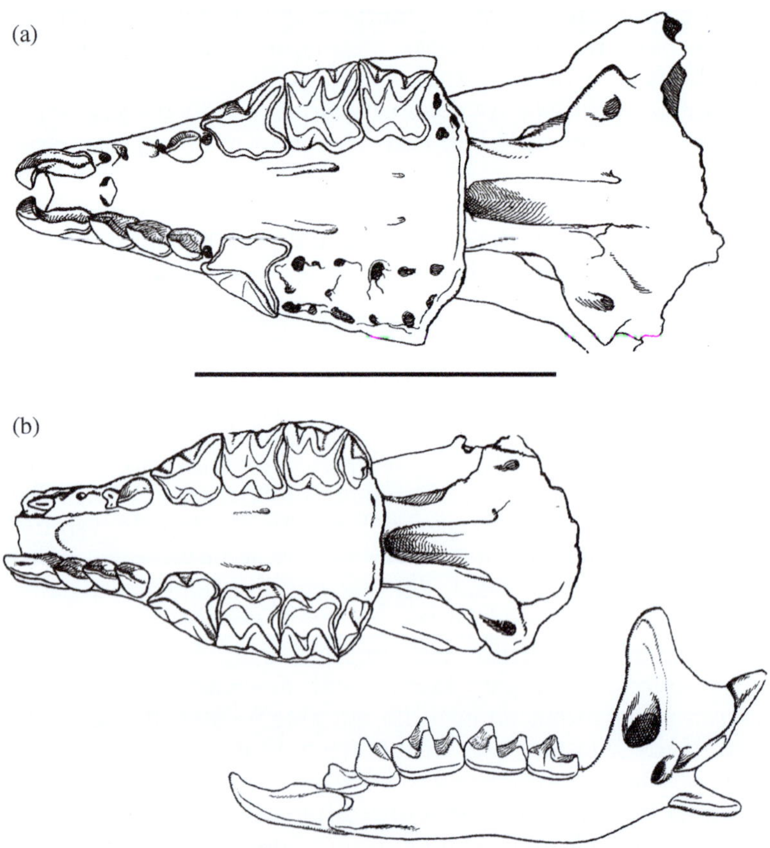

Figure 9.8 (a) Skull (palatal view) of the Corsican shrew (*Asoriculus corsicanus*, originally described as *Nesiotites corsicanus*), Teppa di Lupino, northern Corsica, holotype. Palatal view. (b) Skull (top, palatal view) and left lower jaw (lingual view) of the Sardinian shrew (*Asoriculus similis*). Late Pleistocene. Scale bar: 1 cm. (From 'Pleistocene shrews from the larger Western Mediterranean Islands' by Dorothea Bate in *Journal of Natural History* (1944), reprinted by permission of Taylor & Francis Ltd, http://www.tandfonline.com.)

the genus *Crocidura*, possibly accidentally introduced through human agency, may have caused their decline because this decline coincides with an increase in the population of the white-toothed shrews.

The phylogenetic affinities of the red-toothed shrews are uncertain. Bate (1944) erected the genus name *Nesiotites* to accommodate the Balearic shrew (*Nesiotites hidalgo*), Hensel's Sardinian species, and the new Corsican species which she described in the same paper as *Nesiotites corsicanus*. At present, this attribution is doubtful since the relation between these three species (and the Sicilian species as well) is insufficiently known. The three insular red-toothed shrews are sometimes referred to as *Episoriculus*, *Asoriculus*, and *Soriculus* or grouped into '*Nesiotites*'.

Asoriculus appears to have survived longer on the islands than on the mainland because it disappeared from the Eurasian mainland before the beginning of the Middle Pleistocene.

Otters and Mustelids

The Sardinian giant otter (*Megalenhydris barbaricina*) is one of the most spectacular otter taxa. It is known from a single incomplete skeleton, described by Willemsen and Malatesta (1987), of which not all parts could be recovered from the hard sediment of the cave. Only the lower jaw and some teeth, a femur, part of the axial skeleton, and some phalanges were retrieved, while more vertebrae, the hind limbs, and tiny fragments of the skull, pelvis, and the ribs are visible in the sediment. The lower jaw is much larger than that of the living giant otter (*Pteronura brasiliensis*), which has a maximum head and body length of 1.4 m and a tail of 1 m long. The jaw size thus implies that the Sardinian giant otter was truly a giant. This is confirmed by the larger size of the postcranial elements in relation to the living giant otter. The massive, robust teeth resemble those of the clawless otters (*Aonyx*) and indicate a shellfish diet. The tail was already much flattened right from the tail root, which is a unique feature amongst living otters. This shape is explained as an adaptation to a more aquatic lifestyle than modern river otters. In accordance with this is the flexible backbone, as inferred from parts preserved in the sediment. Whether it was a marine or a river otter cannot be concluded from the remains. The age of the surrounding sediments, and thus of the skeleton of the giant otter, is tentatively estimated to the last Ice Age, about 70,000 to 10,000 years ago.

The earliest otter of Sardinia-Corsica is probably *Lutra castiglionis*, a species recovered from Middle Pleistocene deposits in Castiglione cave near Oletta on Corsica. The remains belong to a single individual, likely an adult male. This otter probably had an aquatic way of life and lived in the rivers as indicated by accompanying faunal elements such as freshwater gastropods. It was a small, robust otter with shortened limbs and a flattened tail. It was originally described as *Cyrnolutra castiglionis* by Pereira and Salotti (2000), but moved to *Lutra* by Willemsen (2006) based on its similarity to *Lutra simplicidens*.

Sardolutra ichnusae was more adapted to aquatic life and swimming than the common river otter, as indicated by its brain morphology. One of the most bizarre features of this otter is its very large baculum, or penis bone. This bone is not only relatively (but not absolutely) larger than in the sea otter (*Enhydra lutris*) but also ends in a blade-like process with two smaller protruding processes. It has been suggested that this is an adaptation to mating in open sea, which means that *Sardolutra* was a marine otter. The age of the single skeleton is probably Late Pleistocene. Originally, this otter was attributed by Malatesta (1977) to the genus *Nesolutra*, the endemic insular otter genus to which also the Sicilian and Maltese otters were thought to belong. However, Willemsen (2006) transferred the type species of *Nesolutra*, *Nesolutra euxena*, to the genus *Lutra*, and thus a new genus name was needed for the Sardinian otter, for which the name *Sardolutra*, literally 'Sardinian otter', was proposed.

Algarolutra majori was not only found on Sardinia (Dragonara cave) but also on Corsica, in Grotta del Margine. It is known only by some dental remains. Originally, the species was attributed to *Cyrnaonyx antiqua* by Helbing (1935), but the dentition of the latter species differs from the Corsican and Sardinian material for which thus a new genus was established by Malatesta and Willemsen (1986). The stratigraphic occurrence of *Algarolutra majori* is Late Pleistocene. There are slight differences between the Sardinian and Corsican specimens, but these differences are small and explained as due to intraspecific variation by Willemsen (1992). For example, the teeth are slightly larger in Corsican specimens and the protocone and protoconid of the upper first molar are slightly longer in the Sardinian specimens.

The phylogenetic relationships of *Algarolutra* are poorly understood, but the other three species are all part of a radiation of the mainland *Lutra simplicidens* on those islands. *Lutra castiglionis* has many features in common with *Sardolutra* but also differs in some features, such as a very different baculum morphology, which cannot be regarded as a more primitive, ancestral condition.

The large terrestrial Sardinian mustelid *Enhydrictis galictoides* is closely related to the Eurasian genus *Pannonictis*, as suggested by Nuria García and colleagues (2008) in their review of the two genera. The Sardinian species is the only of its genus. Its fossils were found in several Sardinian sites (Capo Figari,

San Giovanni, Grotta della Dragonara, and Capo Caccia) but also on Corsica. The stratigraphic range is Middle Pleistocene.

Dwarf Mammoth: Mammuthus

The Sardinian dwarf mammoth (*M. lamarmorae*) stood about 1.3–1.5 m tall at the shoulder, based on the maximal length of 450 mm of the humerus, described by Palombo and colleagues (2012). Compared to its mainland ancestor, the southern mammoth (*Mammuthus meridionalis*), it evolved a body mass reduction of over 90%, according to van der Geer and colleagues (2016). The morphology of the carpal and tarsal bones of the dwarf mammoth resembles more the morphology seen in the southern mammoth than in the straight-tusked elephant, though shape of joints may change considerably during adaptation to a different environment. This means that the shape and proportions of the foot bones of the Sardinian mammoth may simply reflect an adaptation rather than inform about phylogeny.

The molars of the Sardinian mammoth are relatively small, have thick enamel and a relatively low lamellar frequency, in parallel with other insular dwarf proboscideans. Despite these adaptations, the enamel thickness, average lamellar frequency, and degree of hypsodonty are most consistent with the mammoth lineage.

The deposits with mammoth remains from Funtana Morimenta were radiocarbon (^{14}C) dated to about 43,000 years ago (see, however, Palombo and colleagues, 2012). An upper molar, possibly the first molar, from Tramariglio, north-western Sardinia, is attributed to the Late Pleistocene, because it was found in a breccia layer overlying a Tyrrhenian conglomerate. The two mammoth molars from Campu Giavesu were buried in marshland sediments younger than 200,000 years ago, the time of activity of the volcano Monte Annaru.

References

Abbazzi L., Angelone C., Arca M., Barisone G., Bedetti C., Derfino M., Kotsakis T., Morcolini F., Palombo M.R., Pavia M., Piras P., Rook L., Torre D., Tuveri C., Valli A., Wilkens B. 2004. Plio-Pleistocene fossil vertebrates of Monte Tuttavista (Orosei, E. Sardinia, Italy), an overview. *Rivista Italiana di Paleontologia e Stratigrafia*, 110: 681–706.

Abbazzi L., Arca M., Tuveri C. Rook L. 2005. The endemic canid *Cynotherium* (Mammalia: Carnivora) from the Pleistocene deposits of Monte Tuttavista (Nuoro, Eastern Sardinia). *Rivista Italiana di Paleontologia e Stratigrafia*, 111 (3): 493–507.

Abbazzi L., Delfino M., Gallai G., Trebini L., Rook L. 2008. New data on the vertebrate assemblage of Fiume Santo (North-West Sardinia, Italy), and overview on the late Miocene Tusco-Sardinian palaeobioprovince. *Palaeontology*, 51: 425–451.

Acconci L. 1881. Sopra alcove ossa fossili di Elefante rinvenute nel Quaternario della zona di Morimenta in Sardegna. *Processi Verbali Atti Società Toscana di Scienze Naturali*, 2: 266–267.

Angelone A., Stanislav C., Kotsakis T. 2015. The most ancient lagomorphs of Sardinia: An overview. *Geobios*, 48 (4): 287–296.

Angelone C., Čermák S., Moncunill-Solé B., Quintana J., Tuveri C., Arca M., Kotsakis T. 2018. Systematics and paleobiogeography of *Sardolagus obscurus* n. gen. n. sp. (Leporidae, Lagomorpha) from the early Pleistocene of Sardinia. *Journal of Paleontology*, 92 (3): 506–522.

Angelone C., Kotsakis T. 2000. Rodents and lagomorphs from the Pliocene of Mandriola (Sinis Peninsula, Western Sardinia) and their bearing on the palaeogeography of Sardinia. *11th Congress R.C.M.N.S. Abstracts*, 82.

Angelone C., Kotsakis T. 2001. *Rhagapodemus azzarolii* n. sp. (Muridae, Rodentia) from the Pliocene of Mandriola (western Sardinia-Italy). *Bollettino della Società Paleontologica Italiana*, 40(2): 127–132.

Angelone C., Tuveri C., Arca M., López Martíez N., Kotsakis T. 2008. Evolution of *Prolagus sardus* (Ochotonidae, Lagomorpha) in the Quaternary of Sardinia Island (Italy). *Quaternary International*, 182: 109–115.

Antón M., Turner A., Salesa M.J., Morales J. 2006. A complete skull of *Chasmaporthetes lunensis* (Carnivora, Hyaenidae) from the Spanish Pliocene site of La Puebla de Valverde (Teruel). *Estudios Geológicos*, 62 (1): 375–388.

Azzaroli A. 1946. La scimmia fossile della Sardegna. *Rivista di Scienze Preistoriche*, 1: 68–76.

Bate D.M.A. 1944. Pleistocene shrews from the larger Western Mediterranean Islands. *Annals and Magazine of Natural History (Series 11)*, 1944: 738–769.

Bate D.M.A. 1945. The Pleistocene mole of Sardinia. *Annals and Magazine of Natural History (Series 11)*, 12: 448–461.

Bosco C. 1902. Il *Lophiodon sardus* (n.sp.) delle ligniti di Terras de Collu in Sardegna. *Rendiconti della Reale Accademia Nazionale dei Lincei (Serie 5)*, 11: 178–182.

Brandy L.-D. 1978. Données nouvelles sur l'évolution du rongeur endémique fossile corso-sarde Rhagamys F. Major (1905). *Bulletin de la Société Géologique de France*, 7 (20): 831–835.

Burness G.P., Diamond J., Flannery T. 2001. Dinosaurs, dragons, and dwarfs: the evolution of maximal body size. *Proceedings of the National Academy of Sciences of the USA*, 98: 14518–14523.

Carbone C., Mace M., Craig Roberts S., Macdonald D.W. 1999. Energetic constraints on the diet of terrestrial carnivores. *Nature,* 408: 286–287.

Carbone C., Teacher A., Rowcliffe J.M. 2007. The costs of carnivory. *PLoS Biology*, 5: e22.

Casanovas-Vilar I., van Dam J.A., Trebini L., Rook L. 2011. The rodents from the Late Miocene *Oreopithecus*-bearing site of Fiume Santo (Sardinia, Italy). *Geobios*, 44 (2–3): 173–187.
Cetti F. 1774. *Storia Naturale di Sardegna. Vol. 1: I Qadrupedi di Sardegna*. Giuseppe Piattoli, Sassari.
Comaschi Caria I. 1953. L'*Amphitragulus boulangeri* Pomel, primo mammifero terrestre segnalato nel Miocene della Sardegna. *Rivista Iatliana di Paleontologia*, 59: 91–99.
Cordy J.M. Ginesu S. 1994. Fiume Santo (Sassari, Sardaigne, Italie): Un nouveau gisement à Orèopithèque (Oreopithecidae, Primates, Mammalia). *Comptes Rendus de l'Académie des Sciences de Paris (Série 2)*, 318: 679–704.
Dal Piaz G.B. 1929. *Atalonodon*, nuovo genere di perissodattilo dell'Eocene di Gonnesa (Sardegna). *Memorie del Istituto di Geologia della Regia Università di Padova*, 8: 1–9.
Dawson M.R. 1969. Osteology of *Prolagus sardus*. A Quaternary Ochotonid (Mammalia, Lagomorpha). *Paleovertebrata*, 2: 157–190.
de Bruijn H., Rümke C.G. 1974. On a peculiar mammalian association from the Miocene of Oschiri (Sardinia), I and II. *Proceedings of the Koninklijke Akademie van Wetenschappen (Series B)*, 77: 45–79.
Déhaut E.-G. 1911. Animaux fossiles du Cap Figari. *Materiaux pour servir à l'histoire zoologique et paléontologique des îles de Corse et de Sardaigne*, 3: 53–59.
Depéret C. 1897. Etude de quelques gisements nouveaux de vertébrés pléistocènes de l'île de Corse. *Annales de la Société Linnéenne de Lyon* 44: 113–116.
Eisenmann V., van der Geer B. 1999. The *Cynotherium* of Corbeddu (Sardinia): Comparative biometry with extant and fossil canids. *Deinsea*, 7: 147–168.
Engesser B. 1976. *Tyrrhenoglis majori*, ein neuer fossiler Gliride (Rodentia, Mammalia) aus Sardinien. *Eclogae Geologica Helveticae*, 69: 783–793.
Engesser B., 1983. Die jungtertiären Klinsäuger des Gebietes der Maremma (Toskana, Italien. 1.Teil: Gliridae (Rodentia, Mammalia). *Eclogae Geologicae Helvetiae*, 76, 763–780.
Ferrero della Marmora G.A. 1826. *Viaggio in Sardegna dal 1819 al 1825, o Descrizione statistica, fisica e politica di questa isola*. Paris.
Forsyth Major C.I. 1882. L'origine della fauna delle nostre isole. *Atti Società Toscana delle Scienze Naturali, Proceedings*, 3: 113–131.
Forsyth Major C.I. 1883. Die Tyrrhenis: Studien über geographische Verbreitung von Tieren und Pflanzen im westlich Mittelmeergebiet. *Kosmos*, 13: 1–17 and 81–106.
Forsyth Major C.I. 1901. Exhibition of, and remarks upon, the skull of a new fossil mammal (*Enhydrictis galictoides*). *Proceedings of the Zoological Society of London*, 1901 (2): 625–628.
Forsyth Major C.I. 1905. Rodents from the Pleistocene of the western Mediterranean region. *Geological Magazine (Decade 5)*, 2: 462–467 and 501–506.
García N., Arsuaga J.L., Bermudez de Castro J.M., Carbonell E., Rosas A., Huguet R. 2008. The Epivillafranchian carnivore *Pannonictis* (Mammalia, Mustelidae) from Sima del Elefante (Sierra de Atapuerca, Spain) and a revision of the Eurasian occurrences

from a taxonomic perspective. *Quaternary International*, 179 (1): 42–52.

Ginesu S., Cordy J.M. 1997. *Il Monte Tuttavista, Orosei–Galtellì, I riempimenti carstici e la scoperta delle nuove specie del Pleistocene.* Poddighe, Sassari.

Hammond A.S., Rook L., Anaya A.D., Cioppi E., Costeur L., Moyà-solà S., Almécija S. 2020. Insights into the lower torso in late Miocene hominoid *Oreopithecus bambolii. Proceedings of the National Academy of Sciences of the USA*, 17 (1): 278–284.

Harrison T. 1991. The implications of *Oreopithecus bambolii* for the origins of bipedalism. In: Coppens Y., Senut B. (eds), *Origine(s) de la Bipédie chez les Hominidés (Cahiers de Paléoanthropologie)*. Editions du CNRS, Paris, pp. 235–244.

Helbing H., 1935. *Cyrnaonyx antiqua* (Blainv.), ein Lutrinae aus dem europaeischen Pleistocaen. *Eclogae Geologicae Helvetiae*, 27: 443–457.

Hensel R.H. 1855. Beiträge zur Kenntnis fossiler Säugethiere. Insektenfresser und Nagethiere der Diluvialformation. *Zeitschrift der Deutschen Geologischen Gesellschaft*, 7: 458–501.

Hensel R.H. 1856. Beiträge zur Kenntniss fossiler Säugethiere. *Zeitschrift der deutschen geologischen Gesellschaft*, 8: 279–290, 660–704.

Hürzeler J. 1983. Un alcélaphiné aberrant (Bovidé, Mammalia) des "lignites de Grosseto" en Toscane. *Comptes Rendus de l'Académie des Sciences de Paris (Série 2)*, 296: 497–503.

Hürzeler J., Engesser, B. 1976. Les faunes des mammifères néogènes du Bassin de Baccinello (Grosseto, Italie). *Comptes Rendus de l'Académie des Sciences de Paris, (Série 2) D*, 283: 333–336.

Klein Hofmeijer G. 1997. Late Pleistocene deer fossils from Corbeddu Cave. *BAR International Series*, 663: 432–663.

Klein Hofmeijer G., Martini F., Sanges M., Sondaar P.Y., Ulzega A. 1987. La fine del Pleistocene nella Grotta Corbeddu in Sardegna. *Rivista di Scienze Preistoriche*, 41 (1–2): 1–36.

Köhler M., Moyà-Solà S. 1997. Ape-like or hominid-like? The positional behavior of *Oreopithecus bambolii* reconsidered. *Proceedings of the National Academy of Sciences of the United States of America*, 94 (21):11747–11750.

Kotsakis T., Murru M., Palombo M.R. 2008. Terras de Collu. *EuroMam 2008, Fossil Mammalian Biotas of Sardinia, Fieldtrip Guide-Book, Sardinia 16–21 September 2008*, pp. 65–68.

Louchart A., Bedetti C., Pavia M. 2005. A new species of eagle (Aves: Accipitridae) close to the Steppe Eagle, from the Pleistocene of Corsica and Sardinia, France and Italy. *Palaeontographica A*, 272: 121–148.

Lydekker R. 1887. *Catalogue of the Fossil Mammalia in the British Musemn (Nat. Hist.).* Part V. Supplement.

Lyras G.A., van der Geer A.A.E. 2006. Adaptations of the Pleistocene island canid *Cynotherium sardous* (Sardinia, Italy) for hunting small prey. *Cranium*, 23 (1): 51–60.

Lyras G.A., van der Geer A.A.E., Dermitzakis M.D., De Vos J. 2006. *Cynotherium sardous*, an insular canid (Mammalia: Carnivora) from

the Pleistocene of Sardinia (Italy), and its origin. *Journal of Vertebrate Paleontology*, 26 (3): 735–745.
Malatesta A. 1970. *Cynotherium sardous* Studiati. An extinct Canid from the Pleistocene of Sardinia. *Memorie dell' Istituto Italiano di Paleontologia Umana (nuova seria)*, 1: 1–72.
Malatesta A. 1977. The skeleton of *Nesolutra ichnusae* sp. n., a quaternary otter discovered in Sardinia. *Geologica Romana*, 16: 173–209.
Malatesta A., Willemsen G.F. 1986. *Algarolutra* g.n. established for a fossil otter of the Sardinia island. *Geologica Romana* 25: 285–286.
Marcolini F., Arca M., Kotsakis T., Tuveri C. 2005. The endemic vole *Microtus Tyrrhenicola*) (Arvicolidae, Rodentia) from Monte Tuttavista (Sarclinia, Italy) new perspectives for phylogeny and biochronology. In: Alcover J.A., Bover Arbós P. (eds.), *Proceedings of the International Symposium "Insular Vertebrate Evolution: the Palaeontological Approach", September, 16-19 Mallorca*, 185–92.
Marcolini F., Tuveri C., Arca M., Kotsakis T., 2006. *Microtus (Tyrrhenicola) sondaari* n. sp. (Arvicolidae, Rodentia) from Monte Tuttavista (Sardinia, Italy). *Hellenic Journal of Geosciences*, 41: 73–82.
Martini F. 1992. Early human settlement in Sardinia: The palaeolithic industries. In: Tykot R., Andrews T.K. (eds), *Sardinia in the Mediterranean, a Footprint in the Sea. Monographs in Mediterranean Archaeology*, 2: 40–48. Sheffield Academic Press, Sheffield.
Mennecart B., Zoboli D., Costeur L., Pillola G.L. 2019. On the systematic position of the oldest insular ruminant *Sardomeryx oschiriensis* (Mammalia, Ruminantia) and the early evolution of the Giraffomorpha. *Journal of Systematic Palaeontology*, 17 (8): 691–704.
Moncunill-Solé B., Tuveri C., Arca M., Angelone C. 2016. Comparing the body mass variations in endemic insular species of the genus *Prolagus* (Ochotonidae, Lagomorpha) in the Pleistocene of Sardinia (Italy). *Rivista Italiana di Paleontologia e Stratigrafia*, 122 (1): 25–36.
Nehring, A. 1884. Über den Schädel eines zwergenartigen Schweines, *Sus scofa nanus* aus dem Torf Moor von Tripsees in Neu-Vorpommeren, *Sitzungs-Berichten der Gesellschaft naturforschenden Freunde*, 2: 7–14.
Palombo M.R. 2006. Biochronology of the Plio-Pleistocene terrestrial mammals of Sardinia: The state of the art. *Hellenic Journal of Geosciences*, 41: 47–66.
Palombo M.R., Bover P., Valli A.M.F., Alcover J.A. 2006. The Plio-Pleistocene endemic bovids from the Western Mediterranean Islands: Knowledge, problems and perspectives. *Hellenic Journal of Geosciences*, 41 (1): 153–162.
Palombo M.R., Ferretti M.P., Pillola G.L., Chiappini L. 2012. A reappraisal of the dwarfed mammoth *Mammuthus lamarmorai* (Major, 1883) from Gonnesa (south-western Sardinia, Italy). *Quaternary International*, 255: 158–170.
Palombo M.R., Melis R. 2005. Su Fossu de Cannas Cave (Sadali, central-eastern Sardinia, Italy): The oldest deposit holding megalocerine remains in the Pleistocene of Sardinia. In: Alcover J.A., Bover P. (eds), *Proceedings of the International Symposium Insular Vertebrate Evolution: The Palaeontological Approach. Monografies de la Societat d'Història Natural de les Balears*, 12: 265–276.

Passemard E. 1925. Une brèche osseuse à petits mammifères des environs de Corte (Corse). *Bulletin Société Géologique de France*, 25 (4): 347–350.

Pereira E. 2001. *Cervus elaphus rossii* (Mammalia, Artiodactyla), a new endemic sub-species from the middle Pleistocene in Corsica. *Paleovertebrata*, 30 (2–4): 189–213.

Pereira E. 2005. Les deux Cervidés du Pléistocène de Corse. In: Crégut-Bonnoure E. (ed.), *Les ongulés holarctiques du Pliocène et du Pléistocène. Quaternaire*, hors-série, 2: 221–230.

Pereira E., Michaux J., Montuire S. 2005. New data on the extinct endemic rodents *Tyrrhenicola* and *Rhagamys* (Rodentia, Muridae) of Corsica (France) with special emphasis on their dental morphology and adaptation. In: Alcover J.A., Bover P. (eds), *Proceedings of the International Symposium Insular Vertebrate Evolution: The Palaeontological Approach. Monografia de la Societat d'História Natural de les Balears*, 12: 277–286.

Pereira E., Salotti M. 2000. *Cyrnolutra castiglionis*, a new otter (Mustelidae, Lutrinae) from the Middle Pleistocene "Castiglione 3CG" deposit (Oletta, Corsica). *Comptes Rendus de l'Academie des Sciences (Series IIA Earth and Planetary Science)*, 331 (1): 45–52.

Radinsky L.B. 1967. *Hyrachyus, Chasmotherium*, and the early evolution of helaletid tapiroids. *American Museum Novitates*, 2313: 1–23.

Rook L., Abbazzi L., Engesser B. 1999. An overview on the Italian Miocene land mammal faunas. In: Agustì J., Rook L., Andrews P. (eds), *The Evolution of Neogene Terrestrial Ecosystems in Europe*. Cambridge University Press, Cambridge, pp. 191–204.

Rook L., Ferretti M., Arca M., Tuveri C. 2004. *Chasmaporthetes melei* n. sp. an endemic hyaenid (Carnivora, Mammalia) from the Monte Tuttavista fissure fillings (Late Pliocene to Early Pleistocene; Sardinia, Italy). *Rivista Italiana di Paleontologia e Stratigrafia*, 110 (3): 707–714.

Rook L., Harrison T., Engesser B. 1996. The taxonomic status and biochronological implications of new finds of *Oreopithecus* from Baccinello (Tuscany, Italy). *Journal of Human Evolution*, 30 (1): 3–27.

Rook L., Mazza P., Rustioni M., Torre D. 2000. Lands and endemic mammals in the Late Miocene of Italy: Paleogeographic outlines of Tyrrhenian and Adriatic areas between 11–9 and 7–4 Ma. *'State of the Art' Workshop, Lyon, France, 16–18 November 2000. Environments and Ecosystem Dynamics of the Eurasian Neogene (EEDEN)*. http://www.esf.org.

Sondaar P.Y. 2000. Early human exploration and exploitation of islands. *Tropics*, 10 (1): 203–230.

Sondaar P.Y., Elburg, R., Klein Hofmeijer, G., Martini, F., Sanges, M., Spaan, A., De Visser, J. 1995. The human colonization of Sardinia; a Late-Pleistocene human fossil from Corbeddu Cave. *Comptes Rendus de l'Académie des Sciences de Paris (Serie 2a)*, 320: 145–150.

Spoor C.F., Sondaar P.Y. 1986. Human fossils from the endemic island fauna of Sardinia. *Journal of Human Evolution*, 15: 399–408.

Studiati C. 1857. Description des fossiles de la brèche osseuse de Monreale de Bonaria près de Cagliari. In: La Marmora A.F., *Voyage en Sardaigne, ou description statistique, physique et politique de cette île, avec des recherches sur ses productions naturelles et ses antiquités. Troisième partie. Description géologique*, 2: 651–704. 3rd edn., Turin.

Thaler L. 1973. Nanisme et gigantisme insulaires. *La Recherche*, 37 (4): 741–750.

Tobien H., 1935. Über die pleistozänen und postpleistozänen *Prolagus* formen Korsikas und Sardiniens. *Berichte der Naturforschenden Gesellschaft zur Freiburg i. Br*, 34: 253–344.

van den Hoek Ostende L.W., 2001. A revised generic classification of the Galericini (Insectivora, Mammalia) with some remarks on their palaeobiogeography and phylogeny. *Géobios*, 34 (6): 681–695.

van der Geer A.A.E., van den Bergh G., Lyras G.A., Prasetyo U.W., Due R.A., Setiyabudi E., Drinia H. 2016. The effect of area and isolation on insular dwarf proboscideans. *Journal of Biogeography*, 43: 1656–1666.

van der Made J. 1988. *Sus nanus* nov. sp., a Pliocene dwarf pig from Capo Figari (Northeastern Sardinia). *Bolletino della Società Paleontologica Italiana*, 27: 367–378.

van der Made J. 1999. Biogeography and stratigraphy of the Mio-Pleistocene mammals of Sardinia and the description of some fossils. *Deinsea*, 7: 337–360.

van der Made J. 2005. The fossil endemic goat *Nesogoral cenisae* n.sp. from Campidano Sardinia-cursorial adaptations in insular environment. In: Alcover J.A., Bover P. (eds), *Proceedings of the International Symposium Insular Vertebrate Evolution: The Palaeontological Approach. Monografies de la Societat d'Història Natural de les Balears*, 12: 347–368.

van der Made J. 2008. New endemic large mammals from the Lower Miocene of Oschiri (Sardinia): Observations on evolution in insular environment. *Quaternary International*, 182: 116–134.

van der Made J., Palombo M.R. 2006. *Megaloceros sardus* n. sp., a large deer from the Pleistocene of Sardinia. *Hellenic Journal of Geosciences*, 41: 163–176.

Wagner R. 1829. Beiträge zur Geschichte der fossilen Thiere. *Isis von Oken*, 22 (11): 1132–1141.

Willemsen G.F. 1992. A revision of the Pliocene and Quaternary Lutrinae from Europe. *Scripta Geologica*, 101: 1–115.

Willemsen G.F. 2006. *Megalenhydris* and its relationship to *Lutra* reconsidered. *Hellenic Journal of Geosciences*, 41: 83–87.

Willemsen G.F., Malatesta A. 1987. *Megalenhydris barbaricina* sp. nov., a new otter from Sardinia. *Verhandelingen der Koninklijke Akademie van Wetenschappen (Series B)*, 90: 83–92.

CHAPTER TEN

The Balearic Islands

Evolution of Island Mammals: Adaptation and Extinction of Placental Mammals on Islands, Second Edition.
Alexandra van der Geer, George Lyras and John de Vos.
© 2021 John Wiley & Sons Ltd. Published 2021 by John Wiley & Sons Ltd.

The Balearic Islands – consisting of Mallorca (Majorca), Minorca (Menorca), Ibiza (Eivissa), Formentera and a hundred or so smaller islets – are situated between the Iberian and the Sardinia-Corsica block. Mallorca and Minorca were populated by endemic insular faunas since the early mid-Miocene, first by giant pika, dormice, and a giant tortoise, later by a giant hare, a peculiar bovid, giant dormice, and a shrew. The bovid lineage (*Myotragus*) shows a trend towards extreme high-crowned teeth (hypsodonty) in combination with ever-growing incisors (hypselodonty). Ibiza stands apart and was first inhabited by a bovid and small mammals and later by birds, a lizard, and bats only.

Geology and Palaeogeography

The Balearic Islands in the western Mediterranean Sea can be divided into two main groups, the Gymnesics in the east and the Pityusics in the west (Figure 10.1). The first consists of the larger islands Mallorca – or Majorca – and Minorca and nearly 30 surrounding islets, including the natural park Dragonera and the Cabrera archipelago. The second consists of the main islands Ibiza – or Eivissa – and Formentera and about sixty islets.

The Mediterranean Sea did not exist as such in the Eocene and before but was part of the Tethys Ocean. This changed in the Early Oligocene when the eastern part of the Tethys Ocean started to close following the relative movements of the Eurasian and African plates. In the same time, several fragments of the Iberian Peninsula broke off and moved to the east and south-east. This process continued into the Miocene. The major detached blocks are Sardinia-Corsica and the Balearics. Some parts collided later with mainland masses, such as the region of Las Murchas (Granada), now part of southern Spain but an island in the Middle Miocene. The western opening of the former Tethys Ocean closed in the Late Miocene, resulting in a completely closed Mediterranean Sea, eventually leading to the Messinian salinity crisis between 5.33 and 5.96 million years ago. As shown by Guillem Mas and colleagues in 2018, the sea level dropped minimally between 800 and 1200 m below the present level, based on an analysis of the late Neogene faunas of the Balearic Islands, combined with the available geostructural data. When the Strait of Gibraltar opened in the latest Miocene – ending the Messinian salinity crisis of the Mediterranean – the Gymnesics and Pityusics attained more or less their present configuration, and a long-term isolation for their faunas began. Before the Messinian, periods of isolation seem to have existed as well, as evidenced by some elements of the fauna, though the nature and the extent of these endemic

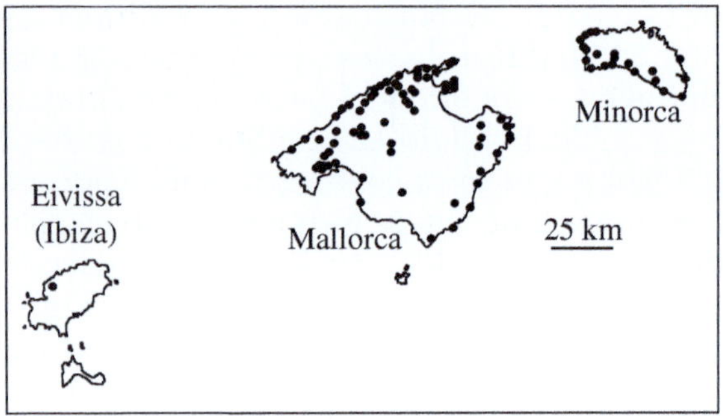

Figure 10.1 Some of the most important fossil localities of the Balearic Islands. More than 140 fossiliferous deposits are known with *Myotragus balearicus* on Mallorca, Minorca, Cabrera and Dragonera (for details, see Bover, 2004). For the other *Myotragus* species, just a few localities are known, restricted to Mallorca and Minorca (*Myotragus batei*, *Myotragus* sp.), or to Mallorca alone (remaining species).

features is not entirely clear. The intra-Messinian connection of the Balearics to each other and to the surrounding mainland was of unknown duration and seems not to have been complete, because the invading faunas were very unbalanced and impoverished. Initially, larger species arrived only in Mallorca and Ibiza, while Minorca was colonised only by hares and rodents, apart from reptiles. Before the onset of the Pleistocene, Mallorca, Minorca, and Ibiza all had their own fauna. The Mallorcan fauna invaded Minorca in the Late Pliocene or Early Pleistocene, and from that moment on, they shared a common fauna. Ibiza, on the other hand, remained isolated and suffered an unexplained extinction of non-flying mammals before the Late Pleistocene.

Historical Palaeontology

The history of fossil findings in the Balearics begins in the middle of the nineteenth century. The Italian general and geologist Giuseppe Alberto Ferrero della Marmora (1846) mentioned a fossiliferous breccia at the base of the hill of Bellver castle at Palma de Mallorca. Here, he retrieved a bone of what he thought to be a '*Lagomys* or rabbit', which subsequently got lost. The deposit could not be relocated either. The loss of both the site and the fossil shed serious doubt on the exact nature of his finding. More than a century later, however, remains of a hare (*Hypolagus*) were found at the early Pliocene site of Caló den Rafelino on Mallorca, as reported by Bover, Quintana and colleagues in 2007, restoring the credibility of della Marmora's report.

The first scientific report of fossils from the Balearics is that of British naturalist Dorothea Bate at the start of the twentieth century. She went to Mallorca, in the hope to be as successful in finding dwarf hippos and elephants as she was in Cyprus and Crete. Unfortunately, she failed in finding such fossils. Instead, after many unsuccessful surveys and excavations in remote places, she found remains of a strange goat with ever-growing rodent-like incisors. Bate (1909) published a short description of the species, based on remains excavated from Cova de Na Barxa at Capdepera, Mallorca. She named it *Myotragus balearicus*, literally the mouse-goat of the Balearics.

She had also found remains of a giant dormouse on Mallorca as well as Minorca, for which she (1918) described the new genus *Hypnomys* and the new species *morpheus*, based on jaws, some limb bones, and skull fragments from several fissures and caves on both islands. In the same paper (Bate, 1918), she erected a species for Minorca (*Hypnomys mahonensis*) and one for Mallorca (*Hypnomys morpheus*). The former is the type species of the genus and distinguished from the latter mainly by its larger size. Bate (1944) finally published the shrew remains that she had discovered at Cap Farrutx at Artà (Figure 10.2) as *Nesiotites hidalgo*, based on the anterior portion of a skull.

In the 1960s, the American artist William 'Bill' Waldren discovered the fossiliferous cave Son Muleta close to his home at Deia. The sediments are sometimes loose, sometimes cemented. During the years, he excavated thousands of bones of *Myotragus* at Muleta and started a local museum with exhibits on the prehistory of Mallorca.

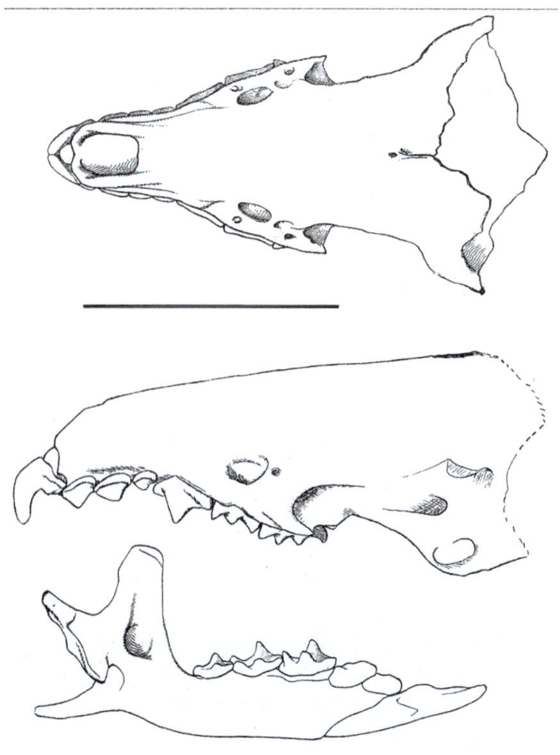

Figure 10.2 Incomplete skull (holotype) and left mandibular ramus (paratype) of the large shrew (*Nesiotites hidalgo*) from Cap Farrutx, Mallorca, Late Pleistocene. From top to bottom, dorsal view, lateral view, and buccal view. Scale bar: 1 cm. Source: From Bate, D. M. A. (1944). LXIII.—Pleistocene shrews from the larger Western Mediterranean Islands. Annals and Magazine of Natural History, 11(83), 738–769. © 1944, Taylor & Francis.

Between 1966 and 1982, several more *Myotragus* species other than Bate's *balearicus* were discovered and described, all belonging to the same lineage but to different geological time frames.

The small dormouse *Hypnomys waldreni* and the small shrew *Nesiotites ponsi* were described by Reumer (1979) based on material he had recovered from Cap Farrutx. These Pliocene forms are ancestral to the Pleistocene forms.

The year 1982 turned out to be an auspicious year for small mammal species because five new species were described from various deposits. These were the giant Minorcan dormouse *Muscardinus cyclopeus* by Jordi Agustí and colleagues, and one of the world's largest known ochotonids *Gymnesicolagus gelaberti*, the giant dormouse *Carbomys sacaresi*, and the two dormice *Margaritamys llulli* and *Peridyromys ordinasi* by Pierre Mein and Rafael Adrover (1982).

Fossil sites continue to be discovered and excavated, amongst which are Cova des Coralloides near Calvià – yielding a complete *Hypnomys* skeleton, described by Bover, Alcover and colleagues (2010) – Caló den Rafelino near Manacor – with the earliest species of *Myotragus*, described by Bover, Quintana and colleagues (2010) – Cova des Pas de Vallgornera near Llucmajor – preserving complete skeletons of three species, described by Bover, Valenzuela and colleagues (2014) – Na Burguesa 1 in the Serra de Tramuntana – the earliest deposit of the second fauna – and Cova Estreta – yielding coprolites of *Myotragus* – all on Mallorca, and the area of Punta Nati on Minorca.

Biozones and Faunal Units

The evolution of the Plio-Pleistocene fauna of Mallorca and Minorca is gradual, lacking turnovers, but experiencing early extinctions and gradual, anagenetic evolution within the surviving lineages. The bovid lineage evolved a smaller body size and acquired extremely high-crowned molars and rodent-like, ever-growing incisors. The dormouse lineage, on the other hand, evolved larger body mass. The stages in evolution of these two elements mark the Pleistocene subzones.

The pre-Pliocene faunas of Mallorca and Minorca, on the other hand, show a clear zonation with distinct faunal turnovers and have no relation to the younger faunal episode (for detailed overview, see Mas and colleagues, 2018).

Mallorca: Middle–Late Oligocene

The oldest biozone of the Balearics is known from Paguera and some other sites and contains only small mammals – for the greater part described by Adrover and colleagues (1977) – reptiles and possibly an amphibian. During this period, the Balearics were still part of the Iberian Peninsula, although the presence of a few endemic species suggests a filtered connection, forming a barrier for some species but not for all. The majority of the small mammals corresponds with European mainland species.

Two of the three dormice, however, differ from mainland relatives, as far as their size and certain morphologies are considered. These dormice are *Moissenetia paguerensis*, described by Marguerite Hugueney and Adrover (1995), and *Bransatoglis adroveri*, described by Hugueney (1997). The third species, *Bransatoglis planus*, on the other hand, is found outside Mallorca as well. Also, the thryonomid rodent (*Sacaresia moyaeponsi*) is an endemic form, considered related to the African cane rats (*Thryonomys*) by Raquel Antoñanzas and Sevket Sen (2005).

The endemic gymnure (Galericinae), *Tetracus daamsi*, has longer and wider anterior teeth than the single other species of the genus *Tetracus nanus* from continental Europe. This difference is probably correlated with an ecological difference, providing further evidence for some sort of geographic isolation. The overall size of Daams' moonrat probably was similar to that of the European species because the dimensions of the posterior teeth of *daamsi* do not differ essentially from those of *nanus*.

Adrover and colleagues (1977) suggested the presence of a primitive gundi (Ctenodactylidae) belonging to the genus *Tataromys* in a preliminary note based on some incomplete fragments, but admitted that this attribution was tentative. Their report was never confirmed by more findings.

Mallorca–Minorca–Las Murchas: Middle–Late Miocene

The second biozone of Mallorca is characterised by several small mammals, apart from large terrestrial tortoises. These are a giant pika (*G. gelaberti*), a giant dormouse (*C. sacaresi*), and two more dormice (*Margaritamys llulli* and *Peridyromys ordinasi*). This very impoverished and unbalanced fauna is known from the coal layers of Santa Margalida and Sant Llorenç, first described by Adrover and colleagues (1983–1984). The pika of this fauna is the largest known species of its family (Ochotonidae).

A similar fauna is known from Minorca, discovered in the karst deposits of Punta Nati 2 (Agustí and Moyà-Solà, 1982), and consisting of an ochotonid closely related to *Gymnesicolagus*, a dormouse (*Margaritamys adroveri*) and a giant tortoise. The dormouse is closely related *Margaritamys llulli* of Mallorca of the same period but is more archaic, according to Josep Quintana and Agustí (2007).

A third potentially similar fauna is known from the Middle Miocene deposit of Las Murchas near Granada, Spain. The two only known species are dormice (*Pseudodryomys granatensis* and an undetermined species related to the mouse-tailed dormouse *Myomimus*), which are considered to have evolved under conditions of insularity, according to Elvira Martín-Suárez and colleagues (1993). The former species is closely related to the *Margaritamys* species of the Gymnesics but is the most archaic of the three.

These three faunas indicate that in the Middle to Late Miocene, so well before the Messinian salinity crisis, an arc of islands existed near the coast of the Iberian Peninsula. Probably, all its islands – Mallorca, Minorca, and Las Murchas – were colonised by the same fauna, with only local differences as suggested by Bover, Quintana and Alcover (2008).

Minorca: Pliocene

The Pliocene fauna of Minorca is highly unbalanced and impoverished, with a mammal fauna consisting of only a giant hare (*Nuralagus rex*), a giant dormouse (*Muscardinus cyclopeus*), and a horseshoe bat (*Rhinolophus* cf. *grivensis*; Figure 10.3). The non-mammal vertebrate fauna is richer and

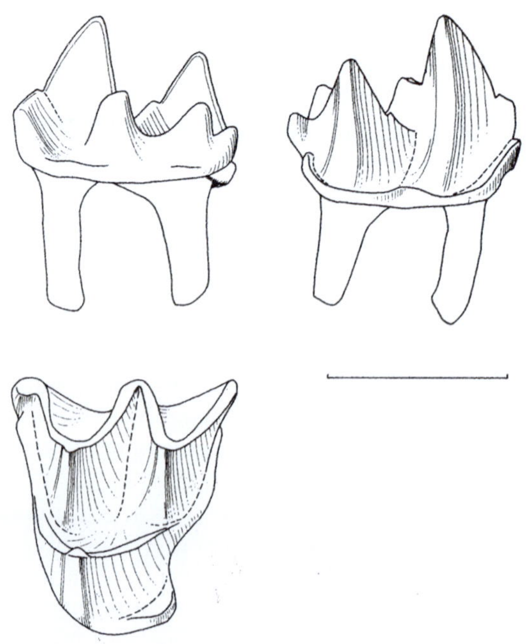

Figure 10.3 Lower molar (top; left lingual view, right buccal view) and upper M2 (below; occlusal view) of the bat from Son Bou 2, Minorca (*Rhinolophus* cf. *grivensis*). Scale bar: 1 mm. (From Reumer, 1982.)

includes a geckonid, a lizard (*Podarcis* sp.), a worm lizard (*Blanus* sp.), at least four snakes (*Coluber*, *Vipera natiensis*, *Vipera* sp., and Viperidae indet.), a tortoise (*Cheirogaster gymnesica*), and a frog (*Latonia* sp.), as summarised by Bover, Quintana and Alcover (2008). The fauna is known from sites between Punta Nati and Cala Es Pous.

At the end of this period, coinciding with the beginning of the Pleistocene Ice Age, the sea level dropped and the Gymnesics constituted one larger island resulting in the invasion of the *Myotragus* fauna from Mallorca to Minorca and the subsequent extinction of the giant rabbit. From this moment on, Minorca and Mallorca share their faunal history.

BOX 10.1

Faunal History of Ibiza

Ibiza followed its own path regarding faunal history, partly independent of that of the Gymnesics. The first colonisation likely took place during the Messinian salinity crisis from eastern Iberia, as in the case of Mallorca and probably the other islands.

The impoverished, unbalanced fauna of the latest Miocene and Early Pliocene (post-Messinian) of Ibiza consists of a bovid, a gerbil (*Debruijnimys* sp.; previously referred to as *Protatera*), a dormouse (*Eliomys* sp.), an insectivore, and a hare (*Hypolagus balearicus*), besides a wall lizard (*Podarcis* sp.) and a tortoise. The fauna is known from one site only, Ses Fontanelles, with unknown stratigraphic position. The hare apparently also reached Mallorca during this period, as reported by Quintana and colleagues (2010). The elements of this biozone are poorly understood and mostly determined only at the genus level due to the paucity of the fossil material. Earlier, two bovids were thought to be present, but Mas and colleagues (2018) show a single species. This bovid was earlier attributed to *Tyrrhenotragus* – the Late Miocene genus of the Tusco-Sardinian palaeo-bioprovince – by Agustí and Moyà-Solà (1990), but this was later questioned by Moyà-Solà and colleagues (1999).

The Late Pliocene fauna of Ibiza is likewise poorly known and reported from only one site, the karst deposit Cova de Ca Na Reia. Recovered elements are two dormice (*Eliomys [Eivissia] canarreiensis* and *Hypnomys* sp.) and some bats, apart from birds and reptiles. From other sites, giant tortoises have been reported. Towards the end of this period, only birds, bats, and reptiles survived, which persisted into the early Holocene until the arrival of humans.

Mallorca: Latest Miocene–Early Holocene

During the Messinian salinity crisis, a new fauna arrived on Mallorca, containing at least six mammal species. Due to the subsequent isolation of this fauna caused by the flooding of the Mediterranean basin, this fauna gradually developed into a strongly impoverished endemic fauna. The Pleistocene mammal fauna of Mallorca only consisted of three surviving species: a bovid (*Myotragus*), a shrew (*Nesiotites*), and a dormouse (*Hypnomys*). Though this fauna shows a long-term stasis regarding composition, the individual lineages show a considerable evolution through time, obviously in response to a progressively changing habitat under influence of, amongst others, climate change.

The earliest deposit – Na Burguesa 1, dating to the earliest Pliocene – was first reported by Bover, Rofes and colleagues (2014). The site yielded the remains of a giant hamster (*Apocricetus darderi*, described by Enric Torres-Roig and colleagues, 2017), a giant mouse (*Paraethomys balearicus*, described by Torres-Roig and colleagues, 2019), a hare (*Hypolagus*), a bovid, a dormouse, and a shrew. The latter three are primitive forms of the three Pleistocene lineages (*Myotragus*, *Hypnomys*, and *Nesiotites*). The hamster is closely related to *Apocricetus alberti* from Spain (MN13), according to Torres-Roig and colleagues (2017), in line with a Messinian arrival of the fauna, but had already evolved a larger size and unique characters, indicating that the isolation of Mallorca had begun. This also applies to the mouse, resembling mainland *Paraethomys meini* (upper Turolian) but with a larger size and with higher-crowned molars. Likely, the relatively rich fauna of Na Burguesa 1 arrived overland, seen the presence of at least six mammal species, twelve reptile species, and four amphibians.

The second stage is represented by the late Early-early Late Pliocene site Caló den Rafelino near Manacor, as reported by Bover, Quintana and colleagues (2007). The fauna consists of a bovid (*Myotragus palomboi*, described by Bover, Quintana and colleagues, 2010), a dormouse (*Hypnomys* sp.), a shrew (*Nesiotites rafelinensis,* described by Juan Rofes and colleagues, 2012), a giant hamster (*Tragomys macpheei*, 'goat-mouse', described by Agustí and colleagues, 2012), a hare (*Hypolagus balearicus*, described by Quintana and colleagues, 2010), and six reptile species including two vipers (*Vipera*) and the endemic lizard *Maioricalacerta rafelinensis.* The hare has also been reported from Ibiza (Ses Fontanelles) (Box 10.1) by Quintana and Blanca Moncunill-Solé (2014). The hamster is much larger than related Cricetinae species and has crescent-shaped ridges on its teeth (selenodonty), which inspired the name *Tragomys*: goats have selenodont teeth as well. The hamster may have evolved from *Apocricetus* from the earlier stage. The hamster and the hare are limited to these earliest stages and went extinct soon after.

From this stage on, the mammal fauna of Mallorca is limited to representatives of three lineages only, being a bovid (*Myotragus*), a shrew (*Nesiotites*) and a dormouse (*Hyponmys*) (for details about the individual lineages, see next section), and shared with Minorca since the late Early Pleistocene (Box 10.2). The spectacular and long-term evolution of these three surviving lineages was possible due to the total absence of mammalian predators and competitors. Not even a mustelid has been recovered from the Balearics. The only known danger was constituted by birds of prey, as attested by *Myotragus* bones found in Cova Murada, Minorca, and identified as food remains of the golden eagle (*Aquila chrysaetos*) by Pere Arnau and colleagues (2000).

A Shared History

BOX 10.2

The late Early-early Middle Pleistocene Mallorcan species of *Myotragus* – *Myotragus batei* or a form intermediate between *Myotragus batei* and *Myotragus kopperi* – extended its range to Minorca. By this time, the sea level had dropped sufficiently to enable a colonisation over land. Initially, the Minorcan form was considered a separate species (*Myotragus binigausensis*), described by Salvador Moyà-Solà and Joan Pons-Moyà (1980), on the basis of fossils retrieved from Barranc de Binigaus at Es Mercadal. At present, however, the species is generally regarded a junior synonym of *Myotragus batei* of Mallorca. The same applies to the Minorcan dormice (*Hypnomys eliomyoides*) and the Minorcan shrew (*Nesiotites meloussae*, described by Pons-Moyà and Moyà-Solà, 1980) of the same deposit; they are often considered synonymous with *Hypnomys onicensis* (if so, *eliomyoides* has priority) and *N. hidalgo* of Mallorca, respectively. From the Early Pleistocene on, Mallorca and Minorca share the same fauna.

Peculiarities and Evolution of Endemic Lineages

The Mouse-goat: Myotragus

During its long evolution of more than 5 million years in isolation, starting during the Messinian salinity crisis (5.6–5.32 Ma ago) when it arrived to the islands, the lineage of *Myotragus* (Figure 10.4) evolved smaller and smaller body size, starting with a body mass of perhaps 60 kg and ending with a mere 23 kg in the smallest specimens of the last species. *Myotragus* did not only shrink through geological time but also evolved a typical lower jaw dentition, with progressive reduction in number of teeth and the evolution of ever-growing incisors like that of rodents, hence the name 'mouse-goat'. Besides, the acquisition of this unique dentition and a dwarf size, the *Myotragus* lineage also shows a gradual shortening of the muzzle, forward-facing eye sockets, a progressive fusion of tarsal bones, and decrease in relative limb length (Bover, 2004). The last stage – represented by the species *Myotragus balearicus* – has the relatively shortest metapodials ever observed in insular ruminants.

Analysis of the complete mitochondrial genome of *Myotragus* by Bover, Llamas and colleagues (2019) shows that the genus is not closely related to the *Ovis* clade – as was earlier concluded by Carles Lalueza-Fox and colleagues (2000) based on short fragments of ancient mitochondrial DNA. Tentatively, Bover, Llamas and colleagues (2019) suggest a potential relationship with the takin (*Budorcas taxicolor*) instead and a split between the two genera around 7.1 million years ago.

Six chronospecies have been described, ranging from the earliest Early Pliocene to the Holocene. The earliest form – *Myotragus palomboi* from Caló den Rafelino – already has short and robust metapodials and phalanges, but is not yet as robust as the geologically younger species. The third incisor is neither ever-growing nor high crowned, and the second premolar is relatively large, though reduced, in respect to related caprine species.

The second earliest form – *Myotragus pepgonellae* of the late Early Pliocene – is a primitive species of the lineage as well, because it still sports an incisiform canine in the lower jaw, the normal condition for bovids. The main difference with the earliest species is the extremely reduced second lower premolar. The species was described by Salvador Moyà-Solà and Joan Pons-Moyà (1982), based on a damaged left jaw from Cala Morlanda at Manacor. The species is named after a Mallorcan folk tale figure, Pep Gonella.

The third in line – *Myotragus antiquus* of the Late Pliocene or earliest Pleistocene (Gelasian) – still had all incisors as well an incisiform canine, but the second lower premolar is now lost, unlike in the two earlier species. There is a tiny alveolus for the second premolar, but it is not clear whether this represents the size-reduced alveolus for the permanent second premolar or the partly obliterated alveolus for the – already lost – deciduous premolar. The two remaining premolars are well developed, and large. *Myotragus antiquus* further has significantly shortened metapodials relative to the earlier species.

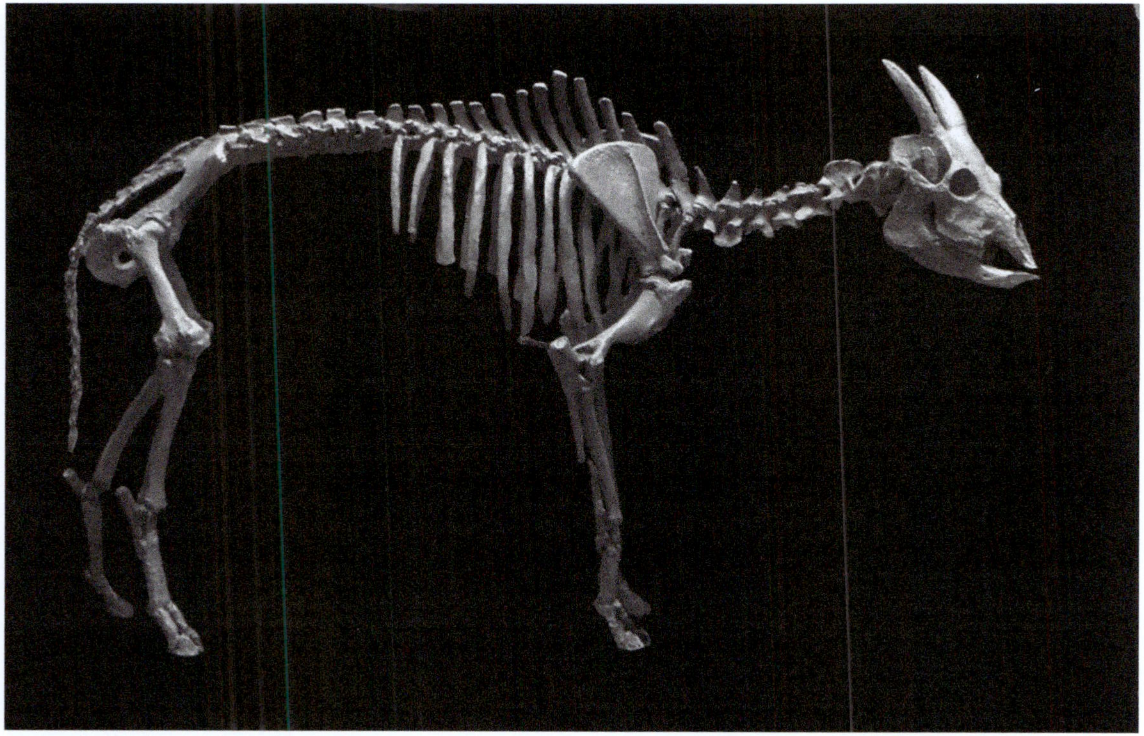

Figure 10.4 Composite skeleton of *Myotragus balearicus*, cast. Natural History Museum, Basel. (Photograph: Loic Costeur.)

The species was described by Joan Pons-Moyà (1977), based on a few fragments from Cap Farrutx.

The next step in evolution – *Myotragus kopperi* of the late Early Pleistocene (early Calabrian) – lacks the second lower premolar. Its incisors are high crowned relative to the previous species but not yet ever-growing. Some later authors seem to have mixed up the various species and attribute the ever-growing incisors also to *kopperi*. The species was described by Moyà-Solà and Pons-Moyà (1981), named after John Kopper from the University of Long Island who did the dating of the sediments. The species is based on a complete mandible from Sa Pedrera de S'Onix at Manacor, Mallorca. There are a two palaeomagnetic dates for this stage of the *Myotragus* lineage from the same site, Canet Cave at Esporles, Mallorca, providing some insight into the speed of evolution. Remains of *Myotragus antiquus* from layer J provided an age of 2.6 million years and those of *Myotragus* aff. *kopperi* from layer E an age of 2.4 million years, according to Pons-Moyà and colleagues (1979).

Evolution progresses with the following species – *Myotragus batei* of the late Early (late Calabrian) and early Middle Pleistocene – the first species with ever-growing incisors, a non-functional third premolar and a much size-reduced fourth premolar. The ever-growing incisors have been suggested by Bover and Alcover (1999a) to be retained milk incisors with open root that were not replaced by permanent incisors with closed root (monophyodonty). However, Moyà-Solà and colleagues (2007) show that this is not the case, and that these incisors are permanent incisors that erupt much earlier than common in ruminants, just after the first molar eruption, in response to resource limitation. The species was named by Miguel Crusafont and Basilio Angel (1966) in honour of Dorothea Bate, the discoverer of the first *Myotragus* fossils. They based their new species on a complete mandible of a young specimen with incompletely erupted third molar from Pedrera de Gènova near Palma, Mallorca. In contrast to *balearicus*, it retained the second and third lower incisor and the third premolar, whereas *balearicus* had lost them all.

The youngest species of the lineage is *Myotragus balearicus* of the Middle Pleistocene to Holocene (Plate 19), with ever-growing first incisors, no second and third incisor, no third premolar, and a small fourth premolar. The molars have extremely high crowns in this stage. The shoulder height of *Myotragus balearicus* was only 0.45–0.50 m, as calculated by Quetglas and Bover (1998). One adult specimen is found of only 0.22 m shoulder height, as reported by Bover and Alcover (1999b).

The brain of *Myotragus balearicus* was suggested to have been smaller than expected by Köhler and Moyà-Solà (2004). However, the lack of an ancestral form undermines such an observation. Besides, as already noted by Colette Dechaseaux (1961) in her detailed description of the bovid brain, the *Myotragus* brain is remarkably convoluted for its size, which means a significant increase in cortical surface. Köhler and Moyà-Solà overlooked this aspect, together with the reference. George Lyras (2018) showed that its brain is normal sized for its body mass when compared to Late Miocene bovids.

The diet of *Myotragus balearicus* was varied, as documented in their fossilised pellets, or coprolites, analysed by Alcover and colleagues (1999). They even included large amounts of the otherwise toxic plant, *Buxus balearica*, in their diet. This was later confirmed using such coprolites by Jordi Bartolomé and colleagues (2011) and Leidy Rivera and colleagues (2014), both based on preserved plant epidermal fragments, and Frido Welker and colleagues (2014), based on plant DNA. These box plants likely contained alkaloid steroids – buxines, cyclobuxines, parabuxines, and others – similar to the closely related boxwood, *Buxus sempervirens*. The consumption of mud and earth is known to act as an antidote to several toxic plants. An explanation behind the hypsodonty is the resulting convenient ability to enlarge the range of consumable plants – and parts of plants such as roots – significantly. *Buxus*, on the other hand, was just one component of *Myotragus'* diet, as shown by Alcover and colleagues (1999). A more likely alternative explanation is that its hypsodonty is not related to diet but to longevity instead, as proposed by Xavier Jordana and colleagues (2012).

The extinction of *Myotragus* likely was not caused by a climatic change but was due to human colonisation of the islands, be it directly through over-hunting or indirectly through human-induced changes to its habitat. The latter cause is suggested by the relatively sudden replacement of *Buxus balearicus* woods by *Olea europaea* maquis shrubland around the time of the first human arrival to the islands, indicative of a dramatic change in landscape use. Based on direct dating of *Myotragus* bones, its last dated occurrence on Mallorca is 4035 ± 32 years before present, as given by Bover, Valenzuela and colleagues (2016). Despite earlier claims, there is no reliable evidence for the presence of *Myotragus* in the human layers of two of the oldest archaeological sites of Mallorca – Cova des Moro near Manacor at the coast and Coval Simó at Escorca in the highlands. This indicates that by that time, about 4000 years ago, the lineage had already gone extinct. The same is valid for Minorca. Cabrera was not inhabited before Punic times (2300 years ago) and can thus not have played any role in the extinction of *Myotragus*. Seeing the lack of evidence for a significant overlap of coexistence of humans and *Myotragus*, in combination with the restricted mobility and the likely ecological naivety of the latter, it is likely to assume a very rapid extinction immediately following the arrival of the first humans leaving no trace in the archaeological record. This is confirmed by Bover, Valenzuela and colleagues (2016), who reduce the gap between the latest occurrence of *Myotragus* and the first occurrence of domestic goats, a proxy of human presence, to between only 350 and 620 years.

Giant Pikas and Hares

The pre-Pleistocene faunas of the Balearics are characterised by giant lagomorphs. The Middle Miocene giant pika of Mallorca and Minorca (*G. gelaberti*) is one of the world's largest known ochotonids, if not the largest on record. The average body weight, estimated from the length of the lower molar row, is 5.4 kg, the size of a large Arctic hare (*Lepus arcticus*). Its teeth are the largest amongst lagomorphs. This exceptionally large size – that is to say, for a pika – is explained as a consequence of insular evolution by Quintana and Agustí (2007).

The giant hare of the Pliocene of Minorca, described as *Nuralagus rex* by Quintana and colleagues (2011, based on the doctoral thesis by Quintana, 2005), reached the amazing body mass of an estimated 14 kg, about double the mass of the European hare (*Lepus europaeus*), as shown by Quintana and colleagues (2005). Initially, it was thought to belong to the Pliocene mainland genus *Alilepus*, but currently it is recognised as an endemic genus and species with unresolved phylogenetic relation to *Alilepus*. Its main characteristics are relatively small eye sockets and tympanic bullae, combined with a small head and short limbs, proportional to its body size.

The Early Pliocene hare of Mallorca and Ibiza (*Hypolagus balearicus*), on the other hand, had not evolved insular gigantism – body mass ranged between 1.3 and 2.7 kg – but has a particular morphology of the forelimb, which has been explained as either an insular trait or adapted to a particular ecological condition.

Giant Dormice: Hypnomys

The dormouse lineage *Hypnomys* of the Gymnesics is most closely related to the mainland genus *Eliomys* – to which the common garden dormouse belongs – and often considered a subgenus of the latter. Analyses of ancient mitochondrial DNA by Bover, Mitchell and colleagues (2020) suggest that *Eliomys* is indeed a sister taxon to *Hypnomys* and reject a subgeneric

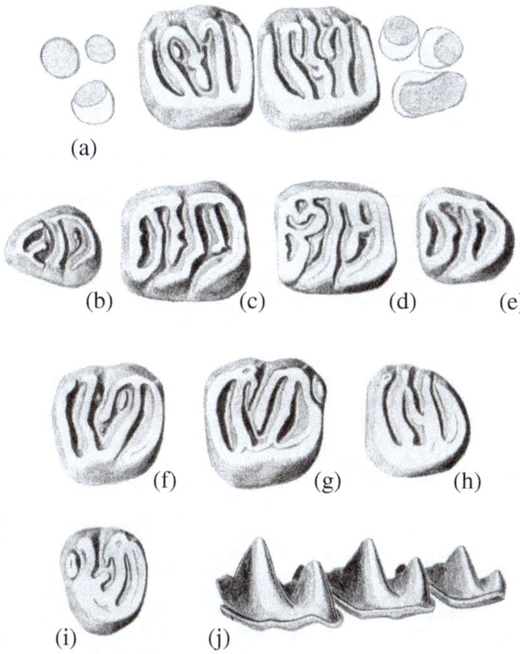

Figure 10.5 Molars of the intermediate dormouse *Hypnomys onicensis* (a–i; occlusal view) and a molar series of an intermediate stage of the shrew *Nesiotites* ex. interc. *ponsi-hidalgo* (j; lingual view). All from Sa Pedrera de S'Onix, Mallorca. (a) left upper M1 and M2, holotype of *Hypnomys intermedius* (now junior synonym of *onicensis*), (b) left lower P4, (c) left lower M1, (d) left lower M2, (e) left lower M3, (f) right upper M1-M2, (g) right upper M1-M2, (h) left upper M3, (i) left upper P4, and (j) left lower M1. Scale bar: 1 mm. (From Reumer, 1981.)

status. They also show that *Hypnomys* may have derived from a species belonging to the *Vasseuromys* lineage in the Middle–Late Miocene.

The earliest and smallest species is *Hypnomys waldreni* of the Early Pliocene of Mallorca. The next step in evolution is represented by *Hypnomys onicensis* (replacement name provided by Reumer in 1994 for the invalid name*Hypnomys intermedius*) (Figure 10.5 (a–i), of the Late Pliocene of Mallorca with intermediate molar size, as its previous name already suggests. *Hypnomys onicensis* had a body mass of about 200 g, calculated by Blanca Moncunill-Solé and colleagues (2014). This means that is already about 2.5 times heavier than its closest mainland relative (*Eliomys quercinus*), according to Bover, Alcover and colleagues (2010). The largest species, *Hypnomys morpheus* of Mallorca and Minorca of the Pleistocene (Figure 10.6), had about the size of the extant edible dormouse (*Glis glis*), a robust build, and a body mass between 173 and 284 g (or between 190 and 276 g in Moncunill-Solé and colleagues, 2014).

Based on a complete skeleton from Cova des Coralloides, Bover and colleagues reconstructed its mode of life. Likely, *Hypnomys* had evolved a more fossorial behaviour relative to *Eliomys* and may have lived on bulbs, corms, and rhizomes (geophytes). In addition, it had a relatively very short tail,

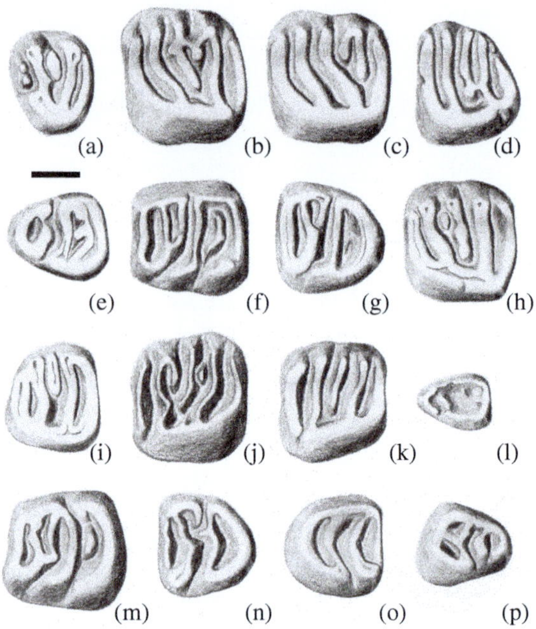

Figure 10.6 Isolated molars of the giant dormouse (*Hypnomys morpheus*), occlusal view. (a) left upper P4, (b) left upper M2, (c) left upper M2, (d) left upper M3, (e) left lower P4, (g) left lower M3, (h) left upper M1–M2, (i) right upper M3, (j) right upper M2, (k) right upper M1, (l) left lower dP4, (m) left lower M1, (n) left lower M3, (o) right lower M3, and (p) right lower P4. (a–g) originally attributed to *Hypnomys mahonensis*. (a–h) Son Bou, Minorca; (i–q) Cala Blanes, Minorca. Scale bar: 1 mm. (From Reumer, 1982.)

comparable to that of Jentink's dormouse (*Graphiurus crassicaudatus*), and elongated zygopodium – the radius, ulna, tibia, and fibula – broader distal humerus, and femur. Such a combination of characters is unique amongst living Sciurognath rodents, complicating the reconstruction of its way of life.

Balearic Shrews: Nesiotites

During the Late Miocene-Pleistocene, an endemic red-toothed shrew genus inhabited the Balearic Islands. This shrew lineage *Nesiotites* hardly shows any change through time, except for a minimal gradual increase in size from the Early Pliocene *N. rafelinensis* and *N. ponsi* to the Middle and Late Pleistocene and Holocene *N. hidalgo* (Figure 10.2). Intermediate sizes are found as well (Figure 10.5 (j)). Furthermore, fossils from Holocene sites (Son Muleta and Sa Bassa Blanca) are somewhat larger than those from Late Pleistocene sites (Porto Cristo, Son Bauzà). Holocene jaws are about 12 mm long, and often, the fourth antemolar is missing, in contrast to older specimens, as shown by Reumer (1981).

Analysis of ancient mitochondrial DNA of *N. hidalgo* by Bover, Mitchell and colleagues (2018) shows that the Balearic shrew is a sister taxon of *Asoriculus gibberodon* with the extant Himalayan shrew (*Soriculus nigrescens*) as the immediate outgroup. The estimated split between *N. hidaldo* and the Himalayan shrew of around 6.44 million years ago is in line with the proposed fauna immigration to the Balearic Islands from mainland Europe during the Messinian salinity crisis, when at least five other mammals, eight reptiles, and four amphibians arrived, as summarised by Torres-Roig and colleagues (2017).

References

Adrover R., Agustí J., Moyà-Solà S., Pons-Moyà J. 1983–1984. Nueva localidad de micromamíferos insulares del Mioceno medio en las proximidades de San Lorenzo en la isla de Mallorca. *Paleontologia i Evolució*, 18: 121–130.

Adrover R., Hugueney M., Mein P. 1977. Fauna africana oligocena y nuevas formas endémicas entre los micromamíferos de Mallorca (Nota preliminar). *Boletín de la Sociedad de Historia Natural de Baleares*, 22: 137–149.

Agustí J., Bover P., Alcover J.A. 2012. A new genus of endemic cricetid (Mammalia, Rodentia) from the Late Neogene of Mallorca (Balearic Islands, Spain). *Journal of Vertebrate Paleontology* 32 (3): 722–726.

Agustí J., Moyà-Solà J. 1982. Une espèce géante de *Muscardinus* Kaup, 1829 (Gliridae, Rodentia, Mammalia) dans le gisement karstique de Cala es Pou (Miocene supérieur de Minorque, Baléares). *Géobios*, 15 (5): 783–789.

Agustí J., Moyà-Solà S. 1990. Neogene-Quaternary mammalian faunas of the Balearics. In: Azzaroli A. (ed.), *International Symposium on Biogeographical Aspects of Insularity, Rome, 18–22 May 1987. Accademia Nazionale dei Lincei, Atti Dei Convegni Lincei*, 85: 459–469.

Alcover J.A., Perez-Obiol R., Yll E.-I., Bover P. 1999. The diet of *Myotragus balearicus* Bate 1909 (Artiodactyla: Caprinae), an extinct bovid from the Balearic Islands: Evidence from coprolites. *Biological Journal of the Linnean Society*, 66: 57–74.

Antoñanzas R.L., Sen S. 2005. New species of *Paraphiomys* (Rodentia, Thyronomyidae) from the Lower Miocene of As-Sarrar, Saudi Arabia. *Palaeontology*, 48: 223–233.

Arnau P., Bover P., Seguí B., Alcover J.A. 2000. Sobre alguns jaciments de *Myotragus balearicus* Bate 1909 (Artiodactyla, Caprinae) de tafonomia infreqüent. *Endins*, 23: 89–100.

Bartolomé J., Retuerto C., Martínez X., Alcover J.A., Bover P., Cassinello J. J., Baraza Y.E. 2011. Sobre la dieta de *Myotragus balearicus*, un bóvido sumamente modificado del Pleistoceno-Holoceno de las Baleares. *Naturaleza Aragonesa*, 27: 4–7.

Bate D.M.A. 1909. Preliminary note on a new artiodactyle from Mallorca *Myotragus balearicus*, gen. et sp. nov. *Geological Magazine (Decade 5)*, 6: 385–389.

Bate D.M.A. 1918. On a new genus of extinct muscardine rodent from the Balearic Islands. *Proceedings of the Zoological Society of London*, 1918: 209–222.

Vicens D. (eds), *Geomorfologia Litoral i Quaternari. Homenatge a Joan Cuerda Barceló. Monografies de la Societat d'Història Natural de les Balears*, 14: 155–180.

Moyà-Solà S., Pons-Moyà J. 1980. Una nueva especie de *Myotragus*, Bate 1909 (Artiodactyla, Rupicaprini), en la Isla de Menorca: *Myotragus binigausensis* n. sp. Implicaciones paleozoogeográficas. *Endins*, 7: 37–47.

Moyà-Solà S., Pons-Moyà J. 1981. *Myotragus kopperi*, une nouvelle espèce de *Myotragus* Bate 1909 (Mammalia, Artiodactyla, Rupicaprini). *Proceedings of the Koninklijke Akademie van Wetenschappen (Series B)*, 84: 57–69.

Moyà-Solà S., Pons-Moyà J. 1982. *Myotragus pepgonellae* nov. sp. un primitivo representante del género *Myotragus* Bate, 1909 (Bovidae, Mammalia) en la isla de Mallorca (Baleares). *Acta Geologica Hispànica*, 12: 77–87.

Moyà-Solà S., Quintana J., Alcover J.A., Köhler M. 1999. Endemic island faunas of the Mediterranean Miocene. In: Rössner G.E., Heissig K. (eds), *The Miocene Land Mammals of Europe*. Pfeil, Munchen, pp. 435–442.

Pons-Moyà J., 1977. La nouvelle espece *Myotragus antiquus* de l'ile de Majorque (Baleares). *Proceedings of the Koninklijke Nederlandse Akademie van Wetenschappen (Series B)*, 80 (3): 215–221.

Pons-Moyà J., Moyà-Solà S. 1980. Nuevo representante del género *Nesiotites* Bate 1944; *Nesiotites meloussae* nov. sp. (Insectivora, Soricidae) de los rellenos cársticos del "Barranc de Binigaus" (Es Mercadal, Menorca). *Endins*, 7: 53–56.

Pons-Moyà J., Moyà-Solà S. Kopper J.S. 1979. La fauna de mamíferos de la Cova de Canet (Espodes) y su cronología. *Endins*, 5–6: 55–58.

Quetglas G., Bover P. 1998. Reconstrucció de l'esquelet de *Myotragus balearicus* Bate, 1909 (Artiodactyla, Caprinae): Noves implicacions morfofuncionals. *Bolleti de la Societat d'Història Natural de les Balears*, 41: 159–172.

Quintana J. 2005. *Estudio morfológico y funcional de Nuralagus rex*. Unpublished PhD thesis. Universitat Autònoma de Barcelona, Barcelona.

Quintana J., Agustí J., 2007. Los mamíferos del Mioceno medio y superior de Menorca (Islas Baleares, Mediterráneo occiental). *Geobios*, 40 (5): 677–687.

Quintana J., Bover P., Alcover J.A., Agustí J., Bailon S. 2010. Presence of *Hypolagus* Dice, 1917 (Mammalia, Lagomorpha) in the Neogene of the Balearic Islands (Western Mediterranean): Description of *Hypolagus balearicus* nov. sp. *Geobios*, 43: 555–567.

Quintana J., Köhler M., Moyà-Solà S. 2011. *Nuralagus rex*, gen. et sp. nov., an endemic insular giant rabbit from the Neogene of Minorca (Balearic Islands, Spain). *Journal of Vertebrate Paleontology*, 31(2): 231–240.

Quintana J., Moncunill-Solé B. 2014. *Hypolagus balearicus* Quintana, Bover, Alcover, Agustí & Bailon, 2010 (Mammalia: Leporidae): New data from the Neogene of Eivissa (Balearic Islands, Western Mediterranean). *Geodiversitas* 36 (2): 283–310.

Quintana J., Moyà S., Köhler M. 2005. El conejo gigante de los depósitos cársticos de Punta Nati – Cala's Pous (Menorca, Illes Balears). In: Alcover J.A., Bover P. (eds), *Proceedings of the International Symposium Insular Vertebrate Evolution: The Palaeontological Approach. Monografies de la Societat d'Història Natural de les Balears*, 12: 297–308.

Reumer J.W.F. 1979. On two new micromammals from the Pleistocene of Mallorca. *Proceedings of the Koninklijke Akademie van Wetenschappen (Series B)*, 82: 473–482.

Reumer J.W.F. 1981. The Pleistocene small mammals from Sa Pedrera de S'Onix, Majorca (Gliridae, Soricidae). *Proceedings of the Koninklijke Akademie van Wetenschappen (Series B)*, 84 (1): 3–11.

Reumer J.W.F. 1982. Some remarks on the fossil vertebrates from Menorca, Spain. *Proceedings of the Koninklijke Nederlandse Akademie van Wetenschappen* (B), 85: 77–87.

Reumer J.W.F. 1994. *Eliomys (Hypnomys) onicensis* nomen novum, to replace the homonym *Hypnomys intermedius* Reumer, 1981 (Rodentia: Gliridae) from Majorca. *Zeitschrift für Säugetierkunde*, 59: 380–381.

Rivera L., Baraza E., Alcover J.A., Bover P., Rovira C.M., Bartolomé J. 2014. Stomatal density and stomatal index of fossil *Buxus* from coprolites of extinct *Myotragus balearicus* Bate (Artiodactyla, Caprinae) as evidence of increased CO2 concentration during the late Holocene. *The Holocene*, 24: 876–880.

Rofes J., Bover P., Cuenca-Bescós G., Alcover J.A. 2012. Nesiotites rafelinensis sp. Nov., the earliest shrew (Mammalia, Soricidae) from the Balearic Islands, Spain. Palaeontologia Electronica 15.1.8A.

Torres-Roig E., Agustí J., Bover P., Alcover J.A. 2017. A new giant cricetine from the basal Pliocene of Mallorca (Balearic Islands, western Mediterranean): Biostratigraphic nexus with continental mammal zones. *Historical Biology*, 31 (5): 559–573.

Torres-Roig E., Piñero P., Agustí J., Bover P., Alcover J. 2019. First evidence of endemic Murinae (Rodentia, Mammalia) in the early Pliocene of the Balearic Islands (western Mediterranean). *Geological Magazine*, 156 (10): 1742–1750.

Welker F., Duijm E., van der Gaag K.J., van Geel B., de Knijff P., van Leeuwen J., Mol D., van der Plicht J., Raes N., Reumer J., Gravendeel B. 2014. Analysis of coprolites from the extinct mountain goat *Myotragus balearicus*. *Quaternary Research*, 81: 106–116.

CHAPTER ELEVEN

Madagascar

Evolution of Island Mammals: Adaptation and Extinction of Placental Mammals on Islands, Second Edition.
Alexandra van der Geer, George Lyras and John de Vos.
© 2021 John Wiley & Sons Ltd. Published 2021 by John Wiley & Sons Ltd.

Today, Madagascar is inhabited by a highly diverse endemic fauna – one of the richest species in the world – including lemurs, tenrecs, fossas, bats, and rodents. This biodiversity was even greater during the Late Pleistocene, with now extinct dwarf hippos, giant lemurs, a large fossa, Malagasy aardvark-like mammals, and giant elephant birds. The oldest mammalian fossils from Madagascar date back to the Mesozoic, but these ancient taxa are unrelated to the mammals of later times. The endemic lemurs, tenrecs, and rodents of Madagascar are all part of impressive adaptive radiations, originating from just three early colonising species.

Geology and Palaeogeography

Madagascar is the world's fourth largest island and the largest oceanic-like one. It is separated from Africa by the Mozambique Channel, a 400–1000 km wide body of water with depths of more than 2 km, exceeding 3 km at some parts. The only prominent feature of this channel is the Davie Ridge, a north–south running ridge that separates Madagascar from Africa.

Geological evidence suggests that Madagascar split from Africa during the Jurassic period. Initially, India and Madagascar were still connected. According to David Krause and colleagues (2019), the separation of Indo-Madagascar from Antarctic took place 124 million years ago. The India–Madagascar separation started some 88 million years ago. From that moment on, Madagascar remained an island, geographically isolated from the rest of the world (Figure 11.1).

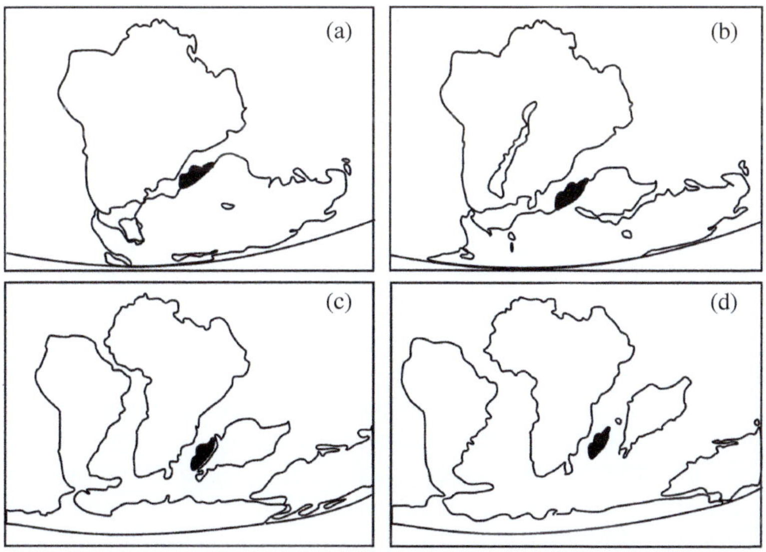

Figure 11.1 Map illustrating the position of Madagascar relative to the other continents during the Mesozoic. (Redrawn from Tattersall, 2008.)

The subsequent geographic isolation resulted in a latest Cretaceous Malagasy fauna that included a range of unusual taxa such as massive predatory frogs, herbivorous crocodyliforms, and specialised theropod dinosaurs. In addition to these, *Vintana sertichi* and *Adalatherium hui*, two Gondwanatherian mammaliaform described by David Krause and colleagues in 2014 and 2020, respectively, have large sizes and a mosaic of primitive and derived features that has been explained as the result of geographic isolation. But exactly how isolated was Madagascar during the Cenozoic? As Ian Tattersall (2008) noted, the absence of Palaeogene and Neogene terrestrial vertebrate fossils from Madagascar makes any hypothesis concerning the timing of immigration or the route followed difficult to check. At present, studies based on phylogenetic topologies and diversification rates of the modern taxa, such as the work of Karen Samonds and her colleagues (2013), provide the only reconstruction of colonisation events of Madagascar.

Three models have been proposed for the colonisation of Madagascar. These are overseas sweepstake dispersal, island-hopping via a series of islands, and ancient vicariance. Gaylord Simpson (1940) was the first to consider overseas sweepstake dispersal via rafting as the way mammals had reached Madagascar. At present, this hypothesis is still by far the most likely one, taking the geology into account, and is followed by most scientists (Box 11.1).

An alternative kind of dispersal is given by Robert McCall (1997), who suggested island-hopping via a series of larger areas of dry land in the Mozambique Channel to Madagascar's northeast margin. These dry areas then subsided in the Early Miocene, isolating the Malagasy mammals since then. In his view, small areas of the Mozambique Channel along the Davie Ridge were exposed above sea level as evidenced by terrestrial sediments in cores (Figure 11.2). According to him, the geotectonic mechanism behind the formation of this land bridge was the uplift of the Davie Ridge due to compression forces related to the collision with India. However, Raymond Rogers and colleagues (2000) noted that the major part of the ridge remained submerged and that these exposures were at most isolated dots of dry land, each with an area of at most a few square kilometres. Therefore, this series of exposed dots could never have constituted a substantial land bridge. The land bridge scheme along the Davie Ridge was raised again by Judith Masters and her colleagues (2006). They suggested a series of alternative migration routes – e.g., along the Davie Ridge, the Antarctic–Africa corridor, and the Deccan hotspot corridor – all of which lack a demonstrated geological basis.

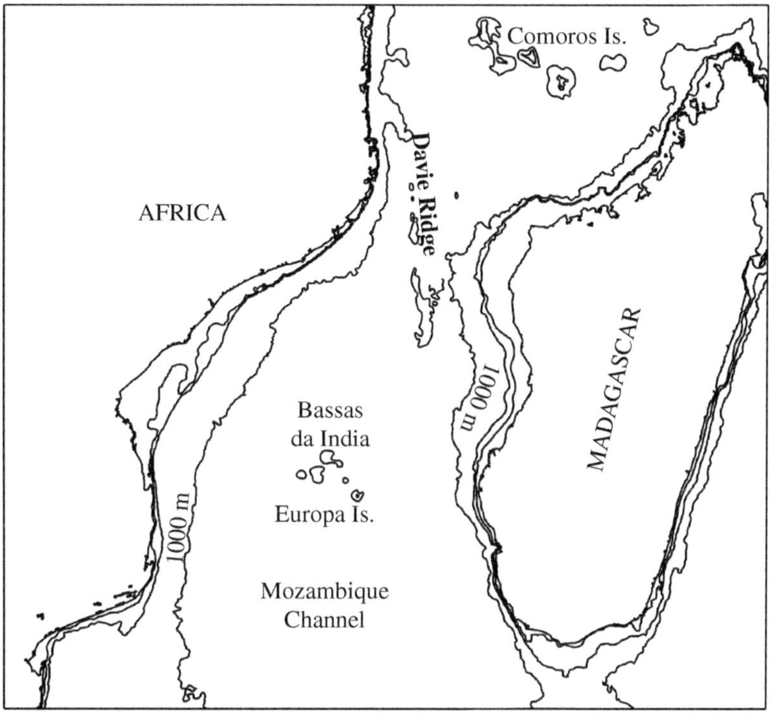

Figure 11.2 Bathymetric map of the Mozambique Channel. (Drawing George Lyras.)

The third alternative explanation is ancient vicariance since the Mesozoic. Such a scenario implies ancient divergence dates, which are not confirmed by DNA studies. The living mammalian groups of Madagascar did not appear before the separation of the island from India, although Ulfur Arnason and colleagues (2000) assumed an age of colonisation older than 84 million years for the lemurs, which is about the time that Madagascar became isolated. However, the resolved lemur phylogeny of the team led by Horvath (2008), based on nuclear as well as mitochondrial DNA, clearly indicates that the time of lemuriform divergence postdates the separation of Madagascar from Africa and India by many millions of years. They calculated the time of dispersal to Madagascar to between 50 and 80 million years ago, confirming earlier calculations by Anne Yoder and Ziheng Yang (2004) and by Céline Poux and colleagues (2005). If this is true, lemurs can have arrived only by overseas dispersal. The same applies to the Malagasy latecomers – the tenrecs, rodents, and fossas – which arrived during the Oligocene, as calculated by Poux and colleagues (2005), based on nuclear genes. Hippopotamuses, which arrived during the Late Pleistocene, are certainly not leftover from a Mesozoic stock nor could they have taken a land bridge in the Early Miocene.

BOX 11.1

A Long Journey

The reason for doubt about overseas rafting, swimming, or floating to Madagascar is found in the immense distance between Africa and Madagascar. This made many researchers question whether such a travel was possible at all, particularly for small mammals like lemurs. There is no evidence that lemurs are able to swim, according to Steven Goodman and Jörg Ganzhorn (2004), and although this leaves the rafting hypothesis open, this possibility was heavily criticised by Jacek Stankiewicz and colleagues (2006). On the basis of a geometric probability model, they calculated the possibility for a raft emerging from an estuary on Africa's east coast and landing on the Malagasy coast as improbably small, taking the prevailing winds and currents into account. Transport enhanced by hurricanes or tornadoes is even less likely than rafting for mammals, in their view. Stankiewicz and colleagues may be certainly right regarding the present-day situation, which explains why no new colonisations of Madagascar took place since the end of the Pleistocene. However, the currents, prevailing winds, and patterns of hurricanes and tornadoes of earlier times are unknown and could, at times, have been different from those observed today. Indeed, using palaeogeographic reconstructions and palaeo-oceanographic modelling, Jason Ali and Matthew Huber (2010) demonstrated that during the Palaeogene period (about 23 to 66 million years ago), strong surface currents flowed from west to east, thus allowing the colonisation of Madagascar by rafting. The hypothesis of colonisation by land bridge was raised again by Paul Mazza (2014), this time for the Malagasy hippopotamuses. Modern hippos have been described as bottom walkers rather than surface swimmers; therefore, according to Mazza, the available geological evidence for the absence of a land bridge to Madagascar should be reconsidered and an alternative explanation to overseas swimming has to be looked for. However, we replied (van der Geer and colleagues, 2015) that hippos are likely able to swim in seawater, which greatly enhances their buoyancy and that geological evidence for the proposed land bridge is lacking. Besides, baby hippos up to weaning age are apt swimmers in freshwater.

At least four periods of significant sea level lowering have been recognised, during which the chances to reach Madagascar by swimming or rafting were considerably better. These are commonly placed during the Early Oligocene (c. 36–35 million years ago), the mid-Miocene (c. 16–13 million years ago), the Late Miocene (c. 7–5 million years ago), and the Late Pliocene or Early Pleistocene (c. 2 million years ago).

Historical Palaeontology

Western scientists heard of the existence of large-sized mammals in Madagascar through the writings of the French colonial governor Etienne de Flacourt, who in 1658 reported stories of locals who had seen large-sized animals. The first European to discover the actual remains of these enigmatic beasts was the French scientist and explorer Alfred Grandidier. He and his brother decided to go for a scientific trip around the world, but Madagascar captured Alfred's interest, and he systematically explored the island and its natural history. In 1865, Grandidier discovered Ambolisatra at the south-western coast, the first Malagasy locality with subfossil mammalian remains (Figure 11.3).

All he had to do was to follow the indications of a village chief to a marsh where, as he was told, he would find the bones of the 'Song'aomby' – literally, the cow that is not a cow. In the marsh, today, covered by a coconut plantation, Grandidier found the fossils of pygmy hippos:

> *Because I asked for information regarding the Song'aomby (previously known to me only through a very poor description that Flacourt had provided under the name Mangarsahoc), the chief of the region indicated the location of a nearby marsh, and informed me that I could find this animal's bones there. On that advice, I hurried to the location – barefooted and barelegged, with pants cut at the knees, as I am prone to do. I entered the marsh, and lowering myself, tapped the bottom where I sensed a large object, and lifted it. After washing it, I found to my surprise and joy that it was a femur – the thighbone of a bird. The bird must have been enormous, like the famous Roc of 1001 Nights. Enthusiastically, I returned to the water and, with some of my men, dug into the mud that carpeted the floor of the marsh. I retrieved more bones of the colossal bird, Aepyornis, known previously only from its 8-liter eggs and a few indeterminable pieces sent by Mr. Abadi and described in 1850 by Isidore Geoffroy Saint-Hilaire. Alongside these bird bones were numerous other bones belonging to an unknown species of hippopotamus that I named Hippopotamus Lemerlei in honor of our odd-job man at Tuléar, as well as bones of other new and interesting animals.* Alfred Grandidier, '*Souvenirs de voyages d'Alfred Grandidier: 1865-1870*' [published as Grandidier (1971); translation from the original French is from Godfrey and Jungers (2003).]

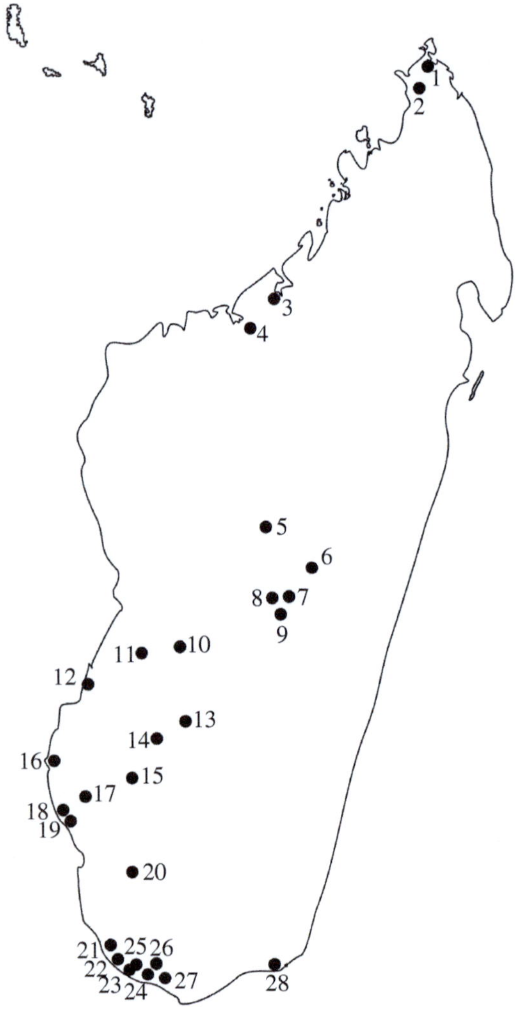

Figure 11.3 Map of Madagascar with the most important subfossil localities. (1) Mont des Français, (2) Ankarana, (3) Anjohibe, (4) Amparihingidro, (5) Ampasambazimba, (6) Sambaina, (7) Antsirabe, (8) Betafo, (9) Morarano, (10) Antanimbaribe, (11) Ambararata, (12) Belo sur Mer, (13) Bemavo, (14) Tsirave, (15) Ampoza, (16) Lamboharana, (17) Ankilitelo, (18) Andranovato, (19) Ambolisatra, (20) Taolambiby, (21) Itampolo, (22) Bemafandry, (23) Bevoalovao-West, (24) Anavoha, (25) Behova, (26) Ampotaka, (27) Beloha, and (28) Andrahomana. Overviews and descriptions of the various sites have been published by Mahé and Sourdat (1972) and Tattersall (1973a). Additional data are provided by MacPhee and colleagues (1985), Gommery and colleagues (2003), Jernvall and colleagues (2003), Godfrey and Jungers (2003), Samonds (2007), and Burney and colleagues (2008).

Grandidier collected the remains of some fifty hippos from the swamp, but only part of this huge collection was sent to Paris, as noted by Solveig Stuenes (1989).

The first lemur subfossil recognised as such was an incomplete skull cap collected in the marsh of Ambolisatra. It was discovered by Mr J.T. Last, a collector working for Lionel Walter Rothschild, a British banker, and politician with a passion for natural history. The specimen was presented by Forsyth Major (1893) of the British Museum to a meeting of the Royal Society. However, Forsyth Major, pending the discovery of new and better preserved material, did not assign a new name to it (it is now known as *Archaeolemur*). The same year, Forsyth Major described another subfossil lemur skull, but this time he assigned a name to it, *Megaladapis madagascariensis* (Box 11.2) He immediately organised an expedition to central-eastern Madagascar, funded by, amongst others, the Royal Society and Lionel Rothschild, in the hope of finding more fossil lemurs and hippos, as Forsyth Major was interested in insular faunas and had collected on various Mediterranean islands. His expedition lasted nearly two years, including the lengthy travel from London to the island (Forsyth Major, 1897). His most important locality was Antsirabe, where he collected subfossils from September till, perhaps, December 1897, according to Paulina Jenkins and Michael Carleton (2005). Apart from fossils, he collected an impressive amount of recent zoological specimens.

BOX 11.2

Taxonomic Confusions

The majority of the subfossil lemur species were named towards the end of the nineteenth century and in the first half of the twentieth century. These included *Megaladapis madagascariensis* by Forsyth Major (1893); *Thaumastolemur grandidieri*, *Archaeolemur majori*, *Archaeolemur edwardsi*, and *Pachylemur insignis* by Henri Filhol (1895); *Palaeopropithecus ingens* and *Megaladapis edwardsi* by Guillaume Grandidier (1899); *Hadropithecus stenognathus* by Ludwig R. Lorenz von Liburnau (1899); *Palaeopropithecus maximus* and *Megaladapis grandidieri* by Herbert Standing (1903); *Mesopropithecus pithecoides* by Standing (1905); *Archaeoindris fontoynontii* by Standing (1909); *Mesopropithecus globiceps* by Charles Lamberton (1936); and *Pachylemur jullyi* by Lamberton (1948).

Some of these names were later synonymised and are not considered valid anymore. A typical example is the case of *Thaumastolemur*. Filhol based his new genus and species on a distal humerus collected in 1865 by Alfred Grandidier at Ambolisatra. Guillaume Grandidier (1902), Alfred Grandidier's son, incorrectly considered the holotype of *Thaumastolemur* a *Megaladapis* and, therefore, synonymised it. Eighty years later, Martine Vuillaume-Randriamanantena (1990) realised his mistake and saw that the holotype of *Thaumastolemur* was in fact a *Palaeopropithecus ingens*. Thus, *Thaumastolemur* should have priority over *Palaeopropithecus*. However, the latter name had been in use for decades, while former name was practically forgotten. For that reason, the International Commission for Zoological Nomenclature suppressed its use in favour of the name *Palaeopropithecus*. Other examples of invalid names are the various synonyms of *Archaeolemur*, a genus based on a humerus and the proximal parts of a radius and an ulna from Belo-sur-Mer at the western coast. Several synonyms were in vogue, e.g., *Lophiolemur, Dinolemur, Nesopithecus, Globilemur, Bradylemur,* and *Protoindris* until Forsyth Major (1900) suggested that they should be grouped together. Guillaume Grandidier (1902, 1905) did so officially a few years later.

A common problem in the early works was the attribution of the isolated postcranial elements to the right taxon. This was partly caused by the absence of detailed field notes and the mixing of bones from several sites as no data were written on the bones themselves. As a result, many elements were attributed to a new or the wrong taxon. This happened, for example, with *T. grandidieri* and *Palaeopropithecus ingens*, now considered synonymous. The former was based on a distal humerus and the latter on a large mandibular fragment. However, the problem of the correct attribution of the postcranial material of many taxa remained for many years, thus complicating later phylogenetic and morpho-functional studies. In recent literature, original finding spots are sometimes given with a question mark, reflecting the uncertainty of provenance of the early fossils.

In 1899, the Austrian pipe-carver and naturalist Franz Sikora excavated Andrahomana, a cave in southern Madagascar. He worked there for 18 days and collected many subfossil specimens, which he sent to the Naturhistorisches Museum in Vienna and to the British Museum of Natural History in London. The Vienna material included a nearly complete skull with mandible and other skeletal elements of a giant 'monkey lemur'. Ludwig Lorenz von Liburnau described it the same year as *Hadropithecus stenognathus.* The London material included the first complete skull and a lower jaw, plus associated postcranial bones, of the earlier described *Archaeolemur majori* and abundant subfossils of the extant Malagasy giant rat, *Hypogeomys antimena.* The cave was visited two years later by Guillaume Grandidier and independently by Charles Alluaud and Lieutenant Gaubert. The material they collected was sent to Paris in 1902 and described by Grandidier (1903) as a new species of Malagasy giant rat, *Hypogeomys australis,* now a junior synonym of the extant species *Hypogeomys antimena,* which was named by his father (Grandidier, 1869).

Perhaps, the leading expert of the extinct mammalian fauna of Madagascar of the first half of the twentieth century was the French palaeontologist Charles Lamberton. Lamberton carried out extensive field work in Madagascar and spend his entire scientific career studying the extinct Malagasy fauna. In 1919, he excavated at Morarana, a locality east of Betafo, where he collected the remains of *Archaeolemur*, *Megaladapis*, and *Palaeopropithecus*. At irregular intervals between 1921 and 1927, he collected subfossils at Ampasambazimba. Here, he also collected fruits and seeds from the peat surrounding the fossiliferous area. This material was analysed and described by Joseph Marie Henry Alfred Perrier de la Bâthie (1927), making Ampasambazimba one of only two subfossil sites for which a published floral list exists. Some other expeditions were in 1924 and 1925 to Ampoza, in 1927 to Sambaina, and in 1931 to Tsirave. Sambaina yielded many hippo bones and only a very few remains of other mammals that included two femoral heads of *Archaeolemur* and a humerus of *Hypogeomys*. Lamberton wrote many scientific works, most of them published in the bulletin of the Malagasy Academy. His most important contributions deal with subfossil lemurs (1934, 1936, 1938, 1947, and 1957) and the Malagasy 'aardvark' (1946).

In the meantime, other naturalists and amateurs also undertook explorations of Madagascar to collect specimens. One of them was the French Raymond Decary, who worked in Madagascar between 1916 and 1944, first as a military officer, later as a colonial administrator, and finally as scientific director. He led several scientific expeditions and collected lemur and micromammal subfossils mainly from Anjohibe and the cave of Andrahomana. Another was Errol White of the British Museum. In 1929, he worked in Madagascar for about six months, as part of an Anglo–French–American expedition, aiming mainly at collecting bird specimens, living as well as extinct. His expedition consisted of two parts, one carried out at Lamboharana under the supervision of Ramamònjy and one carried out at Ampoza, Itampolove, and La Table by White himself (White, 1930). Simultaneously, a Swedish-Norwegian couple, Mr and Mrs Ljungqvist, were collecting fossils at Ampoza. They left their work unfinished, and White continued in their excavation pit.

A new era of excavation started in 1965 when Joël Mahé introduced water pumps to keep the excavation pits of Amparihingidro dry. The locality was discovered by a farmer. The dry pits enabled field observations that had been impossible until then. Around the same time, Alan Walker visited many of the Malagasy sites and reported their status in his (unpublished) doctoral thesis of 1967. Unfortunately, he found several old localities covered by plantations, and, in most cases, nothing could thus be seen.

In the early 1980s, a new series started of expeditions led by, independently, Elwyn Simons, Ross MacPhee, and others.

Excavations of and research on Malagasy subfossils continue to date. One of the more recently named lemurs is *Babakotia radofilai*, described by Laurie Godfrey and colleagues (1990), based on a left upper jaw fragment, a right upper jaw, and some apparently associated midshafts of the long bones from Ankarana and Anjohibe. The first fossils had been discovered two years before at Antsiroandoha, a cave in the Ankara Massif of northern Madagascar, and soon after, many more specimens were found, including a skull and associated skeleton, described by Simons and colleagues (1992). *Babakotia* is the first new fossil lemur genus discovered since 1909. Another new species is *Mesopropithecus dolichobrachion*, named by Simons and colleagues (1995), based on a skull with associated lower jaw, vertebrae, and some hand and foot bones from Ankarana.

In the summer of 2003, subfossil remains of a shrew tenrec were found at Andrahomana cave in extreme south-eastern Madagascar during an excavation led by David Burney. They were described by Goodman and colleagues (2007a) as *Microgale macpheei*, after MacPhee (1987), who had revised the genus. It is the first and only extinct tenrec discovered so far. The same cave and excavation yielded the first fossils of an extant big-footed mouse (*Macrotarsomys petteri*), described by Goodman and Voahangy Soaramalala (2005).

Finally, in the spring of 2009, a new extinct lemur species was discovered in north-western Madagascar by a French-Malagasy team. The new species is described as *Palaeopropithecus kelyus* by Dominique Gommery and colleagues (2009) based on a right half upper jaw with fourth premolar and the first and second molars. The discovery is evidence for larger lemur diversity in the Late Pleistocene and Early Holocene than today.

Biozones and Faunal Units

The oldest mammalian fossils of Madagascar are of Mesozoic age. This Mesozoic mammalian fauna, described by John Flynn and André Wyss (2003), was rather diverse. However, it consisted only of non-placental taxa, and as it was noted by Karen Samonds and her colleagues (2013), none played a role in the ancestry of the extant clades of placental mammals that live on the island today.

The only pre-Quaternary fossil placentals of Madagascar have all been discovered in near-shore marine deposits. Mammalian fossils from the small island of Nosy Makamby, 4 km off the coast of north-western Madagascar, include several sea cow remains and a single lumbar vertebra from a small dolphin. Samonds and her colleagues (2019) reported the Miocene sea cows *Rytiodus heali* (previously attributed incorrectly to the genus *Halitherium* by Collignon and Cottreau, 1927), *Norosiren zazavavindrano*, and *Metaxytherium* cf. *krahuletzi*. On another site, in the Mahajanga Basin of north-western Madagascar, Samonds and her colleagues (2009) described the Middle Eocene sea cow *Eotheroides lambondrano*.

There are more than thirty localities with mammalian remains of Late Pleistocene or Holocene age. Radiocarbon dates indicate that the oldest assemblages are only about 26,000 years old, while the youngest are not older than 630 years before the present. Due to this lack of a fossil record with depth, no separate biozones can be distinguished for Madagascar. Differences between the localities do, though, exist. They can be ascribed to differences in palaeo-environment. The localities can roughly be divided into two main groups, those of the highlands in the volcanic massifs of Itasy and Ankaratra and those of the south-western coast between Morondava and Cape Sainte-Marie.

Most localities yield just a subset of the fossil fauna. The subfossil fauna list of Ampasambazimba, however, is extensive and is the longest of all Malagasy localities, as is the literature list on this site. In addition, it is one of the two subfossil sites for which a published floral list exists, giving some insight into past environmental conditions and ecosystems. The site was discovered accidentally during a survey in 1902 for the economic potential of limestone deposits. Subsequently, several palaeontological expeditions visited there, e.g., Herbert Standing's (1903, 1905) and Charles Lamberton's between 1921 and 1927. Claude Chanudet (1975) composed a review of the early excavations at Ampasambazimba – and other localities – based on all available old documents on these excavations kept at the archive of the Malagasy Academy. Additional data on this locality were published by Tattersall (1973a) and MacPhee and colleagues (1985).

The Late Pleistocene–Early Holocene fossil fauna of Madagascar is known to us by several lemurs (Lemuriformes; Plate 20), three species of dwarf hippos (*Hippopotamus madagascariensis* [Plate 21], *Hippopotamus lemerlei* [Plate 22], *Hippopotamus laloumena*), two species of Malagasy aardvark-like mammals (*Plesiorycteropus germainepetterae* and *Plesiorycteropus madagascariensis*), two tenrecs (*Microgale macpheei, Microgale brevicaudata*), one Malagasy mouse (*Macrotarsomys petteri*), perhaps one Malagasy fossa (*Cryptoprocta spelaea*) and several bats, plus birds – e.g., the elephant birds (*Aepyornis, Mullerornis, Vorombe*) – broad-snouted crocodiles (*Voay robustus*), and other reptiles, including giant Aldabra tortoises (*Aldabrachelys abrupta, Aldabrachelys grandidieri*), with carapace lengths of 115 cm and 125 cm, respectively. The

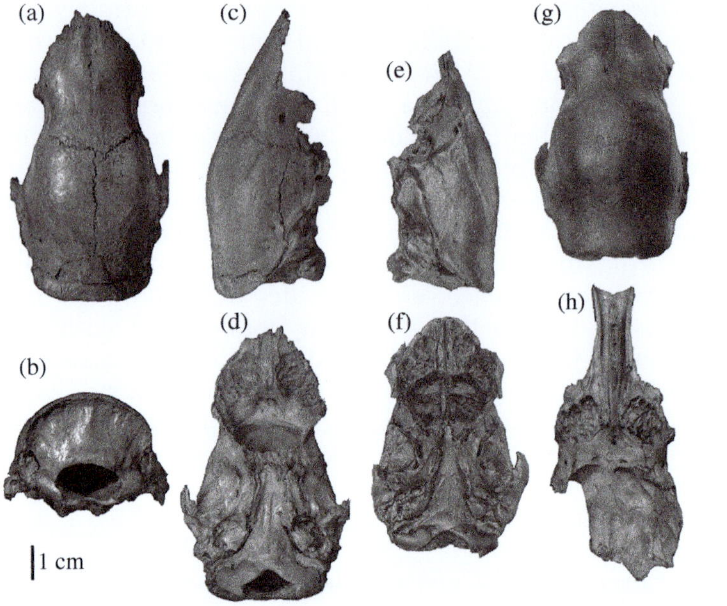

Figure 11.4 Cranial elements of *Plesiorycteropus*. (a) dorsal, (b) occipital, (c) lateral, and (d) ventral views of the holotype of *Plesiorycteropus madagascariensis* (Belo skull). (e) lateral, (f) ventral, and (g) dorsal views of the holotype of *Plesiorycteropus germainepetterae* (Ampasambazimba skull). (h) ventral view of the calvarium preserving nasals of *Plesiorycteropus madagascariensis* (Ambolisatra). National Museum of Natural History, Paris. (Photographs: Ross MacPhee.)

elephant birds were the largest animals, with masses reaching 730 kg (*Vorombe titan*), as calculated by Hansford and Turvey in 2018.

The Malagasy 'Aardvarks'

The Malagasy aardvark-like mammals (*Plesiorycteropus*), also known as Bibymalagasia, are known only from some partial skulls of which none preserves the facial region and many postcranial remains (Figure 11.4). MacPhee's (1994) monograph contains an extensive list of specimens and provides illustrations and descriptions of all available skeletal elements, for the greater part housed in the Muséum National d'Histoire Naturelle in Paris. Two species have been described so far, the larger *Plesiorycteropus madagascariensis* – originally described by Filhol (1895) – and the smaller *Plesiorycteropus germainepetterae* – named by MacPhee (1994).

Since its original description by Filhol, *Plesiorycteropus* has been often considered some sort of aardvark. Charles Lamberton (1946) published a paper in which he recognised that, although *Plesiorycteropus* indeed has many aardvark-like features, its anatomy was rather puzzling and, therefore, its phylogenetic position not easy to access. Instead, it turns out that *Plesiorycteropus* is not an aardvark at all but is the sole representative of a distinct mammalian order called Bibymalagasia – literally 'animals of Madagascar' – erected by MacPhee (1994). He concluded that many of its so-called aardvark features are simply parallel adaptations for digging. A number of anatomical characters are shared with the true aardvarks (Tubulidentata), ungulates and subungulates. Given its phylogenetic uncertain position, MacPhee classifies *Plesiorycteropus* in a family (Plesiorycteropodidae) and order (Bibymalagasia) on its own. Michael Buckley (2013) performs a molecular analysis on bone protein collagen obtained from *Plesiorycteropus*. His results place the 'Malagasy aardvark' as more closely related to tenrecs than aardvarks (Tubulidentata). According to his results, the taxonomic order Bibymalagasia is obsolete, and *Plesiorycteropus* is likely a form of large-sized tenrec. However, two years later, Julien Benoit and his colleagues (2015) indicated that the anatomy of the bony labyrinth of *Plesiorycteropus* supports their ordinal distinctiveness from tenrecs and aardvarks. Therefore, as Benoit and his colleagues pointed out, to wholly solve the puzzle of the phylogenetic affinities of the enigmatic Malagasy 'aardvark' *Plesiorycteropus*, we need to find more fossils.

The larger Malagasy 'aardvark' was probably less than half the size of the living true aardvark (*Orycteropus afer*). Its skull length is estimated to have averaged about 100 mm in length. It had a comparatively long brain case, combined with a short face. Teeth or mandibles were never found, so whether it was toothless like the true aardvark or not is unknown. The lesser Malagasy aardvark-like mammal differs from the larger in a number of anatomical features, mainly in the skull. The body mass has been estimated to have ranged between 6 kg for the smallest *germainepetterae* and 18 kg for the heaviest *madagascariensis*. Not much is known about the habits of the Malagasy aardvark-like mammals. It probably had a very good smell, as indicated by the broad nasals. Its sight, on the other hand, likely was poor, inferred from its tiny eyes. It must have been a proficient digger, seeing its strong and sharp claws. The diet was soft, maybe consisting of insects, as suggested by the shallow mandibular joint and the possible absence of teeth.

The Hippopotamuses

As many as four different species of dwarf hippos were identified by various researchers. In a detailed revision of the fossil record of the Malagasy hippos, Solweig Stuenes (1989) concluded that there were only two valid species, *Hippopotamus lemerlei* (Figure 11.5; Plate 22) and the somewhat larger *Hippopotamus madagascariensis* (Plate 21). A third species was described by Martine Faure and Claude Guerin (1990) as *Hippopotamus laloumena*. It is larger than the other two species and only somewhat smaller than the extant hippopotamus. The similarity with the latter is insufficiently

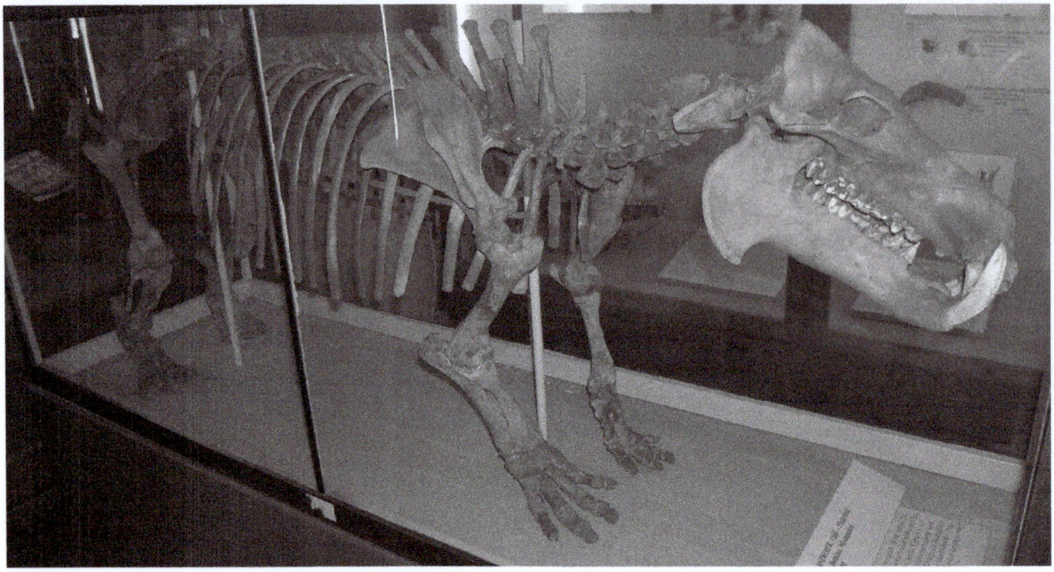

Figure 11.5 Skeleton of *Hippopotamus lemerlei*. Natural History Museum, Berlin. (Photograph: John de Vos.)

resolved, and the species is not widely accepted. According to Burney and colleagues (2004), this third species simply may represent evidence for incidental crossing of the Mozambique Channel by individuals of the extant hippopotamus during the Holocene, including since the European colonisation of Madagascar.

Before Stuenes' revision, the specific names *lemerlei* and *madagascariensis* were used without much reasoning. The name *lemerlei* was given by Alfred Grandidier in a brief description published by Alphonse Milne Edwards (1868). Guldberg (1883) described *Hippopotamus madagascariensis*, and Grandidier and Filhol (1894) suggested the presence of an additional species, *Hippopotamus leptorhynchus*. *Hippopotamus amphibius standini* was established by Monnier and Lamberton (1922). According to Stuenes, however, both *Hippopotamus leptorhynchus* and *Hippopotamus amphibius standini* are junior synonyms of *Hippopotamus lemerlei*. William Fovet and his colleagues (2011) indicated that the name *Hippopotamus madagascariensis* needs a taxonomic revision. According to them, the animal discussed by Guldberg (1883), where he for the first time described *Hippopotamus madagascariensis* is in reality *Hippopotamus lemerlei*. Guldberg used *Hippopotamus madagascariensis* not as a new species, but simply as a name replacement for *Hippopotamus lemerlei*, as the latter does not provide a geographic indication. As the scientific name *Hippopotamus madagascariensis* is a synonym of *Hippopotamus lemerlei*, they thus proposed a new name, *Hippopotamus guldbergi*. The suggestion of Fovet was incidentally followed by Goodman and Jungers (2014), but not anymore by Goodman and Soarimalala (forthcoming), where they continued the use of *Hippopotamus madagascariensis* on the ground of prevalent usage.

The body masses of *Hippopotamus lemerlei* and *Hippopotamus madagascariensis* were estimated by Eleanor Weston and Adrian Lister (2009) as 374 kg and 393 kg, respectively. The differences between the two species are mainly based on skull and teeth morphology (Figure 11.6). The supraorbital margin is thick in *Hippopotamus lemerlei* but thin in *Hippopotamus madagascariensis*, the anterior margin of the orbit lies above

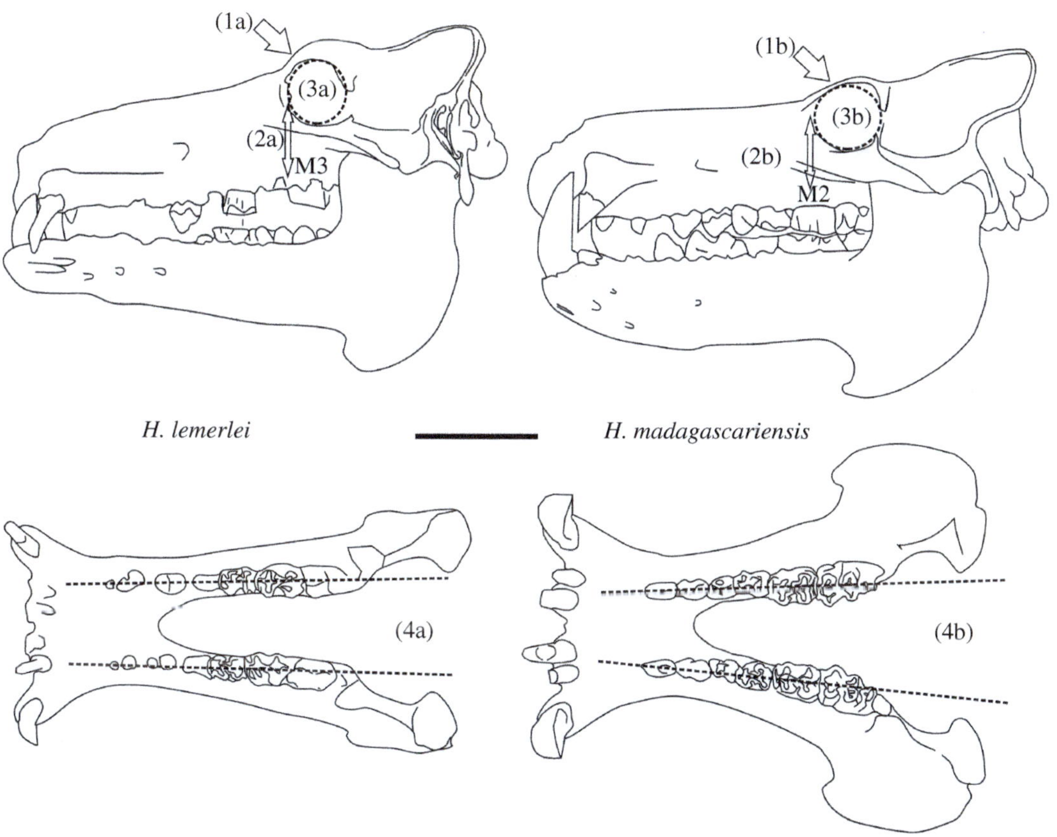

Figure 11.6 Schematic drawing of the skulls and lower jaws of the Malagasy dwarf hippos (*Hippopotamus lemerlei*, top; *Hippopotamus madagascariensis*, below) to show morphological differences between the species as noted by Stuenes (1989). The supraorbital margin is either thick (1a) or thin (1b). The anterior margin of the orbit lies above M3 (2a) or above M2 (2b). The orbit is oval (3a) or round to slightly flattened (3b). The molar rows of the lower jaw run parallel (4a) or converge towards the front (4b). Scale bar: 10 cm. (Drawing by George Lyras based on specimens figured by Stuenes, 1989).

the third molar in *Hippopotamus lemerlei* but above the second molar in *Hippopotamus madagascariensis*, the orbit is oval in *Hippopotamus lemerlei* but round or slightly flattened in *Hippopotamus madagascariensis*, the jugal process lies behind the third molar in *Hippopotamus lemerlei* but above its posterior part in *Hippopotamus madagascariensis*, the glenoid fossa is compact in *Hippopotamus lemerlei* but fairly long and slender in *Hippopotamus madagascariensis*, the molar rows run parallel to each other or diverge anteriorly in *Hippopotamus lemerlei* but converge anteriorly in *Hippopotamus madagascariensis*, the incisors have oblique wear facets as a result of a scissors-like occlusion in *Hippopotamus lemerlei* but terminal wear facets in *Hippopotamus madagascariensis*, and finally, the distance from the wear facet of the canine to the margin of the alveolus is long in *Hippopotamus lemerlei* but short in *Hippopotamus madagascariensis*.

According to Stuenes, the two Malagasy dwarf hippos were adapted to two different environments and had a different way of life, based on anatomical and taphonomical differences. The skull of the smaller species, *Hippopotamus lemerlei*, has many analogies with that of the extant African hippo, *Hippopotamus amphibius*, namely the more elevated eye sockets and skull proportions. She thus suggested that this species led an amphibious life. In contrast, the skull of the larger species, *Hippopotamus madagascariensis*, shows some similarities with that of the extant African dwarf hippo, *Choeropsis liberiensis*, which is mainly terrestrial in its habits (see for its skull shape, van der Geer and colleagues, 2018). Stuenes thus suggested that *madagascariensis* had been more terrestrial than the larger *lemerlei*. In addition, the anatomy of the jaw, the wear facets on the incisors, and the degree of wear of the molars between the two Malagasy hippos indicate differences in mastication and thus probably in diet as well. The larger, more terrestrial species likely ate more abrasive food than the smaller and more amphibious species. Additional support for this hypothesis is provided by the geographic location of the fossil sites. The majority of the fossil remains of *Hippopotamus lemerlei* were found in the coastal lowlands, whereas those of *Hippopotamus madagascariensis* originate from the island's central highlands. There are a couple of overlaps, but according to Stuenes, this is explained as the result of mixing up of specimens in the old collections.

John Harris (1991) suggested that *Hippopotamus madagascariensis* should be placed in the genus *Hexaprotodon*, together with *Choeropsis liberiensis*, based on their anatomical similarity. While acknowledging that hippo phylogeny was unresolved, Harris proposed to maintain the generic names *Hippopotamus* and *Hexaprotodon*, till the publication of a more complete revision. This came when Jean Renaud Boisserie (2005) clearly separated *Choeropsis*, *Hexaprotodon*, and *Hippopotamus* from each other. Two years later (Boisserie, 2007), he classified all three species of Malagasy hippos within the genus *Hippopotamus*.

Hippos survived well into the Holocene on Madagascar. *Hippopotamus lemerlei* lived at least until 980 ± 200 years ago. Early human colonisation of Madagascar may have caused their extinction. MacPhee and Burney (1991) provided some evidence for this in the form of possibly butchered hippos found in south-western Madagascar and dated to as early as the first century. The youngest specimen, unfortunately of unknown provenance, dates to the period of European colonisation. The range of co-occurrence of humans and hippos on Madagascar depends on the time of the first human colonisation of the island. The first colonisation is estimated to have taken place around 2000 years ago. Some works, however, have indicated much earlier dates of colonisation. For example, James Hansford and colleagues (2005) suggested an early Holocene age. Other researchers, such as Atholl Anderson and his colleagues (2018), casted doubt on such an early arrival of humans in Madagascar and proposed a colonisation date younger than 2000 years ago. Historical accounts and folklore recorded in Madagascar mention a creature that could represent a dwarf hippo. The earliest report is that of Etienne de Flacourt, governor of Madagascar appointed by the French East Indian Company. In 1658, he published an extensive natural history of Madagascar describing the many native animals. Along with these descriptions, he provided descriptions of unknown animals as told by the indigenous people. Despite the fact that the natives claimed that these animals were still living, scientists linked them to extinct megafaunal elements such as the elephant bird, *Aepyornis*, the giant lemurs and the dwarf hippos. Since the time of Flacourt, several encounters with mysterious creatures by natives have been reported, the last of which is said to have happened in 1976, published by Burney and Ramilisonina (1998). They noted that, although there is some possibility that hippos might have survived indeed till very late in remote pockets in Madagascar, there is a fair chance that these stories were simply the result of misidentification of extant animals confused by data from reworked old traditions or information which the villagers had gathered from palaeontologists.

The Lemurs

The best-known Malagasy mammals are the lemurs, an infraorder of strepsirrhine ('bend-nosed') primates. The lemurs are restricted to Madagascar (Box 11.3), whereas the other strepsirrhine – lorises, pottos, and galagos – by contrast occur outside Madagascar. The most typical feature of lemurs is the presence of a highly specialised anterior dentition in which the canine and incisors form a procumbent toothcomb. Today, a large number of lemur species inhabit the island but exactly how large that number is, is still an open question. Tattersall (1982) listed 36 species in his monograph on the Malagasy primates but this number raised since then to around different species today. However, as he noted (Tattersall, 2007), this explosion of species is mainly due to the promotion of subspecies to species level plus the recognition of several cryptic species.

The lemurs are divided into five families. Based on phylogenetic analyses of Horvath and colleagues (2008), the most primitive family is that of the aye-ayes (Daubentoniidae). Further, the families of woolly lemurs and sifakas (Indriidae), sportive lemurs (Lepilemuridae), and dwarf lemurs (Cheirogaleidae) appear to form a clade, in which the latter two are more closely related to each other than to the former. This clade is in turn a sister clade to the Lemuridae, to which it is more closely related than to the Daubentoniidae.

Subfossil species described so far (Plate 20) belong to the monkey lemurs (Archaeolemuridae), the sloth lemurs (Palaeopropithecidae), the koala lemurs (Megaladapidae), the aye-aye's (Daubentoniidae), and 'true' lemurs (Lemuridae). The first family is extinct. Two genera are included, namely, *Archaeolemur* and *Hadropithecus*. The second family is extinct as well and includes the genera *Mesopropithecus*, *Babakotia*, *Archaeoindris*, and *Palaeopropithecus* (Plate 23). The same holds for the third family, which is considered the sister taxon of both the living families Lepilemuridae (sportive lemurs) and the Lemuridae. It contains one genus only, *Megaladapis*. The fourth family is known from two species of the genus *Daubentonia*, one extinct and one living. The fifth and last family is extant and includes a single subfossil genus, *Pachylemur*.

BOX 11.3

Lemurs Outside of Madagascar?

The oldest potential lemur fossil (*Bugtilemur mathesonae*) was initially reported by Laurent Marivaux and colleagues (2001) from the Oligocene of Pakistan. The specimens consisted of a few isolated teeth, which were considered to be most similar to those of the Malagasy dwarf lemurs (Cheirogaleidae). However, five years later, Marivaux and colleagues (2006) described another small-bodied adapiform primate (*Muangthanhinius siami*) – this time from the late Eocene of Thailand – which is clearly not related to the Lemuriformes. Considering the dental similarities with the earlier described enigmatic taxon from Pakistan, he questioned his original interpretation of a primitive lemur and considered the dental resemblances with the Malagasy lemurs the result of parallel evolution, testifying to a similar degree of dental specialisation instead.

The Aye-ayes (Daubentoniidae)

This family includes one extinct species, the giant aye-aye (*Daubentonia robusta*) and one living, the aye-aye (*Daubentonia madagascariensis*). The aye-aye is a nocturnal insect-eating primate with ever-growing incisors and a peculiar, long and thin middle finger which it inserts into holes in the tree to pull insects out. It attracts the insects in the wood by tapping on the tree, somewhat similar to the behaviour of woodpeckers and striped opossums. At present, there is no reason to assume a different lifestyle for the extinct giant aye-aye.

The giant aye-aye was named by Lamberton (1934). Several postcranial elements are known, found at Anavoha, Ankilitelo, Lamboharana, Tsirave, and perhaps Ampasambazimba, but no cranial material so far. Its skeleton closely resembles that of its living representative in morphology but is about a third larger in linear dimensions. William Jungers and colleagues (2008) estimated that its size ranged from 11 to 18 kg (their best estimate is 14.2 kg). Some differences in limb proportions and robusticity were noted and discussed by several authors, e.g., Lamberton (1934), Walker (1967), and Godfrey and colleagues (1995). The latter concluded that the similarities in postcranial anatomy strongly suggest similar locomotion to the extant aye-aye, though with more emphasis on slow quadrupedalism and climbing seen in the differences in the limb proportions.

Three large-sized incisors were attributed by MacPhee and Raholimavo (1988) to this extinct species. They were recovered from the collection of the Natural History Museum in Paris and had been collected in 1901 by Grandidier at Lamboharana. They had been drilled, and it is thus reasonable to assume that they were collected by humans and probably strung on necklaces.

The 'monkey lemurs' (Archaeolemuridae)

The family includes two genera and three species (*A. edwardsi*, *Archaeolemur majori*, and *Hadropithecus stenognathus*), all three known only from subfossil remains. The first species has been found in localities in the central part of Madagascar, such as Ampasambazimba, Ampoza, Morarano, and Sambaina, and have been reported from western Madagascar as well (Amparihingidro and Ambolisatra). However, Tattersall (1973b) doubted if the specimens from the latter localities should indeed be attributed to *Archaeolemur edwardsi* and suggested that the range of this species possibly was restricted to the central plateau of the island. *Archaeolemur majori* has been found in central as well as coastal localities, such as Ampoza, Andrahomana, Belo-sur-Mer, Lamboharana, Andrahomana, Itampolo, Manombo, Taolambiby, and Tsirave. *Hadropithecus stenognathus* possibly had the widest range and has been found in southern, western, and central Madagascar (Andrahomana, Belo-sur-Mer, Tsirave, Ampoza, Anavoha, Ampasambazimba, and the lower Menarandra).

In contrast to the rest of the lemurs, *Archaeolemur* and *Hadropithecus* were pronograde terrestrial quadrupeds. They were long-tailed, short-limbed, and stocky animals, neither particularly apt runners nor leapers. They had an extremely short face and a relatively large brain case. Forsyth Major and Standing considered them related to monkeys, but in reality, they are lemurs that developed characters in parallel with monkeys. Jungers and colleagues (2008) estimated the body mass of the *Archaeolemur majori* at about 18 kg (Table 11.1), of *Archaeolemur edwardsi* at about 26 kg, and of *Hadropithecus stenognathus* around 35 kg.

The anatomy and functional morphology of the Archaeolemuridae skull is best known through Tattersall's monograph of 1973 (b). *Archaeolemur* has very large central upper incisors that contact one another at the midline, lacking in this way the interincisal gap typical for strepsirhines and resulting in a monkey-like bite together with the highly modified lower incisors. The premolars form a continuous shearing blade that wears considerably faster than the molars. According to Tattersall, the masticatory apparatus of *Archaeolemur* resembles that of the baboons, to which it is phylogenetically unrelated. Monkey lemurs are considered to have been adapted to a mixed diet of herbs and hard-skinned fruits. *Hadropithecus* probably was dietarily more specialised than *Archaeolemur*.

The postcranial skeleton of Archaeolemuridae indicates that they spend more time on the ground than any other lemur, living or extinct. Based on the limb anatomy, Jungers and colleagues (2005) have shown that *Archaeolemur* had a mixed repertoire of climbing and both arboreal and terrestrial pronograde quadrupedalism and that it was less terrestrial than *Hadropithecus*. A terrestrial lifestyle for *Hadropithecus* was first suggested by Lamberton (1938). Later, Walker (1967) described it as a lemur analogue to the patas monkey (*Erythrocebus patas*) and Jolly (1970a, b) and Tattersall (1973b,

Table 11.1 Body mass estimations in kg of subfossil lemurs, based on Jungers and colleagues (2008). Body mass of the large-sized lemurs was estimated by calculating the femoral and humeral cortical areas at midshaft from radiographs.

Species	*Number of humeral (H) and femoral (F) radiographs*	*Range of individuals (kg)*	*Mean of means (kg)*
Megaladapis edwardsi	6 H, 6 F	58–110	85.1
Megaladapis grandidieri	2 H, 5 F	60–98	74.3
Megaladapis madagascariensis	4 H, 6 F	31–61	46.5
Archaeoindris fontoynontii	1F	151–188	161.2
Palaeopropithecus maximus	4 H, 3 F	27–56	45.8
Palaeopropithecus ingens	4 H, 3 F	22–56	41.5
Babakotia radofilai	4 H, 4 F	12–30	20.7
Mesopropithecus dolichobrachion	1 H, 1 F	9–18	13.7
Mesopropithecus globiceps	5 H, 6 F	6–18	11.3
Archaeolemur edwardsi	7 H, 8 F	15–35	26.5
Archaeolemur majori	2 H, 3 F	13–26	18.2
Hadropithecus stenognathus	3 H	29–41	35.4
Daubentonia robusta	3 H, 2 F	11–18	14.2
Pachylemur jullyi	4 H, 4 F	8–18	13.4
Pachylemur insignis	9 H, 9 F	6–21	11.5

1982) as analogue to the geladas (*Theropithecus gelada*), both ground-dwelling species of open environments.

The Sloth Lemurs (Palaeopropithecidae)

Sloth lemurs had long, curved digits, and most species were specialised hangers like sloths, hence their common name. Four genera have been described so far: *Palaeopropithecus*, *Mesopropithecus*, *Babakotia*, and *Archaeoindris*. Schwartz and colleagues (2002) showed that they had an unusually fast dental development, apparently in order to cope with the processing of fibrous foods, such as leaves, at a young age. The family exhibits a great range in body size, varying from 10 kg for the lightest (*Mesopropithecus*) up to about 200 kg for the heaviest (*Archaeoindris*). Sloth lemurs survived the advent of humans to Madagascar by at least 1500 years. There is evidence of human butchery of sloth lemurs in south-west Madagascar dating to at least 2000 years ago, likely shortly after the first arrival of humans on the island.

Palaeopropithecus includes three species: *Palaeopropithecus ingens*, *Palaeopropithecus maximus* (Figure 11.7), and

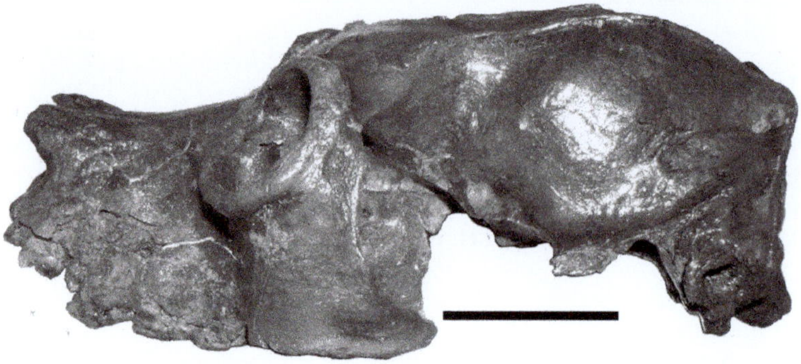

Figure 11.7 Skull of the extinct giant lemur *Palaeopropithecus maximus*, lateral view. Scale bar: 5 cm. Muséum National d'Histoire Naturelle, Paris. (Photograph: George Lyras.)

the recently discovered *Palaeopropithecus kelyus*. The first is known from several southern as well as western localities, such as Ambolisatra, Anavoha, Ankilitelo, Ankomaka, Belo-sur-Mer, and Tsiandroina. The second, on the other hand, is known from central localities such as Ampasambazimba, Antsirabe, and Masinsndraina. The third species has been found at the north-western localities Belobake and Ambongonambakoa. The early morpho-functional interpretations of the skeletal elements of *Palaeopropithecus* were quite fanciful. Herbert Standing believed that it was some sort of aquatic primate. He based that interpretation amongst others on some particular features of the skull such as the elevation of the nasals and the upward orientation of the axes of the orbits. These features enabled the animal to swim with its eyes, nostrils, and ears entirely emergent. This interpretation was elaborated by the Italian palaeontologist Guiseppe Sera (1935). Three years later, he suggested that more fossil lemurs had adopted this mixed arboreal-aquatic lifestyle. However, Sera (1938) based his conclusions on postcranial elements that happened to belong to other lemur genera and on a tarsal bone of a crocodile. Today, we know that *Palaeopropithecus* was a slow vertical climber and a specialised hanger, similar to the living sloths, as revealed by the studies of Brigitte Demes and Jungers (1993) on bone geometry, of Liza Shapiro and colleagues (1994) on vertebral anatomy, and of Godfrey and colleagues (1995) on the ratios of joint surfaces. The smallest species is *Palaeopropithecus kelyus*, which had an estimated body mass of around 35 kg, according to Gommery and colleagues (2009). In addition, it seems to have eaten tougher food, such as seed, than did the two larger species.

Besides the hippopotamus subfossil remains, Alfred Grandidier also found the femur of what he thought to be a sloth. His son Guillaume (Grandidier, 1901) thus named it *Bradytherium*. In the years thereafter, many similar femurs were found at Ambolisatra, but never a skull. Standing (1908) noted that these femora belonged to giant lemurs instead. In later publications, Standing (1910, 1913), Carleton (1936), and Lamberton (1947), the so-called sloth material was correctly attributed to the sloth lemur *Palaeopropithecus.*

Mesopropithecus includes the three species *Mesopropithecus globiceps*, *Mesopropithecus pithecoides*, and *Mesopropithecus dolichobrachion.* The first species has been found in several southern and south-western localities, such as Anavoha, Belo-sur-Mer, Taolambiby, Tsiandroina, and Tsirave. The second and the third are known from the central localities, Ampasambazimba and Antsirabe for the second and Antananarivo for the third.

Babakotia is known by one species only, *B. radofilai.* The morphology of its teeth and its dental formula resemble those of the indriids and the other sloth lemurs. The analysis of Jungers and colleagues (1991) places *Babakotia* as the sister species of the *Palaeopropithecus* – *Archaeoindris* group. Its postcranial anatomy suggests that it was primarily a slow climbing and hanging form, not as specialised as the sloth-like *Palaeopropithecus*, though it was clearly well along this adaptive path, according to Jungers and colleagues.

Archaeoindris includes only one species, the formidable *Archaeoindris fontoynontii.* Subfossils of this species have been found in Ampasambazimba. Due to its huge body mass of approximately 200 kg – coinciding with the upper range of male gorillas – it likely was more terrestrial than the other sloth lemurs.

The Koala Lemurs (Megaladapidae)

Koala lemurs are supposed to have paralleled the Australian koala in locomotion and lifestyle, hence their common name. Three species are known, *Megaladapis edwardsi* (Figure 11.8), *Megaladapis madagascariensis*, and *Megaladapis grandidieri*.

Megaladapis is the most unusual looking lemur. Forsyth Major (1893) noted 'a superficial examination of the skull will certainly not suggest its classification among Lemuroidea'. Indeed, the skull of *Megaladapis* is narrow with a very elongated muzzle, a relatively small braincase, and a very well-developed frontal sinus. The paraoccipital processes are large, elongated, and protruding, facilitating the attachment of enlarged digastric masticatory muscles. The auditory bulla is not inflated. Forsyth Major explained the unusual skull shape of *Megaladapis* as an evolutionary adaptation parallel to the living Australian koalas (*Phascolarctos*). The skull's elongation, the retroflexion of the facial axis relative to the plane of the skull base, and the posteriorly facing occipital condyles and foramen magnum are shared with koalas and are suggestive of a similar lifestyle. The combination of the retroflexed face and

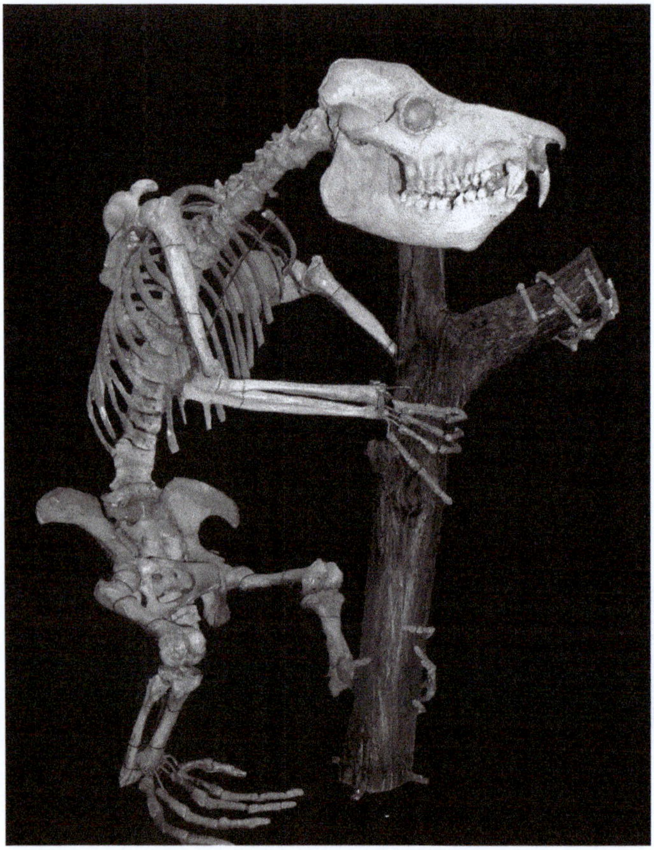

Figure 11.8 Mounted skeleton of the giant koala lemur *Megaladapis edwardsi*. Muséum National d'Histoire Naturelle, Paris. (Photograph: George Lyras.)

the backwardly oriented occipital condyles was explained by Tattersall (1972) as part of a functional complex, merging the head and neck as it were so that the animal could browse a larger area without leaving its resting position. The long nasals, projecting beyond the anterior end of the palate, indicate the presence of a mobile snout. The hands and feet were very large with long, divergent thumb and big toe and moderately curved proximal phalanges, whereas the limbs were short and very robust. This combination indicates that it clung to tree trunks and branches with all four limbs like koalas as shown by Walker (1967). Roshna Wunderlich and colleagues (1996) described some newly discovered foot remains of *Megaladapis* that reinforced this model.

The Common Lemurs (Lemuridae)

The common lemurs include one extinct genus with two species, *Pachylemur insignis* and *Pachylemur jullyi*, though Frederick Szalay and Eric Delson (1979) considered the two species synonymous. The first species is known from several sites, such as Ambolisatra, Anavoha, Andrahomana, Ampoza-Ankazoabo, Belo-sur-Mer, Bemafandry, Lamboharana, Tsiandroina, and Tsirave. The second species is only known from central localities, such as Ampasambazimba, Antsirabe, and Morarano-Betafo. The teeth of *Pachylemur* are similar in morphology to those of the living ruffed lemurs (*Varecia*). It differs from the latter by a much larger body size and far more robust build. Furthermore, its limbs are shorter and have different proportions and its orbits face more forwards. *Pachylemur insignis* is the smaller of the two with a body mass of about 10 kg.

The Tenrecs

The family of the tenrecs, or Malagasy hedgehogs, consists of about 28 living species. They were grouped by Link Olson and Goodman (2003) into three subfamilies, the Tenrecinae, the Geogalinae, and the Oryzorictinae. Despite their amazingly large diversity, they appear to have originated from a single colonisation event, followed by an extensive adaptive radiation. This radiation gave rise to a number of specialised forms, all insect eaters, as shown in detail by Olson and Goodman (2003). The body plans of several kinds of Eurasian insectivores can be recognised in those of the tenrecs. Some resemble true hedgehogs, some mole shrews, some terrestrial shrews, and some the semiaquatic desmans. In stark contrast to the large array of living tenrecs, only one extinct subfossil tenrec (*Microgale macpheei*) is known from Madagascar. It can be distinguished from the other shrew tenrecs by several osteological and dental characters and is most closely related to the extant short-tailed shrew tenrec (*Microgale brevicaudata*), according to Goodman and colleagues (2007a).

The Malagasy Rodents

The rodents of Madagascar form an endemic subfamily (Nesomyinae) within the Muridae. The nine living genera are widely distinct from each other, evidence of a strong adaptive radiation, ranging from the terrestrial gerbil-like *Macrotarsomys*, the arboreal dormouse-like *Eliurus*, the subterranean vole-like *Brachyuromys*, to the rabbit-like giant rat *Hypogeomys*. Based on their distinctiveness, the different genera have been related to several different muroid subfamilies in the past, implying a polyphyletic origin. Jean-Yves Du Bois and colleagues (1996) placed them close to the African Cricetomyinae and suggested a monophyletic origin instead. This is confirmed by their molar pattern, which can be derived from a primitive four-cusped cricetid pattern as shown by Francis Petter (1990).

They further suggested that the adaptive radiation (cladogenesis) within the Nesomyinae started around 12 million years ago, the presumed timing of arrival on the island of the single invader. This period coincides with the Middle Miocene sea level lowering. Poux and colleagues calculated a much earlier colonisation (20–25 million years ago), but Du Bois showed that though the ribosomal RNA clock seems to suggest an age of 20 million years ago for the first intra-nesomyine split, the DNA hybridisation values predict that this happened much later, at c. 8–9 million years ago. It appears that the 12S ribosomal RNA gene evolved faster in the Nesomyinae than in the other tested murids. In addition, such an early date is unlikely seen in the light of the fact that advanced cricetids and murids do not appear in the fossil record before 14 million years ago.

The only fossil Malagasy rodents discovered belong to the extant big-footed mouse *Macrotarsomys petteri*. Fossils of a *Macrotarsomys* were reported from sediments dated to the Late Pliocene or Early Pleistocene by Sabatier and Legendre (1985). However, as commented by Goodman and Soarimalala (2005), the basis of their dating is vague, and the deposits may be much more recent. More sound evidence for subfossil big-footed mice is provided by Goodman and colleagues (2006) from Andrahomana cave, who attribute it to Petter's big-footed mouse.

The Carnivora

There are seven genera of endemic Malagasy carnivores, all members of the endemic family Eupleridae. Recent molecular studies by Yoder and colleagues (2003) and Yoder and Flynn (2003) strongly support the model that the Malagasy carnivores form a monophyletic group that is sister to the cosmopolitan mongoose family Herpestidae. The fossa (*Cryptoprocta ferox*) (Figure 11.9) and the Malagasy civet (*Fossa fossana*) are believed to be the most ancient members within this group. The Malagasy carnivores are unknown from the fossil record, with perhaps an exception for the fossa. Subfossil remains from Andrahomana cave have been attributed to *C. spelaea* by Burney and colleagues (2008). Grandidier (1902), however, mentions a *C. ferox* nov. var. *spelaea* of large size from the same cave, simply indicating a relatively large form of the living fossa found there.

The Bats

At present, about 60% of the bats of Madagascar are endemic, including one endemic family, the sucker-footed bats (Myzopodidae), with two members, *Myzopoda aurita* and *Myzopoda*

Figure 11.9 Lower jaw of the fossa (*Cryptoprocta ferox*). Scale bar: 5 cm. Natural History Museum, London. (Photograph: George Lyras.)

schliemanni, both living in different habitats, the first in the humid eastern forests and the second in the dry western forests, as described by Goodman and colleagues (2007b). The most remarkable feature of these two bats is the presence of suckers, being large and flat adhesive organs attached to the thumb and hind foot (the anatomy of these bats has been extensively studied by Harald Schliemann in the 1970s). The suckers are supposed to help the bats climb onto the slick banana-like leaves of *Ravenala* and other large broad-leaf plants which they use as day roosts, similar to what is observed in the South- and Central-American disc-winged bats (*Thyroptera*). The latter evolved remarkably similar suction discs on wrists and ankles which they use to grip the slick leaves of amongst others *Heliconia*.

The evolutionary history of the Malagasy bat fauna is largely unknown due to the nature of the fossil record. Although bat remains have frequently been recovered from palaeontological sites, they have been rarely identified or described. The oldest bat remains come from Anjohibe cave in north-western Madagascar and are dated to about 80,000 years ago. Karen Samonds (2007) describes the fossil bats from this cave and identifies *Rousettus*, *Eidolon*, *Hipposideros*, *Triaenops*, and *Myotis*. She describes two new subfossil species (*Triaenops goodmani* and *Hipposideros besaoka*), indicating that Malagasy bats experienced a recent species turnover. The latter species was the largest insectivorous bat of Madagascar known so far. Samonds further notes that the subfossil bats tend to be somewhat larger than their extant relatives.

References

Ali J.R., Huber M. 2010. Mammalian biodiversity on Madagascar controlled by ocean currents. *Nature*, 463: 653–656.

Anderson A., Clark G., Haberle S., Higham T., Nowak-Kemp M., Prendergast A., Prendergast A., Radimilahy C., Rakotozafy L.M., Ramilisonina,
Schwenninger J.L., Virah-Sawmy M., Camen M. 2018. New evidence of megafaunal bone damage indicates late colonization of Madagascar. *PLoS ONE*, 13 (10): e0204368.

Arnason U., Gullberg A., Burguete A.S., Janke A. 2000. Molecular estimates of primate divergences and new hypotheses for primate dispersal and the origin of modern humans. *Hereditas*, 133: 217–228.

Benoit J., Lehmann T., Vatter M. Lebrun R., Merigeaud S., Costeur L., Tabuce R. 2015. Comparative anatomy and three-dimensional geometric-morphometric study of the bony labyrinth of Bibymalagasia (Mammalia, Afrotheria). *Journal of Vertebrate Paleontology*, 35 (3): e930043.

Boisserie J.R. 2005. The phylogeny and taxonomy of Hippopotamidae (Mammalia: Artiodactyla): A review based on morphology and cladistic analysis. *Zoological Journal of the Linnean Society*, 143 (1): 1–26.

Boisserie J.R. 2007. Family Hippopotamidae. In: Prothero D.R., Foss S.E. (eds), *The Evolution of Artiodactyls.* Johns Hopkins University Press, Baltimore, pp. 106–119.

Buckley M. 2013. A molecular phylogeny of *Plesiorycteropus* reassigns the extinct mammalian order "Bibymalagasia." *PLoS ONE*, 8: e59614.

Burney D.A., Burney L.P., Godfrey L.R., Jungers W.L., Goodman S.M., Wright H.T., Jull A.J.T. 2004. A chronology for late prehistoric Madagascar. *Journal of Human Evolution*, 47: 25–63.

Burney D.A., Ramilisonina. 1998. The kilopilopitsofy, kidoky, and bokyboky: Accounts of strange animals from Belo-sur-Mer, Madagascar, and the megafaunal "extinction window." *American Anthropologist*, 100 (4): 1–10.

Burney D.A., Vasey N., Godfrey L.R., Ramilisonina, Jungers W.L., Ramarolahy M., Raharivony L. 2008. New findings from Andrahomana Cave, southeastern Madagascar. *Journal of Cave and Karst Studies*, 70 (1): 13–24.

Carleton A. 1936. The limb bones and vertebrae of the extinct lemurs of Madagascar. *Proceedings of the Zoological Society of London*, 106: 281–307.

Chanudet C. 1975. *Conditions géographiques et archéologiques de la disparition des subfossiles malgaches.* Mémoire de Maîtrise, Université de Bretagne Occidentale, Brest.

Collignon M., Cottreau J. 1927. Paléontologie de Madagascar: Fossiles du Miocene marin. *Annales de Paleontologie*, 16: 135–170.

de Flacourt E. 1658. *Histoire de la grande isle de Madagascar composée par le Sieur de Flacourt.* 2 vols. G. Clouzier, Paris.

Demes B., Jungers W.L. 1993. Long bone cross-sectional dimensions, locomotor adaptations and body size in prosimian primates. *Journal of Human Evolution*, 25: 57–74.

Du Bois J.-Y., Rakotondravony D., Hänni C., Sourrouille P., Catzeflis F.F. 1996. Molecular evolutionary relationships of three genera of Nesomyinae, *endemic rodent taxa from Madagascar. Journal of Mammalian Evolution*, 3: 239–260.

Faure M., Guerin C. 1990. *Hippopotamus laloumena* nov. sp., la troisième éspece d'hippopotame holocène de Madagascar. *Comptes Rendus de l'Académie des Sciences de Paris (Série 11)*, 310: 1299–1305.

Filhol H. 1895. Observations concernant les mammifères contemporains des *Aepyornis* à Madagascar. *Bulletin du Muséum National d'Histoire Naturelle*, 1: 12–14.

Flynn J.J., Wyss A.R. 2003. Mesozoic terrestrial vertebrate faunas: The early history of Madagascar's vertebrate diversity. In: Goodman S.M., Benstead J.P. (eds), *The Natural History of Madagascar*. University of Chicago Press, Chicago, pp. 34–40.

Forsyth Major C.I. 1893. On *Megaladapis madagascariensis*, an extinct gigantic lemuroid from Madagascar. *Proceedings of the Royal Society*, 54: 176–179.

Forsyth Major C.I. 1897. On the general results of a zoological expedition to Madagascar in 1894–96. *Proceedings of the Zoological Society of London (for the year 1896)*, pp. 971–981.

Forsyth Major C.I. 1900. A summary of our present knowledge of extinct primates from Madagascar. *Geological Magazine (Decade 4)*, 7: 492–499.

Fovet W., Faure M., Guérin C. 2011. *Hippopotamus guldbergi* n. sp.: Révision du statut d'*Hippopotamus madagascariensis* Guldberg, 1883, après plus d'un siècle de malentendus et de confusions taxonomiques. *Zoosystema*, 33 (1): 61–82.

Godfrey L.R., Jungers W.L. 2003. The extinct sloth lemurs of Madagascar. *Evolutionary Anthropology*, 12: 252–263.

Godfrey L.R., Simons E.L., Chatrath P.S., Rakotosamimanana B. 1990. A new fossil lemur (*Babakotia*, Primates) from northern Madagascar. *Comptes Rendus de l'Académie des Sciences de Paris (Série 2)*, 310: 81–87.

Godfrey L.R., Sutherland M.R., Paine R.R., Williams F.L., Boy D.S., Vuillame-Randriamenantena, M. 1995. Limb joint surface areas and their ratios in Malagasy lemurs and other mammals. *American Journal of Physical Anthropology*, 97: 11–36.

Gommery D., Ramanivosoa B., Tombomiadana-Raveloson S., Randrianantenaina H., Kerloc'h P. 2009. Une nouvelle espèce de lémurien géant subfossile du Nord-Ouest de Madagascar (*Palaeopropithecus kelyus*, Primates). *Comptes Rendus Palevol*, 8 (5): 471–480.

Gommery D., Sénégas F., Mein P., Tombomiadana S., Ramanivosoa B., Cauvin J., Cauvin C. 2003. Les résultats préliminaires des prospections dans la presquîle de Narinda, Madagascar. *Comptes Rendus Palevol*, 2: 639–648.

Goodman S.M., Ganzhorn J.U. 2004. Biogeography of lemurs in the humid forests of Madagascar: The role of elevational distribution and rivers. *Journal of Biogeography*, 31: 47–55.

Goodman S.M., Jungers W. L. 2014. *Extinct Madagascar: Picturing the Island's Past*. University of Chicago Press, Chicago.

Goodman S.M., Soaramalala V. 2005. A new species of *Macrotarsomys* (Rodentia: Muridae: Nesomyinae) from southwestern Madagascar. *Proceedings of the Biological Society of Washington*, 118: 450–464.

Goodman S.M., Vasey N., Burney D.A. 2006. The subfossil occurrence and paleoecological implications of *Macrotarsomys petteri* (Rodentia: Nesomyidae) in extreme southeastern Madagascar. *Comptes Rendus Palevol*, 5(8): 953–962.
Goodman S.M., Vasey N., Burney D.A. 2007 (a). Description of a new species of subfossil shrew tenrec (Afrosoricida: Tenrecidae: *Microgale*) from cave deposits in southeastern Madagascar. *Proceedings of the Biological Society of Washington*, 120 (4): 367–376.
Goodman S.M., Rakotondraparany F., Kofoky A. 2007 (b). The description of a new species of *Myzopoda* (Myzopodidae: Chiroptera) from western Madagascar. *Mammalian Biology - Zeitschrift für Säugetierkunde*, 72 (2): 65–81.
Goodman S.M., Soarimalala V. in press. Introduction to mammals. In: Goodman S.M. (ed.), *The New Natural History of Madagascar*. Princeton University Press, Princeton.
Grandidier A. 1869. Descriptions de quelques animaux nouveaux découverts pendant l'année 1869, sur la côte ouest de Madagascar. *Revue et Magasin de Zoologie, Paris (Série 2)*, 21: 337–342.
Grandidier A. 1971 "*Souvenirs de voyages d'Alfred Grandidier: 1865–1870 (d'après son manuscrit inédit de 1916)*. Association malgache d'archéologie, Tananarive.
Grandidier A., Filhol H. 1894. Observations relatives aux ossements d'hippopotames trouvés dans le marais d'Ampolisatra à Madagascar. *Annales des Sciences Naturelles -Zoologie et Paléontologie*, 16: 151–190.
Grandidier G. 1899. Description d'ossements de lémuriens disparus. *Bulletin du Muséum National d'Histoire Naurelle, Paris*, 5: 344–348.
Grandidier G. 1901. Un nouvel édentè subfossile de Madagascar. *Bulletin du Muséum d'Histoire Naturelle*, 7: 54–56.
Grandidier G. 1902. Observations sur les lémuriens disparus de Madagascar: Collections Alluaud, Gaubert, Grandidier: *Bulletin du Muséum d'Histoire Naturelle*, 8: 497–505, 587–592.
Grandidier G. 1903. Description de l'*Hypogeomys australis*, une nouvelle espèce de rongeur subfossile de Madagascar. *Bulletin du Muséum d'Histoire Naturelle*, 9: 13–15.
Grandidier G. 1905. Recherches sur les lémuriens disparus et en particulier ceux qui vivaient à Madagascar. *Nouvelle Archives du Muséum d'Histoire Naturelle (Série 4)*, 7: 1–142.
Guldberg G.A. 1883. Undersøgelser over en subfossil flodhest fra Madagascar. *Videnskabs-selskabets forhandlinger, Christiania*, 6: 1–24.
Hansford J.P., Turvey S.T., 2018. Unexpected diversity within the extinct elephant birds (Aves: Aepyornithidae) and a new identity for the world's largest bird. *Royal Society Open Science*, 5 (9): 181295.
Hansford J.P., Wright P.C., Rasoamiaramanana A., Pérez V.R., Godfrey L.R. Errickson D., Thompson T., Turvey S.T. 2018. Early Holocene human presence in Madagascar evidenced by exploitation of avian megafauna. *Science Advances*, 4 (9): 6925.
Harris J.M. 1991. Family Hippopotamidae. In: *Koobi Fora Research Project. 3. The Fossil Ungulates: Geology, Fossil Artiodactyls and Paleoenvironments*. Clarendon Press, Oxford, pp. 31–85.

Horvath J.E., Weisrock D.W., Embry S.L. Fiorentino I., Balhoff J.P., Kappeler P., Wray G.A., Willard H.F., Yoder A.D. 2008. Development and application of a phylogenomic toolkit: Resolving the evolutionary history of Madagascar's lemurs. *Genome Research*, 18: 489–499.

Jenkins P.D., Carleton M.D. 2005. Charles Immanuel Forsyth Major's expedition to Madagascar, 1894 to 1896: Beginnings of modern systematic study of the island's mammalian fauna. *Journal of Natural History*, 39 (20): 1779–1818.

Jernvall J., Wright P.C., Ravoavy F.L., Simons E.L. 2003. Report on findings of subfossils at Ampoza and Ampanihy in southwestern Madagascar. *Lemur News*, 8: 21–23.

Jolly C.J. 1970 (a). *Hadropithecus*: A lemuroid small-object feeder. *Man (New Series)*, 5: 619–626.

Jolly C.J. 1970 (b). The seed-eaters: A new model of hominid differentiation based on a baboon analogy. *Man (New Series)*, 5: 5–26.

Jungers W.J., Demes B., Godfrey L.R. 2008. How big were the "giant" extinct lemurs of Madagascar? In: Fleagle J.G., Gilbert C.C. (eds), *Elwyn Simons: A Search for Origins*. Springer, New York, pp. 343–360.

Jungers W.L., Godfrey L.R., Simons E.L., Chatrath P.S., Rakotosamimanana B. 1991. Phylogenetic and functional affinities of *Babakotia radofilai*, a new fossil lemur from Madagascar. *Proceedings of the National Academy of Sciences of the United States of America*, 88: 9082–9086.

Jungers W.L., Lemelin P., Godfrey L.R., Wunderlich R.E., Burney D.A., Simons E.L., Chatrath P.S., James H.F., Randria G.F.N. 2005. The hands and feet of *Archaeolemur*: Metrical affinities and their functional significance. *Journal of Human Evolution*, 49: 36–55.

Krause D.W., Hoffmann S., Hu Y. Wible J.R., Rougier G.W., Kirk E.K., Groenke J.R., Rogers R.R., Rossie J.B., Schultz J.A., Evans A.R., von Koenigswald W., Rahantarisoa L.J. 2020. Skeleton of a Cretaceous mammal from Madagascar reflects long-term insularity. *Nature*, 581: 421–427.

Krause D.W., Hoffmann S., Wible J.R., Kirk E.C., Schultz J.A. von Koenigswald W., Groenke J.R., Rossie J.A., O'Connor P.M., Seiffert E.R., Dumont E.R., Holloway W.L., Rogers R.R., Rahantarisoa L.J., Kemp A.D., Andriamialison H. 2014. First cranial remains of a gondwanatherian mammal reveal remarkable mosaicism. *Nature*, 515: 512–517.

Krause D.W., Sertich J.J.W., O'Connor P.M., Rogers K.C. Rogers R.R. 2019. The Mesozoic biogeographic history of Gondwanan terrestrial vertebrates: Insights from Madagascar's fossil record. *Annual Review of Earth and Planetary Sciences*, 47: 519–553.

Lamberton C. 1934. Contribution à la connaissance de la faune subfossile de Madagascar. Lémuriens et ratites. *Mémoires de l'Académie Malgache (Nouvelle Série)*, 17: 9–39, 40–6, 47–105 and 123–168.

Lamberton C. 1936. Nouveaux lémuriens fossiles du groupe des Propithèques et l'intérêt de leur découverte. *Bulletin du Muséum National d ' Histoire Naturelle (Série 2)*, 8: 370–373.

Lamberton C. 1938. Contribution à la connaissance de la faune subfossile de Madagascar. Note III. Les Hadropithèques. *Bulletin de l'Académie Malgache (Nouvelle Série)*, 20: 127–170.

Lamberton C. 1946. Contribution à la connaissance de la faune subfossile de Madagascar. Note XV. *Plesiorycteropus madagascariensis* Filhol. *Bulletin de l'Académie Malgache (Nouvelle Série)*, 25: 25–53.

Lamberton C. 1947. Contribution à la connaissance de la faune subfossile de Madagascar. Note XVI: *Bradytherium* ou Paleopropithèque? *Bulletin de l'Académie Malgache (Nouvelle Série)*, 26: 89–140.

Lamberton C. 1948. Contribution à la connaissance de la faune subfossile de Madagascar. *Note XVII. Le Pachylemurs. Bulletin de l'Académie Malgache (Nouvelle Série)*, 27: 7–22.

Lamberton C. 1957. Examen de quelques hypothèses de Sera concernant les lémuriens fosssiles et actueles. *Bulletin de l'Académie Malgache (Nouvelle Série)*, 34: 51–65.

Lorenz von Liburnau L.R. 1899. Herr Custor Dr. Ludwig v. Lorenz berichtet über einen fossilen Anthropoiden von Madagascar. *Anzeigen der Kaiserlichen Akademie der Wissenschaften, Mathematisch-naturwissenschaftliche Classe, Wien*, 27: 255–257.

MacPhee R.D.E. 1987. The shrew tenrecs of Madagascar: Systematic revision and Holocene distribution of Microgale (Tenrecidae, Insectivora). *American Museum Novitates*, 2889: 1–45.

MacPhee R.D.E. 1994. Morphology, adaptations, and relationships of *Plesiorycteropus*, and a diagnosis of a new order of eutherian mammals. *Bulletin of the American Museum of Natural History*, 220: 1–214.

MacPhee R.D.E., Burney D. 1991. Dating of modified femora of extinct dwarf *Hippopotamus* from southern Madagascar: Implications for constraining human colonization and vertebrate extinction events. *Journal of Archaeological Science*, 18: 695–706.

MacPhee R.D.E., Burney D.A., Wells N.A. 1985. Early Holocene chronology and environment of Ampasambazimba, a Malagasy subfossil lemur site. *International Journal of Primatology*, 6 (5): 461–487.

MacPhee R.D.E., Raholimavo E.M. 1988. modified subfossil aye-aye incisors from southwestern Madagascar: Species allocation and paleoecological significance. *Folia Primatologica*, 51: 126–142.

Mahé J., Sourdat M., 1972. Sur l'extinction des vertébrés subfossiles et l'aridification du climat dans le Sud-ouest de Madagascar. *Bulletin de la Société Géologique de France*, 14: 295–309.

Marivaux L., Chaimanee Y., Tafforeau P., Jaeger J-J. 2006. New Strepsirrhine primate from the late Eocene of peninsular Thailand (Krabi Basin). *American Journal of Physical Anthropology*, 130: 425–434.

Marivaux L., Welcomme J.L., Antoine P.O., Metais G., Baloch I.M., Bennami M., Chaimanee Y., Ducrocq S., Jaeger J.J. 2001. A fossil lemur from the Oligocene of Pakistan. *Science*, 294: 587–591.

Masters J.C., de Witt M.J., Asher R.J. 2006. Reconciling the origins of Africa, India and Madagascar with vertebrate dispersal scenarios. *Folia Primatologica*, 77 (6): 399–418.

Mazza P.P.A. 2014. If hippopotamuses cannot swim, how did they colonize islands? *Lethaia*, 47: 494–499.

McCall R.A. 1997. Implications of recent geological investigations of the Mozambique Channel for the mammalian colonization of Madagascar. *Proceedings of the Royal Society of London (Series B)*, 264: 663–665.

Milne Edwards A. 1868. Sur des découvertes zoologiques faites récemment à Madagascar par M. Alfred Grandidier. *Commptes rendus hebdomadaires des séances de l'Académie des Sciences, Paris*, 67: 1165–1167; also in *Annales de Sciences Naturelles* (Zoologie et Paléontologie), 10: 375–378.

Monnier L., Lamberton C. 1922. Note sur des ossements subfossiles de la region de Manajary. *Bulletin de l'Académie Malgache (Nouvelle Série)*, 3: 211–213.

Olson L.E., Goodman S.M. 2003. Phylogeny and biogeography of tenrecs. In: Goodman S.M., Benstead J.P. (eds), *The Natural History of Madagascar*. Chicago University Press, Chicago, pp. 1235–1242.

Perrier de la Bâthie H. 1927. Fruits et graines du gisement de subfossiles d'Ampasambazimba. *Bulletin de l'Académie Malgache (Nouvelle Série)*, 10: 24–25.

Petter F. 1990. Relations de parenté des rongeurs de Madagascar. *Atti Convegni Lincei*, 85: 829–837.

Poux C., Madsen O., Marquard E., Vieites D. R., de Jong W.W., Vences M. 2005. Asynchronous colonization of Madagascar by four endemic clades of primates, insectivores, carnivores and rodents as inferred from nuclear genes. *Systematic Biology*, 54: 719–730.

Rogers R.R., Hartman J.H., Krause D.W. 2000. Stratigraphic analysis of Upper Cretaceous rocks in the Mahajanga Basin, northwestern Madagascar: Implications for ancient and modern faunas. *Journal of Geology*, 108: 275–301.

Sabatier M., Legendre S. 1985. Une faune à rongeurs et chiroptères Plio-Pléistocènes de Madagascar. *Actes du 100ème Congrés national des Sociétés savantes, Montpellier, Section des Sciences*, 6: 21–28.

Samonds K.E. 2007. Late Pleistocene bat fossils from Anjohibe Cave, northwestern Madagascar. *Acta Chiropterologica*, 9 (1): 39–65.

Samonds K.E., Ernat R.A., Andrianavalona T., Domning D.P. 2019. New Miocene sirenians from Nosy Makamby, northwestern Madagascar. *Journal of Vertebrute Paleontology*, 39 (1): e1570223.

Samonds K.E., Godfrey L.R., Ali J.R., Goodman S.M., Vences M., Sutherland M.R., Irwin M.T., Krause D.W. 2013. Imperfect isolation: Factors and filters shaping Madagascar's extant vertebrate fauna. *PLoS ONE*, 8 (4): e62086.

Samonds K.E., Zalmout I.S., Irwin M.T., Krause D.W., Rogers R.R., Raharivony L.L. 2009. *Eotheroides lambondrano*, new Middle Eocene seacow (Mammalia, Sirenia) from the Mahajanga Basin, Northwestern Madagascar. *Journal of Vertebrate Paleontology*, 29 (4): 1233–1243.

Schwartz G.T., Samonds K.E., Godfrey L.R., Jungers W.L., Simons E.L. 2002. Dental microstructure and life history in subfossil Malagasy lemurs. *Proceedings of the National Academy of Sciences of the Unites States of America*, 99: 6124–6129.

Sera G.L. 1935. I caratteri morfologici di "*Paleopropithecus*" e l'adattamento acquatico primitivo dei Mammiferi e dei Primati

in particolare. *Contributo alla morphologia, all filogenesi e alla paleobiologia dei Mammifera. Archivio Italiano di Anatomia e di Embriologia*, 35: 229–370.
Sera G.L. 1938. Alcuni cratteri scheletrici di importanza ecologica e filletica nei Lemuri fossili ed attuali. *Studi sulla paleobiologia e sulla filogenesi dei Primati. Paleontographica Italiana (nuova seria 8)*, 38: 1–113.
Shapiro L., Jungers W.L., Godfrey L.R., Simons E.L., 1994. Vertebral morphology of extinct lemurs. *American Journal of Physical Anthropology*, 18 (supplement): 179–180.
Simons E.L., Godfrey L.R., Jungers W.L., Chatrath P.S., Rakotosamimanana B. 1992. A new giant subfossil lemur, *Babakotia*, and the evolution of the sloth lemurs. *Folia Primatologica*, 58: 197–203.
Simons E.L., Godfrey L.R., Jungers W.L., Chatrath P.S., Ravaoarisoa J. 1995. A new species of *Mesopropithecus* (Primates, Palaeopropithecidae) from northern Madagascar. *International Journal of Primatology*, 16: 653–682.
Simpson G.G. 1940. Mammals and land bridges. *Journal of the Washington Academy of Sciences*, 30: 137–163.
Standing H.F. 1903. Rapport sur des ossements sub-fossiles provenant d'Ampasambazimba. *Bulletin de l'Académie Malgache*, 2: 227–235.
Standing H.F. 1905. Rapport sur des ossements sub-fossiles provenant d'Ampasambazimba. *Bulletin de l'Académie Malgache*, 4: 95–100.
Standing H.F. 1908. On recently discovered subfossil primates in Madagascar. *Transactions of the Zoological Society of London*, 18: 69–162.
Standing H.F. 1909. Subfossiles provenant d'Ampasambazimba. *Bulletin de l'Académie Malgache*, 6 (published for the year 1908): 9–11.
Standing H.F. 1910. Note sur les ossements subfossiles provenant des fouilles d'Ampasambazimba. *Bulletin de l'Académie Malgache*, 7 (published for the year 1909): 61–64.
Standing H.F. 1913. Procès-verbal de la séance du 28 novembre 1912. (Reported by M. Fontoynont). *Bulletin de l'Académie Malgache*, 10: 41–44.
Stankiewicz J., Thiart C., Masters J.C., De Wit M.J. 2006. Did lemurs have sweepstake tickets? An exploration of Simpson's model for the colonization of Madagascar by mammals. *Journal of Biogeography*, 33 (2): 221–235.
Stuenes S. 1989. Taxonomy, habits and relationships of the sub-fossil Madagascan hippopotamuses *Hippopotamus lemerlei* and *H. madagascariensis*. *Journal of Vertebrate Paleontology*, 9: 241–268.
Szalay F., Delson E. 1979. *Evolutionary History of the Primates*. Academic Press, New York.
Tattersall I. 1972. The functional significance of airorynchy in *Megaladapis*. *Folia Primatologica*, 18: 20–26.
Tattersall I. 1973 (a). A note on the age of the subfossil site of Ampasambazimba, Miarinarivo Province, Malagasy Republic. *American Museum of Natural History Novitates*, 2520: 1–6.
Tattersall I. 1973 (b). Cranial anatomy of the Archaeolemurinae (Lemuroidea, Primates). *Anthropological papers of the American Museum of Natural History*, 52: 1–110.

Tattersall I. 1982. *The Primates of Madagascar*. Columbia University Press, New York.
Tattersall I. 2007. Madagascar's lemurs: Cryptic diversity or taxonomic inflation? *Evolutionary Anthropology*, 16: 12–23.
Tattersall I. 2008. Vicariance vs. dispersal in the origin of the Malagasy Mammal Fauna. In: Fleagle J.G., Gilbert C.C. (eds), *Elwyn Simons: A Search for Origins*. Springer, New York, pp. 397–408.
van der Geer A.A.E., Lyras G., MacPhee R.D.E., Mitteroecker P. 2018. From Jumbo to Dumbo: Cranial shape changes in elephants and hippos during phyletic dwarfing. *Evolutionary Biology*, 45 (3): 303–317.
van der Geer A.A.E., Anastasakis G., Lyras G.A. 2015. If hippopotamuses cannot swim, how did they colonize islands: A reply to Mazza. *Leithaia*, 48: 147–150.
Vuillaume-Randriamanantena M. 1990. *Palaeopropithecus ingens* Grandidier, 1899 synonyme de *Thaumastolemur grandidieri* Filhol, 1895. *Comptes Rendus de l'Académie des Sciences de Paris (Série 2)*, 310: 1307–1313.
Walker A.C. 1967. *Locomotor Adaptation in Recent and Fossil Madagascar Lemurs*. Unpublished PhD thesis, University of London, London.
Weston E.M., Lister A.M. 2009. Insular dwarfism in hippos and a model for brain size reduction in *Homo floresiensis*. *Nature*, 459: 85–88.
White E.I. 1930. Fossil hunting in Madagascar. *Natural History Magazine*, 15 (2): 209–235.
Wunderlich R.E., Simons E.L., Jungers W.L. 1996. New pedal remains of *Megaladapis* and their functional significance. *American Journal of Physical Anthropology*, 100 (1): 115–139.
Yoder A., Flynn J.J. 2003. Origin of Malagasy Carnivora. In: Goodman S.M., Benstead J.P. (eds), *The Natural History of Madagascar*. University of Chicago Press, Chicago, pp. 1253–1256.
Yoder A., Yang Z. 2004. Divergence dates for Malagasy lemurs estimated from multiple gene loci: Geological and evolutionary context. *Molecular Ecology*, 13: 757–773.
Yoder A.D., Burns M.M., Zehr S., Delefosse T., Veron G., Goodman S.M., Flynn J.J. 2003. Single origin of Malagasy Carnivora from an African ancestor. *Nature*, 421: 734–737.

CHAPTER TWELVE

Java

Evolution of Island Mammals: Adaptation and Extinction of Placental Mammals on Islands, Second Edition.
Alexandra van der Geer, George Lyras and John de Vos.
© 2021 John Wiley & Sons Ltd. Published 2021 by John Wiley & Sons Ltd.

Java is one of the Greater Sunda Islands. In the Early Pleistocene, the island was inhabited by an endemic insular fauna, characterised by a small hippo, deer, a dwarf gomphothere proboscidean, and a giant tortoise. During the Middle Pleistocene, gradually more and more mainland elements succeeded in reaching Java, likely in relation to sea level lowerings of glacial stages. This biozone is characterised by *Homo erectus* and a middle-sized stegodon. At the start of the Late Pleistocene, a tropical mainland fauna arrives with, amongst others, *Homo sapiens*, orangutans, and Asian elephants. This fauna continued only partially into the Holocene fauna, due to increasing aridity, and after the last glacial period, the isolation of Java.

Geology and Palaeogeography

Java, together with Sumatra and Borneo, lies on the continental shelf of Asia that is called the Sunda Shelf. Java started to uplift during the Early Pleistocene after about 2.4 million years ago (Gelasian) due to a combination of tectonic and volcanic processes. At about 1.8 million years ago (Calabrian), Java was still for the greater part below the water level. The uplift of Java was a stepwise process, starting with Proto-Java including the western and central part as far to the east as Sangiran. The uplift of the eastern part was slightly delayed. Throughout its history, Java suffered regularly from volcanic eruptions, which had a varying impact on the island's geography and on the flora and fauna. Pleistocene sea level changes, however, constituted the most important factor in the arrival and subsequent isolation of terrestrial mammals, as shown in detail by Gert van den Bergh (1999). Around 0.8 million years ago, or the start of the Middle Pleistocene, the sea level fluctuations started to become more pronounced, with high amplitudes and minimum sea levels down to 170 m below the present level. During these very low sea level stands of glacial periods, faunal immigrations to Java were easier because a sea level decrease of 40 m is already sufficient to connect Java, Sumatra, and Borneo with the mainland, as demonstrated by Harold Voris (2000). His reconstructed maps should, however, be used with some caution, as he reconstructs ancient shorelines based on present-day depth contours, neglecting the influence of tectonics and volcanism. Prior to the Middle Pleistocene, the amplitude of fluctuations was lower and sea level reached a minimum of 100 m below the present level. Between about 128,000 and 70,000 years ago (Late Pleistocene), a land corridor in the west connected Java, Sumatra, and Borneo with the mainland, permitting overland migration of tropical rain forest taxa.

Historical Palaeontology

Marine fossils had been found by, amongst others, the German physician-naturalist and traveller Franz Junghuhn in the 1850s. Junghuhn's task was to describe the geology, geography, and nature of Java and part of Sumatra by order of the colonial authorities. His richly illustrated work (Junghuhn, 1853) greatly improved the knowledge on the tropics at the time. He also drew the first geological map of Java and a precise topographical map as well. As far as fossils are concerned, Junghuhn mentioned only a fossil shark. After a second visit to Java, he (Junghuhn, 1857) reported a rich site with fossil proboscideans at the slopes of Mount Pati Ayam (Figure 12.1). Junghuhn referred to this site as the 'Battlefield of Giants' and reported four fossil species, *Elephas* (now *Mammuthus*) *primigenius*, *Mastodon* (now *Stegodon*) *elephantoides*, *Elephas* sp., and an unknown larger bovine, *Bos*. Based on the stegodon, the Dutch geologist Winand Staring (1865) concluded that the fauna must be of 'diluvial age', thus dating from before the biblical Flood. Other early findings of Javanese fossils had been reported by Baron Anton Sloet van Oldruitenborgh (1859) from Desa Teguan in the Kendeng Hills, by P.E.C. Schmulling (1864) from Desa Sangiran, and by the painter and naturalist Radèn Saléh (1867) from Mount Pandan and Kedung Lumbu near Kedung Brubus in the Kendeng Hills. These collections were sent to Leiden, the Netherlands.

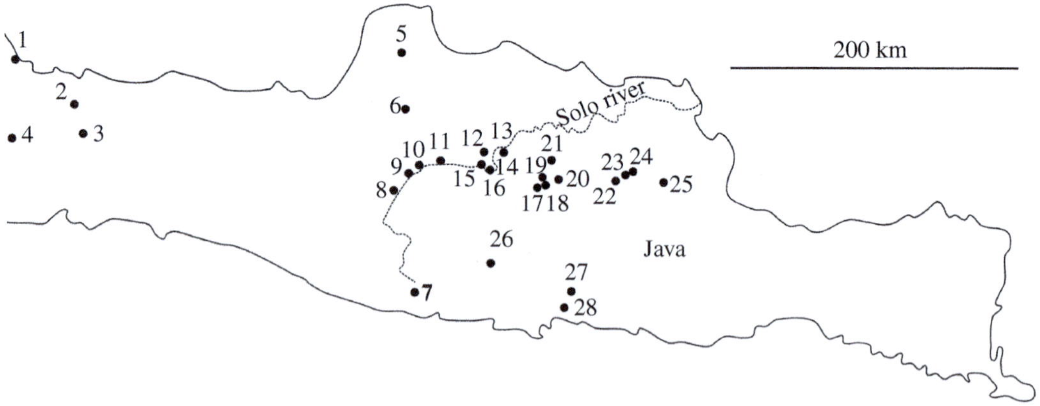

Figure 12.1 Map showing some of the Pleistocene localities of Java. (1) Cheribon, (2) Limbangan, (3) Bumiaju, (4) Tji Djulang, (5) Pati Ajam, (6) Ngembak, (7), Punung, (8) Sangiran, (9) Gesi, (10) Baringinan, (11) Sambunmacan, (12) Randublatung, (13) Tinggang, (14) Ngandong, (15) Trinil, (16) Watualang, (17) Notopuro, (18) Butak, (19) Kedung Brubus, (20) Pandan, (21) Bareng, (22) Kalibeng, (23) Kabuh, (24) Pucangan, (25) Mojokerto, (26) Sampung, (27) Tulung Agung, and (28) Wajak. (Redrawn from Theunissen and colleagues, 1990.)

Karl Martin, then professor of geology and director of the museum in Leiden, described those and other Javanese fossils between 1883 and 1890. He interpreted the mammal remains from the Kendeng Hills, Mount Pati Ayam, and Ngembak as descendants of a Neogene Indian stock. This ancient Indian fauna had been depicted (1846) and described (1868 by Murchison) by Hugh Falconer and Proby Thomas Cautley as the Siwalik fauna, after a mountain range of the outer Himalayas of the Indian subcontinent where this fauna was first found. The fauna may have reached Java and other Greater Sunda islands via Myanmar. Martin (1884, 1886) not only listed several Siwalik species for Java but also recognised two new species that were unique to the Pleistocene of Java. These were Lydekker's deer, *Cervus* (= now *Axis*) *lydekkeri* and the Javan stegodon, *Stegodon trigonocephalus* (Figure 12.2). Martin also actively bought and collected specimens from other Dutch colonies – the Antilles and the Moluccas – to enrich his museum collection.

In 1888, mining engineer B.D. van Rietschoten had found a fossil skull in a cave near the village of Wajak (old spelling Wadjak) in East Java, known as the Wajak Man (now *Homo sapiens*). The skull was passed on to the Dutch anatomist Eugène Dubois – on a quest for a transitional form between Apes and Man – for his opinion. Dubois had searched on Sumatra, where he found thousands of fossils, but only from extant mammals. This Wajak skull was the proof that human fossils could be found on Java. Dubois (1922) regarded it as different from the contemporaneous Javanese skulls, though *Homo sapiens* as well. He immediately asked for a transfer to Java where he took up his new assignment soon after. Hoping for more human fossils, he started exploring the region where van Rietschoten had discovered his Wajak Man. In the autumn of 1890, he found a second, less complete skull in the same rock shelter, as well as animal bones. He described the fossils in 1922 as *Homo wadjakensis*, in his opinion a geologically recent Australian race that was directly related to modern humans. He then extended his exploration to the open countryside of the Kendeng Hills, because earlier fossil discoveries had been made outside caves and rock shelters. One of the promising places was Mount Pati Ayam, or Junghuhn's 'Battlefield of the Giants' with fossil elephant remains. Here, however, his assistant Anthony de Winter suffered unexpected competition from the local people who dug for fossils in order to sell them as 'dragon bones' to Chinese merchants. The latter were not willing to sell, but eventually some locals sold him a few fossils without revealing the actual finding spot. De Winter asked the authorities to prevent this trade, but without much success. The commercial value of the fossils was too high to stop illegal

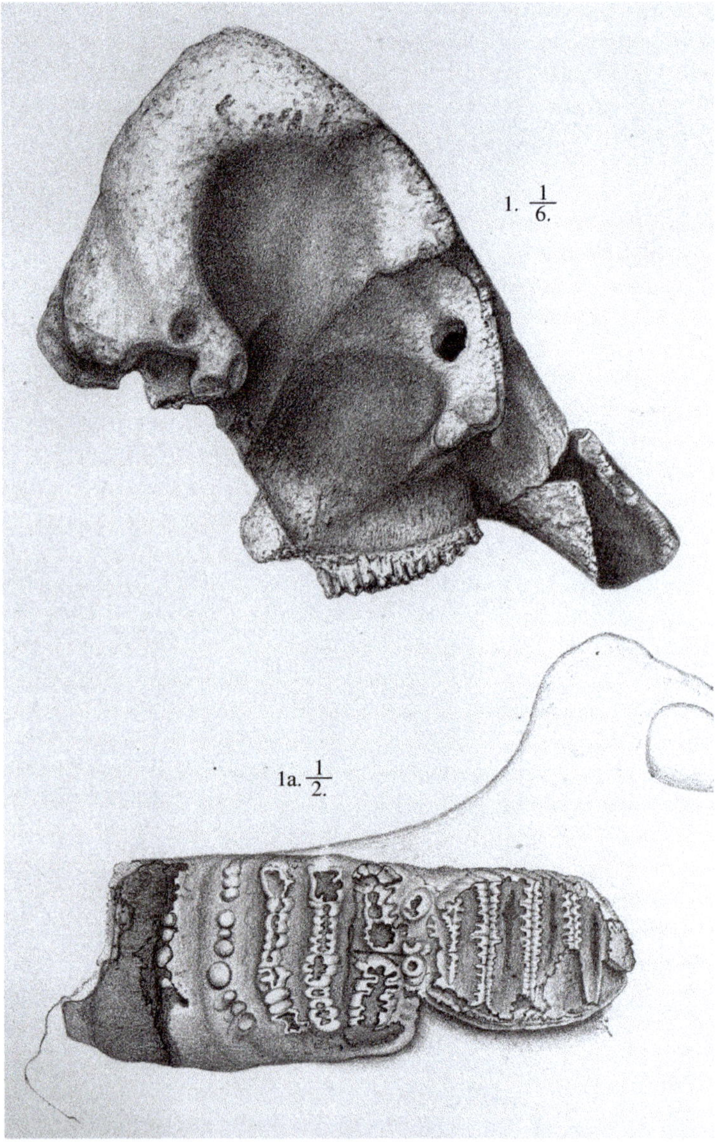

Figure 12.2 Holotype skull of *Stegodon trigonocephalus* from Java. Lateral view of the right side of the skull, and occlusal view of the molars. (From Martin, 1886.)

trade, and several excavated fossils were even stolen from De Winter's collection.

In the meantime, Dubois himself explored the open fields. In November 1890, near Kedung Lumbu at Kedung Brubus in the Kendeng Hills, he found a fragment of a mandible which he considered as belonging to a primitive type of human. Furthermore, he searched in the valley of the Solo River (=Bengawan Solo in Javanese). The riverbanks showed a clear stratigraphy and yielded fossils of extinct species like *Stegodon*, which reminded Dubois of the formations in the

Siwalik Hills of Pakistan and India. He started to collect fossils from the different strata in order to get a better picture of the prehistoric environments from before the Wajak Man. In the autumn of 1891, Dubois started excavations on the banks of the Solo River at Trinil in the Kendeng Hills. Apart from the famous fossils of his 'missing link' *Pithecanthropus erectus*, or Java Man (*Homo erectus*), he and his assistants collected an enormous number of other vertebrate fossils (Figure 12.3), up to more than 40,000 in total from different sites between 1891 and 1900 (Figure 12.4). He referred to the fauna as the Javanese Siwalik fauna, following Martin's original view that the fossil fauna of Java was practically identical to that of the Siwalik Hills. However, Dubois (1907, 1908) changed this view and introduced a new name, the Kendeng or Trinil fauna, which supposedly had a Pliocene (now Early Pleistocene) age. The main difference was that the Trinil fauna was somewhat poorer in species than the Siwalik fauna.

The German zoologist Emil Selenka planned to continue the work of Dubois in Java but unfortunately died before he

Figure 12.3 The veranda of Dubois' house is used as a storage place for the thousands of mammalian fossils excavated at Trinil, Java. (Photograph: Eugène Dubois, 1893–1894; Dubois Archive, Naturalis Biodiversity Center, the Netherlands.)

Figure 12.4 The excavation by Dubois of the site Trinil in 1900. (Photographer unknown; Dubois Archive, Naturalis Biodiversity Center, the Netherlands.)

could even start the excavation. The work was taken over by his wife, zoologist and explorer Margarethe Lenore Selenka, who led the Trinil excavations between 1906 and 1908. Dubois (1907, 1908) hastily described several new Javanese species, before Selenka could do so. He described, amongst others, the two antelopes *Tetraceros kroesenii* and *Anoa santeng*, which are now considered identical to each other and were recombined as *Duboisia santeng*, in honour of Dubois. He named another bovid species *Leptobos groeneveldtii*. One of the more spectacular findings was that of a giant pangolin (*Manis palaeojavanica*, described by Dubois, 1926), three times the size of the extant species (*Manis javanica*). Other Javanese species described by Dubois are two pigs (*Sus brachygnathus* and *Sus macrognathus*), the Javanese hippopotamus (*Hippopotamus* (=*Hexaprotodon*) *sivajavanicus*), a banteng (*Bos (Bibos) palaesondaicus*), a water buffalo (*Bubalus palaeokerabau*), and an otter (*Lutra* (=*Lutrogale*) *palaeoleptonyx*). The Selenka expedition also yielded a vast collection of fossils, but no further remains of *Pithecanthropus*. The fossils were studied and described by various palaeontologists and published in a book edited by Margarethe Selenka and Max Blanckenhorn (1911).

Elbert (1907) mentioned a river terrace at the village of Ngandong on the banks of the Solo River that turned out to

Figure 12.5 The skull of the Mojokerto child (*Homo erectus*) before preparation. (Photographer unknown; Dubois Archive, Naturalis Biodiversity Center, the Netherlands.)

be a rich deposit of fossil vertebrates. Eleven human skulls were excavated here in 1931 and 1932 by the mining engineer Willem Oppenoorth and described as Solo Man, *Homo soloensis*, named after the river (Oppenoorth, 1932). In 1936, Andoyo, a geological assistant with the Geological Survey of the Netherlands, found a skull of a child near the village of Sumbertengah, 3–4 km north of Perning in the Mojokerto district (Figure 12.5). The skull was passed on to the palaeontologist of the survey, Gustav von Koenigswald. He identified it as a juvenile early hominin and published it as the 'Mojokerto child', *Homo modjokertensis* (von Koenigswald, 1936) Although in his view it was a *Pithecanthropus*, he nonetheless published it as *Homo* to prevent a possible conflict with Dubois. He considered its geological age as older than that of the Trinil finds. The skull from Perning, however, was found at a depth of only a metre during a survey rather than in a full-scale excavation, according to Von Koenigswald. Frank Huffman and colleagues (2005) relocated the find spot and concluded that, based upon historical evidence and sediment infilling of the skull, the Mojokerto skull was, however, found *in situ* after all and not a surface find. The fission-track-based dating by Morwood and colleagues (2003) was thus performed on the correct layers.

They dated the layers immediately below and above the supposed find spot to 1.49 and 1.43 million years ago, respectively (note: seen the reworking of volcanic material on Java, these are best considered maximum ages, not absolute ages). Huffman and colleagues (2006) showed that the original dating of the Mojokerto skull to around 1.81 million years old was in error due to stratigraphic and spatial inconsistencies between the actual find spot and the sample localities of the dating. At the time of deposition, Mojokerto was a marine delta, contrary to its inland situation today.

Von Koenigswald also collected a lot of human skulls in the Sangiran area, but these findings are surface finds. He worked with a biostratigraphy that is nowadays difficult to reproduce in the field. It is therefore unknown from which formation his skulls originate. Much later, Ernst Mayr (1950) aggregated all *Pithecanthropus*, *Sinanthropus*, and similar species into the single species *Homo erectus*. At present, also, Solo Man is generally considered to be *Homo erectus*. A few large hominin fossils from Sangiran were attributed by von Koenigswald to *Meganthropus palaeojavanicus*. However, the specimens, the genus, and species are still under discussion; see also, Zanolli and colleagues (2019).

Gustav von Koenigswald (1933) described a new enigmatic mastodon from the Pleistocene, *Cryptomastodon martini*, based upon two anomalous molar fragments from Sangiran and Mount Pati Ayam and a strange partial humerus from Limbangan. Von Koenigswald had not excavated the specimens himself. The molars had already been mentioned and figured as, respectively, '*Mastodon* sp.' by Martin (1888) and '*Mastodon* (?)' by Van Es (1931). Dubois remarked that the first molar resembled the posterior portion of a *Stegodon* molar, not a mastodon molar. The humerus was published by the Swiss palaeontologist Hans Georg Stehlin (1925) as 'problematicum', because it looked basically like a proboscidean humerus but was at the same time quite specific. Von Koenigswald believed that his *Cryptomastodon* was a surviving representative of an ancient proboscidean family, isolated on the island. Soon after the publication, the new genus was under attack by Dietrich (1934), who could not accept the idea of survival of a very ancient lineage together with stegodons and elephants and raised the possibility of some sort of mechanical splitting of the so-called *Cryptomastodon* molars. Dirk Hooijer (1984) followed Dubois and Dietrich and postulated that the molars are best explained as belonging to *S. trigonocephalus*. Hooijer also had correctly identified the enigmatic humerus as belonging not to an elephant but to the giant land tortoise *Geochelone atlas* (now *Megalochelys sivalensis*) a year before. The new genus and species *C. martini* has since been reduced to a junior synonym of *Stegodon*. As pointed out by Hans van Essen and colleagues (2006), the molars are simply isolated malformed supernumerary teeth ('M4') consisting of a cluster of isolated cones and lacking the normally built and parallel transverse ridges. Such surplus molars are found regularly in elephants, and Van Essen figures four specimens belonging to the Javan stegodon.

Josephine Joordens and colleagues (2015) demonstrated that *Homo erectus* at Trinil not only consumed local large freshwater mussels (*Pseudodon vondembuschianus trinilensis*) but also used the discarded shells for tool production. One of these shells was engraved with a geometric zigzag pattern, attributed by them to *Homo erectus*, and likely produced using a shark tooth. Joordens and colleagues conclude that this is the first engraving found in Asia and the oldest in the world. The engraved shell had been discovered accidentally during a study of the century-old fossil shell collection of Dubois. The discovery renewed interest in Dubois' original *Pithecanthropus* site at Trinil, and in 2016, a new series of excavations were started by an Indonesian-Dutch team.

Biozones and Faunal Units

The first biostratigraphic schemes of the Pleistocene of Java, made by von Koenigswald (1934), were based on composite faunal assemblages and reconstructed fauna lists. Fifty years later, these schemes were replaced by John de Vos and colleagues (de Vos et al., 1982; de Vos, 1983), based on the faunal contents from single localities and taking lithostratigraphic data into consideration. This scheme is, in adapted form, still in use and recognises seven subsequent faunas (Figure 12.6). These are, from old to young, the Satir fauna (Early Pleistocene); the Ci Saat Fauna, the Trinil H.K. Fauna, the Kedung Brubus fauna, and the Ngandong Fauna (late Early and Middle Pleistocene); the Punung Fauna (Late Pleistocene); and the Wajak fauna (terminal Pleistocene – Holocene). The transitions between the individual faunas are gradual because new arrivals, speciation, and extinction of the various species are not synchronous. They rather represent time windows when major changes can be detected. The seven faunas can be grouped into three main units with considerable overlaps. Only the oldest unit is truly endemic and unbalanced; humans are lacking. The middle unit is characterised by stegodons and early humans (*Homo erectus*). This unit, the Siva-Malayan fauna, clearly shows affinities with the faunal association from the Indian subcontinent (the Siwaliks) and Myanmar but has endemics mostly on the species level. Through time, the number of species

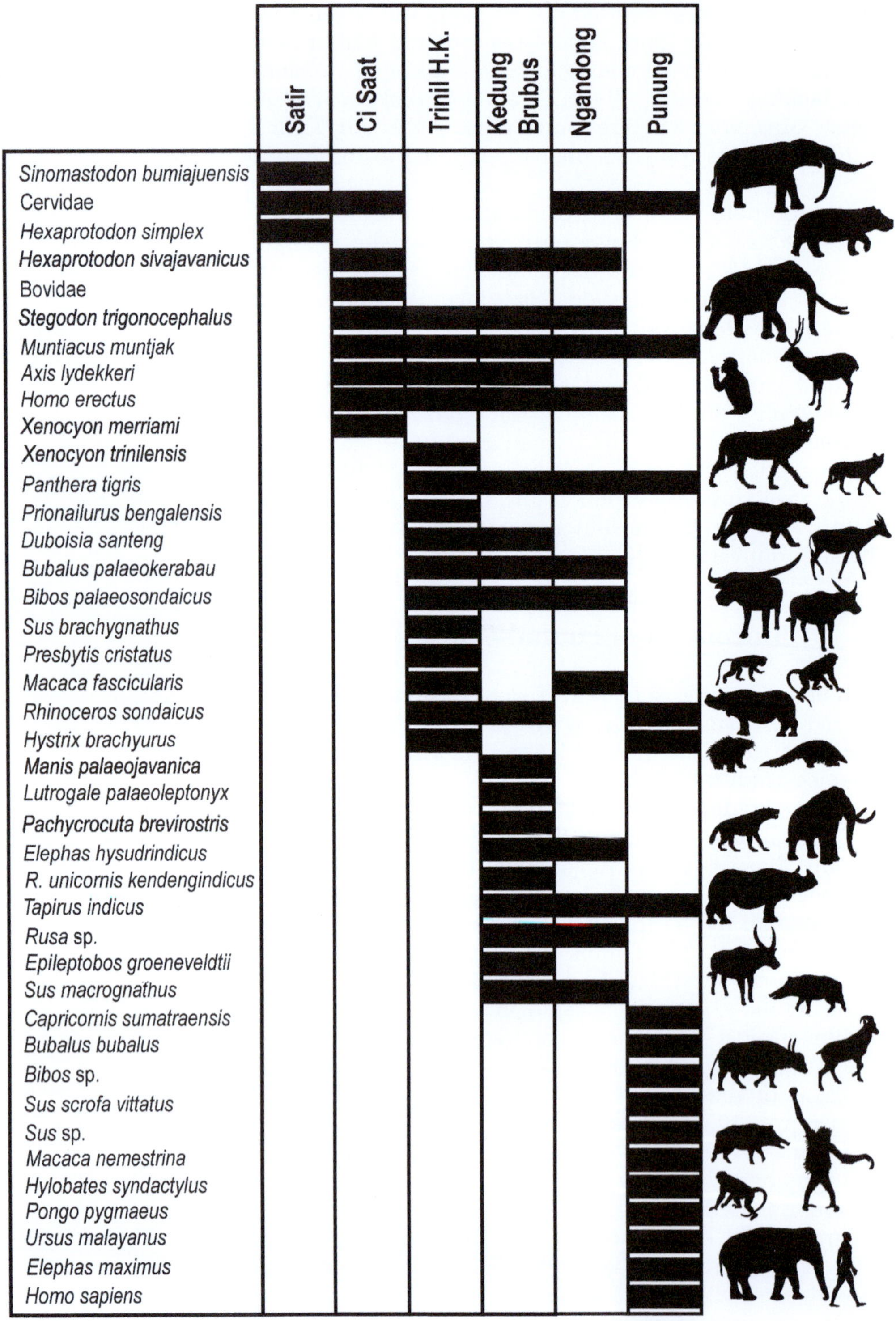

Figure 12.6 Stratigraphic scheme showing the land vertebrate faunal succession of Java. Characteristic elements of the successive faunal units are shown to the right.

increased by new arrivals. The youngest Late Pleistocene unit, the Sino-Malayan fauna, represents a new arrival of a balanced, non-endemic tropical rainforest fauna with modern humans (*Homo sapiens*) and Asian elephants.

The chronostratigraphy of Java is much less clear. At present, there is no consensus, and datings of various authors for the same locality are often in conflict, for example, the published datings for Trinil range between about 1 and 0.5 million years ago. Datings for various sites are provided by, amongst others, Watanabe and Kadar (1985), Suzuki and colleagues (1985), Carl Swisher III and colleagues (1994, 1996), Roy Larick and colleagues (2001), François Sémah and colleagues (2000), Frank Huffman and colleagues (2006), Josephine Joordens and colleagues (2015), Yan Rizal and colleagues (2020), and Matsu'ura and colleagues (2020).

Early Pleistocene

The fauna of this period is impoverished and unbalanced and consists of a gomphothere (an early elephant with primitive teeth) (*Sinomastodon bumiajuensis*), a small hippopotamus (*Hexaprotodon simplex*), deer, a giant tortoise (*Megalochelys* cf. *sivalensis*), and perhaps one or more species of dwarf stegodons. Sondaar (1984) named this fauna the Satir fauna after the village Satir in the Bumiayu area where these elements were first recognised. The same elements had also been found in the lower lahars of the black clays of the Pucangan Formation of Sangiran, which were dated by François Sémah and colleagues (2000) and E. Arthur Bettis and colleagues (2004) to about 1.67 and 1.9 million years ago for, respectively, the lower and upper boundary. The fauna is generally considered to range in age between about 1.5 and 1.8 million years ago.

The environment at the time was swampy with immense coastal mangrove forests and tropical rainforest, as can be inferred from the pollen spectrum given by Anne-Marie Sémah (1984) and Polhaupessy (1996) for Sangiran. The hippopotamus, giant tortoise, and stegodons are closely related to elements of the Siwalik and Burma faunas, indicating a Siva-Malayan origin and migration route for most of the Satir fauna. Probably, several islands existed on the Sunda Shelf during this period, allowing hippos, deer, and proboscideans to reach Java by sweepstake dispersal.

The gomphothere is the earliest proboscidean from Java. The molar cusps stand in three ridges (trilophodont) and form a clover-like pattern when worn, unlike that seen in mastodons. The unbalanced character of the fauna, with only herbivore megafauna and a giant tortoise, is indicative of isolated conditions.

Scattered findings of dwarfed elephantoids are known from Early Pleistocene deposits with unknown stratigraphic position and geological age, likely either belonging to this Satir fauna, the following Ci Saat fauna (see below), or to a contemporaneous fauna from an isolated area. Dwarf stegodons were reported from Sambungmacan (central Java), Cirebon (probably derived from archaic *S. elephantoides*), and Cariang (West Java). From Ci Panggloseran near Bumiayu, a molar of *Stegoloxodon celebensis* has been reported (Hooijer, 1974), a four-tusked dwarf elephant that seems conspecific with the species on Sulawesi.

Finally, fossils of two other Siwalik elements were retrieved from Early Pleistocene layers, also lacking stratigraphic data. These are the anthracothere *Merycopotamus dissimilis* – a hippopotamus-like even-toed ungulate – from Cijulang (East Java) and the chalicothere *Anisodon* (=*Nestoritherium*) cf. *sivalense* – a browsing odd-toed ungulate. They may, like the dwarf stegodons, pertain to the Satir fauna, the earliest stage of the following biozone (Cisaat), or an intermediate stage.

Late Early and Middle Pleistocene

The sites Ci Saat, Trinil H.K., Kedung Brubus, and Ngandong are attributed to the *Stegodon-Homo erectus* fauna association, each representing a successive stage within the biozone. The most highlighted element of the Javanese fauna is doubtlessly *Homo erectus* (see below). Although fossils of *Homo erectus* are missing in the Siwalik Hills themselves, they have been found elsewhere on the Indian subcontinent (Narmada Valley). In Java, on the contrary, several fossils were found. The biozone ranges between about 1.5 million years and 128,000 years ago – sandwiched between the Satir fauna (see above) and the Punung fauna (see below). The analysis of pollen from the Sangiran area by Anne-Marie Sémah and colleagues (2016) indicates a drier, open, seasonal woodland environment for this period with fragmentation, though not complete disappearance, of tropical rainforest and more or less extended grassland patches.

The faunal association of this period clearly shows close affinities with the faunal association from the Indian subcontinent and Myanmar, known as the Siwalik fauna. The species arrived on Java through the so-called Siva-Malayan Route by filter dispersal because not all Siwaliks elements reached Java, and a limited degree of endemism seems to have evolved. The various faunal elements did not arrive in a single event, but likely came in waves, because not all taxa are found in all four faunal stages. Besides new arrivals, there were also local extinctions of previous elements.

Figure 12.7 Skull of the antelope *Duboisia santeng* showing the characteristic horn cores. Scale bar: 5 cm. National Museum of Natural History, Leiden. (Photograph: Eelco Kruidenier.)

The species from this Javan fauna are identical to or closely related with Siwalik species and differ on the subspecies, species, or genus level. These are, to name just a few: the Javanese dwarf stegodon (*S. trigonocephalus*), the Javanese elephant (*Elephas hysudrindicus*), the Kendeng rhinoceros (*Rhinoceros unicornis kendengindicus*), and Dubois' antelope (*D. santeng*, related to *Boselaphus*) (Figure 12.7). Lydekker's deer (*Axis lydekkeri*) has a unique, inward curved antler (Figure 12.8), which makes it difficult to indicate a Siwalik ancestor. In addition, several *Cervus* species were described for the Siwaliks – *sivalensis*, *simplicidens*, and *triplidens* by Richard Lydekker (1879, 1884), *punjabiensis* by Barnum Brown (1926), and *rewati* by Mohammad Arif and colleagues (1992).

Further elements in this Javanese 'Siwalik' fauna are a water buffalo with impressive horns (*Bubalus palaeokerabau*), a banteng (*Bibos palaesondaicus*), two pigs (*Sus brachygnathus* and *Sus macrognathus*; Figure 12.9), a sambar deer (*Rusa* sp.), the Javanese rhinoceros (*Rhinoceros sondaicus*), muntjac (*Muntiacus muntjak*), and the Asian tapir (*Tapirus indicus*). Noteworthy is the first arrival of mammalian carnivores to Java during this biozone, including the tiger (*Panthera tigris*), the fishing cat (*Prionailurus bengalensis*), canids (*Xenocyon*, originally referred to *Megacyon* and *Mececyon*; at a later stage, *Cuon*), two smooth-coated otters (*Lutrogale palaeoleptonyx* [Figure 12.10] and *Lutrogale robusta*), and the short-faced hyena (*Pachycrocuta brevirostris*).

Apart from the above-mentioned taxa, dwarf stegodons seem to have occurred simultaneously in isolated areas. Van den Bergh (1999) mentioned a small-sized *Stegodon* molar (right

Figure 12.8 Complete antler of Lydekker's deer (*Axis lydekkeri*) from Trinil, early Middle Pleistocene. Scale bar: 5 cm. National Museum of Natural History, Leiden. (Photograph: Paul Albers.)

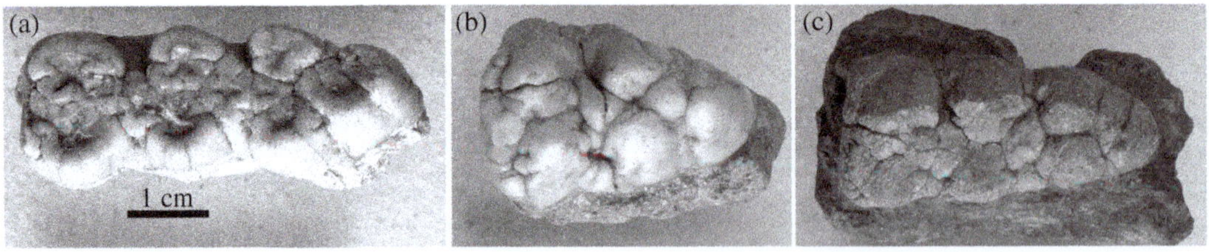

Figure 12.9 Molars of two Pleistocene pigs of Java, (a) *Sus macrognathus* and (b, c) *Sus brachygnathus*; occlusal view. Scale bar: 1 cm. National Museum of Natural History, Leiden. (Photograph: Eelco Kruidenier.)

lower third molar) from Ranji Kares in Cirebon in eastern West Java. The main differences with molars of *S. trigonocephalus* are its smaller size and very low plate number. It is very similar not only to another molar from an unknown locality but also to Von Koenigswald's subspecies *S. trigonocephalus praecursor*. Van den Bergh suggested the existence of a separate palaeo-island in the Cirebon area at the onset of the Pleistocene, inhabited by a dwarf stegodon. Periodic isolations of other

Figure 12.10 Upper P4–M1 of the Pleistocene otter *Lutrogale palaeoleptonyx* from Java. Occlusal (top), lingual (centre), and buccal (bottom) views. Scale bar: 1 cm. National Museum of Natural History, Leiden. (Photograph: Eelco Kruidenier.)

parts of Java may in his view have occurred from time to time as well due to eustatic sea level fluctuations resulting in repeated genetic isolation of *Stegodon* populations. This could have led to the evolution of distinct subspecies, which explains the existence of smaller as well as larger morphotypes.

Ci Saat Fauna

The oldest stage of the *Stegodon-Homo erectus* fauna, that of Ci Saat, is rather hypothetical. This stage is defined as recording the first occurrence of *S. trigonocephalus* on Java. Further new elements of this stage are spotted deer, muntjac, antelope (*Duboisia*), otter, tiger, and *Homo erectus*. The hippopotamus is a survivor from the previous period. The subzone ranges between about 1.5 and 0.9 million years ago, the latter estimated to be the age of Trinil H.K. as inferred from lithological correlation with the Grenzbank layer at Sangiran. Humans first reached Java during this subzone, as attested by c. 1.5 million years old remains discovered in the Pucangan clays of Sangiran (central Java), shown by Francois Sémah and colleagues (2000), flake tools – the 'Sangiran Flakes' – dated to over a million years old by Widianto (2006), and, importantly, the Mojokerto child (*Homo erectus*) dated to about 1.49 and 1.43 million years ago. The Sangiran human fossils of this stage have been redated to be at most between about 1.3 and 1.45 million years old by Matsu'ura and colleagues (2020). Pollen and spores analyses by Robert Morley and colleagues (2020) of the Mojokerto area reconstructed a landscape with a sandy delta, muddy deltas with swamps, extensive open savannah grasslands, and montane broad-leaf and plum pine forests at the volcano flanks. Their data support the age of the Mojokerto child at about 1.43 million years ago.

Apart from the remains of dwarf *S. trigonocephalus*, a mandible of a normal-sized archaic *S. elephantoides* was found near Bukuran in the Sangiran Dome *in situ*, dated at between 1.2 and 1.3 million years ago by van den Bergh (1999), and likely pertaining to the Ci Saat stage. Van den Bergh remarks that the molars of *S. elephantoides* are not easily distinguished from those of *S. trigonocephalus*, while their mandibles differ widely. The *S. elephantoides* mandible is large and has an elongated symphysis, unlike that seen in the Siwalik stegodons. *Stegodon elephantoides* is further only known from the Early Pleistocene Upper Irrawadi beds of Myanmar. The carnivore elements are difficult to place in a biostratigraphy because these are isolated finds and lack precise provenience documentation. These are the sabre-toothed cats (*Homotherium ultimum*, *Megantereon* sp., *Hemimachairodus zwierzyckii*) and Merriam's dog (*Xenocyon merriami*). The latter was found in the so-called Jetis beds, originally attributed to the late Early Pleistocene, so most likely this large canid belongs to the Cisaat faunal level. The composition of the Ci Saat fauna is impoverished though relatively unbalanced, with several cats and a dog as large mammalian predators. This is suggestive of some sort of filter dispersal between the Asian continent and Java via Sumatra.

Figure 12.11 Molar of the Javanese dwarf stegodon *Stegodon trigonocephalus*; occlusal view. Middle Pleistocene. Scale bar: 5 cm. National Museum of Natural History, Leiden. (Photograph: Eelco Kruidenier.)

Trinil H.K. Fauna

In the next stage, Trinil H.K., or 'Hauptknochenschicht' (= main bone layer), is documented in much detail, thanks to the extensive excavations by Eugène Dubois and later by Lenore Selenka. The new arrivals of this stage were, amongst others, the water buffalo, the banteng, the small-jawed pig, the Javan rhinoceros, the Trinil canid, and the Japanese macaque. The stegodon, *S. trigonocephalus* (Figure 12.11), is best explained as a continuation of the Ci Saat stegodon. The same is valid for the Trinil canid (*Xenocyon trinilensis*, originally *Mececyon*), which is likely derived from *X. merriami* (originally *Megacyon*) of the previous biozone. It differs from *Cuon* – to which genus it is often incorrectly attributed – by a series of characters the most typical of which is the presence of a lower third molar and the lesser degree in the specialisation of the m1 talonid (the endoconid is less reduced and it retains a connecting cristid to the hypoconid) (Lyras and colleagues, 2010; van der Geer and colleagues, 2018a, b).

In the Trinil H.K. stage, *Homo erectus* is undoubtedly present. Numerous human fossils were discovered after Dubois' famous findings at Trinil, most of them in the Sangiran dome near the city of Surakarta in central Java. The majority of these fossils likely originate from the early Middle Pleistocene layers. However, as they are often incidentally found by farmers, many of them lack any stratigraphic context. The area is particularly important because the deposits at Sangiran represent a complete stratigraphical sequence, ranging from the Early to the Middle Pleistocene, on the basis of which a biostratigraphy can be reconstructed. From Sangiran, many isolated molars of *Homo erectus* are known, some of which are discussed as potentially belonging to orangutan (*Pongo*). At Sangiran, though, true tropical rainforest seems not to have been present, in contrast with the Late Pleistocene Punung biozone. The stone artefacts of this period include for the first time Acheulean-like tools in sites dated between 0.8 and 1 million years ago, as reported by Francois Sémah and colleagues (1992).

Figure 12.12 Molar of the Javanese elephant *Elephas hysudrindicus*; occlusal view. Scale bar: 5 cm. Middle Pleistocene; National Museum of Natural History, Leiden. (Photograph: Eelco Kruidenier.)

Kedung Brubus Fauna

Elephant, deer, large-jawed pig, hyena, smooth-coated otters, dhole, giant pangolin, Indian rhinoceros, and tapir are first recorded from the next faunal stage, that of Kedung Brubus. The elephant, *E. hysudrindicus* (Figure 12.12), likely is a descendant of *Elephas hysudricus* from the Irrawadi beds. The fauna has an age of roughly 700,000–800,000 years.

The transition between the Trinil and Kedung Brubus Fauna appears to coincide with the opening up of the landscape and a sea level lowering, facilitating faunal exchange between the region of Java and other regions of the Sunda Shelf. The stratigraphically higher findings of *Homo erectus* at the Sangiran dome likely belong to this faunal unit, according to Matsu'ura and colleagues (2020), who dated the human fossil-bearing layers of the Bapang Formation at between about 900,000 and 790,000 years ago.

Ngandong Fauna

Hardly any new arrivals have been recorded for Ngandong, the last stage of the Javanese Siwalik *Homo erectus-Stegodon* biozone. The fauna seems instead to have evolved slightly during this period, indicating a certain degree of isolation. The *Homo erectus* bone bed in a river terrace at Ngandong dates to between 108,000 and 117,000 years ago, based on a combination of uranium-series (U-series), electron spin resonance, and luminescence and argon ($^{40}Ar/^{39}Ar$) dating by Yan Rizal and colleagues (2020). This sets the last occurrence of *Homo erectus* on Java – and likely that of other typical elements of the *Homo erectus-Stegodon* fauna as well – to before 108,000 years ago. An earlier dating by Indriati and colleagues (2011) indicated a minimum age of around 143,000 years ago for the same river terrace. Given that the terrace is 20 m in height, these estimates roughly agree.

A number of species of this stage can be distinguished on the subspecies level from those of the previous stage, such as Ngandong Man for the most-advanced form of *Homo erectus* on Java, and the *S. trigonocephalus ngandongensis*. Body size increase has been demonstrated not only for these two taxa but also for *Bubalus palaeokerabau*, *Panthera tigris soloensis*, and *Sus macrognathus terhaari*. The hippo seems to follow this general pattern as well, as has been demonstrated by Johanna de Visser (2008). Based on measurements and morphology, she concluded that hippopotami immigrated into Java as a single event and that this hippopotamus – *He. sivajavanicus* – underwent a gradual increase in body size through time. In that case, the smallest and earliest form from Satir – *He. simplex* – is a junior synonym of the former. However, the hippo is lacking in Trinil, suggestive of a second dispersal during the Kedung Brubus stage, together with other elements during a sea level low stand.

Late Pleistocene

At the end of the Middle Pleistocene, a significant faunal turnover took place and a complete new fauna migrated into the Indonesian Archipelago, following the Sino-Malayan route. This new fauna, known as the *Pongo-Homo sapiens* fauna, is a typical tropical rainforest fauna of the mainland, consisting of amongst others the Indian elephant (*Elephas maximus*), orangutan (*Pongo pygmaeus*), siamang (*Symphalangus syndactylus*), pig-tailed macaque (*Macaca nemestrina*), the sun bear (*Helarctos malayanus*), the Malay tapir (*T. indicus*), serow (*Capricornis sumatraensis*), muntjac (*Muntiacus muntjac*) and other deer (*Cervus* sp.), wild water buffalo (*Bubalus arnee*), banded pig and bearded pig (*Sus scrofa vittatus* and *Sus barbatus*), long-tailed rat (*Leopoldamys sabanus*), a porcupine (*Hystrix vanbreei* [or *Hystrix brachyura*]), and moon rat (*Echinosorex* sp.), all species which are still extant on the continent or in other regions of the Indonesian Archipelago but went extinct in historical times in Java. A few species were already present during the preceding Ngandong stage, but these belong to different subspecies, recognised by their larger size.

The large quantity of orangutan fossils and the presence of other primates indicate a humid tropical rainforest environment. A pollen and faunal analyses review by Anne-Marie Sémah and colleagues (2016) confirmed that this period, coinciding with Marine Isotope Stage (MIS) 5, was characterised by high sea levels, abundant precipitation, and an overwhelming development of the tropical rainforest. A number of species persisted from the previous biozone(s), such as the tiger, the Javan rhino, the tapir, and the muntjac, while others went extinct, amongst which *Homo erectus*, the stegodon, and the hippopotamus are the most noteworthy.

The fauna is named after two sites – Punung I and Punung II – in the Gunung Sewu limestones of East Java. Punung I, discovered in the 1930s and near Song Agung, was relocated by

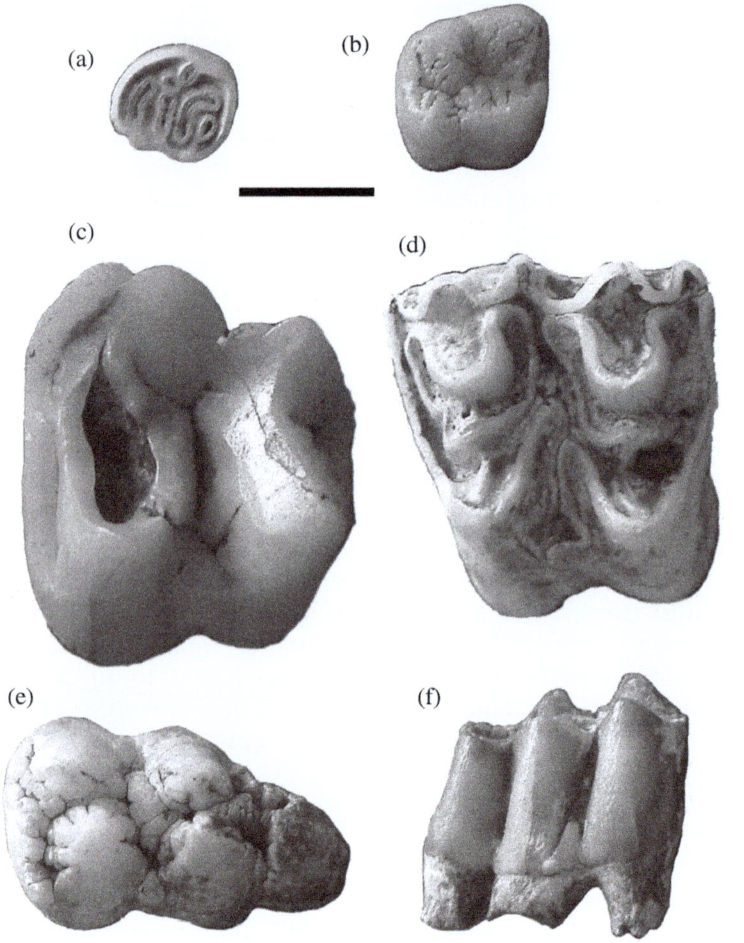

Figure 12.13 Molars from the Late Pleistocene rainforest site Punung, Java, showing the typical gnawing pattern caused by porcupines. The dental roots and the jaw have completely disappeared. Occlusal (a–e) and lingual views (f). (a) Porcupine, *Hystrix brachyurus*; (b) Orangutan, *Pongo pygmaeus*; (c) Asian tapir, *Tapirus indicus*; (d) Banteng, *Bibos banteng*; (e) Wild pig, *Sus* sp.; and (f) Muntjac, *Muntiacus muntjac*. Scale bar: 1 cm. National Museum of Natural History, Leiden. (Photographs: Eelco Kruidenier.)

Paul Storm and colleagues (2005) based on old photographs and notes. The fauna was already known since von Koenigswald had sampled two fissure fillings near Punung, one near Mendolo Kidul and one at the entrance to the Tabuhan cave. At all sites, the fossil vertebrate material consists only of dental elements from which most part of the roots had been gnawed by porcupines (Figure 12.13). Porcupines are known for their habit of collecting carcasses and bones in their dens to eat them entirely (except for the dental enamel). This taphonomic peculiarity is typical for this Late Pleistocene rainforest fauna. In 2005, a new site was found, Punung III at Gunung Dawung, with a fauna similar to Von Koenigswald's fissure fillings, a similar taphonomy, and importantly, with datable layers unlike the other sites. The fossil-bearing breccias of Punung III were dated to between 118,000 and 128,000 years ago by Westaway and colleagues

(2007), using luminescence and U-series dating methods. This age coincides with an early stage of the Last Interglacial.

A similar fauna with similar taphonomic peculiarities – that is, eaten by porcupines – has been found on Sumatra (Jambu caves and Lida Ayer, dated to about 73,000 years ago by Westaway and colleagues, 2017), Malaysia (Niah cave), Vietnam (Lang Trang cave, Duoi U'Oi cave), Cambodia (Phnom Loang), Laos (Tam Hang), and South China (Ho-shang-tung). Apparently, the connection of Java and Sumatra with the mainland had become more continuous. At 128,000 years ago, the sea level was low enough to connect at least large parts of Sumatra, Java, and Borneo to the continent, enabling the tropical rainforest fauna with *Homo sapiens* and orangutan to immigrate.

Late Pleistocene and Holocene

Gradually, the climate changed from humid to dry and the environment changes accordingly from an evergreen tropical forest to the open woodland of today. The Sunda Shelf broke up again and first Java and then Sumatra and Borneo became isolated, based upon the fauna. Every island got its own evolutionary history. In Java and the mainland, for example, the orangutan disappeared, while in Borneo and Sumatra, the taxon managed to survive, although evolving into three different species – *Pongo pygmaeus* of Borneo, *Pongo abelii* of North Sumatra, and *Pongo tapanuliensis* of South Sumatra – due to lack of genetic contact. As a result of Java's isolation, many species today are endemic, such as Kuhl's hog deer (*Hyelaphus kuhlii*), Javan deer (*Rusa timorensis*), the banteng (*Bos javanicus*), warty pig (*Sus verrucosus*), silvery lutung (*Trachypithecus cristatus*), Sunda pangolin (*M. javanica*), and plantain squirrel (*Callosciurus notatus*), while some survived from the previous period(s) (tiger, tapir, muntjac, porcupine, rhino, and humans). This local fauna, described by Louise van den Brink (1982), is known as the Wajak fauna, named after the cave where the first human fossils had been found in 1888. Laser ablation U-series dating results on human and faunal bone fragments from the cave by Paul Storm and colleagues (2013) bracket the age range between 28,500 and 37,400 years for the whole assemblage. This is older than previous absolute datings by Shutler and colleagues (2004), who found ages of between 12,140 and 12,930 years ago for the mammal fauna and only between 7,670 and 7,210 years ago for the humans.

Towards the end of the Pleistocene or during the Early Holocene, small mammals, such as mongoose (*Herpestes javanicus*), fishing cat (*Felis viverrina*), common otter (*Lutra*

lutra), and binturong (*Arctitis binturong*) must have arrived, as they are unknown before the Neolithic period.

Peculiarities and Evolution of Endemic Mammals

Homo erectus

The site Trinil of Java is where the type specimens of *Homo erectus* were found. In 1891, Dubois found here a molar and skullcap of a small-brained primate, which he considered an ape: *Anthropopithecus*, the genus of chimpanzees in those days. A year later, he discovered a femur, which was like that of modern humans, clearly indicating a bipedal locomotion. In 1893, he referred to the molar, the skullcap, and the femur as *Anthropopithecus erectus*, an upright-walking ape. Dubois was in search for a transitional form between Apes and Man. Realising that a species with a small brain yet walking upright might be such a form, he renamed his genus in 1894 into *Pithecanthropus erectus*, now *Homo erectus*. Since the beginning, there has been discussion whether the three fossils actually belong together. The femur is often considered to be identical to that of *Homo sapiens*, while the molar is alternatively assigned to orangutans (*Pongo pygmaeus*), or a large hominid, *Meganthropus*.

Many more human fossils have been discovered over the years at the Sangiran area of central Java. Basically, two groups can be distinguished: a geologically older group – originating from the uppermost Sangiran Formation to the Grenzbank zone – and a younger group – originating from the Bapang Formation. The older forms are more robust, more primitive, and with lower cranial capacity and are sometimes referred to as *M. palaeojavanicus*. The majority of anthropologists, however, consider these findings as within the variation of *Homo erectus*.

Javanese *Homo erectus* left a varied toolkit behind, including the so-called Patjitanian industry (choppers, flakes, [giant] cores), the 'Sangiran Flakes' (small siliceous flakes), Acheulian-like tools, and shell tools. They may have used shark teeth to engrave a shell, as demonstrated by Joordens and colleagues (2015).

Dwarf Elephants

Each biozone of Java is characterised by its own endemic proboscidean. The first arrivals are a gomphothere (*S. bumiajuensis*) of the Early Pleistocene (Satir biozone), followed by the Javan dwarf stegodon (*S. trigonocephalus*) of the late Early and Middle Pleistocene (Cisaat to Ngandong biozones), and finally by *E. maximus* of the Late Pleistocene (Punung). Besides the dwarf stegodon, *E. hysudrindicus* occurs towards the end of the Middle Pleistocene (Ngandong).

Apart from the four common species, scattered findings of other dwarfed elephantoids are known from Early Pleistocene deposits with unknown stratigraphic position and geological age, likely with an Early Pleistocene age (Satir or Ci Saat fauna). They may represent smaller, isolated areas of islands separate from the main island. The dwarf stegodon from Cirebon is probably derived from archaic *S. elephantoides*. Fragments of a pair of upper third molars from Jetis (East Java) were described as the pygmy *Stegodon hypsilophus* by Hooijer (1954). However, he later considered this to be a junior synonym of *Elephas celebensis* from Sulawesi, following Maglio (1973). Van den Bergh and colleagues (1992) argued against this revision, showing that the Jetis molars clearly show stegodon affinities (pygmy *Stegodon* sp.).

The elephant molar found at Ci Panggloseran (Early Pleistocene) near Bumiayu was initially described as *Elephas planifrons* by Van der Maarel (1932), later renamed to *Stegoloxodon indonesicus* by Kretzoi (1950). Van den Bergh (1999) noted a possible relation between this species and the dwarf elephant from Sulawesi. Based on van den Bergh's descriptions and comparisons, Georgi Markov and Haruo Saegusa (2008) renamed the Sulawesi species accordingly into *Stegoloxodon celebensis*.

Stegodon trigonocephalus had evolved a body mass reduction of about 25 % compared to its mainland ancestor *S. ganesa*, and weighed about 2.8 ton, as calculated by Mark Lomolino and colleagues (2013). Based on stable carbon and oxygen isotope analysis of tooth enamel, Mika Puspaningrum and colleagues (2020) reconstructed the diets and habitats of the successive proboscideans of Java. Their results suggest a grassy habitat for *Sinomastodon*, a closed canopy rainforest for *Stegoloxodon*, and a gradual expansion of C_4 plants during the *Homo erectus-Stegodon* biozone. The presence at Ngandong of *E. hysudrindicus*, feeding on both C_3 and C_4 plants, shows the reappearance of fragmented evergreen forests or woodland vegetation and, hence, a shift from a dry to a more humid climate condition. The Punung biozone finally is characterised by a wet, tropical rainforest with *E. maximus*.

Dwarf Canids

Since Stremme (1911) described the small dog from Trinil, excavated by the Selenka team, as *Mececyon* (=*Xenocyon*) *trinilensis*; several other canid fossils were discovered and reported in the early twentieth century from Early and earliest Middle Pleistocene deposits of Java (Plate 24). We revised these fossils in 2018 and showed that none can be attributed to the genus *Cuon* (van der Geer and colleagues, 2018). A hallmark of *Cuon* is the loss of a third lower molar, whereas the Javanese fossils retain this element. Early Cuon with this element, such as *Cuon dubius*, are today considered forms of *Xenocyon*. During the Early Pleistocene, a large-sized species, Merriam's dog (*X. merriami*, previously *Megacyon*), was present, which was replaced by the small-sized Trinil dog (*X. trinilensis*, previously *Mececyon*) during the earliest Middle Pleistocene. Likely, this small form evolved from the larger form of the preceding period. Merriam's dog is similar to the Eurasian continental *Xenocyon lycaonoides* not only in size but also in tooth morphology.

The body size reduction of about 70% within *Xenocyon* on Java may be the result of increased competition within the carnivore guild within the restricted boundaries of the island. In order to meet energetic constraints, *Xenocyon* must have shifted its diet from large-sized prey for Merriam's dog to a much smaller prey for the Trinil dog, in parallel to the evolution of the Sardinian dog (*Cynotherium*, a descendant of *Xenocyon* as well). The presence of Trinil's dog indicates a significant level of endemism of the terrestrial fauna of the Trinil horizon. This implies that *Homo erectus* during this period was isolated to some degree from mainland populations. The following continentalisation enabled the invasion by mainland terrestrial mammals, as is seen in younger faunas of Java, and led to the extinction of *Xenocyon*, and the arrival of *Cuon priscus* (= *Cuon sangiranensis*) in the Middle Pleistocene as part of the Kedung Brubus fauna, as shown by Rebekka Volmer and colleagues (2019). *Cuon* then disappeared, at least from the fossil record, during the Punung biozone, and re-entered at some point during the latest Pleistocene or Holocene (*Cuon alpinus*, earlier described as *Cuon javanicus*).

Lydekker's Deer and Sambar

All endemic faunas of Java contain deer species, generally a medium-sized deer (*A. lydekkeri*), a large-sized species (*Rusa* sp.), and muntjac (*M. muntjak*) in various combinations. The large deer have been variably described as, amongst others, *Cervus kendengensis* (several sites, based on many elements but no holotype), *Cervus stehlini* (Bumiaju, based on three antlers), *Cervus problematicus* (Bumiaju, based on a lower first molar), *Rusa hippelaphus* (Ngandong), and *Rusa* sp. (various Pleistocene localities). A morphometric approach comparing the third molars of *Axis* and the large deer in relation to extant and extinct Southeast Asian deer by Ben Gruwier and colleagues (2015) shows that the large-sized species are most likely species of sambar (*Rusa*), with the former two species closely related to the living Javanese deer (*R. timorensis*). They also demonstrate that *A. lydekkeri* is probably closely related to Bawean hog deer (*Axis kuhlii*). This species, together with Calamian hog deer (*Axis calamianensis*), Indian hog deer (*Axis porcinus*), and Indochinese hog deer (*Axis annamiticus*), have been placed in a genus of their own, *Hyelaphus*, based on genetic analysis by Pitra and colleagues (2004). Their data indicates that *Hyelaphus* is closer related to sambar (*Rusa unicolor*) than to Indian spotted deer (*Axis axis*). These four species, and perhaps *A. lydekkeri* as well, are small-sized forest deer that flee in a typical hog-like manner from predators, with its head held low and going under obstacles instead of leaping over them.

Axis lydekkeri has a characteristic lyre-shaped antler with a brow tine and a forked beam, described and figured by Martin (1886). Contrary to other antlers, the beam turns backwards, then forward, and then inward (Figure 12.8). The surface of the beam is rather smooth, lacking conspicuous pearls. Yahdi Zaim and colleagues (2003) show that the mandibles and isolated third molar from Bumiaju that were earlier ascribed to *Cervus zwaani* belong, in fact, to *A. lydekkeri*, reducing the former species to junior synonym of the latter. Another junior synonym is *Cervus liriocerus*, given by Dubois (1907). *Axis lydekkeri* is similar in size to *A. porcinus*. A larger but otherwise similar form (with often an accessory tine at the base of the brow tine) from Ngandong and other sites in East Java has been described as *Axis javanicus* by von Koenigswald (1933).

Antelopes and Cattle

Throughout the fossil history of Java, several different bovid species arrived, evolved, and went extinct. The earliest species is the small-sized antelope *D. santeng* from the early Middle Pleistocene Trinil fauna. An earlier, ancestral form of *D. santeng* might be represented by a partial skull from the lower vertebrate-bearing horizon of Bumiaju (Early Pleistocene; Ci Saat or Satir biozone), which was erroneously attributed to a deer by von Koenigswald (1934). Gert van den Bergh identified it as a hornless skull fragment of a female member of Boselaphini (in: de Vos and colleagues, 1994). *Duboisia* has been reported by Hooijer (1963) from the Middle Pleistocene of Malaysia as well, but the fossil record of the Pleistocene in Malaysia is poorly known and based on a single locality, Tambun, and the find has not been confirmed. Additionally, Yuichiro Nishioka and Chavalit Vidthayanon (2018) report *Duboisia* fossils from Thailand, but, in our view, the figured specimen differs from *Duboisia*.

Roberto Rozzi and colleagues (2013) reconstructed the ecology of *Duboisia* based on body mass, morphology, and dental analyses and concluded that it was a forest dweller, mostly feeding on leaves with occasionally more abrasive vegetation, similar to extant browsers. This agrees with the palaeo-environmental reconstructions by Anne-Marie Sémah and colleagues (2010). The Early Pleistocene, the likely arrival time of ancestral *Duboisia*, was characterised by closed canopy forests, swamps, and immense mangrove forests in the lower elevations and tropical rainforests during warmer periods and more open forests during cooler periods. The rainforest fragmented and more seasonal forests developed around the transition to the Middle Pleistocene, but although the landscape was gradually dominated by grasslands, relicts of rainforest and swamp forests persisted, and providing a home to forest- dwellers like *Duboisia*.

The other bovids are grazers and are a water buffalo with impressive horns (*Bubalus palaeokerabau*), a banteng (*Bibos palaesondaicus*), and a large bovine (*Epileptobos groeneveldtii*). The former two are present at the Trinil faunal stage, whereas *Epileptobos* is limited to the Kedung Brubus fauna. The long-horned water buffalo is replaced at the transition to the Punung fauna by the extant wild water buffalo (*Bubalus arnee*), while the banteng (*Bibos javanicus*) either arrived during the Holocene or as part of the Punung fauna as well.

References

Arif M., Ibrahim Shah S.M., de Vos J. 1992. *Cervus rewati* sp. nov. (Mammalia, Cervidae) from the upper Siwaliks of Pakistan. *Memoirs of the Geological Survey of Pakistan*, 17: 11–24.

Bettis E.A., Zaim Y., Larick R., Ciochon R.L., Sumito, Rizal Y., Reagan M., Heizler M. 2004. Landscape development preceding *Homo erectus* immigration into Central Java, Indonesia: The Sangiran Formation Lower Lahar. *Palaeogeography, Palaeoclimatology, Palaeoecology*, 206: 115–131.

Brown B. 1926. A new deer from the Siwaliks. *American Museum Novitates*, 242: 1–6.

de Visser J.A. 2008. *The Extinct Genus Hexaprotodon Falconer & Cautley, 1836 (Mammalia, Artiodactyla, Hippopotamidae) in Asia: Paleoecology and Taxonomy*. Unpublished PhD thesis. University of Utrecht, the Netherlands.

de Vos J. 1983. The *Pongo* faunas from Java and Sumatra and their significance for biostratigraphical and paleo-ecological interpretations. *Proceedings of the Koninklijke Akademie van Wetenschappen (Series B)*, 86 (4): 417–425.

de Vos J., Sartono S., Hardja-Sasmita H.S., Sondaar P.Y. 1982. The fauna from Trinil, type locality of *Homo erectus*; a reinterpretation. *Geologie en mijnbouw*, 61: 207–211.

de Vos J., Sondaar P.Y., van den Bergh G.D., Aziz F. 1994. The *Homo* bearing deposits of Java and its ecological context. *Courier Forschungs-Institut Senckenberg*, 171: 129–140.

Dietrich W.O. 1934. Eine neue Proboscidier-Familie? *Centralblatt für Mineralogie, Geologie und Paläontologie (Abteilung B)*, 1934: 141–144.

Dubois E. 1893. Palaeontologische onderzoekingen op Java. *Extra bijvoegsel der Javasche Courant, Verslag van het Mijnwezen over het 3e kwartaal 1892*: 10–14.

Dubois E. 1894. *Pithecanthropus erectus, einen menschenaenliche Uebergangsform aus Java*. Landesdruckerei, Batavia.

Dubois E. 1907. Eenige van Nederlanschen kant verkregen uitkomsten met betrekking tot de kennis der Kendeng-fauna (Fauna van Trinil). *Tijdschrift Koninklijk Nederlansch Aardrijkskundig Genootschap (Series 2)*, 24: 449–458.

Dubois E. 1908. Das geologische Alter der Kendeng- oder Trinil-Fauna. *Tijdschrift Koninklijk Nederlansch Aardrijkskundig Genootschap (Series 2)*, 25: 1235–1270.

Dubois E. 1922. The proto-Australian fossil man of Wadjak, Java. *Verhandelingen der Koninklijke Akademie van Wetenschappen*, 23 (7): 1013–1051.

Dubois E. 1926. *Manis palaeojavanica*, the Giant Pangolin of the Kendeng fauna. *Verhandelingen der Koninklijke Akademie van Wetenschappen*, 29 (9): 1233–1243.

Elbert J. 1907. De nieuwste onderzoekingen over het *Pithecanthropus*-vraagstuk. *Natuurkundig Tijdschrift voor Nederlandsch-Indië*, 67 (3–4): 125–142.

Gruwier B., de Vos J., Kovarovic K. 2015. Exploration of the taxonomy of some Pleistocene Cervini (Mammalia, Artiodactyla, Cervidae) from Java and Sumatra (Indonesia): A geometric- and linear morphometric approach. *Quaternary Science Reviews*, 119: 35–53.

Falconer H., Cautley P.T. 1846. *Fauna Antiqua Sivalensis, Being the Fossil Zoology of the Sewalik Hills, in the North of India*. Smith, Elder, London (Plates only). Notes in Murchison, 1868.

Hooijer D.A. 1954. A pygmy Stegodon from the Middle Pleistocene of Eastern Java. *Zoologische Mededelingen*, 33: 91–102.

Hooijer D.A. 1963. Report upon a collection of Pleistocene mammals from tin-bearing deposits in a limestone cave near Ipoh, Kinta Valley. *Federation Museum Journal*, 7: 1–5.

Hooijer D.A. 1974. *Elephas celebensis* (Hooijer) from the Pleistocene of Java. *Zoologische Mededelingen*, 48: 85–93.

Hooijer D.A. 1984. The solution of the *Cryptomastodon* problem. *Netherlands Journal of Zoology*, 34 (2): 228–231.

Huffman O.F., Shipman P., Hertler C., de Vos J., Aziz F. 2005. Historical evidence of the 1936 Mojokerto skull discovery, East Java. *Journal of Human Evolution*, 48: 321–363.

Huffman O.F., Zaim Y., Kappelman J., Ruez D.R., Jr,, de Vos J., Rizal Y., Aziz F., Hertler C. 2006. Relocation of the 1936 Mojokerto skull discovery site near Perning, East Java. *Journal of Human Evolution*, 50: 431–451.

Indriati E., Swisher C.C. III, Lepre C., Quinn R.L., Suriyanto R.A., Hascaryo A.T., Grün R., Feibel C.S., Pobiner B.L., Aubert M., Lees W., Antón S.C. 2011. The age of the 20 meter Solo River terrace, Java, Indonesia and the survival of *Homo erectus* in Asia. *PLoS ONE*, 6 (6): e21562.

Joordens J.C., d'Errico F., Wesselingh F.P., Munro S., de Vos J., Wallinga J., Ankjærgaard C., Reimann T., Wijbrans J.R., Kuiper K.F., Mücher H.J., Coqueugniot H., Prié V., Joosten I., van Os B., Schulp A.S., Panuel M., van der Haas V., Lustenhouwer W., Reijmer J.J., Roebroeks W. 2015. *Homo erectus* at Trinil on Java used shells for tool production and engraving. *Nature*, 518: 228–231.

Junghuhn F. 1853. *Java, deszelfs gedaante, bekleeding en inwendige structuur*. Van Kampen, Amsterdam.

Junghuhn F. 1857. Over de fossiele zoogdierbeenderen te Patihajam, in de resientie Djapara, eiland Java. *Natuurkundig Tijdschrift voor Nederlandsch-Indië*, 14: 215–219.

Kretzoi M. 1950. *Stegoloxodon* nov. gen., a possible Asiatic ancestor of true loxodonts. *Földtani Közlöny*, 80: 405–408.

Larick R., Ciochon R.L., Zaim Y., Sudijono, Suminto, Rizal Y., Aziz F., Reagani M., Heizler M. 2001. Early Pleistocene $^{40}Ar/^{39}Ar$ ages for Bapang Formation hominins, Central Jawa, Indonesia. *Proceedings of the National Academy of Science of the United States of America*, 98 (9): 4866–4871.

Lomolino M.V., van der Geer A.A.E., Lyras G.A., Palombo M.R., Sax D., Rozzi R. 2013. Of mice and mammoths: Generality and antiquity of the island rule. *Journal of Biogeography*, 40: 1427–1439.

Lydekker R. 1879. Notices od Siwalik mammals. *Records of the Geological Survey of India*, 12: 33–52.

Lydekker R. 1884. Indian Tertiary and Post-Tertiary Vertebrata. Rodents and new ruminants from the Siwaliks, and synopsis of Mammalia. *Palaeontologia Indica (Series 3)*, 10 (3): 105–134.

Lyras G.A., van der Geer A., Rook L. 2010. Body size of insular carnivores: Evidence from the fossil record. *Journal of Biogeography*, 37: 1007–1021.

Maglio V.J. 1973. Origin and evolution of the Elephantidae. *Transactions of the American Philosophical Society (New Series)*, 63 (3): 1–149.

Markov G.N., Saegusa H., 2008. On the validity of *Stegoloxodon* Kretzoi, 1950 (Mammalia: Proboscidea). *Zootaxa*, 1861: 55–56.

Martin K. 1884. Ueberreste vorweltlicher Proboscidier von Javaund Banka. *Leiden, Sammlungen des Geologischen Reichs-Museums* 4, 1884–1889: 1–24.

Martin K. 1886. Fossiele Saugethierreste von Java und Japan. *Leiden, Sammlungen des Geologischen Reichs-Museums* 4, 1884–1889: 25–69.

Martin K. 1888. Neue Wirbelthierreste vom Pati-Ajam auf Java. *Leiden, Sammlungen des Geologischen Reichs-Museums* 4, 1884–1889: 87–116.

Matsu'ura S., Kondo M., Danhara T., Sakata S., Iwano H., Hirata T., Kurniawan I., Setiyabudi E., Takeshita Y., Hyodo M., Kitaba I., Sudo M.,

Danhara Y., Aziz F. 2020. Age control of the first appearance datum for Javanese *Homo erectus* in the Sangiran area. *Science*, 367 (6474): 210–214.

Mayr E. 1950. Taxonomic categories in fossil hominids. *Cold Spring Harbor Symposia on Quantitative Biology*, 15 (1951): 109–118.

Morley R.J., Morley H.P., Zaim Y., Huffman O.F. 2020. Palaeoenvironmental setting of Mojokerto *Homo erectus*, the palynological expressions of Pleistocene marine deltas, open grasslands and volcanic mountains in East Java. *Journal of Biogeography*, 47: 566–583.

Morwood M.J., O'Sullivan P., Susanto E.E., Aziz F. 2003. Revised age for Mojokerto 1, an early *Homo erectus* cranium from East Java, Indonesia. *Australian Archaeology*, 57: 1–4.

Nishioka Y., Vidthayanon C. 2018. First occurrence of *Duboisia* (Bovidae, Artiodactyla, Mammalia) from Thailand. *Fossil Record*, 21: 291–299.

Oppenoorth W.F.F. 1932. Ein neuer diluvialer Urmensch von Java. *Natur und Museum*, 62: 269–279.

Pitra C., Fickel J., Meijaard E., Groves P.C. 2004. Evolution and phylogeny of old world deer. *Molecular Phylogenetics and Evolution*, 33 (3): 880–895.

Polhaupessy A.A. 1996. Palynological study on the vertebrate bearing horizon of Java. *Odontologie, Association for Comparative Odontology*, 1: 32–36.

Puspaningrum M.R., van den Bergh G.D., Chivas A.R., Setiabudi E., Kurniawan I. 2020. Isotopic reconstruction of Proboscidean habitats and diets on Java since the Early Pleistocene: Implications for adaptation and extinction. *Quaternary Science Reviews*, 228: 106007.

Radèn Saléh 1867. Over fossiele beenderen van den Pandan. *Natuurkundig Tijdschrift voor Nederlandsch-Indië*, 29: 422–423, 426–429, 433–437, 448–541, 455–459.

Rizal Y., Westaway K.E., Zaim Y., van den Bergh G.D., Bettis E.A., Morwood M.J., Huffman O.F., Grün R., Joannes-Boyau R., Bailey R.M., Sidarto, Westaway M.C., Kurniawan I., Moore M.W., Storey M., Aziz F., Suminto, Zhao J.X., Aswan, Sipola M.E., Larick R., Zonneveld J.P., Scott R., Putt S., Ciochon R.L. 2020. Last appearance of *Homo erectus* at Ngandong, Java, 117,000-108,000 years ago. *Nature*, 577: 381–385.

Rozzi R., Winkler D.E., de Vos J., Schulz E., Palombo M.R. 2013. The enigmatic bovid *Duboisia santeng* (Dubois, 1891) from the Early–Middle Pleistocene of Java: A multiproxy approach to its paleoecology. *Palaeogeography, Palaeoclimatology, Palaeoecology*, 377: 73–85.

Schmulling P.E.C. 1864. Fossiele beenderen van zoogdieren, bij Sanggiran gevonden. *Natuurkundig Tijdschrift voor Nederlandsch-Indië*, 27: 399–400.

Selenka M.I., Blanckenhorn M. (eds) 1911. *Die Pithecanthropus-Schichten auf Java; Geologische und paläontologische Ergebnisse der Trinil-Expedition (1907–1908)* Leipzig, Engelmann.

Sémah A.M. 1984. Paleoenvironnement des sites a hominides de Java. *Resumes des Communications de Table-Ronde; L'Homme fossile et son Environnement a Java; Marseille, 12-14 Janvier*, 1984: 29–32.

Sémah A.-M., Sémah F., Djubiantono T., Brasseur B. 2010. Landscapes and Hominids' environments: Changes between the Lower and the Early Middle Pleistocene in Java (Indonesia). *Quaternary International*, 223–224: 451–454.

Sémah A.-M., Sémah F., Moigne A.-M., Ingicco T., Purnomo A., Simanjuntak T., Widianto H. 2016. The palaeoenvironmental context of the Palaeolithic of Java: A brief review. *Quaternary International*, 416: 38–45.

Sémah F., Féraud G., Saleki H., Falguères C., Djubiantono T. 2000. Did early man reach Java during the Late Pliocene? *Journal of Archaeological Science*, 27: 763–769.

Sémah F., Sémah A.-M., Djubiantono T., Simanjuntak H.T. 1992. Did they also make stone tools? *Journal of Human Evolution*, 23: 439–446.

Shutler R., Head J.M., Donahue D.J., Jull A.J.T., Barbetti M.F., Matsu'ura S., de Vos J., Storm P. 2004. AMS radiocarbon dates on bone from cave sites in southeast Java, Indonesia including Wajak. *Modern Quaternary Research Southeast Asia*, 18: 1–5.

Sloet van Oldruitenborgh A. 1859. Fossiele beenderengevonden op de helling van den Goenung Pandan. *Natuurkundig Tijdschrift voor Nederlandsch-Indië*, 16: 70–71.

Sondaar P.Y. 1984. Faunal evolution and the mammalian biostratigraphy of Java. *Courier Forschungsinstitut Senckenberg*, 69: 219–235.

Staring W.C.H. 1865. Het voorkomen van diluviale gronden op Java. *Verslagen en Mededeelingen der Koninklijke Akademie*, 17: 52.

Stehlin H.G. 1925. Fossile Saügetiere aus der Gegend von Limbangen (Java). Dienst van den Mijnbouw in Nederlandsch-Indië. *Wetenschappelijke mededeelingen*, 3: 1–10.

Storm P., Aziz F., de Vos J., Kosasih D., Baskoro S., Ngaliman, van den Hoek Ostende L.W. 2005. Late Pleistocene *Homo sapiens* in a tropical rainforest fauna in East Java. *Journal of Human Evolution*, 49: 536–545.

Storm P., Wood R., Stringer C., Bartsiokas A., de Vos J., Aubert M., Kinsley L., Grün R. 2013. U-series and radiocarbon analyses of human and faunal remains from Wajak, Indonesia. *Journal of Human Evolution*, 64: 356–365.

Stremme H., 1911. Die Säugetiere mit Ausnahme der Proboscidier. *In*: Selenka L., Blankenhorn M. (eds.), *Die Pithecanthropus-Schichten auf Java, Geologische und Paläontologische Ergebnisse der Trinil-Expedition (1907 und 1908): 82–150*. Engelmann, Leipzig.

Suzuki M., Wikarno, Budisantoso, Saefudin I., Itihara M. 1985. Fission Track ages of Pumice tuff, Tuff layers and Javites of Hominid fossil bearing Formations in Sangiran Area, Central Java. In: Watanabe N., Kadar D. (eds), Quaternary Geology of the Hominid fossil bearing formations in Java. *Geological Research and Development Centre, special publication*, 4: 309–331.

Swisher III C.C., Curtis G.H., Jacob T., Getty A.G., Suprijo A., Widiasmoro. 1994. Age of the earliest known hominids in Java, Indonesia. *Science*, 263: 1118–1121.

Swisher III C.C., Rink W.J., Antón S.C., Schwarcz H.P., Curtis G.H., Suprijo A., Widiasmoro. 1996. Latest *Homo erectus* of Java: Potential Contemporaneity with *Homo sapiens* in Southeast Asia. *Science*, 274: 1870–1874.

Theunissen B., de Vos J., Sondaar P.Y., Aziz F. 1990. The establishment of a chronological framework for the hominid-bearing deposits of Java: A historical survey. *Geological Society of America Special Papers*, 242: 39–53

van den Bergh G.D. 1999. The Late Neogene elephantoid-bearing faunas of Indonesia and their palaeozoogeographic implications; A study of the terrestrial faunal succession of Sulawesi, Flores and Java, including evidence for early hominid dispersal east of Wallace's line. *Scripta Geologica*, 117: 1–419.

van den Bergh G.D., Aziz F., Sondaar P.Y., Hussain S.T. 1992. Taxonomy, stratigraphy, and paleozoogeography of Plio-Pleistocene Proboscideans from the Indonesian Islands. *Bulletin of the Geological Research and Development Center Bandung, Paleontology*, 7: 28–58.

van den Brink L.M. 1982. On the mammal fauna of the Wajak Cave, Java (Indonesia). *Modern Quaternary Research in Southeast Asia*, 7: 177–193.

van der Geer A.A.E., Lyras G., MacPhee R.D.E., Mitteroecker P. 2018 (a). From Jumbo to Dumbo: Cranial shape changes in elephants and hippos during phyletic dwarfing. *Evolutionary Biology*, 45 (3): 303–317.

van der Geer A.A.E., Lyras G.A., Volmer R. 2018 (b). Insular dwarfism in canids on Java (Indonesia) and its implication for the environment of Homo erectus during the Early and earliest Middle Pleistocene. *Palaeogeography, Palaeoclimatology, Palaeoecology*, 507: 168–179.

van der Maarel F.H. 1932. Contributions to the knowledge of the fossil mammalian fauna of Java. *Wetenschappelijke Mededelingen Dienst van den Mijnbouw in Nederlandsch-Indië*, 15: 1–208.

van Es L.J.C. 1931. *The Age of Pithecanthropus*. 's-Gravenhage: Martinus Nijhoff.

van Essen H., van den Bergh G., de Vos J. 2006. The final solution of the *Cryptomastodon* problem. Morphological correlations between supernumerary teeth in stegodonts and elephants (Proboscidea, Mammalia). *Courier Forschungsinstitut Senckenberg*, 256: 29–41.

Volmer R., van der Geer A.A.E., Cabrera P.A., Wibowo U.P., Kurniawan I. 2019. When did *Cuon* reach Java? – Reinvestigation of canid fossils from *Homo erectus* faunas. *Geobios*, 55: 89–102.

von Koenigswald G.H.R. 1933. Beitrag zur Kenntnis der fossilen Wirbeltiere Javas. *Wetenschappelijke Mededeelingen van den Dienst van den Mijnbouw in Nederlandsch Indië*, 23: 1–127.

von Koenigswald G.H.R. 1934. Zur stratigraphie des javanischen Pleistocän. *De Ingenieur in Nederlandsch-Indië* 1 (4): 185–201.

von Koenigswald G.H.R. 1936. Erste Mitteilung über einen fossilen Hominiden aus dem Altpleistocän Ostjavas. *Verhandelingen der Koninklijke Akademie van Wetenschappen, Amsterdam*, 39: 1000–1009.

Voris H.K. 2000. Maps of Pleistocene sea levels in Southeast Asia: Shorelines, river systems and time durations. *Journal of Biogeography*, 27: 1153–1167.

Watanabe N., Kadar D. (eds) 1985. Quaternary Geology of the hominid fossil bearing formations in Java. *Special Publication of the Geological Research and Development Centre*, 4: 1–378.

Westaway K.E., Louys J., Due Awe R., Morwood M.J., Price G.J., Zhao J.X., Aubert M., Joannes-Boyau R., Smith T.M., Skinner M.M., Compton T., Bailey R.M., van den Bergh G.D., de Vos J., Pike A.G.W., Stringer C., Saptomo E.W., Rizal Y., Zaim J., Santoso W.D., Trihascaryo A., Kinsley L., Sulistyanto B. 2017. An early modern human presence in Sumatra 73,000–63,000 years ago. *Nature*, 548: 322–325.

Westaway K.E., Morwood M.J., Roberts R.G., Rokus A.D., Zhao J.X., Storm P., Aziz F., van den Bergh G., Hadi P., Jatmiko, de Vos J. 2007. Age and biostratigraphic significance of the Punung rainforest fauna, East Java, Indonesia, and implications for *Pongo* and *Homo*. *Journal of Human Evolution*, 53 (6): 709–717.

Widianto H. 2006. The oldest Homo erectus stone tools in Java: From the Lower Pleistocene Pucangan formation in Sangiran. Paper presented at the 18th Congress of the Indo-Pacific Prehistory Association, Manila, 20–26 March 2006.

Zaim Y., de Vos J., Huffman O.F., Aziz F., Kappelman J., Rizal Y. 2003. A new antler specimen from the 1936 Perning hominid site, East Jawa, Indonesia, attributable to *Axis lydekkeri* (MARTIN, 1886). *Journal of Mineral Technology*, 10 (2): 1–10.

Zanolli C., Kullmer O., Kelley J., Aacon A.-M., Demeter F., Dumoncel J., Fiorenza L., Grine F.E., Hublin J.-J., Nguyen A.T., Nguyen T.M.H., Pan L., Schillinger B., Schrenk F., Skinner M.M., Ji X., Macchiarelli R. 2019. Evidence for increased hominid diversity in the Early to Middle Pleistocene of Indonesia. *Nature Ecology & Evolution*, 3: 755–764.

CHAPTER THIRTEEN

Flores

Evolution of Island Mammals: Adaptation and Extinction of Placental Mammals on Islands, Second Edition.
Alexandra van der Geer, George Lyras and John de Vos.
© 2021 John Wiley & Sons Ltd. Published 2021 by John Wiley & Sons Ltd.

Flores is one of the Lesser Sunda Islands east of Java – Bali, Lombok, Sumba, Sumbawa, Flores, and Timor. The island was inhabited during the Pleistocene by dwarf stegodons; small, medium, and giant rats; Komodo dragons; and an enigmatic species of dwarf human. The presence of this 1-m tall hominin (*Homo floresiensis*) tends to support the view that an early Asian *Homo* evolved dwarf size on the island. Only one of the giant rats survived till today, as do the dragons on the nearby isle of Komodo.

Geology and Palaeogeography

The island forms part of a series of active volcanoes between Java and Timor, also known as the Inner Banda Arc, and consists mainly of young volcanic rocks. At Flores' west end, there are some small islands – Komodo and Rinca being the largest – which were connected to Flores during periods of low sea level. Further to the west, the island of Sumbawa is separated from these islands by an over 200 m deep sea strait, first reported by Jacqueline Mammerickx and colleagues (1976). Colonisation overland to Flores is otherwise hampered by an over 4 km deep basin in the north and the deep Savu Sea in the south-east. An inactive arch lies south and south-east of Flores, known as the Outer Banda Arc, consisting of Timor, Sumba, and several islands in between, such as Savu and Roti. Based on the bathymetry, we can state that Flores, with or without the smaller islands attached, was an island throughout the Pleistocene.

The active volcanism lies behind the very rugged topography of Flores – with peaks of 2 km – and the irregular coastline of the island. Flores emerged above the sea during the Early Miocene, about 15–21 million years ago, first attested by Hans Ehrat (1925) and dated by Nishimura and colleagues (1981). Since the end of the Pliocene, Flores has been uplifted due the advancing Australian plate, as summarised by Gerrit van den Bergh (1999). The area of active volcanism moved relatively southwards, caused by tectonics. As a result, the northern half of Flores consists of extinct Miocene volcanoes; the southern half of Plio–Pleistocene, extinct volcanoes; and in the extreme south-east and south-west young, still active volcanoes, sloping steeply into the deep Savu Sea. A relatively flat depression in between various active and inactive volcanoes in west Central Flores is known as the So'a Basin, which is known for its Pleistocene fossils. Van den Bergh (1999) suggested that the sediments may have deposited after the Ae Sissa river outlet was blocked with volcanic products from the now inactive Kelilambo volcano. Other important fossil localities consist of

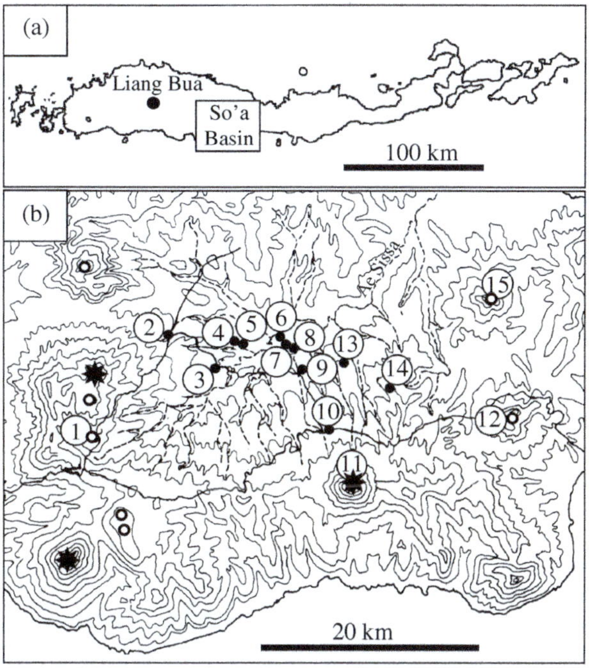

Figure 13.1 Some Pleistocene localities of Flores. (a) Map showing the location of Liang Bua and So'a Basin. (b) Topographic map of the So'a Basin showing various localities and volcanos: (1) Bajawa, capital of the Ngada District, (2) Soa, (3) Menge Ruda, (4) Mata Menge, (5) Boa Leza, (6) Ola Bula, (7) Tangi Talo, (8) Bhisu Sau, (9) Ola Kile (abandoned village), (10) Boawae, (11) Ambulobo Volcano, (12), Kelindora Volcano, (13) Dozo Dhalu, (14) Gero (abandoned village), and (15) Kelilambo Volcano. Contour interval: 200 m. (Adapted from Van den Bergh and colleagues, 2008 and Van den Bergh, 1999.)

caves in the Miocene limestone regions of the islands, particularly in the northern parts. The sediments in these caves derive from weathering of the limestone and are of a Pleistocene age.

Historical Palaeontology

The name Flores is derived from the Portuguese Cabo de Flores, or Cape of Flowers, the translation of the Malay name *Tanjung Bunga* for the north-east cape of the island. In 1956, the Dutch missionary and archaeologist Theodor Verhoeven started excavations in the So'a Basin near the abandoned village Ola Bula (Figure 13.1), Nage Keo, west Central Flores, after the Raja of the region had given him a stegodon molar, found in the neighbourhood. Collaborators of the zoological museum at Bogor (then Buitenzorg) of western Java, including Rokus Due Awe, assisted Father Verhoeven in the field. They collected remains of stegodons, giant rats, monitor lizards, and lithic artefacts from various sites. Verhoeven named the two most important sites Mata Menge and Ola Bula. Apart from the So'a Basin, he also excavated in caves, including Liang Toge, Liang Bua, Liang Momer, and Liang Michael, where they found mainly Neolithic material.

The fossils were sent to Dirk Hooijer of the Dutch Rijksmuseum van Natuurlijke Historie, now Naturalis. Hooijer (1957a) described the subfossil rodent material from Liang Toge and distinguished two species of giant cave rats and a smaller subspecies of the extant giant rat of Flores as, respectively, *Papagomys verhoeveni*, *Spelaeomys florensis*, and *Papagomys armandvillei besar*. Guy Musser (1981) demonstrated that the holotype of Hooijer's first species actually belongs to the living *Papagomys armandvillei*, while the rest of the material represented a separate species indeed. He renamed the smaller giant cave rat *Papagomys theodorverhoeveni*, in this way maintaining the reference to the discoverer of the material.

Musser (1981) described two right upper jaws and an isolated upper molar of a large rat as *Hooijeromys nusatenggara*, in honour of Dirk Hooijer. The species name is inspired by the Indonesian name for the Lesser Sunda Islands. The material came from the collection of Verhoeven, collected in the So'a Basin of central Flores. In the same paper, he also addressed the material from Liang Bua collected by Verhoeven. Middle-sized rats, first mistaken for ship rats (*Rattus rattus*), were shown to represent two more endemic genera, *Komodomys* and *Floresomys*. As the latter name was preoccupied, Musser and colleagues (1986) changed the name to *Paulamys*, after Verhoeven's wife. *Komodomys* was known to have survived on nearby Komodo, but *Paulamys* was considered extinct until a live rat that was collected in 1989 in tropical rainforest from south-central Flores was referred to this species by Darrel John Kitchener and colleagues (1991a), who suggested it is closely related to *Bunomys* of Sulawesi.

Maringer and Verhoeven (1971) considered the co-occurrence of primitive lithic artefacts and remains of large stegodons evidence of the presence of early humans on the island. The fossils of the stegodon dated from the Middle Pleistocene. The early humans are thus likely members of the *Homo erectus* lineage. The suggestion that *Homo erectus* had already crossed the Wallace Line was quite unheard of at the time, and few paid heed. Among those that did were Sondaar and his Indonesian-Dutch team (Figure 13.2), who had access to the Verhoeven collection; in addition, Sondaar knew Verhoeven personally since the 1970s. Sondaar and his colleagues (1994) found artefacts in association with fossils of the larger stegodon at Mata Menge. Based on palaeomagnetic dates for these findings, they came up with a date of 0.73 Ma, coinciding with the Brunhes-Matuyama reversal of the Earth's magnetic field. Three years later, they described the artefacts in collaboration with the Australian archaeologist Mike Morwood (Morwood and colleagues, 1997). By then, funding of the Indonesian-Dutch excavations had stopped, and an Indonesian-Australian team led by Morwood continued the work on Flores. A year later, his team used fission-track dating to date the artefacts to between 800,000 and 880,000 years ago. This early age and the primitive

Figure 13.2 The Indonesian–Dutch excavation team of 1991, directed by Paul Yves Sondaar (centre). (Photograph: John de Vos.)

nature of the artefacts prompted their search for early humans in the early 2000s. The attention went to the cave Liang Bua, not only because human remains have a far higher change to be preserved in caves than elsewhere, but mainly because Verhoeven had found Palaeolithic artefacts at this site (Figure 13.3). In addition, the top layers of Liang Bua had yielded sub-recent skeletons of modern humans (*Homo sapiens*) of a Negrito type (about 1.50 m tall), and radiocarbon (C^{14}) dated to 3,550 years ago. The team of Morwood, led by Indonesian archaeologist Wahyu Saptomo, dug down to bedrock, in the course of which they found stone tools and, in 2003, a partial skeleton (LB1) of a Palaeolithic dwarf human, just a metre tall in stature, which

Figure 13.3 Stone tools found by Father Theodor Verhoeven at Liang Bua in 1965. Scale bar: 5 cm. National Museum of Natural History, Leiden. (Photograph: Eelco Kruidenier.)

Figure 13.4 The cave Liang Bua, type locality of *Homo floresiensis*, discovered in 2003, in July 2012. (Photograph: Gert van den Bergh.)

they, with physical anthropologist Peter Brown as first author, described as the new species *Homo floresiensis* in *Nature* (2004).

The debate on the identity of *Homo floresiensis*, either a pathological modern human or an endemic species of *Homo* (see below), somewhat turned the attention away from another wealth of information that resulted from the excavations in Liang Bua (Figure 13.4). The various sections, which are well dated, provide an excellent record of the development of an island fauna in the Late Pleistocene and Holocene, including the pristine Pleistocene fauna, the arrival of both Palaeolithic and Neolithic *Homo sapiens*, and the arrival of Europeans. In a special 2009 issue of the *Journal of Human Evolution* dedicated to the Flores finds, van den Berg, Meijer and colleagues (2009) give a complete overview of this fauna. Apart from the mammals, the Liang Bua section is also rich in fossil birds, among which is a species – the giant flightless marabou *Leptoptilos robustus*, described by Hanneke Meijer and Rokus Awe Due (2010).

After the findings in Liang Bua cave, the team explored the island further and came across large stone artefacts at the surface at Wolo Sege, which seemed to have eroded from an exposed conglomerate lens stratigraphically below the early Middle Pleistocene site Mata Menge. Wolo Sege proved to contain the oldest *in situ* stone tools of Flores, dated to about a million years ago, as announced by Adam Brumm and colleagues (2010). The tools include large Acheulean pick-like implements, according to Brumm and Moore (2012), which are generally considered evidence for complex tool-making abilities.

Excavations in 2014 of river sediments at the early Middle Pleistocene site of Mata Menge (Plate 25) (van den Bergh and colleagues, 2014), another site of Verhoeven of the 1950s, yielded the oldest *in situ* hominid fossils of Flores. Two years later, they were described by van den Bergh and colleagues and dated to about 700,000 years ago by Brumm. The lithic artefacts found with the fossils are similar in size and of relatively less-advanced typology as those found in the older layers of Liang Bua of about 600,000 years ago and later. The loss of complexity in tool-making technology between Wolo Sege and Mata Menge has been explained as a response to either more arid conditions or resource limitation under isolation.

Biozones and Faunal Units

Three biozones with considerable overlap can be distinguished on Flores (Figure 13.5). The earliest two consisted entirely of endemic faunas while the youngest is mainly anthropogenic. The overseas arrival by sweepstake dispersal of late Early and early Middle Pleistocene species, including an early human, coincides with lower sea levels. Towards the end of the Late Pleistocene, both *Homo floresiensis* and *Stegodon* disappeared from the record. Notably, the small mammals as well as the monitor survived and remain present in the anthropogenic biozone. Among these, *Papagomys theodorverhoeveni* and *Spelaeomys* became extinct later on.

Early Pleistocene

The earliest known fossil fauna of Flores is characterised by a very small dwarf stegodon (*Stegodon sondaari*). This fauna is clearly unbalanced and impoverished. Mammalian carnivores are lacking, and the still living large Komodo dragon (*Varanus komodoensis*) occupied the niche of top predator. Fossils of this early fauna are rare and limited to the locality Tangi Talo in the So'a Basin of central Flores. Tuffaceous layers of the site have been fission-track dated at 0.9 million years ago by Mike Morwood and colleagues (1998). At some point in time, a hominid arrived, as evidenced by the presence of *in situ* stone artefacts at the site Wolo Sege. These are dated to at least 1.02 million years ago by Brumm and colleagues (2010; later adjusted to at least 1.01 million years, 2016), based on a hardened tuff layer (ignimbrite) immediately above the artefacts.

Apart from the stegodon and the monitor lizard, a giant tortoise was element of this early fauna. Erick Setiyabudi (2006) ascribed it to *Collossochelys azizi* (earlier reported as

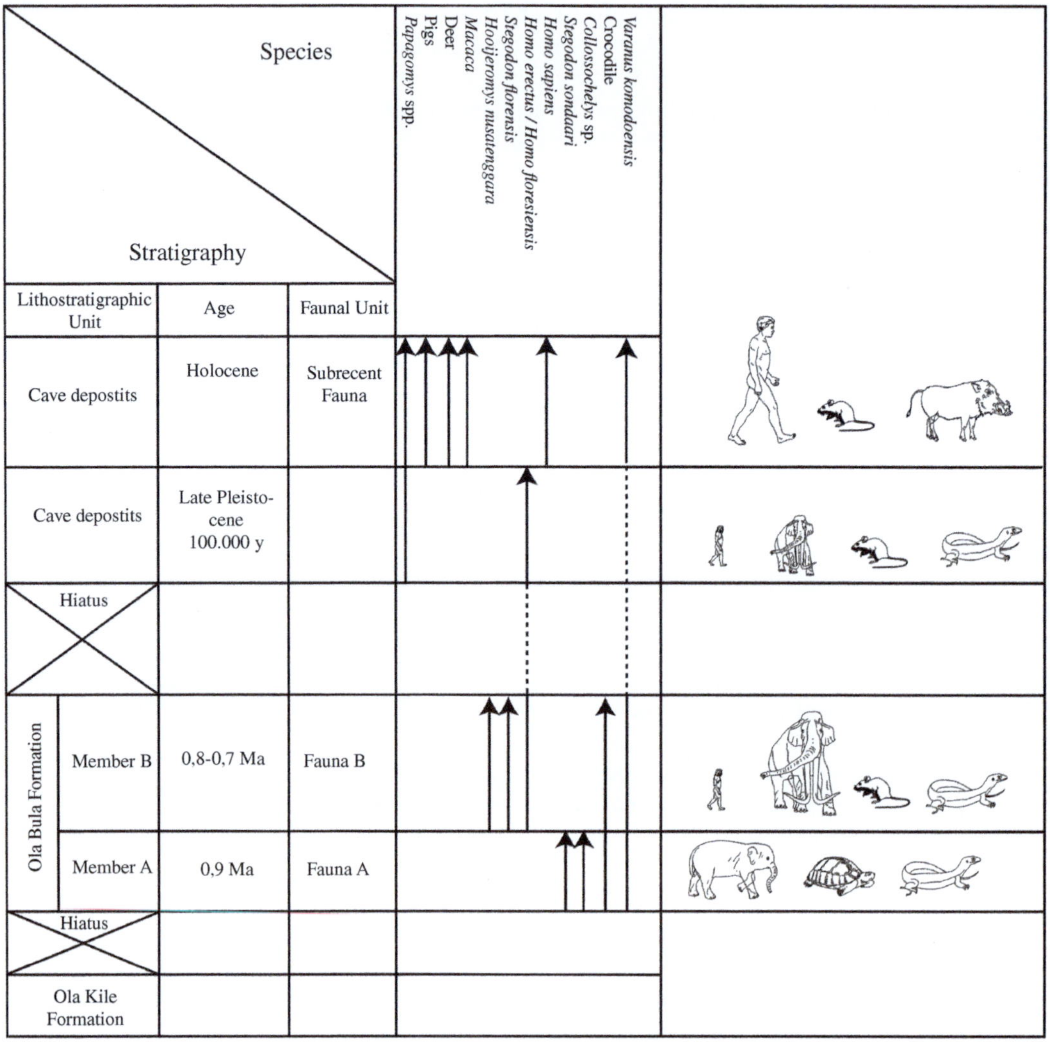

Figure 13.5 Stratigraphic scheme showing the land vertebrate faunal succession of Flores. Characteristic elements of the successive faunal units are shown to the right. The subrecent to recent fauna is omitted.

Geochelone) and related it to the giant tortoise known from the Siwaliks (*Megalochelys sivalensis*). Both reptiles arrived on the island already large bodied, as concluded by Setiabudi. The monitor lizard, however, is related to giant Pliocene forms from Australia, so may either represent a case of insular dwarfism, despite its still impressive size, or a relict taxon that survived unchanged for at least a million years. It is plausible that the first colonisers of the virgin volcanic island of Flores were only large reptiles, a general pattern seen on oceanic islands, and that the stegodon and human arrived later.

Giant Dragons of Timor

Timor, like Flores, was home to a giant monitor. Father Verhoeven had collected three very large varanid vertebrae (Figure Box 13.1) from Middle Pleistocene deposits at Raebia in the Atambua Basin of central Timor. Hooijer (1972) had described them as *Varanus komodoensis*, thus assuming that they belonged to the same species as the Flores dragon. Although some features indeed remind of the latter species, other features are more like those of the Australian *Varanus priscus*, whereas others are unique. Scott Hocknull and colleagues (2009) thus suggest that the Timor monitor should be assigned to a new species, intermediate in size between the Flores monitor and *Varanus priscus*.

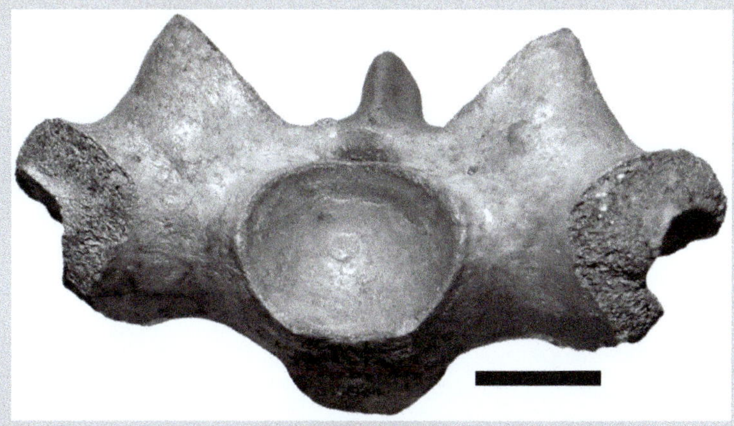

Figure Box 13.1 A vertebra of the Timor monitor (*Varanus timorensis*); proximal view. Scale bar: 2.5 cm. National Museum of Natural History, Leiden. (Photograph: Eelco Kruidenier.)

BOX 13.1

Middle Pleistocene

Around the Early to Middle Pleistocene transition, a second immigration of mammals to Flores took place. The new mammalian elements include a middle-sized stegodon (*Stegodon florensis florensis*) (Plate 26) and a middle-sized cave rat (*Hooijeromys nusatenggara*). The monitor lizard and the hominin from the previous biozone apparently survived the impact, while the dwarf *Stegodon sondaari* disappeared from the record. Important localities of this biozone are Dhozo Dhalu (Figure 13.6), Ola Bula, Mata Menge (Plate 25), and Boa Lesa, all in the So'a Basin. Importantly, primitive stone artefacts have been found in association with fossils of the stegodon in the same area where tektites belonging to the Australasian

Figure 13.6 Excavations in Flores. Top: Excavation of *Stegodon florensis* fossils at Dhozo Dhalu. Below: Excavation details of giant tortoise at Tangi Talo. (Photograps: John de Vos.)

strewn field, dated to about 790,000 years ago. Dating of the volcanic layers here by O'Sullivan of the team of Morwood (1998) support this view with ages between about 680,000 and 880,000 years ago for this biozone. The oldest *in situ* hominid fossils of Flores, excavated at Mata Menge, belong to this biozone. They were dated to about 700,000 years ago by Brumm and colleagues (2016). This small-sized hominid might be ancestral to the Late Pleistocene *Homo floresiensis*.

The environment during the early Middle Pleistocene at Mata Menge consisted of open, savannah-like grasslands with wetland habitats and forest patches, as indicated by the combined analyses of phytoliths by van den Bergh, Aziz and colleagues (2009), stable carbon and oxygen isotopes from stegodon teeth by Mika Puspaningrum and colleagues (2014; in more detail in Brumm and colleagues, 2016), and the bird fossils by Meijer and colleagues (2015).

The colonisation of elements of this biozone is likely related to the low sea levels at the onset of this biozone, which significantly decreased the distances between the islands. The source for the Flores stegodon might be nearby Sulawesi, from where a still poorly known *Stegodon* species of the same size was identified. Flores is easily reached from Sulawesi because of the prevailing southward directed ocean currents, according to Sondaar (1989) and Kuhnt and colleagues (2004), whereas strong surface currents constitute an effective barrier against overseas dispersal along the Inner Banda Arc. In this respect, it is noteworthy that the Flores murids, with the exception of *Spelaeomys*, have their closest relatives on Sulawesi. As in the case of the stegodon, also the hominids might have come through Sulawesi, where abundant artefacts have been found, in association with fossils of megafauna. Furthermore, Middle Pleistocene lithic artefacts in association with rhino fossils, some with butchery marks as shown by Thomas Ingicco and colleagues (2018), have been recovered from Luzon, north of Sulawesi. This could, tentatively, be evidence for a Middle Pleistocene dispersal route from Taiwan through the Philippines and Sulawesi into Flores and the islands further to the east, as first postulated by Sondaar (1989).

Late Pleistocene

Likely, a gradual progressive evolution of the species of the Middle Pleistocene fauna resulted in those of the Late Pleistocene fauna, characterised by a dwarf human (*Homo floresiensis*), giant rats (*Papagomys*) (Plate 27), middle-sized rats (*Komodomys, Paulamys, Hooijeromys*), small rats (*Rattus hainaldi*), dwarf stegodons (*Stegodon florensis insularis*), shrews (*Crocidura* or *Suncus* sp. A and sp. B, and presumably *Suncus mertensis* or its precursor), giant marabou storks (*Leptoptilos robustus*), and several bats (see below). The Komodo dragon and the middle-sized rat *Hooijeromys* are survivors since the Early Pleistocene, whereas *Papagomys* likely evolved from *Hooijeromys*. There is no recorded new arrival, and the distinction between the biozones is artificial, based on evolutionary stages of the species.

The most important site is the cave Liang Bua. Here, stone tools and flaking debris were found in the same layer as dwarf stegodon bones showing cut marks, indicating that they had been eaten, either hunted or scavenged. Also, remains of rats, bats, and the Komodo dragon were present at the same depths, evidence that no faunal turnover had taken place. Instead, a gradual local evolution took place of humans and stegodons and an adaptive radiation of murids. Giant rat remains are abundant and continue into the Holocene layers (Locatelli and colleagues, 2012). Grace Veatch and colleagues (2019) reported for the first time the presence of the middle-sized *Hooijeromys* in Liang Bua throughout the depositional sequence. The genus was until then considered extinct since the Middle Pleistocene.

The extinction of the stegodon and the human at the end of this biozone was initially associated by van den Bergh and colleagues (2008) with a volcanic eruption around 17,000 years ago. However, the revised stratigraphy and updated chronology of the cave sediments by Thomas Sutikna and colleagues (2016) showed otherwise. They dated the skeletal remains of the dwarf hominid and the containing deposits to between about 60,000–100,000 years ago and the Palaeolithic stone artefacts to between about 50,000–190,000 years ago. Previous estimates dated the Liang Bua hominid to between about 12,000–95,000 years ago, so surviving long after modern humans reached insular Southeast Asia and Australia. New stratigraphic data revealed during later excavations showed that this is not the case. It remains thus an open question whether the Palaeolithic dwarf humans ever met modern humans on Flores.

Interestingly, practically all stegodons in Liang Bua were juvenile. Likely, the Flores human favoured to hunt these instead of the more ferocious adults. A similar case is provided by the Middle Pleistocene site Panxian Cave in southern China. Here, the stegodon assemblage (*Stegodon orientalis*) was dominated by juvenile remains, associated with stone artefacts and hominin remains, as reported by Lynne Schepartz and colleagues (2005). The most recent evidence of stegodon hunting from Panxian comes from the layer dated to 17,000 years ago.

Holocene

At the start of the Holocene, a mainland fauna arrived, for the greater part the result of human introductions. The only survivors from the Pleistocene are the Flores giant rat (*Papagomys armandvillei*), and three smaller rats (*Rattus hainaldi*, *Komodomys rintjanus*, and *Paulamys naso*), the Flores shrew (*Suncus mertensi*), and the Komodo monitor (*Varanus komodoensis*).

The presence of modern humans (*Homo sapiens*) is attested by archaeological remains from caves like Liang Bua, Liang Toge, and Liang Momer (Figure 13.7). The caves Liang Toge and Liang Momer were also excavated by Verhoeven, like Liang Bua. The human skeleton from Liang Toge has been ^{14}C dated to about 3,550 years ago, as reported by Teuku Jacob (1967), and belonged to a Negrito, whereas Liang Momer contained

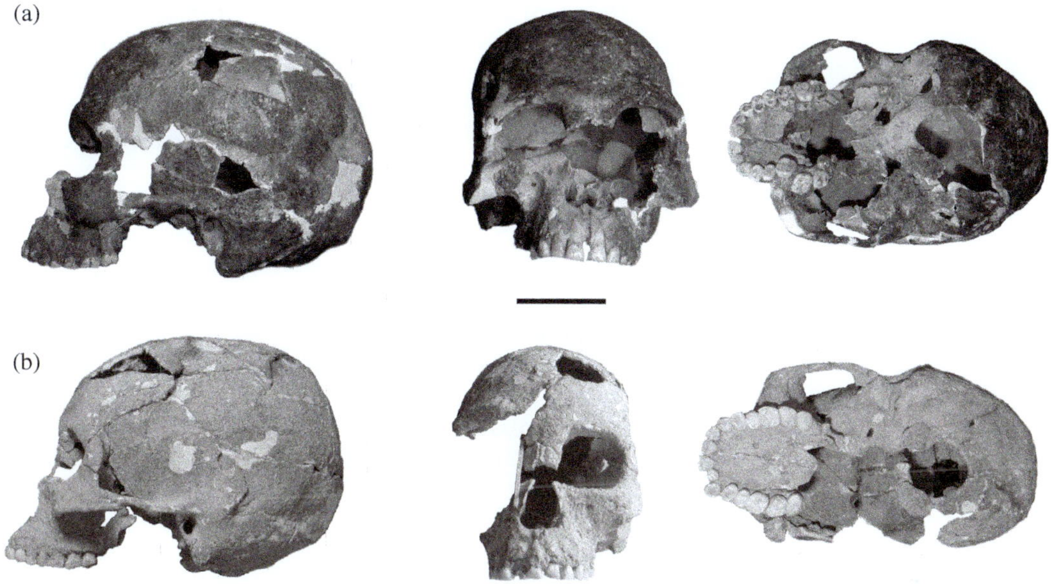

Figure 13.7 Skulls of *Homo sapiens* from (a) Liang Momer and (b) Liang Togè in left, front, and palatal views. Scale bar: 5 cm. National Museum of Natural History, Leiden. (Photograph: George Lyras.)

a normal-sized human skeleton. The material from Liang Momer could not be dated, but Jacob estimated a Mesolithic period at about 3,000–5,000 years ago. The archaeological findings present evidence for the use of shells as tools. An early introduction is the Sulawesi warty pig (*Sus celebensis)* at 7,000 years ago. Later, around 4,000 years ago, more mammals were introduced, such as wild boar (*Sus scrofa*), long-tailed macaque (*Macaca fascicularis*), Asian palm civet (*Paradoxurus hermaphroditus*), and Sunda porcupine (*Hystrix javanica*). They entered the archaeological sequence together with pottery, evidence for a Neolithic culture. The pig, macaque, and porcupine were likely introduced as food items – the latter two are still extensively hunted for food on the island – while the civet may have been imported as pest controller. The giant murids at the archaeological sites – including the still extant *Papagomys armandvillei* – are all survivors from the Late Pleistocene fauna. The same is likely the case for the critically endangered Flores shrew (*Suncus mertensi*).

Subfossil remains of a second, smaller monitor lizard (*Varanus hooijeri*) have been found in several caves (e.g. Liang Michael, Liang Toge), as reported by Leo Daniël Brongersma (1958), and recently from the archaeological sequence of Liang Bua as well. This varanid had blunt, barrel-shaped teeth similar to those of the extant Nile monitor (*Varanus niloticus*) and

Gray's monitor from Luzon, Philippines (*Varanus olivaceus*). These latter two species live on fruits and land snails, and likely this was valid for Hooijer's monitor as well. The smaller lizard was contemporaneous with the large form, though the latter had declined in number, likely due to the disappearance of the small stegodons.

Even more recent arrivals are the white-toothed shrew (*Crocidura monticola*), mice (*Mus musculus, Mus caroli*), and commensal rats (*Rattus exulans, Rattus argentiventer, Rattus rattus*, and *Rattus norvegicus*). The Polynesian rat (*Rattus exulans*), however, may have its origin on the island or the region instead of on the South Asian mainland, as suggested based on the white-bellied stock on Flores populations contrary to all other populations, and largest genetic diversity on Flores, as found by Thomson and colleagues (2014). The fossil record, though, does not yet support this hypothesis. Finally, with the late invasion of domestics and commensals, likely also the extant small monitor arrived (*Varanus salvator*), which is widely distributed throughout Southeast Asia. Deer were not introduced before 500 years ago.

Peculiarities and Evolution of Endemic Mammals

Dwarf Humans: Homo floresiensis

The Flores dwarf hominid is the smallest adult human ever found with a length of about 1 m only, estimated based on the size of the femur. The skull of this diminutive human species is very small and shows some primitive features, such as a receding forehead, prominent eyebrows, and the lack of a chin, compared with modern humans (Figure 13.8). Using high-resolution micro-CT scans, Kubo and colleagues (2013) estimated its endocranial volume, and indirectly its brain volume, to 426 cc (previous volume estimates range between 380 and 430 cc). Such an endocranial volume is about a quarter of that of the average modern human.

To date, remains of at least nine individuals were recovered. The skeletal remains from the cave Liang Bua are dated to about 60,000–100,000 years ago. These are not, however, the oldest remains. Gerrit van den Bergh and colleagues (2016) described hominin fossils from Mata Menge. The latter specimens comprise a mandible fragment and six isolated teeth belonging to at least three small-jawed, small-toothed individuals. Dating to about 700,000 years ago, these fossils now constitute the oldest hominin remains from Flores. The mandible and teeth excavated at Mata Menge are similar in size as well as morphological features to those of *Homo floresiensis*

Figure 13.8 Skulls of *Homo sapiens* (right) and *Homo floresiensis* (left). The latter is a stereolithographic replica. National Museum of Natural History, Leiden. (Photograph: Eelco Kruidenier.)

from Liang Bua, except that the lower first molar retains a more primitive condition. According to the authors, the Mata Menge fossils are derived compared with *Australopithecus* and *Homo habilis* and so tend to support the view that *Homo floresiensis* is a dwarfed descendent of early Asian *Homo erectus*. The environment of Mata Menge at the time was a savannah-like open grassland habitat with a wetland component according to Adam Brumm and colleagues (2016).

The first fossils were described in the time (2004) in which the epic Tolkien saga *Lord of the Rings* was made into a movie trilogy, and *Homo floresiensis* quickly received the nickname 'the Hobbit' in newspaper articles. Immediately after the discovery, a fierce debate started on the true nature of this human. Was it a mentally disabled, microcephalic modern human, as Teuku Jacob and colleagues (2006) suggested, or indeed a small-brained dwarf species, as Debbie Argue and team suggested in the same year? And if so, was its ancestor *Homo erectus* or *Homo sapiens* or even an unknown *Australopithecus* or *Homo habilis*?

The primitive features of the skull and brain were considered closely resembling those of *Homo erectus* in the broader sense according Brown (2004), Falk (2005, 2007), Argue (2006), Gordon (2008), Lyras (2009) (Figure 13.9), and Zeitoun (2016), with colleagues, but *Homo habilis* according to Morwood and Van Oosterzee (2007), or just an early *Homo* of about 1.5 million years ago according to Baab and McNulty (2009). There are some puzzling differences, however, between the Flores 'hobbit' and other early hominids, including *Homo erectus*. The face of

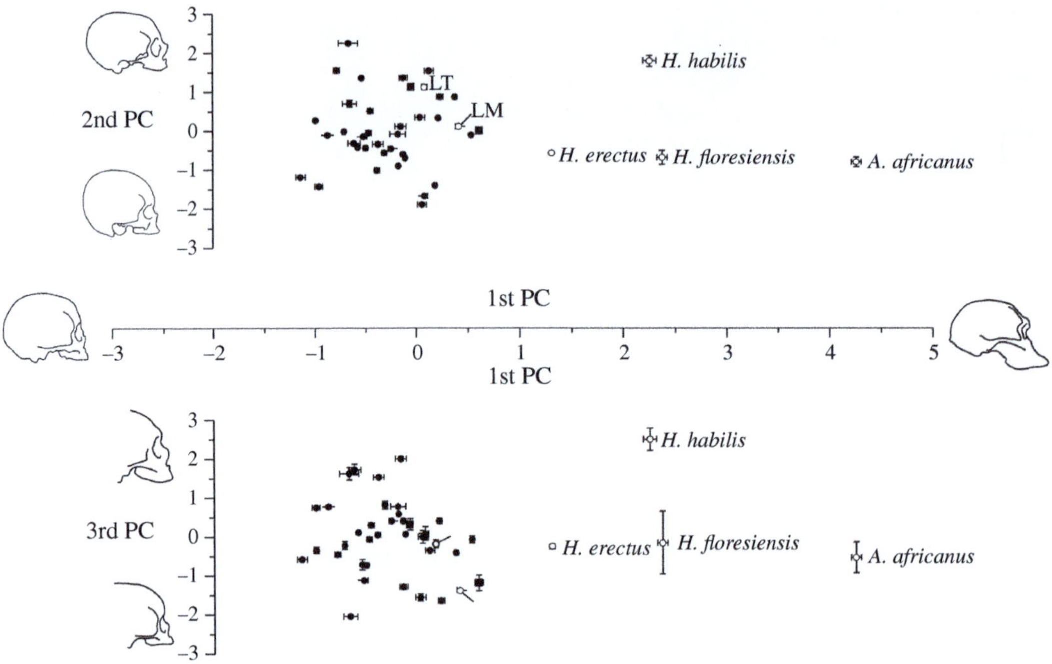

Figure 13.9 Plot of the first three principal components of a principal component analysis performed on 13 three-dimensional landmarks collected on recent and fossil hominid skulls to illuminate the phylogenetic position of *Homo floresiensis*. Error bars explain the uncertainty concerning the exact location of the bregma, which could not be precisely placed on *H. floresiensis*. The first principal component describes the height of the cranial vault, the degree of prognathism, and the development of the supraorbital region. The second principal component describes the changes in the midsagittal contours of the neurocranium (e.g. elongation of the vault), while the third mainly describes facial length. Abbreviations used in the plot: H. = *Homo*, A. = *Australopithecus*, LM = Liang Momer, LT = Liang Toge. The latter two are prehistoric caves on Flores with *Homo sapiens* remains (Redrawn from Lyras and colleagues, 2009.)

the 'hobbit' is rather flat, while *Homo erectus* has a pronounced muzzle. The brain size furthermore is amazingly small, comparable with that of *Australopithecus*, a much more primitive form of our lineage. One explanation is that this small brain is related to insularity and follows brain–body scaling patterns in parallel with some other island forms. This is in line with Falk and colleagues (2005), Weston and Lister (2009), and Kubo and colleagues (2013) who calculated that the observed small brain can indeed be explained for the greater part by normal body–brain size scaling relationships in *Homo*, with additional loss under predator release. Another possibility is that *Homo floresiensis* evolved from a hitherto unknown primitive Asian human of an australopithecine evolutionary stage, a possibility that was already hinted at by Argue and colleagues (2006) and again raised by Matt Tocheri and colleagues (2007), based on the primitive wrist of *Homo floresiensis* and by Jungers and colleagues (2009), based on the morphology of the navicular bone of the foot, which they compare with australopithecines and OH8 by lack of tarsal bones of *Homo erectus*. Argue and colleagues (2017) delved much deeper into this matter using a dataset comprising 50 skulls, 26 lower jaw, 24 dental, and 33 postcranial skeletal characters. Their detailed phylogenetic analysis revealed that *Homo floresiensis* is either closest related to *Homo habilis* or represents a sister group to a clade consisting of the latter species, *Homo erectus*, *Homo ergaster*, and *Homo sapiens*. If that is the case, once on the island, this ancestral human evolved dwarf size, under resource limitation in combination with release from predator pressure, just like other megafauna.

No matter the scenario, the founder population, representing an early offshoot of the genus *Homo* and an early Out-of-Africa migration, survived on Flores long after the source population on the Asian mainland was replaced by more advanced hominids. This implies that the dwarf humans of Flores are another relict taxon surviving on an island.

Dwarf Stegodons: Stegodon sondaari, Stegodon florensis

Hooijer (1964a, b) recognised an extremely small *Stegodon* form from Flores. The material, two milk molars, was collected at the surface in 1960 by Hartono of the geological survey of Indonesia, who described the geology of that area in 1961. The size reminded Hooijer of the smallest dwarf forms from Java (*Stegodon hypsilophus*) and Sulawesi (*Stegodon sompoensis*), which he had described in 1955 and 1964, respectively. More than thirty years later, van den Bergh (1999) described the Flores dwarf form as *Stegodon sondaari*, based on cranial, dental, and skeletal material excavated at Tangi Talo in the lower layer of the Ola Bula formation, dated to the Early Pleistocene. *Stegodon sondaari* is about half as tall as the Javan stegodon (*Stegodon trigonocephalus*) and has a body mass of about 15% of its presumed mainland ancestor (*Stegodon elephantoides*). It exhibits a mixture of primitive and advanced traits: the molars have a relatively low number of ridges, but are relatively high crowned. They are also relatively large in proportion to the jaws. The lower jaw symphysis is further very short and lacks a protruding rostrum, in contrast to the jaw of the Timor dwarf stegodon (Box 13.2). Its diet consisted primarily of C_3 plants, meaning that it was a browser of leaves rather than a grazer, as indicated by analysis of stable carbon and oxygen isotopes from its molars by Mika Puspaningrum and colleagues (2014).

Hooijer (1957b) described larger, but still small, stegodon remains from the So'a Basin as a subspecies of the Javan stegodon, *Stegodon trigonocephalus florensis*. Van den Bergh recognised it as an independent species, *Stegodon florensis*. This Middle Pleistocene dwarf stegodon (Plate 26) is some 20–30% smaller than the Javan stegodon and half the size of its presumed mainland ancestor (*Stegodon ganesa*) but had higher crowned (hypsodont) molars and a more advanced molar ridge formula. In contrast with the smaller stegodon from the previous biozone, this middle-sized stegodon was a grazer of C_4 grasses, as shown by stable carbon and oxygen isotope analysis of tooth enamel samples collected for several *Stegodon florensis* individuals by Puspaningrum and colleagues (2014) and Brumm and colleagues (2016). The dietary shift from C_3 browse to C_4 grasses reflects an increasingly drier climate in the region from the Early to Middle Pleistocene across the region with open grasslands present in the So'a Basin.

Through time, the middle-sized stegodon species evolved smaller body size as well, to such an extent that the form from Liang Bua shows a similar body size reduction relative to its continental ancestor of about 83% as the Early Pleistocene stegodon (van der Geer and colleagues, 2016). The size difference is reflected in separate subspecies as proposed by van den Bergh and colleagues (2008), with *Stegodon florensis insularis* for the smallest and youngest form and *Stegodon florensis florensis* for the larger and older form.

BOX 13.2

Dwarf Stegodon from Timor and Sumba

The dwarf stegodon from Timor is the easternmost stegodon species in insular Southeast Asia, separated from Australia only by the Timor Strait, which was less wide during glacial periods. The history of the Timor stegodons begins in 1960, when the priest and archaeologist Theodor Verhoeven found its fossils in a dry riverbed near Atambua, announced four years later (Plate 28) (Verhoeven, 1964). Sartono (1969) named this new dwarf species *Stegodon timorensis* (Figure Box 13.2) In the same year, Dirk Hooijer also recognised a larger species among the Timor stegodon material, reminding him of the situation he knew from Flores. He referred to the larger Timor stegodon as *Stegodon timorensis* subspecies D, obviously not knowing what to think of its ancestry and geological age. Sartono (1969) proposed eastward migration of stegodons from Java along the island chain of the Lesser Sunda Islands to Flores and further to Timor, evolving into different forms on the various islands. Two migration waves then took place, one eventually leading to the 'pygmy' Early Pleistocene species, such as those of Java (Sambungmacan form) and Flores (*S. sondaari*), followed by a second wave replacing these very small forms, and recognised as the middle-sized dwarfs of the Middle Pleistocene, as on Sulawesi (*Stegodon* sp. B), Luzon (*Stegodon luzonensis*), Java (*Stegodon trigonocephalus*), and Flores (*Stegodon florensis florensis*). The first wave may represent a form close to *Stegodon elephantoides*, as proposed by van den Bergh (1999) for the Flores pygmy form; the second *Stegodon ganesha*. Lack of a sound stratigraphic context of the findings, however, hampers the phylogeographic and temporal reconstruction of these migrations. The only available absolute dating is an Uranium-Thorium dating (U-Th) on a stegodon tusk fragment from landslide deposits at the site Wéaiwé near Atambua by Julien Louys and colleagues (2016), which yielded an age range between about 130,000–168,000 years ago (late Middle Pleistocene). The earliest archaeological assemblages with stone tools – none of which contains stegodon – date back to about 40,000 years ago, indicating a long gap between the last stegodon and the first human on Timor.

Figure Box 13.2 Partial mandible of *Stegodon timorensis*, cast of holotype; occlusal view. Scale bar: 5 cm. National Museum of Natural History, Leiden. (Photograph: Eelco Kruidenier.)

Possibly Sumba had a similar fauna as Flores and Timor, as indicated by the finding of a partial left lower jaw with molars of a very small stegodon at Watu Mbaka, southeast of Waingapu. It was described as *Stegodon sumbaensis* by Sartono (1979), who noted that this stegodon is about the same size as the smallest species of Flores. This was confirmed by van der Geer and colleagues (2016), who calculated a body mass reduction of over 90% for the Sumba stegodon relative to mainland *Stegodon ganesa*. Samuel Turvey and colleagues (2017) reported more dwarf stegodon material from Early Pleistocene sediments at Lewapaku, together with fossils of the monitor *Varanus komodoensis*, further confirming the faunal similarity between Flores, Timor (Box 13.1), and Sumba. They further reported the small monitor *Varanus* cf. *hooijeri* from Holocene sites, as is the case on Flores.

Giant, Medium, and Small Rats

Today, the largest endemic rat of Flores (*Papagomys armandvillei*) has a body length of just below half a metre with a 70 cm long tail. The other giant rat (*Papagomys theodorverhoeveni*), however, seems to be extinct, despite a claim to the contrary of Suyanto and Watts (2002). Jelle Zijlstra and colleagues (2008) showed that this rediscovered specimen is a *Papagomys armandvillei* instead.

The genera of Pleistocene and Holocene giant rats (*Papagomys*; Plate 27) and medium-sized rats (*Komodomys* and *Paulamys*) are more closely related to one another than to any other murid, according to Musser and Carlton (2005). This suggests an adaptive radiation driven by competition and facilitated by ecological release. The Middle to Late Pleistocene giant rat (*Hooijeromys*) likely represents one lineage in this radiation. It had a body head length of about 0.28–0.35 m according to Musser (1981) and resembled the recent giant rat *Papagomys* in several characters. *Hooijeromys* may well be ancestral to the later *Papagomys* species. The other giant rat (*Spelaeomys*; Plate 27), on the other hand, originates clearly from a different stock. Its complex molar pattern resembles that of the giant rat of Timor (Box 13.3).

The taxonomic status of the middle-sized rats is debated. They are endemic to the area and still live on Komodo and Rinca. Musser and colleagues (1986) renamed the preoccupied name *Floresomys* to *Paulamys*, while Kitchener and Mohamad Yani (1997) moved the single species to the genus *Bunomys*, the hill rats of Sulawesi. Van den Bergh and colleagues (2008) treated the two middle-sized rats as a single group, *Komodomys*/*Paulamys*, not only because of the taxonomic uncertainties but also because the mandibles are practically indistinguishable from each other. In parallel with the giant size of the murids, also a barn owl (*Tyto* sp.) seems to have evolved into a large-bodied endemic form.

Hainald's rat, which is a small rat, was provisionally placed under *Rattus* by Kitchener and colleagues (1991b). Molecular data, analysed by Vicki Thomson and colleagues (2018), places *R. hainaldi* close to *Rattus macleari* of Christmas Island, south of Java. This pair is then closest related to *Bandicota*, *Nesokia* and one step further removed to *Diplothrix* of the Ryukyu Islands, but further analyses is needed to establish this relationship and remove *R. hainaldi* and *R. macleari* from the genus *Rattus*.

BOX 13.3

Giant Rats of Timor and Sumba

The extinct giant rats from Timor – *Coryphomys buehleri* and *Coryphomys musseri* of similar size – are only known by subfossil fragments from prehistoric sites in limestone caves. Like on Flores, Timor housed several species of giant rats, next to medium-sized and smaller murids, mentioned by Ken Aplin and Kristopher Helgen (2010). *Coryphomys* has a complex molar pattern (Figure Box 13.3), resembling that of *Spelaeomys* of Flores. The skull features a high degree of anatomical specialisation, likely the result of long-term isolation. Aplin and Helgen (2010) tentatively suggested a potential relation with the cloud rats of the Philippines (Phloeomyini) and the Australo-Papuan water rats (Hydromyini). They estimated that *Coryphomys* weighted in the range of 1.2–2.6 kg or even more, based on skeletal remains. Justyna Miszkiewicz and colleagues (2020) showed that the giant rats had a relatively slow pace of bone metabolic activity, relative to the smaller rats of Timor, consistent with predictions made by the ecological interpretation of the island rule (low predator and competitor pressure), but also consistent with scaling models that predict a slower pace of life in larger species.

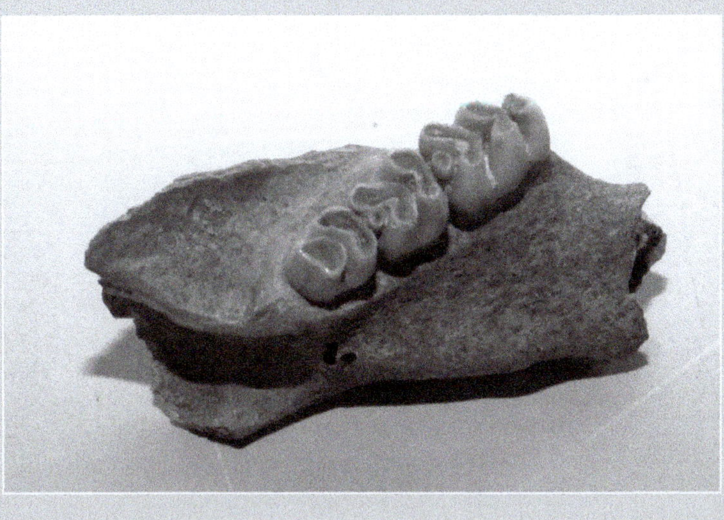

Figure Box 13.3 Maxilla fragment with M1–3 of the giant rat of Timor (*Coryphomys buhleri*) from Liang Leluat; occlusal view. Collection: Verhoeven, National Museum of Natural History, Leiden. (Photograph: Eelco Kruidenier.)

Other endemic rats from Timor, first noted by Jack Mahoney but in part unpublished, are Mahoney's genus A (one species), Mahoney's genus B (two or three species), Mahoney's genus C (two or three species), four small-sized species, and the still extant *Rattus timorensis*. Already Kitchener and colleagues (1991) noticed its peculiar nature and that it is different from *Rattus* and more closer to *Bunomys* of Sulawesi and *Paulamys* of Flores. Molecular data (cytochrome b) obtained by Vicki Thomson and colleagues (2018) does not support a close relationship to the former, while *Paulamys* was not sampled, so remains a potential close relative.

The fossil murids from Sumba are even less known, but included a giant rat (*Raksasamys tikusbesar*) with a body mass estimate of between about 470 and 680 g, calculated by Turvey and colleagues (2017). Its teeth are very high crowned with a simplified crown morphology and peculiar features, unlike that of other eastern Lesser Sunda murids. A second giant rat (*Milimonggamys juliae*) is much smaller than *Raksasamys*, with a body mass between about 250 and 335 g. The two Sumba giant rats show closer molar morphology to the murids from Flores – excluding *Spelaeomys* – than to *Spelaeomys* and the giant rats Timor with their distinctive complex molars with many cusps. The material was retrieved from Holocene cave sediments. Much older material, excavated from a Pleistocene deposit at Lewpaku of about 1 million years old, is too fragmentary to identify other than that it belongs to a giant murid. The same deposit yielded remains of *Stegodon sumbaensis* and a large monitor lizard (*Varanus komodoensis*, but more likely a new species, see above).

Insectivores: Shrews and Bats

Van den Hoek Ostende and colleagues (2007) described the first fossil insectivores from the Late Pleistocene of Flores. They identified three mandibles from the cave Liang Bua as belonging to two different shrew species, which they named *Crocidura* or *Suncus* sp. A and sp. B. The two shrew genera – white-tooted and red-toothed shrews, respectively – differ from each other in the number of antemolars in the upper jaw, but in Liang Bua, only mandibles were recovered, hampering further classification. The new shrew species seem unrelated to the extant endemic Flores shrew (*Suncus mertensis*) or to the widespread Sunda shrew (*Crocidura monticola*) and represent in all likelihood extinct endemic shrews. Regarding size, species A is larger than species B, and both are larger than the Flores shrew.

Liang Bua also yielded one of the best insular records of fossil bats. The bat assemblage consists of small bats only and is described by Esther Ouwendijk and colleagues (2014). Two different stratigraphical groups are present, a Late Pleistocene level, dated to between 61,000 and 74,000 years ago, and an Early Holocene, archaeological level. The Pleistocene assemblage includes a flute-nosed bat (*Murina* sp. aff. *florium*) and the broad-eared horseshoe bat (*Rhinolophus euryotis*). The Holocene assemblage is much richer and more diverse and is dominated by the two bat species of the Late Pleistocene, followed by newcomers, including the diadem leaf-nosed bat (*Hipposideros diadema*), the Lombok horseshoe bat (*Rhinolophus simplex*), the Sumba roundleaf bat (*Hipposideros* sp. cf. *sumbae*), a sheath-tailed bat (*Taphozous* sp.), a woolly bat (*Kerivoula* sp.), and a bent-winged bat (*Miniopterus* sp.) that have also been recovered. The diadem leaf-nosed bat, the most abundant species in population numbers, and the bent-winged bat are today extirpated from Flores. All recovered bats that were determined on the species level have a much wider distribution, most of them including Indonesia, New Guinea, and Australia, and none of them are endemic to Flores, confirming their much better dispersal abilities compared to rodents.

References

Aplin K.P., Helgen K.M. 2010. Quaternary murid rodents of Timor Part I: New material of *Coryphomys buehleri* Schaub, 1937, and description of a second species of the genus. *Bulletin of the American Museum of Natural History*, 341: 1–80.

Argue D., Donlon D., Groves C., Wright R., 2006. *Homo floresiensis*: Microcephalic, pygmoid, *Australopithecus*, or *Homo*? *Journal of Human Evolution*, 51: 360–374.

Argue D., Groves C.P., Lee M.S.Y., Jungers W.L. 2017. The affinities of *Homo floresiensis* based on phylogenetic analyses of cranial, dental, and postcranial characters. *Journal of Human Evolution*, 107: 107–133.

Baab C., McNulty K.P. 2009. Size, shape, and asymmetry in fossil hominins: The status of the LB1 cranium based on 3D morphometric analyses. *Journal of Human Evolution*, 57: 608–622.

Brongersma L.D. 1958. On an extinct species of the genus *Varanus* (Reptilia, Sauria) from the island of Flores. *Zoologische Mededelingen*, 36: 113–125.

Brown P., Sutikna T., Morwood M.J., Soejono R.P., Jatmiko, Wayhu Saptomo E., Awe Due, R. 2004. A new small-bodied hominin from the Late Pleistocene of Flores, Indonesia. *Nature*, 431: 1055–1061.

Brumm A., Jensen G.M., van den Bergh G.D., Morwood M.J., Kurniawan I., Aziz F., Storey M. 2010. Hominins on Flores, Indonesia, by one million years ago. *Nature*, 464: 748–752.

Brumm A., Moore, M.W. 2012. Biface distributions and the Movius Line: A Southeast Asian perspective. *Australian Archaeology* 74: 32–46.

Brumm A., van den Bergh G.D., Storey M., Kurniawan I., Alloway B.V., Setiawan R., Setiyabuti E., Grün R., Moore M.W., Puspaningrum M.R., Wibowo U.P., Insani H., Sutisna I., Westgate J.A., Pearce

N.J.G., Duval M., Meijer H.J.M., Aziz F., Sutikna T., van der Kaars S., Flude S., Morwood M.J. 2016. Age and context of the oldest known hominin fossils from Flores. *Nature*, 534: 249–253.

Ehrat H. 1925. Geologische mijnbouwkundige onderzoekingen op Flores. Geologische schetskaart, 1:250.000. *Jaarboek van het Mijnwezen in Nederlandsch Oost-Indië, Verhandelingen*, 1925 (2): 208–226.

Falk D., Hildebolt C., Smith K., Morwood M., Sutikna T., Brown P., Jatmiko E., Saptomo W., Brunsden B., Prior F. 2005. The brain of *Homo floresiensis*. *Science*, 308: 242–245.

Falk D., Hildebolt C., Smith K., Morwood M. J., Sutikna T., Jatmiko, Wahyu Saptomo E., Imhof H., Seidler H., Prior F. 2007. Brain shape in human microcephalics and *Homo floresiensis*. *Proceedings of the National Academy of Sciences of the United States of America*, 104: 2513–2518.

Gordon A. D., Nevell L., Wood B. 2008. The *Homo floresiensis* cranium (LB1): Size, scaling, and early *Homo* affinities. *Proceedings of the National Academy of Sciences of the United States of America*, 105 (12): 4650–4655.

Hocknull S.A., Piper P.J., van den Bergh G.D., Due R.A., Morwood M.J., Kurniawan I. 2009. Dragon's paradise lost: Palaeobiogeography, evolution and extinction of the largest-ever terrestrial lizards (Varanidae). *PLoS ONE*, 4 (9): e7241.

Hooijer D.A. 1955. Fossil Proboscidea from the Malay Archipelago and India. *Zoologische Verhandelingen*, 28: 1–146.

Hooijer D.A. 1957 (a). Three new giant prehistoric rats from Flores Lesser Sunda Islands. *Zoologische Mededelingen*, 35: 299–313.

Hooijer D.A. 1957 (b). A *Stegodon* from Flores. *Treubia*, 24: 119–129.

Hooijer D.A. 1964 (a). Pleistocene vertebrates from Celebes XII. Notes on pygmy Stegodonts. *Zoologische Mededelingen*, 40: 37–44.

Hooijer D.A. 1964 (b). On two milkmolars of a pygmy stegodont from Ola Bula, Flores. *Bulletin Geological Survey of Indonesia*, 1: 49–52.

Hooijer D.A. 1969. The Stegodon from Timor. *Proceedings of the Koninklijke Akademie van Wetenschappen (Series B)*, 72: 201–210.

Hooijer D.A. 1972. *Varanus* (Reptilia, Sauria) from the Pleistocene of Timor. *Zoologische Mededelingen*, 47: 445–447.

Ingicco T., van den Bergh G.D., Jago-on C., Bahain J.-J., Chacón M.G., Amano N., Forestier H., King C., Manalo K., Nomade S., Pereira A., Reyes M.C., Sémah A.-M., Shao Q., Voinchet P., Falguères C., Albers P.C.H., Lising M., Lyras G., Yurnadi D., Rochette P., Bautista A., de Vos J. 2018. Earliest known hominin activity in the Philippines by 709 thousand years ago. *Nature*, 557: 233–237.

Jacob T. 1967. *Some Problems Pertaining to the Racial History of the Indonesian Region*. Neerlandia, Utrecht.

Jacob T., Soejono R.P., Hsü K., Frayer D.W., Eckhardt R.B., Kuperavaga A.J., Thorne A., Henneberg M. 2006. Pygmoid Australomelanesian *Homo sapiens* skeletal remains from Liang Bua, Flores: Population affinities and pathological abnormalities. *Proceedings of the National Academy of Sciences of the United States of America*, 103: 13421–13426.

Jungers W.L., Harcourt-Smith W.E.H., Wunderlich R.E., Tocheri M.W., Larson S.G., Sutikna T., Awe Due, R., Morwood M.J. 2009. The foot of *Homo floresiensis*. *Nature*, 459: 81–84.

Kitchener D. J., Yani M. 1997. Small non-volant terrestrial mammal diversity along an altitudinal gradient at Gunung Ranaka, Flores Island, *Indonesia. Tropical Biodiversity*, 5 (2): 155–159.

Kitchener D.J., Aplin K.P., Boeadi. 1991. A new species of *Rattus* from Gunung Mutis, South West Timor Island, Indonesia. *Records of the Western Australian Museum*, 15: 445–461.

Kitchener D.J., How R.A., Maharadatunkamsi. 1991 (a). *Paulamys* sp. cf. *P. naso* (Musser, 1981) (Rodentia: Muridae) from Flores Island, Nusa Tenggara, Indonesia: Description from a modern specimen and a consideration of its phylogenetic affinities. *Records of the Western Australian Museum*, 15: 171–189.

Kitchener D.J., How R.A., Maharadatunkamsi. 1991 (b). A new species of *Rattus* from the mountains of West Flores, Indonesia. *Records of the Western Australian Museum*, 15: 611–626.

Kubo D., Kono R.T., Kaifu Y. 2013. Brain size of *Homo floresiensis* and its evolutionary implications. *Proceedings of the Royal Society B*, 280: 20130338.

Kuhnt W., Holbourn A., Hall R., Zuvela M., Käse R., 2004. Neogene history of the Indonesian throughflow. In: Clift P., Hayes D., Kuhnt W., Wang P. (eds), *Continent-Ocean Interactions in the East Asian Marginal Seas*. American Geophysical Union Monographs 149, pp. 299–320.

Locatelli E., Awe Due R., van den Bergh G.D., van den Hoek Ostende L.W. 2012. Pleistocene survivors and Holocene extinctions: The giant rats from Liang Bua (Flores, Indonesia). *Quaternary International*, 281: 47–57.

Louys J., Price G.J., O'Connor S. 2016. Direct dating of Pleistocene stegodon from Timor Island, East Nusa Tenggara. *PeerJ*, 4: e1788

Lyras G.A., Dermitzakis M.D., van der Geer A.A.E., van der Geer S.B., de Vos J. 2009. The origin of *Homo floresiensis* and its relation to evolutionary processes under isolation. *Anthropological Science*, 117: 33–43.

Mammerickx J., Fisher R.L., Emmel F.J., Smith S. M. 1976. *Bathymetry of the East and Southeast Asian Seas*. Scripps Institute of Oceanography, La Jolla.

Maringer J., Verhoeven T. H. 1970. Die Steinartefacte aus der *Stegodon*-Fossil-Schicht von Mengeruda auf Flores, Indonesien. *Anthropos*, 65: 229–247.

Meijer H.J.M., Due R.A. 2010. A new species of giant marabou stork (Aves: Ciconiiformes) from the Pleistocene of Liang Bua, Flores (Indonesia). *Zoological Journal of the Linnean Society*, 160, 707–724.

Meijer H.J.M., Kurniawanc I., Setiabudi E., Brumm A., Sutikna T., Setiawan R., van den Bergh G.D. 2015. Avian remains from the Early/Middle Pleistocene of the So'a Basin, central Flores, Indonesia, and their palaeoenvironmental significance. *Palaeogeography, Palaeoclimatology, Palaeoecology*, 440: 161–171.

Miszkiewicz J.J., Louys J., Beck R.M.D., Mahoney P., Aplin K., O'Connor S. 2020. Island rule and bone metabolism in fossil murines from Timor. *Biological Journal of the Linnean Society*, 129: 570–586.

Morwood M., van Oosterzee P. 2007. *The Discovery of the Hobbit. The Scientific Breakthrough that Changed the Face of Human History*. Random House, New York.

Morwood M.J., Aziz F., van den Bergh G.D., Sondaar P.Y., de Vos J. 1997. Stone artefacts from the 1994 excavations at Mata Menge, West Central Flores, Indonesia. *Australian Archaeology*, 44: 26–34.

Morwood M.J., O'Sullivan P.B., Aziz F., Raza A. 1998. Fission-track ages of stone tools and fossils on the east Indonesian island of Flores. *Nature*, 392: 173–176.

Musser G.G. 1981. The giant rat of Flores and its relatives eat of Borneo and Bali. *Bulletin American Museum Natural History*, 169: 69–175.

Musser G.G., Carlton M.D., 2005. Superfamily Muroidea. In: Wilson D.E., Reeder D.M. (eds.), *Mammal Species of the World. A Taxonomic and Geographic Reference*, 3rd edn., Vol. 2. Johns Hopkins University Press, Baltimore, pp. 894–1531.

Musser G.G., van de Weerd A., Strasser E. 1986. *Paulamys*, a replacement name for *Floresomys* Musser, 1981 (Muridae), and new material of that taxon from Flores, Indonesia. *American Museum Novitates*, 2850: 1–10.

Nishimura S., Otofuji Y., Ikeda T., Abe E., Yokoyama T., Kobayashi Y., Hadiwisatra S., Sopohaluwakan J., Hehuwat F. 1981. Physical geology of the Sumba, Sumbawa and Flores Islands. In: Barber A.J., Wiryosujono S. (eds), *The Geology and Tectonics of Eastern Indonesia. Geological Research Development Center Bandung, Special Publication*, 2: 105–113.

Ouwendijk E.M., Due R.A., Locatelli E., Jatmiko, van den Hoek Ostende L.W. 2014. Bat cave and Hobbit hole, microbats of Liang Bua (Flores, Indonesia). *Alcheringa: An Australasian Journal of Palaeontology*, 38: 422–433.

Puspaningrum M.R., van den Bergh G.D., Chivas A., Setiabudi E., Kurniawan I., Brumm A., Sutikna T. 2014. Preliminary results of dietary and environmental reconstructions of Early to Middle Pleistocene Stegodons from the So'a Basin of Flores, Indonesia, based on enamel stable isotope records. In: Kostopoulos D.S., Vlachos E., Tsoukala E. (eds), *Scientific Annals, School of Geology, Aristotle University of Thessaloniki, Special Volume*, 102: 164–165.

Sartono S. 1969. *Stegodon timorensis*: A pygmy species from Timor (Indonesia). *Proceedings of the Koninklijke Nederlandse Akademie van Wetnschappen (Series B)*, 72: 192–200.

Sartono S. 1979. The discovery of a pygmy *Stegodon* from Sumba, East Indonesia: An announcement. *Modern Quaternary Research in SE Asia*, 5: 57–63.

Schepartz L.A., Stoutamire S., Bekken D.A. 2005. *Stegdon orientalis* from Panxian Dadong, a Middle Pleistocene archaeological site in Guizhou, South China: Taphonomy, population structure and evidence for human interactions. In: Palombo M.R., Mussi M., Gioia P., Cavarretta A. (eds), *Studying Proboscideans: Knowledge, Problems and Perspectives. Quaternary International*, 126–128: 271–282.

Setiyabudi E. 2006. *Palaeontological study of Fossil Giant Tortoises from the Indonesian Islands.* Unpublished MSc thesis, Department of Earth and Environmental Science, University of Kagoshima, Japan.

Sondaar P.Y. 1989. Did Man reach Australia via the giant rat and dingo route? *Publications of the Geological Research and Development Centre, Paleontology*, 5: 76–83.

Sondaar P.Y., van den Bergh G.D., Mubroto B., Aziz F., de Vos J., Batu U.L. 1994. Middle Pleistocene faunal turnover and colonization of Flores (Indonesia) by *Homo erectus. Comptes Rendus de l'Académie des Sciences de Paris (Serie 2)*, 319: 1255–1262.

Sutikna T., Tocheri M.W., Morwood M.J., Saptomo E.W., Jatmiko, Awe Due R., Wasisto S., Westaway K.E., Aubert M., Li B., Zhao J.-X., Storey M., Alloway B.V., Morley M.W., Meijer H.J.M., van den Bergh G.D., Grün R., Dosseto A., Brumm A.R., Jungers W.L., Roberts R.G. 2016. Revised stratigraphy and chronology for *Homo floresiensis* at Liang Bua in Indonesia. *Nature*, 532: 366–369.

Suyanto A., Watts C.H.S. 2002. Verhoeven's Giant rat of Flores, Indonesia (*Papagomys theodorverhoeveni* Musser, 1981; Muridae) is a modern species. *Treubia*, 32: 87–93.

Thomson V., Aplin K. P., Cooper A., Hisheh S., Suzuki H., Maryanto I., Yap G., Donnellan S.C. 2014. Molecular genetic evidence for the place of origin of the Pacific rat, *Rattus exulans. PLoS ONE*, 9 (3), e91356: 1–11.

Thomson V., Wiewel A., Chinen A., Maryanto I., Sinaga M.H., How R., Aplin K., Suzuki H. 2018. A perspective for resolving the systematics of *Rattus*, the vertebrates with the most influence on human welfare. *Zootaxa*, 4459: 431–452.

Tocheri M.W., Orr C.M., Larson S.G., Sutikna T., Jatmiko, Wahyu Saptomo E., Awe Due R., Djubiantono T., Morwood M.J., Jungers W.L., 2007. The primitive wrist of *H. floresiensis* and its implications for hominin evolution. *Science*, 317: 1743–1745.

Turvey S.T., Crees J.J., Hansford J., Jeffree T.E., Crumpton N., Kurniawan I., Setiyabudi E., Guillerme T., Paranggarimu U., Dosseto A., van den Bergh G.D. 2017. Quaternary vertebrate faunas from Sumba, Indonesia: Implications for Wallacean biogeography and evolution. *Proceedings of the Royal Society B*, 284: 20171278.

van den Bergh G.D. 1999. The Late Neogene elephantoid-bearing faunas of Indonesia and their palaeozoogeographic implications; A study of the terrestrial faunal succession of Sulawesi, Flores and Java, including evidence for early hominid dispersal east of Wallace's line. *Scripta Geologica*, 117: 1–419.

van den Bergh G.D., Awe Due R., Morwood M.J., Sutikna T., Jatmiko P., Wahyu Saptomo E. 2008. The youngest *Stegodon* remains in south-east Asia from the Late Pleistocene archaeological site Liang Bua, Flores, Indonesia. *Quaternary International*, 181 (1): 16–48.

van den Bergh G.D., Aziz F., Kurniawan I., Morwood M.J., Lentfer C.J., Suyone, Setiawan R. 2009. Environmental reconstruction of the Middle Pleistocene archaeological and palaeontological site of Mata Menge, Central Flores, Indonesia. In: Aziz F., Morwood M.J., van den Bergh G.D. (eds), *Pleistocene Geology, Palaeontology and*

Archaeology of the Soa Basin, Central Flores, Indonesia. Pusat Servei Geologi, Bandung, Special Publication, 36: 58–94.

van den Bergh G.D., Brumm A., Puspaningrum M.R., Setiabudi E., Kurniawan I., Prasetyo U.W. 2014. Taphonomy of *Stegodon florensis* remains from the early Middle Pleistocene archaeological site Mata Menge, Flores, Indonesia. In: Kostopoulos D.S., Vlachos E., Tsoukala E. (eds), *Scientific Annals, School of Geology, Aristotle University of Thessaloniki, Special Volume*, 102: 207–208.

van den Bergh G.D., Meijer H.J.M., Awe Due R., Morwood M.J., Szabó K., van den Hoek Ostende L.W., Sutika T., Saptomo E.W., Piper P.J., Dobney K.M. 2009. The Liang Bua faunal remains: A 95k.yr. sequence from Flores, East Indonesia. *Journal of Human Evolution*, 57: 527–537.

van den Bergh G.D., Kaifu Y., Kurniawan I., Kono R.T., Brumm A., Setiyabudi E., Aziz F., Morwood M.J. 2016. *Homo floresiensis*-like fossils from the early Middle Pleistocene of Flores. *Nature*, 534: 245–248.

van den Hoek Ostende L.W., van den Bergh G.D., Awe Due R. 2007. First fossil insectivores from Flores. *Hellenic Journal of Geosciences*, 41: 67–72.

van der Geer A.A.E., van den Bergh G., Lyras G.A., Prasetyo U.W., Due R.A., Setiyabudi E., Drinia H. 2016. The effect of area and isolation on insular dwarf proboscideans. *Journal of Biogeography*, 43: 1656–1666.

Veatch E.G., Tocheri M.W., Sutikna T., McGrath K., Saptomo E.W., Jatmiko, Helgen K.M. 2019. Temporal shifts in the distribution of murine rodent body size classes at Liang Bua (Flores, Indonesia) reveal new insights into the paleoecology of *Homo floresiensis* and associated fauna. *Journal of Human Evolution*, 130: 45–60.

Verhoeven T. 1964. Stegodon-Fossilien auf der Insel Timor. *Anthropos*, 59: 634–646.

Weston E.M., Lister A.M. 2009. Insular dwarfism in hippos and a model for brain size reduction in *Homo floresiensis*. *Nature*, 459: 85–88.

Zeitoun V., Barriel V., Widianto H. 2016. Phylogenetic analysis of the calvaria of *Homo floresiensis*. *Comptes Rendus Palevol*, 15: 555–568.

Zijlstra J.S., van den Hoek Ostende L.W., Awe Due R. 2008. Verhoeven's giant rat of Flores (*Papagomys theodorverhoeveni*, Muridae) extinct after all? *Contributions to Zoology*, 77 (1): 25–31.

CHAPTER FOURTEEN

Sulawesi

Evolution of Island Mammals: Adaptation and Extinction of Placental Mammals on Islands, Second Edition.
Alexandra van der Geer, George Lyras and John de Vos.
© 2021 John Wiley & Sons Ltd. Published 2021 by John Wiley & Sons Ltd.

The Indonesian island Sulawesi is renowned for its extraordinarily high level of endemism; over 90% of its terrestrial mammal species are endemic to the island. The same applies to fossil faunas, although much less species are known. The Early Pleistocene fauna is characterised by dwarf elephants and stegodons, large pigs, giant Sulawesi rat, and giant tortoises. The Middle Pleistocene fauna is insufficiently known but seems to have contained small elephants, stegodons, and a derived form of the large pig. During the Late Pleistocene, ancestors of the endemic mammals of today arrived, such as the babirusa and the anoas. A number of mainland taxa were introduced through human agency in the Holocene.

Geology and Palaeogeography

Sulawesi – or Celebes in the older literature – is one of the Greater Sunda Islands and lies east of Borneo. The island has an irregular shape and resembles some sort of mutilated brittle star with its four large peninsulas arranged around a relatively small centre. Because of this shape, the Portuguese initially considered the island an archipelago. The central part is very mountainous, and the best connection between the peninsulas is still overseas. A few island chains belong to Sulawesi, such as the Sangihe Islands extending between the northern peninsula towards Mindanao of the Philippines and the Peleng or Banggai Islands east of central Sulawesi.

Alfred Wallace, who travelled within the Malay Archipelago between 1854 and 1862, was the first to realise that only long-term isolation and local evolution could explain the present-day faunal distribution on the various islands. Especially Sulawesi in the centre of the archipelago awoke his interest, because paradoxically, this island was one of the poorest in number of species and the most isolated as based on the features of the faunas of all the large islands in the archipelago. Wallace (1863) drew a zoogeographic boundary between ‘Sundaland’, consisting of the large islands Borneo, Sumatra, Java, and Bali in the west and Sulawesi and Lombok in the east (Figure 14.1). Five years later, Thomas Henry Huxley extended the line between Borneo and the Sulu Island arc through the Sulu Sea and further through the Strait of Mindanao west of the Philippines. He named the line Wallace’s Line, even though Wallace himself disagreed with the extension; in the latter’s view, the Philippines belonged to the Asiatic realm. Later, a second line was drawn between Australia and the Indonesian islands, named Lydekker’s Line. The area between this line and Wallace’s is referred to as *Wallacea*. The extension made

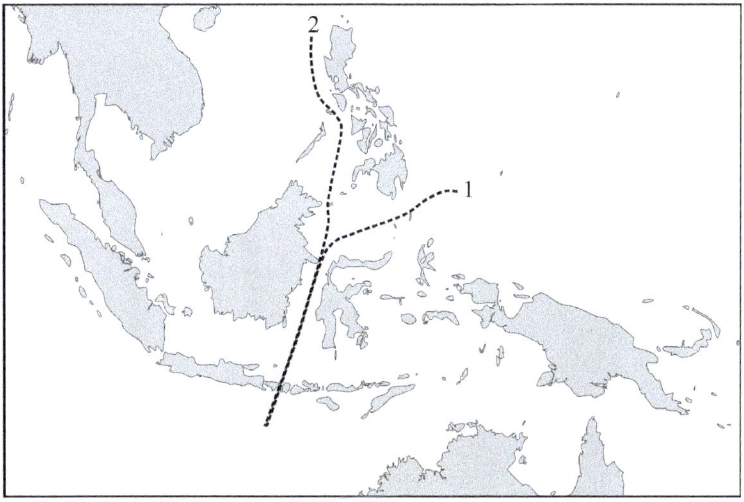

Figure 14.1 Map showing the Oriental and Australian faunal regions. The transitional region between the zoogeographic boundaries, known as *Huxley's line* (2) in the west and *Lydekker's line* (1) in the east, is also referred to as *Wallacea*. (Redrawn from Van den Bergh, 1999.)

by Huxley is now known as Huxley's Line. Back in England, Wallace (1869) wrote:

> Judged by this standard, Celebes must be one of the oldest parts of the Archipelago. It probably dates from a period not only anterior to that when Borneo, Java, and Sumatra were separated from the continent, but from that still more remote epoch when the land that now constitutes these islands had not risen above the ocean. Such an antiquity is necessary, to account for the number of animal forms it possesses, which show no relation to those of India or Australia . . . [Wallace, 1869: 217 (Chapter 18).]

Wallace realised that most fauna elements of Sulawesi had been isolated a long time and had followed their own evolution, such as the babirusa or pig-deer (Figure 14.2), a still living enigmatic inhabitant of the island with huge, backward swept upper tusks penetrating the skin and so different from all other pigs in the world. The observations and interpretations of Wallace can be applied to the fossil faunas as well.

The geologist Audley-Charles (1981) proposed the existence of a land connection during the Late Pleistocene between south-eastern Asia and Australia through Sulawesi, enabling the dispersal of modern humans (*Homo sapiens*) to Australia. A similar connection, or only with Asia, supposedly existed during the Late Pliocene as well, as a result of the collision between Sulawesi and the Sula Peninsula (Banggai-Sula Spur), a continental fragment of Australia-New Guinea. In Audley-Charles' view, Banggai, Sula, Seram, and Buru were exposed by the Late Pliocene or already earlier, and the western

The subfossils of the so-called Toalian cave sites were described by Hooijer (1950). The caves are important for their Neolithic artefacts and known since the Sarasin brothers (1905) published their travel accounts of southern Sulawesi. The Toalian culture is characterised by stone flake tools, pottery, bone tools, and rock paintings, but Hooijer was more interested in the food remnants, among which are several subfossils of recent Sulawesi taxa, such as the babirusa and the anoas, but no new taxa.

Unfortunately, the stratigraphical position of the material collected by Hooijer and Van Heekeren remains unknown. In the late 1980s, Fachroel Aziz of the Geological Museum at Bandung started a project to reconstruct the stratigraphical context of the fossil findings. In the early 1990s, Iwan Kurniawan of the museum at Bandung, Paul Sondaar of the University of Utrecht, his then PhD student Gert van den Bergh, and John de Vos (one of the authors) from Naturalis, the Netherlands, joined the team. They discovered over 30 sites, mainly in the Walanae Valley, while collecting new material and stratigraphic data, as published by Van den Bergh and colleagues (1994). They noticed that, as on other islands, a biozone with a very small dwarf proboscidean preceded a biozone with a larger dwarf proboscidean – in their terminology, pygmy versus dwarf. Later, an Indonesian-international team, including Gert van den Bergh, returned to the Walanae Valley to conduct surveys between 2007 and 2012 to resolve the stratigraphic context of the lithic artefacts found with the vertebrate fossils. They discovered four new sites with *in situ* stone artefacts in a stratigraphic context of which they excavated one, the Talepu site, revealing the first evidence of stone artefacts in dated contexts (van den Bergh and colleagues, 2016).

Biozones and Faunal Units

The biostratigraphy of Sulawesi shows only two Pleistocene faunas, separated from each other by a gap (Figure 14.4). The first biozone – recognised in the Walanae Formation – is much better known than the second biozone – recognised in the Tanrung Formation – because of the relative abundance of its fossils. The relation between the two zones is not entirely clear, but the continuation of the occurrence of a form of the endemic pig *Celebochoerus* into the second biozone strongly suggests that the turnover was incomplete, as seen in most island faunas. The fossil faunas are depauperate and lacking carnivores, characterised by a high degree of endemism and indicating an arrival by a sweepstake route across a sea barrier.

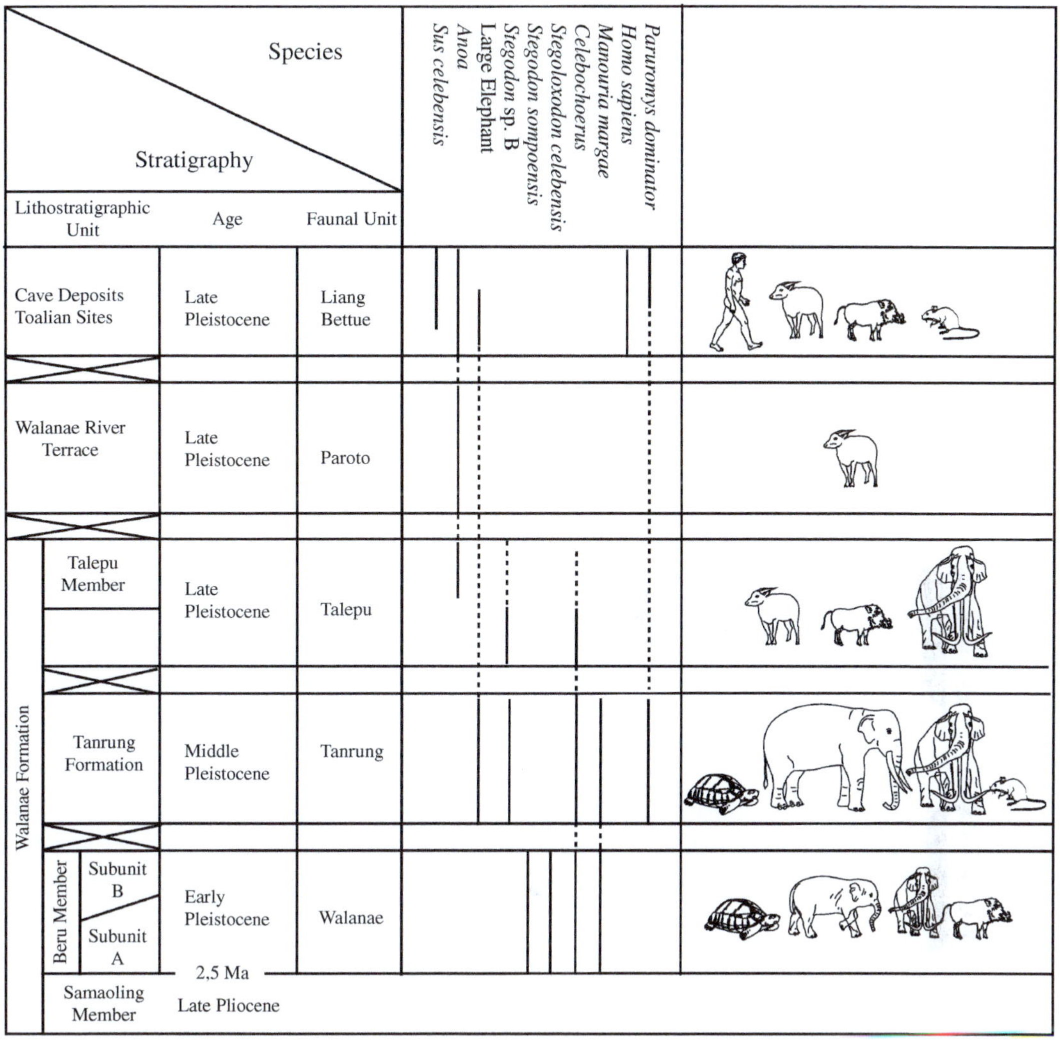

Figure 14.4 Stratigraphic scheme showing the land vertebrate faunal succession of Sulawesi. Characteristic elements of the successive faunal units are shown to the right. The subrecent to recent fauna is omitted.

Early Pleistocene

The fauna of this period, also known as *Walanae faunal unit*, consists mainly of the Sulawesi dwarf elephant (*Stegoloxodon celebensis*=*'Elephas' celebensis*; Plate 29), the giant Sulawesi pig (*Celebochoerus heekereni*), a very small dwarf stegodon (*Stegodon sompoensis*), and the giant Sulawesi rat (*Paruromys dominator*). Apart from these endemic mammals, a giant tortoise (*Manouria margae*) – with a carapace of about 1.2–1.5 m long – described by Hooijer (1949b) is unique to Sulawesi. The

Figure 14.5 Upper dM4 of the Sulawesi dwarf stegodon (*Stegodon sompoensis*), holotype; occlusal view. Scale bar: 5 cm. National Museum of Natural History, Leiden. (Photograph: Eelco Kruidenier.)

oldest occurrence of this fauna is at about 2.5 million years ago, close to the start of the Early Pleistocene.

The dwarf stegodon (*Stegodon sompoensis*) (Figure 14.5) is about half the size of the Javanese *Stegodon trigonocephalus*. In a younger layer (subunit B) of this biozone, a larger stegodon with lower-crowned molars occurs, as already noted by Hooijer (1972a, b) in his publications on Pleistocene Celebes. However, its relation to the rest of the fauna and the dwarf stegodon is unclear. It might represent the first arrival of the subsequent fauna.

Fossils of the giant Sulawesi rat are rare and limited to two isolated lower molars, reported by Kevin Downing and colleagues (1998). The rat is the only species that is still extant and indicates the long-term isolation of the island.

Middle Pleistocene

The fauna of this biozone, known as the *Tanrung faunal unit*, is poorly represented and thus poorly known as well. Recorded elements are a large elephant (probably *Palaeoloxodon namadicus*), a medium-sized dwarf stegodon (*Stegodon* sp. B), and a giant pig (*Celebochoerus* sp.).

The stegodon of this biozone is larger than the dwarf species from the previous biozone, suggestive of a new invasion from the mainland. The first upper molar has one ridge extra, compared to other medium-sized stegodons (*Stegodon trigonocephalus*, *Stegodon orientalis*, and *Stegodon florensis florensis*). Fossils were also found in the Pintareng formation on the Sangihe Islands, a group of islands half-way between the northern arm of Sulawesi Island and Mindanao Island (the Philippines). Initially, the Sangihe material was tentatively associated with the Javan stegodon, and thus named *Stegodon* species B cf. *trigonocephalus* by Fachroel Aziz (1990), but two years later, he removed the addition cf. *trigonocephalus* in a paper with Van den Bergh as first author with other colleagues. The dental material from Tanrung and Pintareng is very similar in morphology and differs from that of the earlier stegodon (*Stegodon sompoensis*). Unfortunately, the designation species B was also given to a primitive stegodon from Liucheng cave in the Guangxi Province of China, but there is no direct relation between the Chinese and Sulawesi stegodons.

A single surface-collected molar fragment consisting of two posterior plates found at the Tanrung riverbed evidences the presence of an elephant. Probably, some tusk fragments from the Tanrung Formation belong to this elephant as well. In the same riverbed, two large-sized metacarpals were found, which might be attributed to the elephant as well. The elephant molar has a very high crown, as in *Palaeoloxodon namadicus* – the Asian species of straight-tusked elephants – to which size it fits and to which species it may probably be attributed.

Van den Bergh and colleagues (2016) report *in situ* stone artefacts associated with fossil remains of large vertebrates (*Bubalus* sp., *Stegodon*, *Celebochoerus*, and giant tortoise). They were recovered at new excavations at Talepu in the Walanae Basin north-east of Maros from stratified deposits. These deposits accumulated between circa 118,000 and 194,000 years ago and are evidence that Sulawesi, like Flores, was host to a long-established population of Palaeolithic hominins.

Late Pleistocene–Recent

Towards the end of the Tanrung faunal unit, newcomers gradually replaced the endemic species, resulting in the present-day or recently extinct fauna. Also, this fauna underwent the effects of long-term isolation and evolved into endemic forms, as already noticed by Wallace (1869). In Toalian sites of the Late Pleistocene and Holocene caves, fossils and subfossils of these unique animals are present, such as the babirusa or deer-pig (*Babyrousa babyrussa*) (Figure 14.2), the Sulawesi warty pig (*Sus celebensis*), the lowland anoa (*Bubalus depressicornis*), the mountain anoa (*Bubalus quarlesi*), the Sulawesi black macaque (*Macaca nigra*), the moor macaque (*Macaca maura*), the booted macaque (*Macaca ochreata*), the Sulawesi or spectral tarsier (*Tarsius tarsier*), Sulawesi giant palm civet (*Macrogalidia musschenbroekii*), the trefoil-toothed giant rat (*Lenomys meyeri*), and *Homo sapiens*. The Asian palm civet (*Paradoxurus hermaphroditus*), the Javanese porcupine (*Hystrix javanica*), the Malay civet (*Viverra tangalunga*), and the Javanese deer (*Rusa timorensis*) were introduced by humans in the Holocene.

The cave sites of Sulawesi yielded more than faunal remains. Early rock paintings were discovered here already by Van Heekeren (1952) in caves in the Maros karsts of southern Sulawesi. Aubert and colleagues (2014) dated them to at least 35,400 years ago and described them in detail. Five years later, they described another rock art panel from the cave Leang Bulu'Sipong 4 in the Pangkep karsts, also in southern Sulawesi. They dated the art to at least 43,900 years ago based on uranium-series dating of the overlying cave deposits (speleothem) (Aubert and colleagues, 2019). Here, Palaeolithic humans carved a hunting scene with pigs and anoas.

Peculiarities and Evolution of Endemic Mammals

Four-tusked Dwarf Elephants: Stegoloxodon

Hooijer (1953a) described, amongst others, a lower jaw with an alveolus for a tusk as well as a tusk fragment that perfectly fits this alveolus. This was evidence that the Sulawesi dwarf elephant occasionally bore lower as well as upper tusks. It is best considered a retention of a primitive character, as seen in the African early elephantids *Primelephas* and *Stegotetrabelodon* and, possibly, the earliest forms of *Elephas planifrons*. The retention of functional lower tusks is not seen in juveniles of otherwise single-paired tuskers, so cannot be considered a paedomorphical feature in relation to its dwarf size. Another primitive character of the Sulawesi dwarf elephant is the development of premolars (Hooijer, 1953b). Between the late 1980s and early 1990s, the Indonesian-Dutch team excavated more material of this elephant, including a fairly complete but rather distorted skull. All material, new as well as old, is described and revised in Van den Bergh's doctoral thesis of 1999 on the Indonesian elephantoids, with a discussion on taxonomy. He puts question marks, '*Elephas*', to indicate the then uncertain taxonomic position, following Sondaar's (1984) approach. Van den Bergh considers a possible relation with '*Elephas*' *indonesicus* from Ci Pangglosoran near Bumiayu on Java, dated to the Early Pleistocene as well. Also, this specimen was originally assigned to *Elephas planifrons* but was later renamed into *Stegoloxodon indonesicus* by Kretzoi (1950). Markov and Saegusa (2008) made a step further and synonymised '*Elephas*' with *Stegoloxodon*.

Anoas: Dwarf Bubalus

The anoas, endemic to Sulawesi, are the smallest of the extant wild cattle species with a shoulder height of 0.6–1.0 m and a body mass of around 88 kg for the lowland species (*Bubalus depressicornis*) and 76 kg for the mountain species (*Bubalus quarlesi*), as calculated by Roberto Rozzi (2017). They have a stocky build, short limbs, and massive necks (Plate 5). These dwarf water buffaloes are closely related to the wild Asian water buffalo (*Bubalus arnee*) and the Philippine tamaraw, or Mindoro dwarf buffalo (*Bubalus mindorensis*), and diverged about 2 million years ago, based on molecular data analysed by Tanaka and colleagues (1996). Fossil and subfossil remains of anoas have been found in Late Pleistocene and Holocene deposits but not in earlier deposits as suggested by Hooijer (1950). Several paintings of anoas in prehistoric rock shelters, some of which are dated to 35–40 thousand years ago, confirm the contemporaneity of these animals with humans (overview in Rozzi, 2017). The taxonomic relation between the two species is unclear despite a number of theories that have been proposed. These include ecological differences, clinal variation (altitudinal), two separate immigrations, geographical separation before Sulawesi was unified, vicariance after Sulawesi was split into smaller islands by sea level rises, and habitat fragmentation followed by divergence. The lack of sufficient data hampers any conclusion, as explained by Van den Bergh and colleagues (2001). To add to the complexity, a third, somewhat larger species (*Bubalus grovesi*) has been described from the Late Pleistocene to Holocene rock shelters in South Sulawesi by Rozzi (2017). According to Rozzi, the body mass reductions of the three anoas are 89% for the extinct Groves' anoa, 92% for the lowland anoa, and 93% for the mountain anoa, relative to the wild mainland water buffalo of South East Asia (*Bubalus arnee*).

Shrew Rats, Tree Rats, and Allies

Sulawesi today is home to amazing adaptive radiations of 12 endemic genera of rats and mice that evolved in response to ecological conditions and opportunities into an array of species adapted to a multitude of ecological niches. The Sulawesi rats and mice include long-nosed shrew rats, adapted to a diet parallel to that of shrews elsewhere, small arboreal tree mice, amphibious water rats, and large terrestrial rats. The bizarre and sometimes extreme evolution, such as that of the extremely long-faced shrew rat without cheek teeth (*Paucidentomys vermidax*, described by Jacob Esselstyn and colleagues, 2012), specialised in a diet of earthworms and grubs, is evidence for millions of years of evolution in isolation. However, the only fossils discovered thus far are two isolated molars of the giant Sulawesi rat (*Paruromys dominator*) from the Early Pleistocene Walanae Formation, apart from subfossils found at archaeological sites. Likely, it is just a matter of time before more Pleistocene fossils of murines will be found, considering the long-term isolation of the present rat and mice species.

Pigs with Enigmatic Tusks

Giant Pigs: Celebochoerus

The Sulawesi giant pig *Celebochoerus heekereni* (Figure 14.6) is, apart from its large size, characterised by two pairs of prominent and large tusks in the males. The upper tusks were even larger than those of the living warthog of Africa. Its limbs were relatively short, indicative for an island form. The relation between the giant pig of the older Walanae Formation and younger Tanreg Formation is unclear but is likely to be explained as a local evolution trend towards further size reduction. This pig genus may have arrived from the Philippines, where a related species (*Celebochoerus cagayanensis*) has been reported. *Celebochoerus* is neither related to extant babirusa (*Babyrousa*) nor to common pigs (*Sus*).

Babirusa, or Pig-Deer: Babyrousa

The babirusa (*Babyrousa*) is an extant enigmatic endemic species of the island with huge, backward swept upper tusks penetrating the skin and so different from all other pigs in the world. This extant genus is restricted to Sulawesi and the nearby Togian Islands, whereas the subspecies from the Sula and Buru Islands is likely the result of ancient introduction by human agency (see Colin Groves, 1980, and Umberto Albarella and colleagues, 2007). In this wild swine (Figure 14.2), the upper tusks grow through the skin of the muzzle and then curve backward towards the forehead, sometimes even touching it. The tusks do not enter the mouth cavity. The lower tusks follow more or less the same direction but do not come into contact

(a) (b)

Figure 14.6 Mandible with molars (a; occlusal view) and upper tusk (b; lateral view) of the Sulawesi pig (*Celebochoerus heekereni*). Scale bar: 5 cm. (a) National Museum of Natural History, Leiden; (b) Geological Museum Bandung, Java. (Photographs: (a) Eelco Kruidenier and (b) Gert van den Bergh.)

with the upper tusks, hampering honing. To do so, the animal uses trees to sharpen the tusks. It differs in many other features as well from *Sus*, such as the lack of the prenasal bone in the snout disc and the very small litter size consisting of only one or two piglets, which are more precocious than those of *Sus* and lack stripes. The babirusa may have reached Sulawesi by swimming, because this species is known to swim often in the sea to reach small islands, as reported by Melisch (1994). Babirusa seems not to be derived from the endemic Sulawesi pig of the Middle Pleistocene (*Celebochoerus*). Fossils of an extinct subspecies (*B. b. beruensis*) were found at Beru and Sompo in south-western Sulawesi in association with fossils of the other endemic pig, *Celebochoerus heekereni*, and are much larger than living specimens, as noted by Hooijer (1948).

The babirusa was known to the Late Pleistocene humans, as revealed by an early rock painting in one of the caves of the Maros karst area, dated to at least 35,400 years ago. This naturalistic painting, possibly the earliest dated figurative depictions worldwide, indicates that this pig species may have played an important role in the Palaeolithic culture as it does today.

References

Albarella U., Dobney K., Ervynck A. 2007. *Pigs and Humans: 10,000 years of Interaction*. Oxford University Press, Oxford.

Aubert M., Brumm A., Ramli M., Sutikna T., Saptomo E.W., Hakim B., Morwood M.J., van den Bergh G.D., Kinsley L., Dosseto A. 2014. Pleistocene cave art from Sulawesi, Indonesia. *Nature*, 514: 223–227.

Aubert M., Lebe R., Oktaviana A.A., Tang M., Burhan B., Hamrullah, Jusdi A., Abdullah, Hakim B., Zhao J.-X., Gerla I.M., Sulistyarto P.H., Sardi R., Brumm A. 2019. Earliest hunting scene in prehistoric art. *Nature*, 576: 442–445.

Audley-Charles M.G. 1981. Geological History of the region of Wallace's line. In: Whitmore T.C. (ed.), *Wallace's Line and Plate Tectonics*. Clarendon Press, Gloucestershire, pp. 24–35.

Audley-Charles M.G., Hooijer D.A. 1973. Relation of Pleistocene migrations of pygmy stegodonts to island arc and tectonics in Eastern Indonesia. *Nature*, 24: 197–198.

Aziz F. 1990. *Pleistocene Mammal Faunas of Sulawesi and Their Bearings to Paleozoogeography*. Unpublished PhD thesis, Kyoto University.

Downing K.F., Mussser G.G., Park L.E. 1998. The first fossil record of small mammals from Sulawesi, Indonesia: The large murid, *Paruromys dominator*, from the Late (?) Pliocene Walanae Formation. In: Tomida Y., Flynn L.J., Jacobs L.L. (eds), *Advances in Vertebrate Paleontology and Geochronology. National Science Museum Monographs*, 14. National Science Museum, Tokyo, pp. 105–121.

Esselstyn J.A., Achmadi A.S., Rowe K.C. 2012. Evolutionary novelty in a rat with no molars. *Biology Letters*, 8: 990–993.

Groves C.P. 1980. Notes on the systematics of *Babyrousa* (Artiodactyla, Suidae). *Zoologische Mededelingen*, 55 (3): 29–46.

Hooijer D.A. 1948. Pleistocene vertebrates from Celebes. I. *Celebochoerus heekereni nov. spec. Verhandelingen der Koninklijke Akademie van Wetenschappen*, 51 (8): 1024–1032.

Hooijer D.A. 1949 (a). Pleistocene vertebrates from Celebes. IV. *Archidiskodon celebensis* nov. spec. *Zoologische Mededelingen*, 30: 205–226.

Hooijer D.A. 1949 (b). Pleistocene vertebrates from Celebes. II. *Testudo margae* nov. spec. *Verhandelingen der Koninklijke Akademie van Wetenschappen*, 51 (9): 1169–1182.

Hooijer D.A. 1950. Man and other mammals from Toalian sites in south-western Celebes. *Verhandelingen der Koninklijke Akademie van Wetenschappen, Afdeling Natuurkunde* 2, 46: 1–164.

Hooijer D.A. 1953 (a). Pleistocene vertebrates from Celebes. V. Lower molars of *Archidiskodon celebensis* Hooijer. *Zoologische Mededelingen*, 31: 311–318.

Hooijer D.A. 1953 (b). Pleistocene vertebrates from Celebes. VII. Milk molars and premolars of *Archidiskodon celebensis* Hooijer. *Zoologische Mededelingen*, 32: 221–230.

Hooijer D.A. 1954. Pleistocene vertebrates from Celebes VIII . Dentition and skeleton of *Celebochoerus heekereni. Zoologische Verhandelingen*, 24: 3–46.

Hooijer D.A. 1964 (a). Pleistocene vertebrates from Celebes XII. Notes on pygmy Stegodonts. *Zoologische Mededelingen*, 40: 37–44.

Hooijer D.A. 1964 (b). On two milkmolars of a pygmy stegodont from Ola Bula, Flores. *Bulletin Geological Survey of Indonesia*, 1: 49–52.

Hooijer D.A. 1972 (a). *Varanus* (Reptilia, Sauria) from the Pleistocene of Timor. *Zoologische Mededelingen*, 47: 445–447.

Hooijer D.A. 1972 (b). Pleistocene vertebrates from Celebes. XIV. Additions to the *Archidiskodon-Celebochoerus* fauna. *Zoologische Mededelingen*, 46: 1–16.

Huxley T.H. 1868. *An Introduction to the Classification of Animals*. Churchill, London.

Kretzoi M. 1950. *Stegoloxodon* nov. gen., a possible Asiatic ancestor of true loxodonts. *Földtani Közlöny*, 80: 405–408.

Markov G.N., Saegusa H., 2008. On the validity of *Stegoloxodon* Kretzoi, 1950 (Mammalia: Proboscidea). *Zootaxa*, 1861: 55–56.

Melisch R. 1994. Observation of swimming babirusa in Lake Poso, central Sulawesi, *Indonesia. Malayan Nature Journal*, 47: 431–432.

Rozzi R. 2017. A new extinct dwarfed buffalo from Sulawesi and the evolution of the subgenus *Anoa*: An interdisciplinary perspective. *Quaternary Science Reviews*, 157: 188–205.

Sarasin P., Sarasin F. 1905. *Reisen in Celebes*. C.W. Kreidel's Verlag, Wiesbaden.

Sondaar P.Y. 1984. Faunal evolution and the mammalian biostratigraphy of Java. *Courier Forschungsinstitut Senckenberg*, 69: 219–235.

Tanaka T., Solis C.D., Masangkay J.S., Maeda K., Kawamoto Y., Namikawa T. 1996. Phylogenetic relationship among all living species of the genus *Bubalus* based on DNA sequences of cytochrome b gene. *Biochemical Genetics*, 34: 443–452.

van den Bergh G.D. 1999. The Late Neogene elephantoid-bearing faunas of Indonesia and their palaeozoogeographic implications; A study of the terrestrial faunal succession of Sulawesi, Flores and Java, including evidence for early hominid dispersal east of Wallace's line. *Scripta Geologica*, 117: 1–419.

van den Bergh G.D., Aziz F., Sondaar P.Y., de Vos J. 1994. The first *Stegodon* from Central Sulawesi and a new advanced *Elephas* species from South Sulawesi. *Bulletin of the Geological Research and Development Centre Bandung*, 17: 22–39.

van den Bergh G.D., Aziz F., Sondaar P.Y., Hussain S.T. 1992. Taxonomy, stratigraphy, and paleozoogeography of Plio-Pleistocene Proboscideans from the Indonesian Islands. *Bulletin of the Geological Research and Development Center Bandung, Paleontology*, 7: 28–58.

van den Bergh G.D., de Vos J., Sondaar P.Y. 2001. The late quaternary palaeogeography of mammal evolution in the Indonesian Archipelago. *Palaeontology, Palaeoclimatology, Palaeoecology*, 171: 385–408.

van den Bergh G.D., Li B., Brumm A., Grün R., Yurnaldi D., Moore M.W., Kurniawan I., Setiawan R., Aziz F., Roberts R.G., Suyono, Storey M., Setiabudi E., Morwood M.J. 2016. Earliest hominin occupation of Sulawesi, Indonesia. *Nature*, 529: 208–211.

van Heekeren H.R. 1949. Early man and fossil vertebrates on the island of Celebes. *Nature*, 163: 492.

van Heekeren, H.R. 1952. Rock-paintings and other prehistoric discoveries near Maros (South West Celebes). *Laporan Tahunan Dinas Purbakala*, 1950: 22–35.

Wallace A.R. 1863. On the physical geography of the Malay Archipelago. *Journal of the Royal Geographical Society*, 33: 217–234.

Wallace A.R. 1869. *The Malay Archipelago: The Land of the Orang-Utan and the Bird of Paradise: A Narrative of Travel with Studies of Man and Nature*. Tyron Press, Scotland.

CHAPTER FIFTEEN

The Philippines

Evolution of Island Mammals: Adaptation and Extinction of Placental Mammals on Islands, Second Edition.
Alexandra van der Geer, George Lyras and John de Vos.

© 2021 John Wiley & Sons Ltd. Published 2021 by John Wiley & Sons Ltd.

Today, the Philippine island arch is a biodiversity hotspot. The fossil record of the history of this biodiversity is, however, rather poor but indicates that past biodiversity was even larger. A number of fossil endemic species from the Pleistocene are described from the various islands, indicating that the Philippines consisted of several palaeo-islands, each harbouring its own endemic fauna. Amongst these fossil mammal species are dwarf stegodons, dwarf buffaloes, dwarf deer, a rhinoceros, and giant rats. In addition, the colonisation date of several of the endemic rodents of today can be traced back to the Miocene, evidence of a high murid biodiversity throughout the entire Quaternary.

Geology and Palaeogeography

The Philippines forms an island group situated between Borneo and Taiwan in the Malay Archipelago. Today, it consists of over 7,000 islands, but this number was much less during periods of low sea stands in the past. On the basis of the current geography in combination with current sea depths below 120 m, Darin Croft and colleagues (2006) reconstructed the palaeogeography of the Philippines for the last glacial maximum, which we follow here. They recognise the following six large palaeo-islands: Greater Luzon, Mindoro, Greater Palawan, Greater Negros-Panay, Greater Mindanao, and Greater Sulu (Figure 15.1). The configuration of the various islands before the Late Pleistocene is much less known. For the middle Miocene, Robert Hall (1998) reconstructs the palaeography of the Philippines and recognises two distinct subaerial land masses, one corresponding to northern Luzon and one roughly corresponding to eastern Mindanao, with deep sea separating them, namely shallow seas separated northern Luzon from Kalimantan (Borneo) and mainland South East Asia and eastern Mindanao from Sulawesi and New Guinea. Hall's map for the middle Pliocene shows a larger amount of subaerial land masses, more or less corresponding to the modern situation but with most islands considerably smaller than today. Eastern Mindanao is now connected to western Mindanao plus south-eastern Luzon and is separated from northern Luzon by a shallow sea only. Smaller current islands and some peninsulas of the larger islands were uplifted during the Pliocene.

The current configuration of the Philippine islands with their characteristic fractured geography resulted mainly from tectonic interactions among the Eurasian continental plate and the Philippine oceanic plate since the early Miocene, resulting in volcanic activities, tectonic uplift and tectonic motion. The various islands are the result of underwater volcanic activities

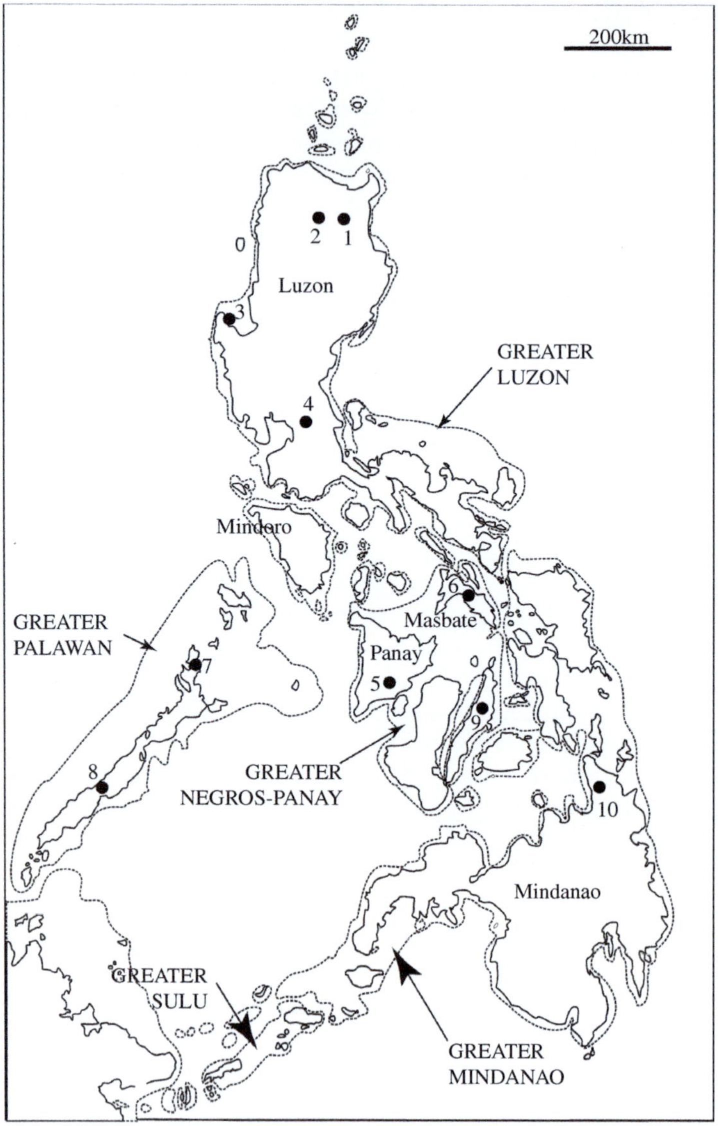

Figure 15.1 Map of the Philippines showing current islands, Late Pleistocene islands and Pleistocene localities. (1) Callao cave, Luzon; (2) Cagayan Valley, Luzon; (3) Pangasinan or Cabarruyan Island; (4) San Juan, Rizal, Luzon; (5) Panay Island (Visayas); (6) Bat-Ongan, Masbate; (7) Ille cave, Palawan; (8) Tabon cave, Palawan; (9) Balamban, Cebu; and (10) Agusan del Norte, Mindanao. (Redrawn from Croft and colleagues, 2006.)

and thus of oceanic origin, with the exception of Mindoro and Palawan, which originated as portions of the Asian continental shelf that rifted away during the mid-Oligocene. They remained below sea level until they were uplifted, Mindoro in the Late Miocene and Palawan in the Early Pliocene. They were never connected by dry land to each other nor with any other Philippine islands. Palawan is thought to have been connected to northern Borneo during the penultimate glacial event and perhaps during previous glacial events as well, according to Scott Steppan and colleagues (2003), based on molecular phylogeny of murids.

The extraordinarily complex configuration of the Philippines, together with sea level changes which caused the joining and fragmentation of subaerial landmasses, gave rise to isolation of local faunas. The Philippines shows at present the one but highest degree of endemism, surpassed only by Madagascar.

Historical Palaeontology

Fossils from the Philippines are hardly known. Several findings have been reported from different islands, but the majority was collected without stratigraphic data. In addition, several of these fossils are lost or cannot be traced in the known collections.

The first reported fossil findings are two fragments of a *Stegodon* obtained from local people on Mindanao Island (Figure 15.1), found somewhere on their island. Edmund Naumann (1887) described a fragment of a lower molar as belonging to the Javanese stegodon, *Stegodon trigonocephalus*. The fossil originated from somewhere from the north-west of Mindanao and had been used as a talisman, representing a tooth of the warrior god Tagbusau. After discussing the fragment with Karl Martin, who at that time was studying the Javanese fossils, Naumann (1890) later changed his mind and presented it as a new species, *Stegodon mindanensis*. It is one of the rare fossils in the world that is not kept in a palaeontological or natural history collection but in an ethnographical collection (Dresden, Germany) because of its ritual use. The fragment has been studied many times, and given accordingly many names, including cf. *Elephas*, *Parastegodon*, or a primitive *Mammuthus*. At present, it is generally supposed to be a middle-sized *Stegodon*.

G.J. Adams (1910) referred to a tooth and some tooth fragments as cf. *Antelope*. The tooth, a worn upper molar, probably the first one, was found at a depth of between 81 and 85 m while drilling a well at Pasig in the neighbourhood of Manila. Ralph von Koenigswald (1956) suggested that it might belong to a small bovine rather than to an antelope. At present, nothing definite can be said because the original specimen is lost, but it might very well have been a dwarf buffalo. A similar finding was done in 1920 by Charles Wingh, who found two teeth while boring a well near San Juan in Rizal on Luzon, which were indicated later as belonging to a water buffalo (cf. *Bubalus*) by Otley Beyer (1949). Apart from these 'carabao' teeth, also a 'prehistoric deer' tooth was found, tentatively named cf. *Cervus* by von Koenigswald (1956), as they are still labelled. The same label was applied to the tooth fragments found earlier by Adams.

Von Koenigswald also described three new endemic species from Luzon, all three of Middle Pleistocene age. His *Rhinoceros philippinensis* is based on two molars found in the Cagayan Valley by Alfonso Bagunu and Rodolfo Albano back in 1936. From Pangasinan or Cabarruyan Island, he described a fragmentary lower molar of a small elephant as *Elephas beyeri*. Fragmentary remains of a stegodon found near Fort MacKinley (= Fort Bonifacio) in Rizal, Luzon, were described by him as *Stegodon luzonensis*. The molars of this new species are a little smaller than those of the Javan stegodon. Unfortunately, these specimens – the holotypes, on which his three descriptions are based – are all lost at present.

In the 1970s, the National Museum of the Philippines in Manila explored the so-called 'Open Sites' on the hills in the centre and towards the eastern side of the Cagayan Valley, Luzon, in order to find archaeological remains. Almost 70 localities with stone tools, and, in some cases, tektites and vertebrate fossils, were found, as reported by Lopez (1972). The age and context of the archaeological materials were since then debated. Robert Fox and Jesus Peralta (1974) argued that at least some of the tools could be coeval with the vertebrate fossils, which in turn have a Middle Pleistocene age, according to Von Koenigswald.

Angel Bautista (1988, published 1995) reported two proboscidean findings from Cabalwan, Cagayan valley of unknown affinity, an *Elephas* sp. and a *Palaeoloxodon* sp. belonging to the *Palaeoloxodon namadicus* group. He also described rhinoceros material from Butuan City, Agusan del Norte, Mindanao as *Rhinoceros philippinensis*. The specimens though were recovered from a disturbed protohistoric cultural level along with traded materials from South East Asia and China. Evidence of human modification of the specimens is seen in the presence of holes for stringing purpose on the root portion of the fourth premolar and third molar. Likely, the Butuan specimens were a part of materials that were brought by Neolithic people to the Philippines from South East Asia.

Three years later, Bautista (1991) mentioned six extinct species from Pangasinan or Cabarruyan Island. He tentatively assigned two of them to mainland species, the Chinese stegodon (*Stegodon* cf. *sinensis*) and the straight-tusked elephant (*Palaeoloxodon* cf. *namadicus*). The four other species were only determined on a genus level (*Stegodon* sp., *Elephas* sp., *Cervus* sp., and *Bubalus* sp.). From site A in the Novaliches-Marilo District of Rizal, he reported a large piece of a stegodon tooth, presumably of the Javan stegodon (*Stegodon* cf. *trigonocephalus*) together with a fossil deer antler. From another site, a few kilometres from site A to site M, another large piece of a large fossil stegodon tooth was found. This was a surface find in a freshly ploughed field. It was found together with four complete tektites and a few probable pre-Neolithic stone implements. From another field, again a few kilometres from the previous sites, site X, two large parts and 12 fragments of fossil ivory from a large stegodon tusk were found. Besides the remains of the large stegodon, Bautista also reported bone fragments of a large tortoise and an elephant (*Elephas* sp.), both of unknown relation to the stegodon. At Fort Bonifacio, Rizal, he found rhinoceros material, which he ascribed to the Philippine rhinoceros (*Rhinoceros philippinensis*). In 1999, Bautista went again to the open sites of Cagayan Valley, this time with one of us (John de Vos) and found bones of elephant and stegodon, stone tools, and tektites, a situation similar to Flores. Plans were made to excavate the open sites systematically, with the hope of finding evidence for Middle Pleistocene human occupation. These had, however, to wait at least a decade, when in 2013 such excavations finally took place (see below).

In the meantime, in 2003, archaeological excavations started at Callao cave, Cagayan Valley, northern Luzon by a Philippine-French team to investigate subsidence strategies during the Holocene (Figure 15.2). Upon discovery of a Late Pleistocene layer near the cave entrance with flake tools and burnt animal bones, the team included Australian colleagues and explored this Palaeolithic occupation floor in 2007. They excavated a foot bone of a small-sized hominid, dated to 67,000 years ago, and in association with a fossil fauna including warty pigs (*Sus philippensis*), Philippine deer (*Rusa marianna*), and murids (*Apomys* sp. and *Batomys* sp.), as reported by Mijares and colleagues (2010). After this discovery, twelve additional hominin elements – representing at least three

Figure 15.2 Callao cave, northern Luzon, where fossils of Late Pleistocene hominins (*Homo luzonensis*) were recovered. (Photograph: George Lyras.)

individuals – were recovered from the same stratigraphic layer. The hominid remains were described as *Homo luzonensis* by Florent Détroit and colleagues (2019).

In 2005, two of us (John de Vos and George Lyras) went to Masbate to relocate the site where the archaeologist William (Bill) Solheim II had found goat-like bones. He thought these were evidence for the earliest domestication of goat or sheep on the island and sent them in 1952 to Dirk Hooijer of Naturalis for verification. The latter, however, identified them as deer with shortened metapodials. We were interested because they strongly resemble the bones of dwarf deer from Crete. After meeting Solheim II, we found the steep hill with almost vertical sites and retrieved subfossil material of deer and pigs as well as a fossil giant rat (Plate 30) in a limestone breccia.

In 2013, a team of the National Museum in Manila, the Archaeological Survey, and one of us (John de Vos) finally undertook a prospecting visit to the Cagayan Valley to identify the most promising site for finding evidence for hominin presence. A year after, an international team led by John de Vos and Thomas Ingicco in collaboration with the National Museum of the Philippines started a large-scale excavation in the Cagayan Valley, Luzon, near the open sites where in the past, fossil vertebrates, tektites, and stone tools had been found, with the goal to understand the stratigraphy and the contemporaneity of the three types (see below). Finally, they reported the first evidence (Plate 31) of the contemporaneity of lithic artefacts and extinct megafauna in Middle Pleistocene sediments (Ingicco and colleagues, 2015).

Biozones and Faunal Units

No reliable biostratigraphy of the Philippines can be made at present, because the fossil record is poor and a clear stratigraphical context is lacking. No fossils were recorded from Greater Sulu, Mindoro, and Greater Palawan, with the exception of (sub)fossils from Ille and Tabon caves (Box 15.1).

However, some observations can be made, though. First of all, the fossil fauna or faunas seem unbalanced and depauperate, with larger as well as smaller proboscideans. Furthermore, fossils of the larger stegodon and the rhinoceros are coeval with stone artefacts. The associated tektites belong to the Australasian strewnfield, dated to between 700,000 and 800,000 years ago. This situation reminds of Flores, and it is thus tempting to assume that first there was a fauna with a small stegodon, followed by a fauna with a larger stegodon and Palaeolithic humans. Excavations since 1999 by a Philippine–Dutch–French–Greek team in Cagayan Valley are aimed at finding evidence for such early human occupation. In 2014, during excavations at Kalinga in the Cagayan Valley of northern Luzon (Plate 31), the team recovered over 50 stone tools associated with an almost-complete disarticulated skeleton of *Rhinoceros philippinensis*, which shows clear signs of butchery, together with other fossil vertebrate remains, attributed to stegodon, Philippine brown deer, freshwater turtle, and monitor lizard. All finds originate from a clay-rich bone bed that was dated to between 777,000 and 631,000 years ago using electron-spin resonance methods that were applied to tooth enamel and fluvial quartz, as reported by Ingicco and colleagues (2018).

Greater Luzon–Greater Negros–Panay–Greater Mindanao: Middle–Late Pleistocene

Remains of molars and molar fragments of a large stegodon were found at several sites in Luzon (*Stegodon luzonensis*) (Plate 32), one site in Visayas (Greater Negros-Panay), and one site in Mindanao (*Stegodon mindanensis*). The material of the latter species, erected by Naumann, falls within the range of Von Koenigswald's species *luzonensis*, so rules of priority would dictate that the latter is a junior synonym of the former. However, it is appropriate to retain – as long as the biostratigraphy of the different islands is unsolved – both names as valid species, because both species are based on remains from different islands, respectively, Mindanao and Luzon.

Fossils of *Stegodon luzonensis* are, like those of the rhinoceros, found in association with tektites of the Australasian strewnfield and stone artefacts.

Remains of a large elephant were found at several sites in Luzon and one in Visayas (Greater Negros-Panay) and attributed to either *Elephas* sp., *Elephas* cf. *namadicus* (= *Palaeoloxodon*) or *Elephas beyeri*. The specific status of these elephants is unclear, not only because the holotype of *Elephas beyeri* from Cabarruyan Island (Greater Luzon) is lost but also because the elephant material is very fragmented. However, as in the case of the stegodon, it might be that Luzon harboured its own elephant race, distinct from that of Greater Negros-Panay, though likely with the same ancestral form.

Fossil bovid material is extremely scanty and limited to Luzon and Cebu (Greater Negros-Panay). De Vos and Bautista (2003) showed that the Luzon material belongs to a small water buffalo (*Bubalus* sp.), probably related to the living tamarau (*Bubalus mindorensis*) from Mindoro. The bovid from Cebu (*Bubalus cebuensis*) was much smaller.

Fossil pig material is even scantier than bovid material and was until 2013 restricted to three lower molars from Luzon,

Figure 15.3 Skull of the Philippine deer (*Rusa marianna*), Pleistocene–Holocene. Scale bar: 4 cm. National Museum of the United States, Washington. (Photograph: George Lyras.)

labelled as *Celebochoerus cagayanensis*, but without description. In 2013, a large upper tusk was found referable to *Celebochoerus* from the Cagayan Valley of Luzon.

Several deer fossils have been reported since 1910 from Luzon and Palawan, but none of these was given a specific name, other than just *Cervus* sp., cf. *Cervus*, *Cervus* sp. 1, and *Cervus* sp. II or a 'small *Rusa* sp.', except for Callao cave from where *Rusa marianna* has been reported from Late Pleistocene sediments. Descriptions and depictions of the other material are entirely lacking. Measurements indicate that all fossil deer are small, and the slightly pearled beam of retrieved antlers resembles that of a sambar deer (*Rusa unicolor*) most. Its relation to the living Philippine sambar (*Rusa marianna*) (Figure 15.3) remains unknown.

Apart from mammalian fossils, also postcranial remains of a giant tortoise were found in Early Pleistocene sediments of Luzon. They were described as *Manouria sondaari* by Hans-Volker Karl and Ulrich Staesche (2007) but later moved to the genus *Megalochelys* by Anders Rhodin and colleagues (2015).

Masbate: Late Pleistocene

From Masbate (Greater Negros-Panay), a dwarf deer (*Cervus* spp.) and a skull of a giant rat (*Rattus* cf. *everetti*) are known. The rat skull is unique because it is one of the rare complete

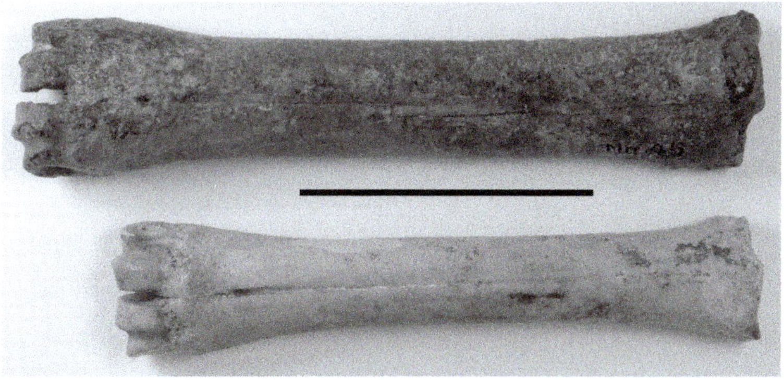

Figure 15.4 Metacarpals of two deer (*Cervus* sp.) of different sizes; anterior (dorsal) view. Scale bar: 5 cm. National Museum of Natural History, Leiden. (Photograph: Eelco Kruidenier.)

skulls of a fossil rodent (Plate 30) found in South East Asia. The deer had shortened, stocky limbs as typical for deer on islands (Figure 15.4) but is further hardly described. Hooijer referred to the deer as *Cervus* spec. div. (=various species of deer), aware of the unusual large variation within the material. De Vos (2006) suggested that the material represents a new species, as the metacarpals are much more massive than those of extant Philippine deer (*Rusa marianna*, *Rusa alfredi*). Both taxa – the deer and the rat – were found together by two of us (George Lyras and John de Vos) in a breccia of a cave at the hill Bat-ongan, but not in the same layers as the archaeological remains and the remains of pigs and domestic animals as was suggested earlier by Hooijer and Solheim in their correspondences. The age of the deer and rat fossils is estimated at Late Pleistocene. Since similar fossils were not found on other parts of the Philippines, it is reasonable to assume that Masbate formed a separate island that was not connected to the rest of Greater Negros-Panay.

Latest Pleistocene and Holocene

From the end of the Pleistocene or the beginning of the Holocene, the species of the present-day fauna are known. Most of these represent much earlier arrivals, but without a fossil record, their time spent in isolation is unknown. The Philippines harbour one of the most unique faunas in the world, regarding the level of endemic species, indicating an isolation of at least millions of years. Regularly, new species are being discovered, and likely, the number of currently described species is only half the actual number of species, as calculated by Lawrence Heaney and colleagues (2016). Characteristic mammal species, as given by Heaney and colleagues (1998), and regularly updated, include the Philippine sambar (*Rusa marianna*) of Luzon, Prince Alfred's deer (*Rusa alfredi*) of Visayas, Philippine tarsier (*Tarsius syrichta*), Philippine flying lemur (*Cynocephalus volans*) – which is a colugo and not a lemur – Philippine long-tailed macaque (*Macaca fascicularis philippinensis*), several bat species, and many endemic rodents (Muridae). Doubtlessly adding to the insular flavour are the three or four subspecies of water monitor (*Varanus salvator*) endemic to various islands of the Philippines. Together with the pythons and crocodiles, they occupy the rank of large terrestrial predators. With their enormous size – they may even reach a length of 3 m – they constitute a real danger for deer and primates.

Remains of the earliest *Homo sapiens* in the Philippines are found in the Tabon cave on Palawan. The cave sediments are dated to 47,000 ± 11,000 years ago as reported by Eusebio Dizon and colleagues (2002). The human material consists of a skull cap, a mandible fragment with a twisted molar and a cast of a complete mandible of which the original is lost. The mandible fragment with the twisted molar and new finds are described by Florent Détroit and colleagues (2004).

BOX 15.1

Tigers in the Philippines?

In northern Palawan, two tiger subfossils have been found during archaeological excavations at Ille cave near the village of New Ibajay, as reported by Philip Piper and colleagues (2008). The material consists of two articulating phalanges, found amidst an animal bone assemblage overlain by a deposit dated to the early thirteenth millennium before present but which was itself dated to about 12,000–9,000 years ago, as given by Helen Lewis and colleagues (2008). The other faunal elements are ascribed to macaques (*Macaca fascicularis*); deer (*Axis calamianensis*, *Rusa* sp.), bearded pigs (*Sus ahoenobarbus*); and a variety of small mammals, turtles, snakes, and lizards. The presence of stone tools and evidence of the use of fire, combined with the presence of cut marks on some bones, implies that the bone accumulation is human-caused. Piper and colleagues take the tiger bones as evidence of a viable tiger population in Palawan. However, tiger claws are widely used as amulet in the whole of South and South East Asia where tigers occur and already in the early centuries as pointed out by van der Geer (2008). Furthermore, the condition of the tiger subfossils differs from that of the other fossils, as they show longitudinal cracking of the cortical bone surfaces due to weathering, indicating a post-mortem exposure to air and light. In addition, slight damages are present at several places, and some concretion adheres to one of the condyles. Piper is aware of the 'ornamental' (magic) value of tiger skins and teeth and admits that these can be transported over long distances, but overlooks tiger claws and jugular bones, which are widely used as well. The tiger subfossils might thus very well have been imported by the human settlers. Such was the case with the tiger canine found in tenth to twelfth century Ambangan sites in Butuan, Mindanao, as explained by Elenita Alba (1994). On the other hand, Palawan is not far away from Borneo, and tigers are known to be excellent swimmers. A few tiger remains were reported by Kitchener (1999) from Borneo as well, being two sub-adult canines and a metacarpal fragment (Niah caves) and a navicular bone (Madai cave), from large latest Pleistocene and Early Holocene vertebrate assemblages. The conclusion that there was a viable tiger population in Palawan during the latest Pleistocene and Early Holocene can thus not be dismissed entirely.

Peculiarities and Evolution of Endemic Mammals

Diminutive Humans

From the Callao cave of northern Luzon (Figure 15.2), a spectacular find was reported in 2007 of a third metatarsal of a small hominid, dated to 67,000 years ago. Analysis of this foot bone suggested that it belonged to the genus *Homo*, but to which species was unclear. Initially, it was described as *Homo sapiens* by Mijares and colleagues (2010). After this discovery, twelve additional hominin elements – representing at least three individuals – were recovered from the same stratigraphic layer of Callao cave as the previously discovered metatarsal. These specimens display a unique combination of primitive and derived morphological features that distinguishes it from all other species of the genus *Homo* – including *Homo floresiensis* and *Homo sapiens*. This warrants their attribution to a new species, named *Homo luzonensis* by Florent Détroit and colleagues (2019). This provides a parallel with the Pleistocene of Flores, which also had a small hominin.

Dwarf Buffaloes and Dwarf Rhinoceros

Croft and colleagues (2006) described a dwarf buffalo, *Bubalus cebuensis* – based on associated postcranial material and two teeth. Michael Armas had found the remains during exploration for phosphate in 1958 in a mining tunnel near Balamban at Cebu, part of Greater Negros-Panay. The Cebu dwarf buffalo was similar in size to the lowland anoa (*Bubalus depressicornis*) of Sulawesi, but with more robust limb bones. Its body weight is estimated to approximately 150–165 kg, which is heavier than the anoa due to its increased robusticity. It further had larger teeth relative to body size, compared to other *Bubalus* species – a feature not uncommon in island dwarfs. Some characters of its postcranial are unique, but have not yet been explained in terms of functional morphology. The age of the fossils of the Cebu dwarf buffalo is Pleistocene but may even be Holocene.

According to Roberto Rozzi (2017), the body mass reduction of the Philippine dwarf buffaloes is 79% for the Mindoro dwarf buffalo (*Bubalus mindorensis*) and 85% for the Cebu dwarf buffalo (*B. cebuensis*), relative to the wild mainland water buffalo of South East Asia (*B. arnee*). The dwarf buffalo remains from Luzon (*Bubalus* sp.) seem similar in size and morphology to the Mindoro species. Due to the fragmented fossil record, however, the relation between the three Philippine buffaloes is unresolved.

The fossil rhinoceros from Luzon, *Rhinoceros philippinensis,* is known from an almost-complete skeleton excavated at the Kalinga site of Rizal province (Plate 31). De Vos and Bautista (2003) reported a scaphoid – a carpal bone – from a *Rhinoceros* of Pangasinan or Cabarruyan Island, likely belonging to the same species. The Philippine rhinoceros was small with the size of a Sumatran rhinoceros (*Dicerorhinus sumatrensis*). It is, however, not closely related to the latter species, but likely a descendant from a southern Chinese form.

Cloud Rats, Earthworm Mice, Moss Mice, and Allies

The impressive present-day biodiversity of endemic murid rodents of the Philippines with over seventy species left hardly any trace in the fossil record. The three exceptions are a Pleistocene rat from Masbate (perhaps *Rattus everetti*), a cloud rat (*Batomys* sp.), and an earthworm mouse (*Apomys* sp.) discovered at Callao cave, dated to between about 50,000 and 68,000 years ago, reported by Heaney and colleagues (2011). Molecular and morphological data, however, indicate that today's murids were present already much earlier and must therefore have been part of the previous biozones. Musser and Heaney (1992) noticed that the murids can be grouped, based on morphological characters, into three distinct clusters, which they name the Old Endemics, the New Endemics, and the single genus *Anonymomys*, assuming multiple invasions. The first group comprises species that retain primitive features yet exhibit various unique specialisations as well, some of them spectacular, indicating a long-term isolation and adaptive radiation *in situ*. To this group belong the arboreal and folivorous cloud rats (*Phloeomys*, *Crateromys*, *Musseromys*, *Carpomys*, and *Batomys*) and the insectivorous and vermivorous earthworm mice (*Archboldomys*, *Chrotomys*, *Rhynchomys*, *Soricomys*, and *Apomys*). The Late Miocene is the likely time of arrival of these Old Endemics, based on calibrated molecular phylogenies and summarised by Lawrence Heaney and colleagues (2016), with about 10–14 million years ago for the cloud rats and about 6–8 million years ago for the earthworm mice. The second group is more '*Rattus*-like', and indeed, several species were initially assigned to *Rattus*. To these New Endemics belong the genera *Tryphomys* and *Abditomys*, *Bullimus*, *Crunomys*, *Rattus everetti* (Plate 30), and perhaps *Tarsomys* and *Limnomys*. The Late Pliocene is the likely time of arrival of this group, calibrated to about 3–4 million years ago. The genus *Crunomys* was initially included in the earthworm mice group of the Old Endemics, but Eric Rickart and Heaney (2002) found substantial DNA (karyotypic) differences between *Crunomys* and *Archboldomys*, which is confirmed by evidence from mitochondrial DNA and nuclear IRBP data, as given by Sharon Jansa and colleagues (2006). They place *Crunomys* in the group of the New Endemics. The genus *Anonymomys* seems difficult to place in the existing phylogenetic framework. Musser and Heaney could not link it with any other Philippine murine, regarding morphology. Jansa and colleagues (2006) did not include the genus in their molecular analysis, because no fresh tissue could be obtained. This leaves *Anonymomys* for so long in a division on its own.

Jansa and colleagues (2006) divided the three groups in smaller clusters, based on combined molecular data, and give an estimation for the time and geographic origin of each cluster using a relaxed molecular clock and taking the appearance of the first murine (*Progonomys*, twelve million years ago) as a minimum date of murine origin. These dates are about 18.7 million years ago for the cloud rats, 15.8 or 13.4 million years ago for the earthworm mice, 7.1 million years ago for the shrew mouse *Crunomys*, though with hesitation, and 3.1 million years ago for the forest rats (*Bullimus* and *Rattus*) and moss mice (*Limnomys* and *Tarsomys*). Leaving out the insufficiently supported date of arrival of *Crunomys*, two main colonisation events can be traced back, one between 15 and 20 million years ago and one around 3 million years ago.

Warty Pigs and Sulawesi Pigs

Celebochoerus is a unique genus of pigs having extremely large upper tusks, and which was to date only known from the Pliocene-Pleistocene of Sulawesi in Indonesia, hence the name – Celebes is an old name for Sulawesi, and *choerus* means pig in Classical Greek. The early Middle Pleistocene findings from Cagayan Valley, Luzon, are ascribed to *Celebochoerus cagayanensis*, which differs from the Sulawesi species *Celebochoerus heekereni* in having mesial and distal enamel bands on the upper canines. Ingicco and colleagues (2016) considered these characteristics as symplesiomorphic in suids and propose a migration route from the Philippines to Sulawesi. Already von Koenigswald (1958) mentioned the existence in the National Museum collections of pig fossils ('*Sus*'), along with elephant fossils and tektites from Anda (Pangasinan province), but without further details. These may be *Celebochoerus* as well.

Fossils of another endemic pig, the Philippine warty pig (*Sus philippensis*), were discovered at Callao cave of Luzon in association with a human metatarsal in a sediment layer dated to between 50,000 and 68,000 years ago. Today, the warty pig has a wide distribution and also occurs on Mindanao, Leyte, Samar, and nearby islands. The most characteristic feature is its 'warts', projections of skin-covered bone on the muzzle. A similar, but smaller warty species (*Sus cebifrons*), is present on Negros, Panay, and Masbate; it is extinct on Cebu and Guimaras. The pig subfossils from a cave at the hill Bat-ongan, Masbate, may belong to this species. A third small-sized warty species is endemic to Mindoro (*Sus oliveri*). It seems that each faunal region, corresponding to the Pleistocene palaeography, harbours its own variety of warty pig.

Elephants and Stegodons

Stegodon fossils from Luzon have been attributed to several species in the past, including *Stegodon orientalis*, *Stegodon sinensis*, *Stegodon trigonocephalus*, *Stegodon luzonensis*, and *Stegodon mindanensis*. Van der Geer and colleagues (2016) calculated a body mass decrease of about 60%, compared to mainland *Stegodon ganesha*, so smaller than the Javanese stegodon. Surprisingly, the third molars of the Luzon dwarf stegodon are in comparison much smaller than first and second molars. Based on this discrepancy, two different large stegodons were mentioned in the past for the Philippines. De Vos and Bautista (2001), however, showed that the second molar of the largest form fits the third molar of the presumably somewhat smaller form. It seems therefore that there was only one species of large stegodon, *Stegodon luzonensis* (Plate 32).

The size of the Philippine elephants is similar to that of the Asian elephant (*Elephas maximus*), based upon molar size only. In case the ancestor of the elephant of Luzon is related to the large-sized *Palaeoloxodon namadicus*, it had evolved a considerably smaller size. If so, the correct name should be

Palaeoloxodon beyeri; alternatively, it may have derived from a form close to *Elephas hysudricus*.

With only postcranial material at hand, it is difficult and often impossible to determine if a specimen is from an elephant or a stegodon. In 2001, a Philippine-Dutch team excavated proboscidean remains from the Espinosa Tusk Site (Cagayan Valley, Luzon), including a substantial part of the right foot, consisting of the second to fifth metacarpal, the first phalange of the fifth metacarpal, and cuneiform and magnum of a right front leg individual. The withers height could be around 90–200 cm, which is in the range of a dwarf proboscidean, according to Nike Liscaljet (2012), whether it evolved from a mainland *Stegodon* or *Elephas*.

References

Adams G.J. 1910. Geologic Reconnaissance of Southwestern Luzon. *Philippines Journal of Science*, 5 (A): 57–117.

Alba E. 1994. Archaeological evidences of animals as trade goods: A preliminary survey. *National Museum Papers*, 4 (2): 1–66.

Bautista A.P. 1991. Recent zooarchaeological researches in the Philippines. *Journal Arkeologi*, 4: 45–58.

Bautista A.P. 1995. Fossil remains of Rhinoceros from the Philippines. *National Museum Papers*, 5: 1–15.

Beyer H.O. 1949. Outline review of Philippine archaeology by island and provinces. *Philippine Journal of Science*, 77: 205–347.

Croft D.A., Heaney L.R., Flynn J.J., Bautista A. 2006. Fossil remains of a new, diminutive *Bubalus* (Artiodactyla: Bovidae: Bovini) from Cebu Island, Philippines. *Journal of Mammology* 87 (5): 1037–1051.

de Vos J. 2006. Notes about parallels in the evolution of the Pleistocene cervids from Greece (Crete, Kassos and Karpathos), Japan (the Ryukyu-islands) and Philippines (Masbate). *Hellenic Journal of Geosciences*, 41 (1): 127–140.

de Vos J., Bautista A. 2001. An update on the vertebrate fossils from the Philippines. *National Museum Papers*: 11 (1): 37.

de Vos, J., Bautista A. 2003. Preliminary notes on the vertebrate fossils from the Philippines. *Proceedings of the Society of Philippine Archaeologists, Semantic and Systematics: Philippine Archaeology*, 1: 42–62.

Détroit F., Dizon E., Falguères C., Hameau S., Ronquillo W., Sémah F. 2004. Upper Pleistocene *Homo sapiens* from the Tabon cave (Palawan, The Philippines): Description and dating of new discoveries. *Comptes Rendus Palevol*, 3 (8): 705–712.

Détroit F., Mijares A.S., Corny J., Daver G., Zanolli C., Dizon E., Robles E., Grün R., Piper Ph.J. 2019. A new species of *Homo* from the Late Pleistocene of the Philippines. *Nature*, 568: 181–186.

Dizon E., Détroit F., Sémah F., Falguères C., Hameau S., Ronquillo W., Cabanis E. 2002. Notes on the morphology and age of the Tabon Cave fossil *Homo sapiens*. *Current Anthropology*, 43 (4): 660–666.

Fox R.B., Peralta J.T. 1974. Preliminary report on the Palaeolithic archaeology of Cagayan Valley, Philippines and the Cabalwanian Industry. *Proceedings of the First Regional Seminar on South East Asian Prehistory and Archaeology, National Museum of the Philippines, Manila*, pp. 100–147.

Hall R. 1998. The plate tectonics of Cenozoic SE Asia and the distribution of land and sea. In: Hall R., Holloway J. D. (eds), *Biogeography and Geological Evolution of SE Asia.* Backhuys Publishers, Leiden, pp. 99–131.

Heaney L.R., Balete D.S., Dolar M.L., Alcala A.C., Dans A.T.L., Gonzales P.C., Ingle N.R., Lepiten M.V., Oliver W.L.R., Ong P.S., Rickart E.A., Tabaranza B.R. Jr., Utzurrum R.C.B. 1998. A synopsis of the mammalian fauna of the Philippine Islands. *Fieldiana Zoology (New Series)*, 88: 1–61.

Heaney L.R., Balete D.S., Duya M.R., Duya M.V., Jansa S.A., Steppan S.J., Rickart E.A. 2016. Doubling diversity: A cautionary tale of previously unsuspected mammalian diversity on a tropical oceanic island. *Frontiers of Biogeography*, 8 (2): e29667.

Heaney L.R., Piper P.J., Mijares A.S.B. 2011. The first fossil record of endemic murid rodents from the Philippines: A late Pleistocene cave fauna from northern Luzon. *Proceedings of the Biological Society of Washington*, 124 (3): 234–247.

Ingicco T., Castro A., de Vos J., van den Bergh G., Manalo K., Amano N., Jago-On C., Lyras G. 2015. Re-evaluation of the association of stone-tools with the Philippine megafauna from new fieldwork in the Kalinga-Apayao region (Luzon Island). *SEASIA Conference, Kyoto, 12–13 décembre* 2015: 92.

Ingicco T., van den Bergh G.D., de Vos J., Castro A., Amano N., Bautista A. 2016. A new species of *Celebochoerus* (Suidae, Mammalia) from the Philippines and the paleobiogeography of the genus *Celebochoerus* Hooijer. 1948. *Geobios*, 49 (4): 285–291.

Ingicco T., van den Bergh G.D., Jago-on C., Bahain J.-J., Chacón M.G., Amano N., Forestier H., King C., Manalo K., Nomade S., Pereira A., Reyes M.C., Sémah A.-M., Shao Q., Voinchet P., Falguères C., Albers P.C.H., Lising M., Lyras G., Yurnadi D., Rochette P., Bautista A., de Vos J. 2018. Earliest known hominin activity in the Philippines by 709 thousand years ago. *Nature*, 557: 233–237.

Jansa S.A., Barker F.K., Heaney L.R. 2006. The pattern and timing of diversification of Philippine endemic rodents: Evidence from mitochondrial and nuclear gene sequences. *Systematic Biology*, 55 (1): 73–88.

Karl H.-V., Staesche U. 2007. Fossile Riesen-Landschildkröten von den Philippinen und ihre paläogeographische Bedeutung. *Geologisches Jahrbuch A*, 160: 171–197.

Kitchener A.C. 1999. Tiger distribution, phenotypic variation and conservation issues. In: Seidensticker J., Christie S., Jackson P. (eds), *Riding the Tiger: Tiger Conservation in Human-Dominated Landscapes.* Cambridge University Press, Cambridge, pp. 19–39.

Lewis H., Paz V., Lara M., Barton H., Piper P., Ochoa J., Vitales T., Carlos J., Neri L., Robles E.C., Padilla R., Ragragio A., Solheim W. II., Ronquillo W. 2008. Dating and interpreting Early to Middle Holocene

cave occupation and an early cremation burial from Palawan, *Philippines. Antiquity*, 82: 318–335.
Liscaljet N. 2012. Napakaliit trompa: New pygmy proboscidean remains from the Cagayan Valley (Philippines). *Quaternary International*, 276–277: 278–286.
Lopez S.M. 1972. Contribution to the Pleistocene Geology of Cagayan Valley, Philippines. *Seminar on Southeast Asian Prehistory & Archaeology*. National Museum of the Philippines, Manila, pp. 1–16.
Mijares A.S., Détroit F., Piper Ph., Grün R., Bellwood P., Aubert M., Champion G., Cuevas N., De Leon A., Dizon E. 2010. New evidence for a 67,000-year-old human presence at Callao Cave, Luzon, Philippines. *Journal of Human Evolution* 59: 123–132.
Musser G.G., Heaney L.R. 1992. Philippine rodents: Definitions of *Tarsomys* and *Limnomys* plus a preliminary assessment of phylogenetic patterns among native Philippine murines (Murinae, Muridae). *Bulletin of the American Museum of Natural History*, 211: 1–138.
Naumann E. 1887. Fossile Elephantenreste von Mindanao, Sumatra und Malakka. *Abhandlungen und Berichte des Königlichen Geologischen und Anthropologisch-Ethnographischen Museums zu Dresden*, 6: 1–11.
Naumann E. 1890. *Stegodon mindanensis*, eine Art von Uebergangs-Mastodonten. *Zeitschrift der Deutschen Geologischen Gesellschaft*, 42: 166–169.
Piper P.J., Ochoa J., Lewis H., Paz V., Ronquillo W.P. 2008. The first evidence for the past presence of the tiger *Panthera tigris* (L.) on the island of Palawan, Philippines: Extinction in an island population. *Palaeography, Palaeoclimatology, Palaeoecology*, 264: 123–127.
Rhodin A.G.J., Thomson S., Georgalis G.L., Karl, H.-V., Danilov I.G., Takahashi A., de la Fuente M.S., Bourque, J.R., Delfino M., Bour R., Iverson J.B., Shaffer H.B., van Dijk P.P. 2015. Turtles and tortoises of the world during the rise and global spread of humanity: First checklist and review of extinct Pleistocene and Holocene chelonians". *Chelonian Research Monographs*, 5 (8): 000e.1–66
Rickart E.A., Heaney L.R. 2002. Further studies on the chromosomes of the Philippine rodents (Muridae: Murinae). *Proceedings of the Biological Society of Washington*, 115: 473–487.
Rozzi R. 2017. A new extinct dwarfed buffalo from Sulawesi and the evolution of the subgenus *Anoa*: An interdisciplinary perspective. *Quaternary Science Reviews*, 157: 188–205.
Steppan S.J., Zawadzki C., Heaney L.R. 2003. Molecular phylogeny of the endemic Philippine rodent *Apomys* (Muridae) and the dynamics of diversification in an oceanic archipelago. *Biological Journal of the Linnean Society*, 80: 699–715.
van der Geer A.A.E. 2008. *Animals in Stone, Indian Mammals Sculptured through Time.* Handbook of Oriental Studies, Section 2 South Asia 21. Brill, Leiden.

van der Geer A.A.E., van den Bergh G., Lyras G.A., Prasetyo U.W., Due R.A., Setiyabudi E., Drinia H. 2016. The effect of area and isolation on insular dwarf proboscideans. *Journal of Biogeography*, 43: 1656–1666.

von Koenigswald G.H.R. 1956. Fossil mammals from the Philippines. *National Research Council of the Philippines, University of the Philippines, Diliman, Quezon City*, pp. 1–24.

von Koenigswald G.H.R. 1958. Preliminary report on a newly-discovered Stone Age culture from Northern Luzon, Philippine Islands. *Asian Perspectives*, 2: 69–71.

CHAPTER SIXTEEN

Japan: Honshu, Shikoku, and Kyushu

Evolution of Island Mammals: Adaptation and Extinction of Placental Mammals on Islands, Second Edition.
Alexandra van der Geer, George Lyras and John de Vos.

© 2021 John Wiley & Sons Ltd. Published 2021 by John Wiley & Sons Ltd.

The three Japanese main islands Honshu, Shikoku, and Kyushu form today a zoogeographical unit together with the northern Ryukyu Islands. The fossil biozones of this part of Japan are based on proboscideans. The late Early Miocene fauna is known by a lineage of four-tusked dwarf stegodons. A large stegodon characterises the Pliocene, followed again by a biozone with a dwarf stegodon and several deer in the Early Pleistocene. The early Middle Pleistocene fauna contains a small mammoth and red deer, followed in the late Middle and Late Pleistocene by a middle-sized elephant and several deer species. During the Last Glacial Maximum, a mainland fauna entered Japan. The Japanese faunas are impoverished and endemic, except during the Last Glacial Maximum, and perhaps during the Early Miocene with a gomphothere.

Geology and Palaeogeography

The Japanese Archipelago is located along the eastern margin of the Asian continent, with Hokkaido in the north and the southern Ryukyu Islands in the south (Figure 16.1). Japan is separated from the Eurasian mainland by the Sea of Japan (Hokkaido and Honshu), the Strait of Korea and the Tsushima Strait (western Honshu and Kyushu), and the East China Sea (Kyushu and the Ryukyu Islands). In the north, Hokkaido is separated from Sakhalin Island by the Soya Strait.

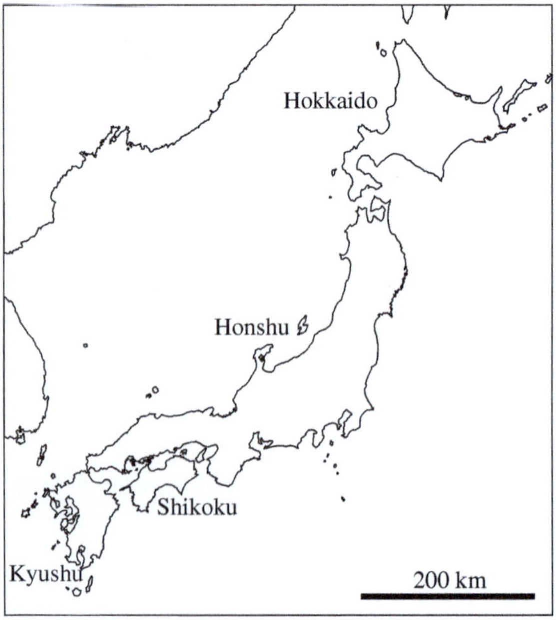

Figure 16.1 Map of Japan. (Redrawn from http://d-maps.com.)

The island arc can be divided into three different zoogeographic units, of which the northernmost one is Hokkaido. Hokkaido was connected with the Asian mainland throughout the Pleistocene and is therefore not considered an island as far as its palaeontology is concerned. The central zoogeographic unit consists of Honshu, Shikoku, and Kyushu – including some 3000 smaller islands and islets – and the northern Ryukyu Islands Tanega, Yaku, and the Tokara Islands. At times of low sea levels, the main islands of Honshu, Shikoku, and Kyushu formed a single landmass, referred to as Hondo. The southern zoogeographic unit is formed by the central and southern Ryukyu Islands. This unit is separated from the central one by the Tokara Strait, which functions as a zoogeographical boundary between the Oriental and the Palaearctic region. This boundary can be traced back at least to the Late Miocene and is confirmed by the lack of conspecific amphibians on either side of the line, as demonstrated by Hidetoshi Ota (2000). In this chapter, only the central unit is considered. The southern unit is treated in the next chapter.

The Early Miocene coincides with the opening of the Sea of Japan. Around 18 million years ago, large parts of the proto-Japanese islands are supposed to have been above sea level, presumably with dry land connections between the different islands and the Asian continent. Subsequently, the islands gradually submerged, resulting in their isolation. Around 15 million years ago, the transgression ceased and uplift began in the south-west, while the transgression continued until the middle Late Miocene in the north-east of the island arc. As a result, the exposed land area reduced significantly, and many small islands were formed.

During periods of low sea level in the Plio–Pleistocene, the Strait of Korea and the East China Sea might have been exposed as dry land, thus playing an essential role in the dispersal of mammals to Japan. The Strait of Korea has at most places a depth of about 100 m, which means that an equivalent drop in sea level is sufficient to bring Japan within reach for non-swimming mammals. Japanese fossil faunas, though, show a high degree of endemism, as was already recognised by Wallace in 1880. During the Pleistocene, the sea level was lowest between 20,000 and 17,000 years ago. Estimations of this drop differ between the various scientists and range from maximal 140 m below the present level according to Iseki (1975) to minimal 75 m according to Ohshima (1980). This seems to coincide with an invasion of a mainland fauna to Japan and the extinction of the dwarf elephant.

The problem of dispersal of mammals to Japan, either by land bridge or by overseas sweepstake dispersal, has been extensively discussed by several authors, e.g. Ryuji Tada (1994); Mike

Dobson and Yoshinari Kawamura (1998); Kawamura (1998); Konishi and Yoshikawa (1999); and Yoshikawa, Kawamura and Taruno (2007), based on information from fossil faunas, sea-level changes, sea-floor topography, and salinity changes in the East China Sea and the Sea of Japan. According to Dobson and Kawamura (1998), Hondo was connected to the Asian continent at about 1 million years ago, at 630,000 years ago, and at 430,000 years ago, evidenced by the immigration of, respectively, *Mammuthus shigensis* from China, *Stegodon orientalis* from the south via the East China Sea continental shelf, and *Palaeoloxodon naumanni* from northern China via the Korean peninsula.

The condition of the Strait of Korea and the Tsushima Strait during the Last Glacial Maximum was discussed by Jin-Oh Park and colleagues (1996) and Hiroyuki Matsui and colleagues (1998), who argued that this problem is based on the Salinity Balance Model of the Sea of Japan. The $\delta^{18}O$ values in the Sea of Japan throughout the Late Pleistocene, as reported by them, indicate that the Tsushima Strait was only about 10 m deep at about 19,000–15,000 years ago. Immigration across the Tsushima Strait during the Last Glacial Maximum has also been suggested by Satoshi Ohdachi and colleagues (2004) and Yoshiki Yasukochi and colleagues (2009), based on phylogeography of white-toothed shrews and black bears, respectively. Conditions of the Tsugaru Strait were discussed by Keiichi Takahashi and colleagues (2004, 2006). Akihisa Kitamura and colleagues (2001) and Kitamura and Kimoto (2004, 2006) reported that the southern channel of the Sea of Japan was open between 3.9 and 1 million years ago (middle Pliocene–late Early Pleistocene), which period coincides with an impoverished fauna. Kazutaka Amano and colleagues (2008) suggested that this channel may have opened longer than Kitamura estimated.

Sweepstake dispersal to Japan is not generally considered, though the composition of the Japanese impoverished faunas – consisting of mainly proboscideans and deer and lacking bovids – is suggestive for such dispersal in our view.

Historical Palaeontology

Karl Martin, then professor of geology and director of the museum in Leiden, described in 1886 some molars of elephants from the Late Pleistocene of Japan (Figure 16.2). Martin recognised similarities with the Siwalik fauna of British India and with the fossil fauna of Java and assigned the molar from Yasiro to *Euelephas* sff. *antiquus* and the one from an unidentified site to *Euelephas namadicus* (now both *Palaeoloxodon naumanni*).

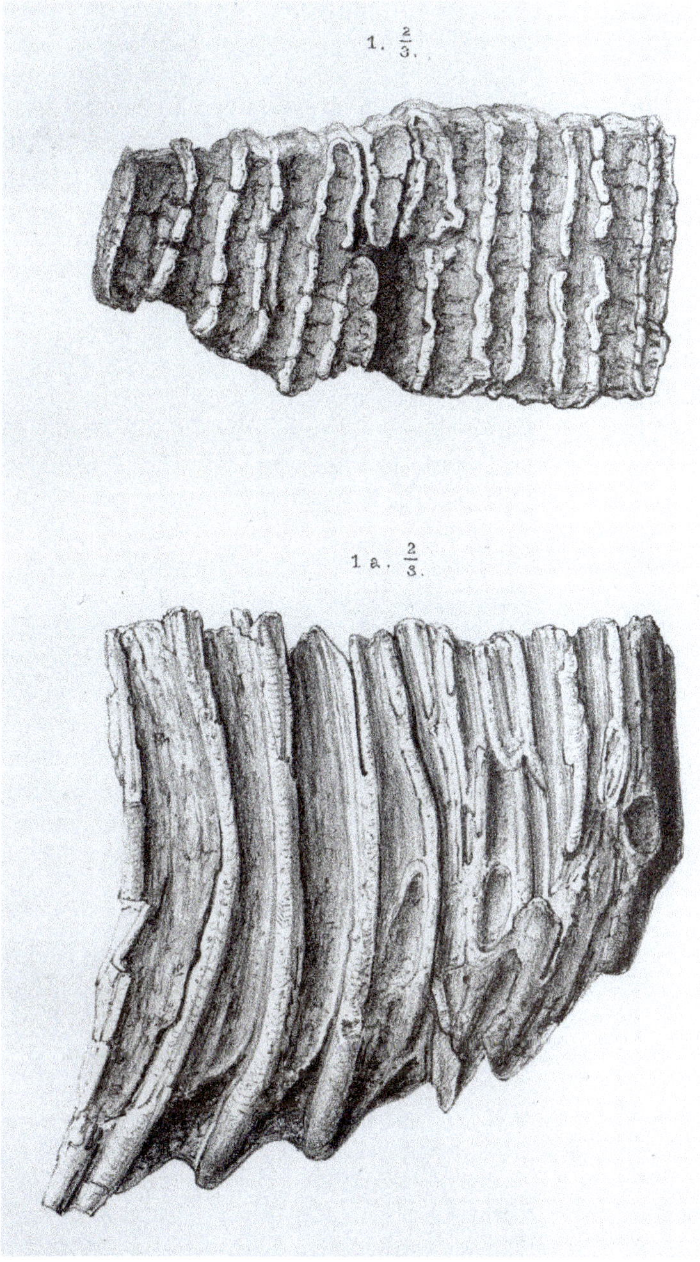

Figure 16.2 Molars of Naumann's elephant (*Palaeoloxodon naumanni*) from Japan, Late Pleistocene. Top: occlusal view; below: lateral view. (From Martin, 1886.)

Hikoshichiro Matsumoto recognised 14 proboscideans for Japan between 1918 and 1941, of which he himself had named no less than five species. He also erected two new stegodon genera (*Parastegodon*, now *Stegodon*, and *Prostegodon*, now *Stegolophodon*) and one elephant subgenus (*Palaeoloxodon*). His profusion of proboscidean taxa has been reduced to a more convenient number, and most of Matsumoto's species are now treated as junior synonyms. His subgenus *Palaeoloxodon*, however, survived the heavy pruning and is still widely used on a generic level, recently re-established by Jeheskel Shoshani and colleagues (2007). In Matsumoto's view, molars of *Palaeoloxodon* resemble those of *Loxodonta*, the African elephant. *Palaeoloxodon* then are the more primitive members of the *Loxodonta* lineage, which explains the name of the subgenus. *Palaeoloxodon* was in the past regularly considered a junior synonym of *Elephas*, the genus of Asiatic elephants.

The first proboscidean described by Matsumoto (1918) was *Elephas aurorae* (now *Stegodon aurorae*), based on a left upper third molar – described as a right upper second molar, see Taruno, 1991 – from Mount Tomuro, Ishikawa Prefecture, Honshu, from an unknown horizon, but possibly Pliocene. A decade later, Matsumoto (1929) erected a new genus, *Parastegodon*, to accommodate not only this new species but also *Elephas planifrons* from India and *Stegodon mindanensis* from Mindanao, the Philippines, as well.

Makiyama (1924) described two subspecies of *Elephas namadicus* in Japan, being *naumanni* and *namadi*. Matsumoto (1924) immediately seized the opportunity and erected the new subgenus *Palaeoloxodon* to accommodate Makiyama's two new subspecies and the Indian variety as well. He assigned the Japanese species *naumanni* as type species and added a fourth species, described by him as *tokunagai*. The type specimen came from Soyama in the Higashitonami district, Toyama Prefecture, Honshu, from an unknown horizon.

Matsumoto (1924) also described *Euelephas protomammonteus*. This species is smaller and more archetypal than *Elephas* (=*Mammuthus*) *trogontherii*. Fossils of this new species were all found in the Kimitsu district of Kazusa, Chiba Prefecture, Honshu (at Nagahama, Seki-mura, Uwehata, Kokubo, and Otomi). Two years later, Matsumoto (1926a) described a new gomphothere as *Hemimastodon* (= *Gomphotherium*) *annectens*, based on a crushed skull with teeth. The specimen had been found at Banjobora in Mino, Giyu Prefecture, Honshu, in the Hiramaki Formation of a late Early Miocene age. The molars bear three ridges and resemble those of the primitive African genus *Phiomia*, but are more hypsodont. According to Matsumoto's (1929) later diagram, this new species is sister taxon to *Gomphotherium angustidens* and *Zygolophodon*.

In the same 1926 paper, Matsumoto described *Trilophodon (= Sinomastodon) sendaicus*, based on two complete third molars, two fragments of a third specimen and a left astragal from Kitayama near Sendai at Rikuzen, Miyaki Prefecture, Honshu, probably from the Tatsunokuchi Formation of a Late Miocene or

Early Pliocene age. In his view, the three-ridged molars resemble those of the European *Trilophodon* (=*Gomphotherium*) *angustidens*, but the tips of unworn cusps are more acute. He thought that his new species was a geographical and geological variety of this species.

More than a decade later, Matsumoto (1941) described *Stegodon clifti miensis*, based on a left lower jaw with a third molar from Mie Prefecture, Honshu. It is a large stegodon, dating to the Late Pliocene. Much later, Sawamura and colleagues (1979) described another large stegodon as *Stegolophodon shinshuensis*, based on fossils from Nagano Prefecture, Honshu. Hiroyuki Taruno (1991) moved this new species to the genus *Stegodon*. However, Taru Hajime and Kono Naoki (2002) synonymised *shinshuensis* with Matsumoto's *miensis*. The latter name has priority, and thus the large Japanese stegodon is now known as *Stegodon miensis*.

Hatai (1959) described the third molar of a primitive stegodon as *Stegolophodon miyakoae*. The specimen came from the Tsukinoki Formation of Shibata, Miyagi Prefecture, Honshu. Haruo Saegusa (2008) considered the new species as a junior synonym of *Stegolophodon pseudolatidens*, because a skull and a mandible of the latter species were found in a nearby locality in a nearly identical horizon.

Biozones and Faunal Units

The biozonation of Japan is mainly based upon proboscidean remains. Fortunately, the horizon and the age of several proboscidean fossils are known accurately since the late 1990s, because the Plio–Pleistocene strata have been well studied by tephrostratigraphy, magnetostratigraphy, and radiometric dating. Correlation of the horizons is possible by dating of the volcanic ash deposits, as was done by amongst others Yasufumi Satoguchi and team (1999). The Miocene biozonation (Figure 16.3) on the other hand is based on biostratigraphy only. No fossils were reported from Shikoku and the northern Ryukyu Islands. Most of the fossil faunas of Japan are impoverished endemic faunas, indicating insular conditions.

Early Miocene

The earliest Japanese vertebrate fauna is hardly known. The primitive proboscidean of this biozone is a primitive type of elephant, a four-tusked gomphothere (*Gomphotherium annectens*; earlier also *Trilophodon sendaicus*, *Trilophodon palaeindicus*, and *Bunolophodon yokotii*), described by Matsumoto (1926a, b). Apart from the type specimen – a partial

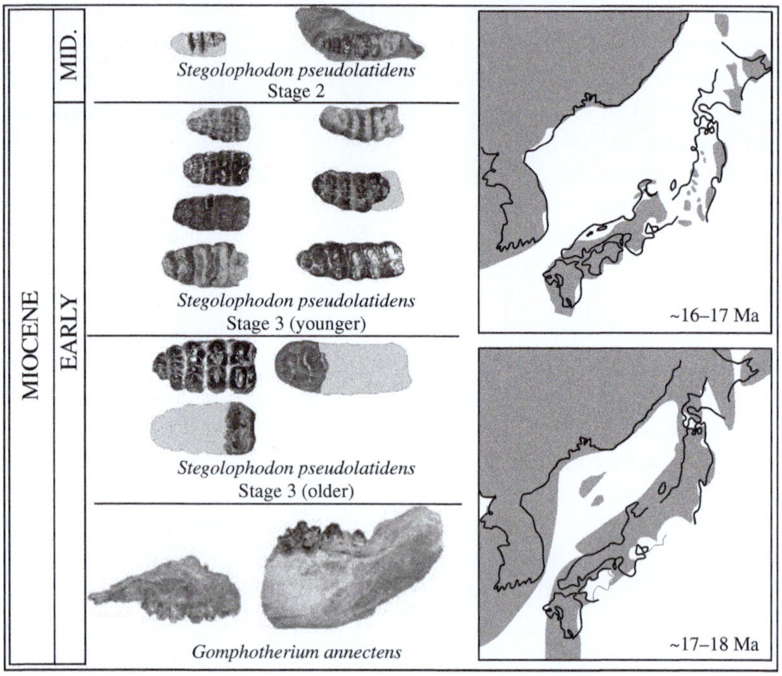

Figure 16.3 Stratigraphic scheme showing the four intervals of proboscideans during the Miocene of Japan. (Based on Saegusa, 2008.)

skull – from Banjobora at Mino, Gifu Prefecture, Honshu, an almost complete lower mandible with teeth and fragments of the lower tusks was found at the same locality, which appears to belong to the same individual, as shown by Tadao Kamei and colleagues (1977). Endemic characters are not known for the Japanese gomphothere, indicating that Japan might very well have been connected to the mainland during this period. The age of the fossiliferous formation is estimated at 18 million years. Takai (1954) reports that the gomphothere is found in association with the anthracothere *Brachyodus* – a widespread Eurasian species like the gomphothere.

Late Early Miocene

The second biozone is known mainly for its primitive four-tusked stegodon (*Stegolophodon pseudolatidens*). A tapir and an anthracothere (*Brachyodus*) have been mentioned as well, but their relation to the stegodon is unclear.

Three different sizes of the Japanese *Stegolophodon* are recognised, which form successive stages of dwarfism. The smallest, and geologically youngest, is represented by an extremely small hemi-mandible and fragmentary second and third molars from the Asakawa Formation of Ibaragi Prefecture, Honshu. These molars are the smallest of all known *Stegolophodon* molars. The smallest stage is dated to 16.4–16 million years ago by Yoshiki

Koda and colleagues (2003), based on fission-track dating of a tuff layer, while the largest stage is dated to 18.3–17 million years ago.

The two ways of dispersal of the ancestral *Stegolophodon* to proto-Japan – land bridge versus sweepstake – both remain open. A northern land bridge would have connected Honshu via Hokkaido-Sakhalin to Siberia, but this route seems less likely because stegodon fossils were not reported from Siberia. A southern land bridge is favoured by the presence of *Stegolophodon* fossils at Early Miocene deposits of coastal China, Thailand, and Myanmar but is contradicted by deep marine basins in the Tsushima Strait – the eastern channel of the Korea Strait – and the offshore area west of Kyushu. Thus, an overseas dispersal via the more central part of the emerging Japan Sea seems the most reasonable alternative. The subsequent isolation of *Stegolophodon* fits the geological data of gradual submersion of the northeast of the island arc. At the onset of the Middle Miocene, Japan, consisted of several islands, and no contact area with the mainland existed.

Pliocene

After a long gap in the fossil record, a next biozone is recognised, characterised by a large stegodon, *Stegodon miensis* (=*shinshuensis*) (Figure 16.4). Fossils were found at over ten localities. The youngest specimen is assigned an age of about 2.93 million years ago, based on fission-track analysis of the tuff layer by Hisao Baba and colleagues (2005), whereas the oldest specimen is dated to about 3.5 million years ago. An isolated upper third molar that is indistinguishable from those of the Chinese species *Stegodon zdanskyi* was dated to around 5 million years ago. This indicates that there was a gradual

Figure 16.4 Lower third molar of *Stegodon miensis* (occlusal view) found at Hattori River, Iga city, Mie Prefecture, Honshu. (Photograph: John de Vos.)

evolution from the ancestral form *zdanskyi* to the endemic *miensis* during this period.

A second element of this biozone is a colobine monkey (*Kanagawapithecus leptopostorbitalis*), reported by Mituo Iwamoto and colleagues (2005) from the Kosawa locality, Kanawaga Prefecture, Honshu. The age of the sediments in which its partial skull was found is estimated as between at least 2.5 million years and 2.9 million years old. The taxonomy and phylogenetic relations of this monkey have been debated by various scholars, and a close relationship to *Dolichopithecus* and *Parapresbytis eohanuman* has been suggested. Nishimura and colleagues (2012) reviewed its characteristics and concluded that the Japanese colobine is a genus on its own, unrelated to all modern Asian colobines, based amongst others on the presence of a maxillary sinus in *Kanagawapithecus*. Among colobines, a maxillary sinus is only present in *Libypithecus* and *Cercopithecoides*. Otherwise, *Kanagawapithecus* shares more important facial and dental features with *Dolichopithecus* than with *Parapresbytis*.

Early Pleistocene

The next biozone is characterised by a dwarf stegodon (*Stegodon aurorae*). The rest of the endemic megafauna consists of two species of Père David's deer (*Elaphurus shikamai* and *Elaphurus tamaensis*) and a third species of deer (*Cervus* sp. or *Cervus kyushuensis*). Fossil remains of a rhinoceros and a canid (*Xenocyon falconeri*) were reported by Yoshikazu Hasegawa and colleagues (1991), and Akihiro Koizumi (2003), respectively. The sediments in which their fossils were found are 1.8 million years old.

The dwarf stegodon (Figure 16.5) of this biozone evolved from the larger stegodon (*Stegodon miensis*) of the previous biozone, on the basis of a cladistic analysis of cranial characters by Saegusa (1987). This implies that this biozone is a mere continuation of the previous one, with probably new arrivals. Fossils of this species, including eight skeletons, have been reported from more than 45 localities, mainly from central Honshu.

A similar or identical dwarf stegodon has also been reported from Taiwan. Initially, it was named *Stegodon akashiensis* by Tokio Shikama and colleagues (1975), but this was considered a junior synonym of *Stegodon aurorae* by Hiroyuki Taruno (1991) who has shown that the holotype of *akashiensis* is actually a worn upper first or second molar of the latter species. Fuyoji Takai (1936), who erected the species, had based his description on a worn molar fragment found at Akashi, Hyogo Prefecture, in Early Pleistocene deposits.

The two species of Père David's deer are endemic to Japan and might have reached the island overseas since the extant

Figure 16.5 Lower and upper molars of the Japanese dwarf stegodon (*Stegodon aurorae*); occlusal view. Geological Museum, Tsukuba, Honshu. (Photograph: Jan van der Made.)

species (*Elaphurus davidianus*) is an excellent swimmer and grazes on grasses and water plants. Probably, the Japanese fossil species had a similar preference for marshlands and river sides as is evidenced by a site at Lake Biwa near Kyoto, Honshu (Plate 33). In the past, this was a muddy river side, where deer and stegodons left their imprints. Volcanic ashes that buried the herbivore tracks and trunks of giant Chinese sequoias (*Metasequoia*) were dated to 1.8 million years ago. During the Early Pleistocene, the area must have been covered with sequoia forests, as described by Yamakawa and colleagues (2009), inhabited by deer and dwarf stegodons. The co-occurrence of small proboscideans and several deer, combined with the lack of other megafauna, is typical for insular faunas with sweepstake origin. The canid, rhinoceros, and monkey may represent vicariant taxa though overseas dispersal is another possibility, as evidenced by the rhinoceros on the Philippines, a descendant of *Xenocyon* on Sardinia, and monkeys on the West Indies.

Taiwan, too, harboured an endemic Père David's deer (*Elaphurus formosanus*) during this period, together with two or perhaps four other species or subspecies. The composition of the Taiwanese fauna with only a dwarf stegodon and Chinese deer thus closely resembles that of Japan and indicates a similar ancestral fauna arriving from China.

Early Middle Pleistocene

The next biozone is characterised by another impoverished, unbalanced fauna, consisting of a steppe mammoth (*Mammuthus trogontherii*), deer (*Cervus kazusensis*), a large bovid ('*Bubalus*' sp.), a field mouse (*Apodemus argenteus*), and a polecat (*Oriensictis nipponica*). Fossils of the extant Japanese dormouse (*Glirulus japonicus*) and Japanese hare (*Lepus brachyurus*) from Middle Pleistocene sediments were reported by Kawamura (1989), and fossils of a wild boar were reported by Fujita and colleagues (2000) from NT cave, Nilmi, Okayama Prefecture, Honshu. An early age for the hare is in agreement with genomic data (mitochondrial DNA) from Yamada and colleagues (2002), who estimated a divergence from its ancestor during the Pliocene. The fauna indicates an increase in temperate grasslands. Also, the dormouse DNA dates back to the Pliocene, as reported by Shumpei Yasuda and colleagues (2007).

The Japanese mustelid was attributed to the new genus *Oriensictis* within the subfamily of polecats (Galictinae) by Shintaro Ogino and Hiroyuki Otsuka (2008), based on tooth morpohology. The presence of a groove on the canine is in favour of a galictine ancestry.

In fissure fillings at Isa, Yamaguchi Prefecture, Honshu, skull remains of a small tiger (*Panthera tigris*) were found together with large *Stegodon* fossils, as reported by Tokio Shikama and Goro Okafuji (1963). Large *Stegodon* is restricted there to the horizon of 0.60–0.65 million years ago of dated fluvial sediments in the Osaka group, as given by Yoshikawa and colleagues (2007), which narrows the tiger window to the early Middle Pleistocene.

In Taiwan, a few thousands year earlier, around 1 million years ago, the dwarf stegodons disappeared as well and new elements are the steppe mammoth (*Mammuthus trogontherii taiwanicus*), a water buffalo (*Bubalus* sp.), the Chinese rhinoceros (*Rhinoceros sinensis*), a pig (*Sus* sp.), and a large felid (*Panthera* sp.). The Taiwanese form of the mammoth differs in molar size and number of plates. The distance between Taiwan and the mainland was relatively short, and likely the fauna had arrived through filter dispersal.

Late Middle–Early Late Pleistocene

During this period, new invasion events occurred, as described by Kawamura (1998), Konishi and Yoshikawa (1999), and Yoshikawa and colleagues (2007). The megafauna of Honshu of this biozone consists mainly of Naumann's elephant (*Palaeoloxodon naumanni*; Figure 16.6), Chinese giant deer (*Sinomegaceros yabei*), and several other species of deer (six

Figure 16.6 Upper molars of Naumann's elephant (*Palaeoloxodon naumanni)*; occlusal and lateral views. Geological Museum, Tsukuba, Honshu. (Photograph: Jan van der Made.)

different *Cervus* species or subspecies, including sika deer, *Cervus nippon*; *Cervus kazusensis* is a relic from the previous biozone). Small mammals include the Japanese mole-shrew (*Anourosorex japonicas*) and a grass vole (*Microtus epiratticepoides*). Palaeontological and molecular data, given by Nagata and colleagues (1999), Takakuwa (2006), and Kawamura (2009), suggest that sika deer may have immigrated together with Naumann's elephant.

Fossils of Naumann's elephant were also reported from Taiwan and the Penghu Islands by Shikama and colleagues (1975), but these appear to differ markedly from the Japanese form in size and wear pattern of the molars. In fact, according to Keiichi Takahashi and colleagues (2001) in their revision of Japanese and Taiwanese proboscideans, the Taiwanese Late Pleistocene form is rather similar to the Chinese mainland form, *Palaeoloxodon huaihoensis*.

Late Pleistocene

Gradually, coinciding with a cooling of the climate, the ancestors of the extant species settle the area, replacing the older endemics. Sites dating to between 29,000 and 57,000 years ago have yielded the fossils of several extant species, including the Japanese hare (*Lepus brachyurus*), Japanese macaque (*Macaca fuscata*), large Japanese field mouse (*Apodemus speciosus*), red fox (*Vulpes vulpes*), raccoon-dog (*Nyctereutes procyonoides*), Asian black bear (*Ursus thibetanus*), leopard (*Panthera pardus*), and red fox (*Vulpes vulpes*), as reported by Yoshinari Kawamura and Ryohei Nakagawa (2012). Naumann's elephant, Chinese giant deer, sika deer, *Cervus kazusensis*, Japanese dormouse, the mole-shrew, and the grass vole are survivors from the previous period. A number of Late Pleistocene species, such as the Japanese dormouse (*Glirulus japonicus*) and the Japanese hare (*Lepus brachyurus*), are relic species dating from 3 to 5 million years ago (Pliocene). The composition of the Late Pleistocene fauna indicates a temperate forest environment.

Around about 30,000 years ago, close to or at the onset of the Last Glacial Maximum, a balanced fauna gradually invades Hondo from the north across the Tsugaru Strait, which is supposed to have formed an 'ice bridge'. The dispersal coincides with a glacial period and a significant sea-level drop. The new fauna, listed in Kawamura and colleagues (1989), includes amongst others, the grey wolf (*Canis lupus*), the brown bear (*Ursus arctos*), Young's lion (*Panthera youngi*), the elk (*Alces alces*), the aurochs (*Bos primigenius*), bison (*Bison priscus*), a wild horse (*Equus hemionus*), and small mammals, e.g. a lemming, a shrew (*Crocidura disinezumi*), and the Japanese mole (*Mogera wogura*). Also, the woolly mammoth (*Mammuthus primigenius*) entered Hokkaido from the north during the late Late Pleistocene, as reported by Kawamura (1998) and Takahashi and colleagues (2004), but did not disperse into Hondo.

The ancestor of the extant Smith's red-backed vole (*Myodes smithii*) may have arrived during this period as well. Earlier, the vole was placed in a genus on its own, *Phaulomys*, because of the presence of ever-growing molars. Genomic data, however, have shown that Smith's vole is closely related to extant Asian *Myodes* species. The acquisition of ever-growing molars likely evolved on the island, reminding of the ever-growing incisors in the Balearic *Myotragus* and the Tuscany *Maremmia*. Based on analysis of mitochondrial cytochrome b gene sequence variation, Asuka Honda and colleagues (2019) reconstructed a rapid expansion event during warmer periods following a severe cold period (population bottleneck) during the Late Pleistocene. They did not detect such a rapid expansion event in the other red-backed vole (*Myodes andersoni*).

Small differences with mainland species have been reported for the Japanese late Late Pleistocene species. For example, *Macaca fuscata* from Matsugae (north-eastern Kyushu) and from early Holocene sites of the Jōmon Period tends to have larger teeth than any modern East Asian form from the '*fascicularis* group' (in the sense of Fooden, 1976), including modern *Macaca fuscata*, as reported by Ogino and Otsuka (2005). Earlier, Mitsuo Iwamoto (1975) noticed that the fossil species had a much broader face than extant Japanese macaques, based on a skull from Shikimizu quarry, Shikoku (latest Pleistocene–Holocene). The earlier described horse *Equus nipponicus* from Late Pleistocene beds had been re-investigated and radiometrically dated by Ann Forsten (1997), who concluded that it was a domestic small-sized horse (*Equus caballus*) of about 1,530 years ago. Other small-sized horse remains were dated to about 28,000 years ago, confirming the arrival of a small-sized horse at the onset of the Last Glacial Maximum.

Naumann's elephant initially survived the invasion, concluding from its co-occurrence with the new fauna in the Upper Kuzuü Formation, Tochigi Prefecture, Honshu. It nevertheless went extinct some 16,000 years ago when the ancestors of the Neolithic Jōmon people came to Japan. The last fossils of Naumann's elephant are radiocarbon dated to 16,720 ± 880 years ago by Koji Okumara and colleagues (1982). By then, the sea level had reached its lowest level, coinciding with the coldest vegetation in the region as analysed by Keiji Suzuki and Tadao Kamei (1981).

After the Last Ice age, the Japanese fauna suffered local extinctions, unbalanced by new colonisations. This resulted in a certain degree of endemism, as shown by Dobson and Kawamura (1998) and Kawamura (2007), for example, the Japanese or Honshu wolf (*Canis lupus hodophilax*), the Japanese serow (*Capricornis crispus*), and the Japanese fox (*Vulpes vulpes japonica*). The Japanese wolf (*Canis lupus hodophilax*) – not to be confused with the mountain dog – was the smallest known wolf variety with a shoulder height of about 40 cm and was raised to specific status by some authors. It was officially extinct in 1905, mainly due to an epidemic of contagious diseases including rabies.

Peculiarities and Evolution of Endemic Mammals

Dwarf Elephants

Stegolophodon lived in Japan during the early Middle Miocene and is represented by three distinct size classes that follow each other in time during the evolution of dwarfism on the island. All specimens show the same derived morphological traits confirming that they indeed belong to a single lineage. The reduction in body size is best explained by insular dwarfism and does not represent successive immigration from the mainland of Asia. To date, the Japanese stegolophodons represent the oldest geological record of insular dwarfism in elephantoids.

The different stages were in the past contributed to different species – *tsudai, miyakoae, latidens, elephantoides* and sp. – or even genera – *Pentalophodon* (= *Anancus*), *Bunolophodon* (= *Gomphotherium*), and *Rhynchotherium*. Saegusa (2008) revised the Japanese species and reduced them to three developmental stages of a single stegodon lineage. The size differences were initially attributed to sexual dimorphism of a single species. However, the range of variation throughout the stratigraphic sequence exceeds that of other elephantoid samples. In addition, a trend towards smaller body size can be observed from the oldest to the youngest layer.

The genus is now considered ancestral to the stegodons, the sister taxon of elephants and mammoths. Generally, members of this genus bear only upper tusks, with the exception of the Thai species *Stegolophodon nasaiensis* and the Japanese species. The same was claimed for the Chinese *Stegolophodon hueiheensis*, but the only material known of this species is a pair of heavily worn third molars. A mandible, showing the alveolus of the lower tusk, was found together with the molars, as described by Minchen Chow (1959), but this mandible appears to be missing now.

The Pliocene proboscidean is *Stegodon miensis*. The molars are similar in morphology to those of *Stegodon zdanskyi* of northern China, which might be its ancestor. The Japanese species is considered to be more derived than the Chinese species. The main difference is found in the number of lophs and the structure of the enamel layer. In the literature, the names *bombifrons* and *insignis* – either with or without the addition *confer* – are sometimes used for this stegodon. This implies a direct relationship with the Siwalik fauna, but this is in conflict with a Chinese origin of the stegodon.

The dwarf stegodon of the Early Pleistocene, *Stegodon aurorae*, is a direct descendant of the larger stegodon (*Stegodon miensis*) of the previous biozone, on the basis of a cladistic analysis of cranial characters by Saegusa (1987). The dwarf stegodon was relatively small with a shoulder height of almost 2 m (Plate 1) and had relatively hypsodont molars (Figure 16.5). The limbs were relatively short in relation to the body, compared to the Chinese stegodon (*Stegodon zdanskyi*), befitting an insular species. This can also be inferred from the nature of the preserved footprints (Plate 33). The Japanese dwarf stegodon had evolved a body mass reduction of about 75% relative to its ultimate mainland ancestor *Stegodon zdanskyi*, as calculated by us in 2016.

Initially, the Japanese mammoth was known as *Mammuthus protomammonteus* and considered smaller than its ancestor, a form of the steppe mammoth *Mammuthus trogontherii* (= *armeniacus*), which stood about 4.5 m tall. The large inter-individual variation observed in the molars of this mammoth gave rise to a lot of confusion and led to a division into several species or subspecies since 1926: *protomammonteus*, *meridionalis* subsp., cf. *meridionalis*, *paramammonteus*, *shigensis*, *paramammonteus shigensis*, *proximus*, and *armeniacus proximus*. However, Kawamura and colleagues (2007) and Taruno and Kawamura (2007) revised the early Middle Pleistocene *Mammuthus* from the Japanese islands and concluded that they all belong to a single species, *Mammuthus trogontherii*, based on features of the molars (Figure 16.7).

Naumann's elephant (*Palaeoloxodon naumanni*; Figure 16.8) is the type species of the genus *Palaeoloxodon*, which is an unusual situation: an island form as reference for a mainland genus. Naumann's elephant has strongly curved and twisted tusks, unlike the straight tusks as typical for *Palaeoloxodon antiquus*. It had a shoulder height less than 3 m, which is a mild decrease in height compared to its ancestor, *Palaeoloxodon namadicus*. Its remains are the most abundant proboscidean fossils known from Japan. Naumann's elephant may have dispersed twice from Hondo to Hokkaido, once at 430,000 years ago, as given by Kawamura (1998), and again at about 30,000 years ago across the so-called Tsugaru ice bridge or by swimming, according to Takahashi and colleagues (2004).

Deer

The combination of deer and dwarf elephant as main megafauna elements is typical for isolated islands. For Japan, this is the case as well during the Early Pleistocene – with *Stegodon aurorae*, *Elaphurus shikamai*, *Elaphurus tamaensis*, *Cervus*

Figure 16.7 Left lower jaw with molar of the Japanese mammoth, *Mammuthus trogontherii* (= *Mammuthus protomammonteus*); occlusal view. Geological Museum, Tsukuba, Honshu. (Photograph: Jan van der Made.)

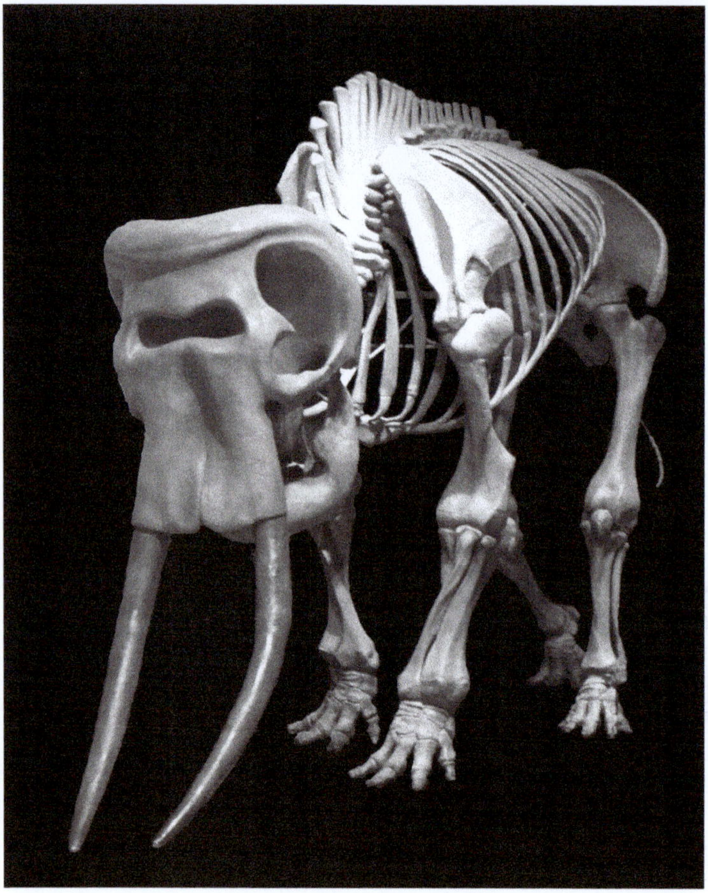

Figure 16.8 Reconstruction of the Japanese dwarf elephant (*Palaeoloxodon naumanni*). Iberaki Nature Museum, Osaki, Honshu. (Photograph: Jan van der Made.)

sp., or *Cervus kyushuensis*, the early Middle Pleistocene – with *Mammuthus trogontherii* and *Cervus kazusensis*, and the late Middle to Late Pleistocene – with *Palaeoloxodon naumanni*, *Sinomegaceros yabei*, *Cervus nippon*, and several more *Cervus*.

The fossil record of the today's endemic Japanese sika deer (*Cervus nippon*) goes back to the Middle Pleistocene. Today, six subspecies are recognised, endemic to different regions. Based on molecular phylogenetic analyses, these six subspecies belong to two different lineages – a northern and a southern lineage – which diverged around 300,000–500,000 years ago in the late Middle Pleistocene, according to Nagata and colleagues (1999). This is confirmed by the fossil record. The lineages differ in teeth and lower jaw morphology, with the northern lineage having higher crowned molars than the southern lineage, as reported by Ozaki and colleagues (2007). The subspecies vary in morphology as well as body mass – ranging between 40 kg and 100 kg – mainly reflecting latitudinal variations in climate (Bergmann's rule) as shown by Mugino Kubo and Takatsuki (2015) based on a large sample, covering most of the Japanese archipelago populations.

Yoshinara Kawamura (2009) infers its arrival time to about 430,000 years ago through a presumed short-lived land bridge formed in the present-day Korea Strait. On the island, it lived in sympatry with *Cervus kazusensis*, a survivor from the Early Pleistocene fauna, and Chinese giant deer (*Sinomegaceros yabei*). The latter two species went extinct in the Late Pleistocene, leaving sika deer as the only endemic deer today.

References

Amano K., Hamuro M., Sato T. 2008. Influx of warm-water current to Japan Sea during the Pliocene-Based on analyses of molluscan fauna from the Mita Formation in Yatsuo-machi of Toyama City. *Journal of the Geological Society of Japan*, 114 (10): 516–531.

Baba K., Ohira H., Aiba H., Matsukawa M. 2005. The stratigraphic level of the specimen of *Stegodon miensis* (Proboscidea, Mammalia) in Akiruna City, Tokyo and its fission-track age. *Bulletin of Toyo Gakugei University (Series 4)*, 57: 185–193. [in Japanese with English abstract.]

Chow M.M. 1959. New species of fossil Proboscidea from South China. *Acta Palaeontologica Sinica*, 7: 251–258.

Dobson M., Kawamura Y. 1998. Origin of the Japanese land mammal fauna: Allocation of extant species to historically-based categories. *Quaternary Research, Tokyo*, 37: 385–395.

Fooden J. 1976. Provisional classifications and key to living species of macaques (Primates: Macaca). *Folia Primatologica*, 25 (2–3): 225–236.

Forsten A. 1997. A note on the Jinomori horse, *Equus nipponicus* Shikama and Onuki (Mammalia, Equidae), and some other equine remains from Japan. *Paleontological Research*, 1 (4): 316–317.

Fujita M., Kawamura Y., Murase N. 2000. Middle Pleistocene wild boar remains from NT cave, Niimi, Okayama prefecture, west Japan. *Journal of Geosciences, Osaka City University*, 43: 57–93.

Hajime T., Naoki K. 2002. Redescription and identification of Stegodon's tooth from Akiruno-shi, Tokyo, with a reference to the Specific Name of the large type *Stegodon* from Pliocene of Japan. *Memoirs of the National Science Museum*, 38: 33–41. [In Japanese, with English abstract.]

Hasegawa Y., Koizumi A., Matsushima Y., Imanaga I., Hirata D. 1991. Fossil remains from the Nakatsu Group. *Research Report of the Kanagawa Prefectural Museum Natural History*, 6: 1–98.

Hatai K. 1959. Discovery of a Miocene elephant molar from the Sen-nan district, Miyagi Prefecture, northeast Japan. *Saito Ho-on Kai Museum Research Bulletin*, 28: 1–4.

Honda A., Murakami S., Harada M., Tsuchiya K., Kinoshita G., Suzuki H. 2019. Late Pleistocene climate change and population dynamics of Japanese *Myodes* voles inferred from mitochondrial cytochrome b sequences. *Journal of Mammalogy*, 100 (4): 1156–1168.

Iseki H. 1975. On the time indices in the geomorphic histories of the coastal sand dunes. *Daiyonki Kenkyu*, 14: 183–188. [in Japanese with English abstract.]

Iwamoto M. 1975. On a skull of a fossil macaque from the Shikimizu limestone quarry in the Shikoku District, Japan. *Primates*, 16 (1): 83–94.

Iwamoto M., Hasegawa Y., Koizumi A. 2005. A Pliocene colobine from the Nakatsu Group, Kanagawa, Japan. *Anthropological Science*, 113: 123–127.

Kamei T., Okazaki Y., Nonogaki I. 1977. On some new material of *Gomphotherium annectens* (Matsumoto) from the Mizunami Group, Central Japan. *Bulletin of the Mizunami Fossil Museum*, 4: 1–8. [in Japanese.]

Kawamura Y. 1989. Quaternary rodent faunas in the Japanese Islands, Part 2. *Memoirs of the Faculty of Sciences of Kyoto*, 54: 1–235.

Kawamura Y. 1998. Immigration of mammals into the Japanese Islands during the Quaternary. *Daiyonki Kenkyu (The Quaternary Research)*, 37: 251–257. [in Japanese with English abstract.]

Kawamura Y. 2007. Last glacial and Holocene land mammals of the Japanese Islands: Their fauna, extinction and immigration. *Daiyonki Kenkyu (The Quaternary Research)*, 46: 171–177.

Kawamura Y. 2009. Fossil record of sika deer in Japan. In: Dale R., McCullough D.R., Takatsuki S., Kaji K. (eds), *Sika Deer: Biology and Management of Native and Introduced Populations.* Springer, pp. 11–25.

Kawamura Y., Kamei T., Taruno H. 1989. Middle and Late Pleistocene mammalian faunas in Japan. *Daiyonki Kenkyu (The Quaternary Research)*, 28: 317–326. [in Japanese with English abstract.]

Kawamura Y., Nakagawa R. 2012. Terrestrial mammal faunas in the Japanese Islands during OIS3 and OIS2. In: Ono A., Izuho M. (eds), *Environmental Changes and Human Occupation in East Asia during OIS3 and OIS2. BAR International Series* 2352. Archaeopress, Oxford, England, pp. 33–54.

Kawamura Y., Taruno H., Inada T. 2007. A primitive mammoth molar from Himeshima Island, Oita Prefecture, Western Japan. *Daiyonki Kenkyu (The Quaternary Research)*, 46: 399–411.

Kitamura A., Kimoto K. 2004. Reconstruction of the southern channel of the Japan Sea at 3.5–0.8 Ma. *Daiyonki Kenkyu (The Quaternary Research)*, 43: 417–434. [in Japanese with English abstract.]

Kitamura A., Kimoto K. 2006. History of the inflow of the warm Tsushima Current into the Sea of Japan between 3.5 and 0.8 Ma. *Palaeogeography, Palaeoclimatology, Palaeoecology*, 236: 355–366.

Kitamura A., Takano O., Takata H., Omote H. 2001. Late Pliocene–early Pleistocene paleoceanograhic evolution of the Sea of Japan. *Palaeogeography, Palaeoclimatology, Palaeoecology*, 172: 81–98.

Koda Y., Yanagisawa Y., Hasegawa Y., Otsuka H., Aizawa M. 2003. A middle Miocene mandible of *Stegolophodon* (Proboscidea, Mammalia) discovered in Katsura Village, Ibaraki Prefecture, eastern Japan. *Chikyu Kagaku (Earth Science)*, 57: 49–59. [in Japanese.]

Koizumi A. 2003. The first record of Plio–Pleistocene hypercarnivorous canid, *Canis (Xenocyon) falconeri* (Mammalia; Carnivora), from Tama River, Akishma City, western Tokyo, Japan. *Daiyonki Kenkyu (The Quaternary Research)*, 42: 105–111. [in Japanese with English abstract.]

Konishi S., Yoshikawa S. 1999. Immigration times of the two proboscidean species, *Stegodon orientalis* and *Palaeoloxodon naumanni*, into the Japanese Islands and the formation of land bridge). *Chikyu Kagaku (Earth Science)*, 53: 125–134.

Kubo M.O., Takatsuki S. 2015. Geographical body size clines in Sika Deer: Path analysis to discern amongst environmental influences. *Evolutionary Biology*, 42: 115–127.

Makiyama J. 1924. Notes on a fossil elephant from Sahamma, Tôtômi. *Memoirs of the College of Science, Kyoto Imperial University (Series B)*, 1 (2): 255–264.

Martin K. 1886. Fossiele Saugethierreste von Java und Japan. *Leiden, Sammlungen des Geologischen Reichs-Museums* 4, 1884–1889: 25–69.

Matsui H., Tada R., Oba T. 1998. Low-salinity isolation event in the Japan Sea in response to eustatic sea-level drop during LGM: Reconstruction based on salinity–balance model. *Daiyonki-kenkyu (The Quaternary Research)*, 37 (3): 221–233. [in Japanese with English abstract.]

Matsumoto H. 1918. On a new archetypal fossil elephant from Mt. Tomuro, Kaga. *Science Reports of the Tôhoku Imperial University (2nd Series)*, 3: 51–56.

Matsumoto H. 1924. On *Loxodonta* (*Palaeoloxodon*) *namadica* (Falconer and Cautley) in Japan. *Journal of the Geological Society of Tokyo*, 31: 2–5.

Matsumoto H. 1926 (a). On two new mastodonts and an archetypal stegodon of Japan. *Science Reports of the Tôhoku Imperial University* (2nd Series), 10: 1–11.

Matsumoto H. 1926 (b). On some new fossil Cervicorns from Kazusa and Liukiu. *Science Reports of the Tôhoku Imperial University* (2nd Series), 10: 21–23.

Matsumoto H. 1929. On *Parastegodon* Matsumoto and its bearing on the descent of earlier elephants. *Science Reports of the Tôhoku Imperial University (2nd Series)*, 13: 13–15.

Matsumoto H. 1941. On Japanese stegodonts and parastegodonts. *Zoological Magazine, Tokyo*, 53 (8): 385–395.

Nagata J., Masuda R., Tamate H.B., Hamasaki S-I., Ochiai K., Asada M., Tatsuzawa S., Suda K., Tado H., Yoshida M.C. 1999. Two genetically distinct lineages of the Sika Deer, *Cervus nippon*, in Japanese Islands: Comparison of Mitochondrial D-Loop region sequences. *Molecular Phylogenetics and Evolution*, 13 (3): 511–519.

Nishimura T.D., Takai M., Senut B., Taru H., Maschenko E.N., Prieur A. 2012. Reassessment of *Dolichopithecus* (*Kanagawapithecus*) *leptopostorbitalis*, a colobine monkey from the Late Pliocene of Japan. *Journal of Human Evolution*, 62: 548–561.

Ogino S., Otsuka H. 2005. Morphological Study of Fossil *Macaca* cf. *fuscata* detected in the Middle Pleistocene Matsugae fauna excavated from the cave deposits in northeastern Kyushu, Japan. *Primate Research*, 21 (1): 1–9.

Ogino S., Otsuka H. 2008. New middle Pleistocene Galictini (Mustelidae, Carnivora) from the Matsugae cave deposits, northern Kyushu, West Japan. *Paleontological Research*, 12 (2): 159–166.

Ohdachi S.D., Iwasa M.A., Nesterenko V.A., Abe H., Masuda R., Haberl W. 2004. Molecular phylogenetics of Crocidura shrews (Insectivora) in east and central Asia. *Journal of Mammalogy*, 85 (3): 396–403.

Ohshima K. 1980. Recording the late Quaternary sea-level change on the topographic feature of straits of the Japanese Islands. *Daiyonki Kenkyu*, 19: 23–37. [in Japanese with English abstract.]

Okumara K., Ishida I., Kawamura Y., Kumada M., Tamiya S. 1982. Latest Pleistocene Mammalian assemblage of Kumaishi-do Cave, Gifu prefecture and the significance of its 14C age. *Earth Science*, 36: 214–218.

Ota H. 2000. The current geographic faunal pattern of reptiles and amphibians of the Ryukyu Archipelago and adjacent regions. *Tropics*, 10 (1): 51–62.

Ozaki M., Suwa G., Kaji K., Ohba T., Hosoi E., Koizumi T., Takatsuki S. 2007. Correlations between feeding type and mandibular morphology in the sika deer. *Journal of Zoology (London)*, 272: 244–257.

Park J.O., Kimura M., Taira A. 1996. Late Pleistocene unconformity of the Tsushima and Korea Straits revealed by seismic reflection profiles. *Chigaku Zasshi (Journal of Geography)*, 105: 297–305.

Saegusa H. 1987. Cranial morphology and phylogeny of the stegodonts. *The Compass*, 64: 221–243.

Saegusa H. 2008. Dwarf *Stegolophodon* from the Miocene of Japan: Passengers on sinking boats. *Quaternary International*, 182: 49–62.

Satoguchi Y., Nagahashi Y., Kurokawa K., Yoshikawa S. 1999. Tephrostratigraphy of the Pliocene to Lower Pleistocene formations in central Honshu, Japan. *Earth Science*, 53: 275–290. [In Japanese with English abstract.]

Sawamura H., Sugiyama T., Tanaka T., Yoshida M., Suzuki K. 1979. New species of *Stegolophodon* found from the Shigarami Formation, northern part of Nagano Prefecture, Central Japan. *Earth Science (Chikyu Kagaku)*, 33: 11–25.

Shikama T., Okafuji G. 1963. On some Choukoutien mammals from Isa, Yamaguchi Prefecture. *Science Reports of the Yokohama National University (Section 2)*, 7: 59–67.

Shikama T., Otsuka H., Tomida Y. 1975. Fossil proboscidea from Taiwan. *Science Reports of the Yokohama National University (Section 2, Biological and Geological Sciences)*, 22: 13–62.

Shoshani J., Ferretti M.P., Lister A.M., Agenbroad L.D., Saegusa H., Mol D., Takahashi K. 2007. Relationships within the Elephantinae using hyoid characters. *Quaternary International*, 169–170: 174–185.

Suzuki K., Kamei T. 1981. Flora and Fauna. *Recent Progress of Natural Sciences in Japan*, 6: 203–211.

Tada R. 1994. Paleoceanographic evolution of the Japan Sea. *Palaeogeography, Palaeoclimatology, Palaeoecology*, 108: 487–508.

Takahashi K., Chang C.H., Cheng Y.N. 2001. Proboscidean fossils from the Japanese Archipelago and Taiwan Islands and their relationship with the Chinese mainland. In: Cavaretta G., Gioia P., Mussi M., Palombo M.R. (eds), *The World of Elephants. Proceedings of the 1st International Congress, Roma, 16–20 October, 2001*, pp. 148–151.

Takahashi K., Soeda Y., Izuho M., Aoki K., Yamada G., Akamatsu M. 2004. A new specimen of *Palaeoloxodon naumanni* from Hokkaido and its significance. *Daiyonki Kenkyu (The Quaternary Research)*, 43: 169–180.

Takahashi K., Soeda Y., Izuho M., Yamada G., Akamatsu M., Chang C.-H. 2006. The chronological record of the woolly mammoth (Mammuthus primigenius) in Japan, and its temporary replacement by *Palaeoloxodon naumanni* during MIS 3 in Hokkaido (northern Japan). *Palaeogeography, Palaeoclimatology, Palaeoecology*, 233: 1–10.

Takai F. 1936. On a new fossil elephant from Okubo-mura, Akashi-gun, Hyogo Prefecture, Japan. *Proceedings of the Imperial Academy of Japan*, 12 (1): 19–21.

Takai F. 1954. An addition to the mammalian fauna of the Japanese Miocene. *Journal of the Faculty of Sciences*, University of Tokyo, section II, 9: 331–335.

Takakuwa Y. 2006. Cervid fossils from the Kiyokawa Formation of Shimosa Group, Sodegaura, Chiba Prefecture, Japan. *Daiyonki Kenkyu (The Quaternary Research)*, 45: 197–206.

Taruno H. 1991. Systematic revision of genus "*Parastegodon*" from the Japanese Islands (Mammalia: Proboscidea). *Bulletin of the Osaka Museum of Natural History*, 45: 5–16. [in Japanese with English abstract.]

Taruno H., Kawamura Y. 2007. Mammoths of East Asia: A revision of their taxonomy, chromospatial distributions, evolution, and immigration into Japan. *Jubilee Publication in Commemoration of Prof. T. Kamei's 80th Birthday*, pp. 59–78. [in Japanese with English abstract.]

van der Geer A.A.E., van den Bergh G., Lyras G.A., Prasetyo U.W., Due R.A., Setiyabudi E., Drinia H. 2016. The effect of area and isolation on insular dwarf proboscideans. *Journal of Biogeography*, 43: 1656–1666.

Wallace A.R. 1880. *Island Life, or the Phenomena and Causes of Insular Faunas and Floras, Including a Revision and Attempted Solution of the Problem of Geological Climates.* Macmillan, London. Reprinted 1911, AMS Press.

Yamada, F., Takaki, M., Suzuki, H. 2002. Molecular phylogeny of Japanese Leporidae, the Amami rabbit *Pentalagus furnessi*, the Japanese hare *Lepus brachyurus*, and the mountain hare *Lepus timidus*, inferred from mitochondrial DNA sequences. *Genes and Genetic Systems*, 77 (2): 107–116.

Yamakawa C., Momohara A., Nunotani T., Matsumoto M., Watano Y. 2009. Paleovegetation reconstruction of fossil forests dominated by Metasequoia and *Glyptostrobus* from the late Pliocene Kobiwako Group, central Japan. *Paleontological Research*, 12: 167–180.

Yasuda S.P., Minato S., Tsuchiya K., Suzuki H. 2007. Onset of cryptic vicariance in the Japanese dormouse Glirulus japonicus (Mammalia, Rodentia) in the Late Tertiary, inferred from mitochondrial and nuclear DNA analysis. *Journal of Zoological Systematics and Evolutionary Research*, 45 (2): 155–162.

Yasukochi Y., Nishida S., Han S.-H., Kurosaki T., Yoneda M., Koike H. 2009. Genetic structure of the Asiatic Black Bear in Japan using Mitochondrial DNA analysis. *Journal of Heredity*, 100 (3): 297–308.

Yoshikawa S., Kawamura Y., Taruno H. 2007. Land bridge formation and proboscidean immigration into the Japanese Island during the Quaternary. *Journal of Geosciences of Osaka City University*, 50 (1): 1–6.

CHAPTER SEVENTEEN

Japan: The Southern and Central Ryukyu Islands

Evolution of Island Mammals: Adaptation and Extinction of Placental Mammals on Islands, Second Edition.
Alexandra van der Geer, George Lyras and John de Vos.
© 2021 John Wiley & Sons Ltd. Published 2021 by John Wiley & Sons Ltd.

The Japanese Ryukyu Islands form an island chain between Kyushu (Japan) and Taiwan. The southern and central islands form a zoogeographical unit and are known for their endemic faunas. Since the Early Pleistocene, they were inhabited by dwarf mammoths, several deer, spiny rats, giant Ryukyu rats, Ryukyu shrews, Amami rabbits, and normal-sized tortoises. Towards the end of the Pleistocene, Palaeolithic humans and wild boar arrived on the central islands and roe deer, cats, rats, and voles on the southern islands. Only the small mammals of the Pleistocene survived until today. The northern islands belong to the main islands of Japan, from a biogeographical point of view.

Geology and Palaeogeography

The Ryukyu Islands – also known as the Nansei Islands – form part of the Japanese archipelago situated between Kyushu in the north and Taiwan in the south. They form an island arch of more than a hundred islands and islets, divided into three main groups, a northern, a central, and a southern group (Figure 17.1). Deep-sea straits – the Tokara Gap between the northern and the central groups and the Kerama Gap between the central and the southern groups – separate these three groups from each other. Most small islands consist of coral reefs, others are of volcanic origin, and again others are continental shelf islands. The nearest distance to the edge of the continental shelf of mainland Asia is about 170 km, where the Senkaku or Pinnacle Islands, a group of uninhabited rocks, lie. The islands are generally included in the Ryukyu Archipelago, which further consists of the Ryukyu Islands proper and the Daito Islands, as shown by Hidetoshi Ota and colleagues (1993). At present, the East China Sea separates the Senkaku Islands from the mainland of Asia, but until the end of the Early Pleistocene and intermittently during the Late Pleistocene and then during the Last Glacial Maximum, the connection consisted of dry land. A deep-sea trench, the Okinawa Trough, separates the Senkaku Islands from the Ryukyu Islands proper since the Pliocene.

Within the Japanese archipelago, two main island groups can be discerned, each with their own zoogeographical history. One group consists of the southern and central Ryukyu Islands – including the main island Okinawa – the other of the northern Ryukyu Islands and Kyushu, Honshu, and Shikoku, the main islands of Japan. The zoogeographical boundary between the two groups coincides with the Tokara Strait and is indicated by the Watase Line. Alfred Wallace (1876) was the first who noted the zoogeographic boundary between the Australasian and Oriental regions in Indonesia. His line was

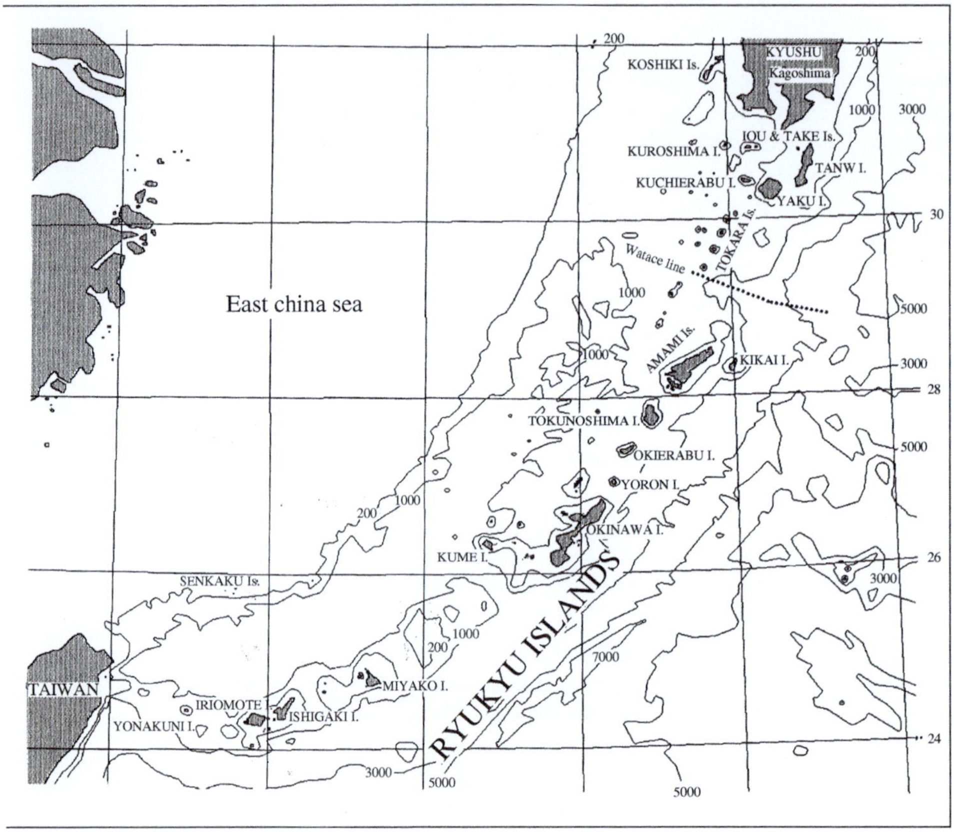

Figure 17.1 Map of the Ryukyu Islands. (From Sondaar, 2000.)

coined as Wallace's Line by Thomas Henry Huxley (1868). David Brauns (1884) recognised a similar phenomenon in the Japanese Island arc, which was elaborated by the Japanese cell-biologist Shosaburo Watase (1912) in a footnote to a paper. To honour his Japanese colleague, Yaichiro Okada (1927) named this line the Watase Line in a paper about the distribution of anurans. In this chapter, we refer to the southern and central Ryukyu Islands as the Ryukyu Islands, while the northern Ryukyu Islands are treated in the previous chapter. The southern Ryukyu Islands are further faunistically and phylogeographically closer to Taiwan than to the central Ryukyu Islands, as shown by Ota (2000).

The traditional view on the zoogeographical history of the Ryukyu Islands, based on the model of Kizaki and Oshiro (1977) (Figure 17.2) and modifications thereof, assumes four occasional land bridge connections of the island arc with the mainland and almost complete submergence of the whole Ryukyu Islands during and after the Pliocene, as summarised

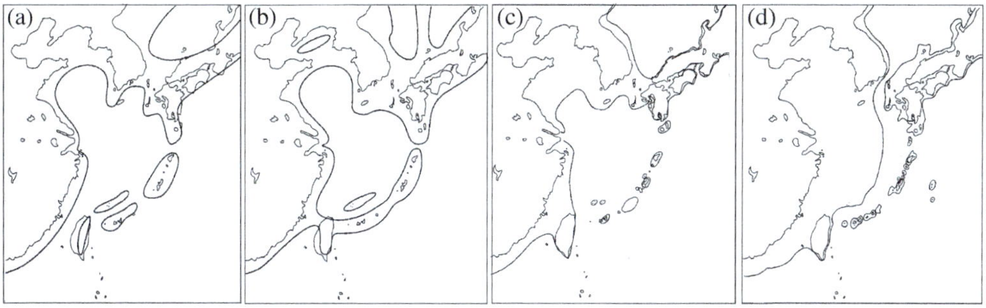

Figure 17.2 Maps showing Kizaki and Oshiro's (1977) hypothesis on the land configurations of the Ryukyu Islands and adjacent regions during the (a) Pliocene, (b) Early, (c) Middle, and (d) Late Pleistocene.

by Hidetoshi Otsuka and Akio Takahashi (2000). However, studies on the distribution and phylogenetic relations of the herpetofauna of the island are in conflict with this hypothesis and indicate instead longer periods of isolation and overseas sweepstake dispersals. These results are in agreement with the palaeontological evidence and are therefore followed here.

The geological history of the Ryukyu Islands is rather complex, involving tectonic movements and volcanism. Starting somewhere in the Late Miocene, the proto-island arc drifted eastward and rotated clockwise with respect to Eurasia as a result of the widening of the East China Sea. Based on the rarity of Late Miocene marine deposits, it is suggested that the proto-island arc and its surrounding areas were exposed above sea level and were possibly connected to the south-eastern part of Asia. This is confirmed by zoogeographical evidence provided by the endemic frogs *Odorrana ishikawae* from Okinawa of the central islands as shown by Masafumi Matsui and colleagues (2005). At this time, several central islands (Kume, Okinawa, Toku, Amami, etc.) and southern islands (Iriomote, Ishigaki, etc.) formed part of the continental shelf. Due to the extension of the Okinawa Trough and the opening of the East China Sea, starting in the Early Pliocene, the connection with the mainland was gradually lost, and these parts became continental shelf islands. The emerging Okinawa Trough also separated the Senkaku Islands from the rest of the Ryukyu Archipelago. The connection between the southernmost Ryukyu Islands and Taiwan also probably got lost in the Early or middle Pliocene, based on the phylogenetic distance between the endemic frogs *Odorrana swinhoana* of Taiwan and *Odorrana utsunomiyaorum* from Ishigaki and Iriomote of the Yaeyama Group, according to Matsui and colleagues. During the remainder of the Pliocene, the Ryukyu Islands were isolated, not only due to the progressive widening of the

East China but also because of a high Pliocene sea level. The Tokara Strait, which separates the central Ryukyu Islands from the northern Ryukyu Islands and Japan proper and coincides with the Watase Line, originated in the Pliocene as well, due to tectonics, as estimated by Masanao Honda and colleagues (2008).

Since the Middle Pleistocene, the Ryukyu Islands gradually became more isolated by both crustal movements and tremendous sea-level rise. At the time, some relatively small land areas of the Ryukyu Islands were under water and coral reefs were formed, but the arch never submerged completely, in contrast to the general opinion. During the Last Glacial Maximum, a land connection between Taiwan and the southernmost Ryukyu Islands may have existed, based on the abundant fossil land mammals recovered from the bottom of the Pengfu Channel of the Taiwan Straits dated to about 20,000–15,000 years ago. The remaining Ryukyu Islands were not connected to the mainland during this period, though some may have been connected to each other. This was likely the case with Ishigaki and Iriomote of the southern group, as evidenced by differences in mtDNA between the two island populations of the endemic frog *Odorrana supranarina*, shown by Matsui and colleagues (2005). The degree of isolation is thus much higher for the central islands than for the southern islands, as concluded by Akio Takahashi and colleagues (2008). After the Last Glacial Maximum, the islands gradually submerged partly as is evidenced by submarine archaeological finds.

Historical Palaeontology

The first scientific paper on fossils from the Ryukyu Islands dates back to 1926. In this paper, Hikoshichiro Matsumoto (1926) reported fossils of a diminutive 'cervicorn' from Liukiu, as the islands were known at that time. Yosezato, who had found them at the southern coast of Shimajiri district, Okinawa, forwarded the fossils to him. Matsumoto recognised two different species amongst the material, a form of muntjac and a form of very small deer, related to the rucervine group (the brow-antlered deer, *Rucervus*). From this, he concluded that 'The discovery of these two species may indicate that this island or archipelago might have been a part of the Oriental Region zoögeographically since those ages (this is, Early Pleistocene)', and further that the Ryukyu Islands 'might probably already in those ages have been isolated from the continent or larger islands'. Matsumoto was the first to point out the isolation of these islands during the Pleistocene.

He describes the 'muntjac' under the new name of *Muntiacus astylodon* based on a fragment of a left lower jaw bearing two molars and the roots of three premolars. The molars differ from those of other muntjac, amongst others, in having very high crowns – hypsodonty, a common feature for island herbivores as we know now – and the lack of basal cingula and accessory columns, hence the name *astylodon*, teeth without columns (stylids). Matsumoto further describes the small deer as *Cervus* (cf. *Rucervus*) *riukiuënsis*, based on a fragment of a very basal part of the right antler and its pedicle and two more tiny antler fragments. He notes that the antler and the pedicle are extremely small, smaller than those in all subgenera of *Cervus* known to him. This very short pedicle excludes its attribution to muntjac and thus his former species. Later, Shigeyasu Tokunaga and Fuyuji Takai (1939) combined the molars and the antlers of his two species into a single species, *Cervus astylodon*.

To date, only three mammoth fossils have been found on the Ryukyu Islands. The first two were upper molars, reported by Tokunaga (1940) as similar to *Palaeoloxodon namadicus*. They originated from cave deposits at Tanabaru near Ohnogoshi, Miyako. The third fossil is an incomplete and much worn upper molar, discovered by Kiichi Maja at the seashore south-west of Kyan village, southern Okinawa (Cape Kyan). Tomohide Nohara and Yoshikazu Hasegawa (1973) described this molar as *Palaeoloxodon* with a question mark and considered it very similar in shape and plate frequency to the Japanese Naumann's elephant (*Palaeoloxodon naumanni*). Tadao Kamei (1970) assigned the three fossils to the *Mammuthus* group, and Otsuka (1996) identified one of the molars from Miyako as *Mammuthus paramonteus shigensis*, the species of the Early Pleistocene of Japan, now a junior synonym of *Mammuthus protomammonteus* or *Mammuthus trogontherii*.

A fragmentary mastodon molar was collected by K. Okamoto at the roadside near the Shimajiri coast of northern Miyako, but no information about its stratigraphic horizon is available. Hasegawa and colleagues (1973) described the molar as *Trilophodon*, a junior synonym of *Gomphotherium*, and assigned it a Pliocene age because of its similarity to Pliocene taxa. They further concluded that it must have come from the Shimajiri Formation, based upon the attached sandstone containing fossil molluscs. Hence, the Shimajiri Formation was of a Pliocene age. In the same paper, the authors described a new roe deer, *Capreolus miyakoensis* (now junior synonym of *Capreolus tokunagai*), from cave deposits of Amagawa-do, Miyako.

In 1962, Tawada discovered a large number of deer fossils in association with bone artefacts and ash seam in the Yamashita cave at Naha City, Okinawa, as reported by Takamiya (1968). A concentrated search for human fossils was started, and soon juvenile human remains were found, attributed to a

7-year-old child. The bones, described by Hisashi Suzuki (1983), were dated to about 32,100 years ago by Kobayashi and colleagues (1971). A year before, Mr Seiho Ohyama had found fossil human remains in a fissure of the Minatogawa limestone quarry at Gushikami-son, Okinawa. Subsequent excavation yielded four partial skeletons including skulls and some isolated bones, together with abundant deer remains. Suzuki and Kazuro Hanihara (1982) edited a special volume about the remains of the so-called Minatogawa Man. In this publication, Suzuki described the skulls, while Hisao Baba and Banri Endo did so for the post-cranial elements in much detail. Much less known are the fragmentary human fossil remains from Oyama cave and Tobaru cave on Okinawa and Kadabaru cave and Gohezu cave on Ie, found between 1962 and 1976 and attributed to the Late Pleistocene based on degree of fossilisation and results of fluorine analysis, as summarised by Matsu'ura Shuji (1999).

Biozones and Faunal Units

The reconstruction of the biozones of the Ryukyu Islands is hampered by the uncertainty concerning the taxonomic status and the phylogenetic affinities of most mammalian fossils, mainly due to the lack of appropriate comparison material. At present, four faunal units can be recognised ranging in age from the Late Miocene to the latest Pleistocene. Most authors prior to 2008 held the opinion that the Ryukyu Island arc had been repeatedly connected to the mainland of Asia and that each land connection had been followed by a migration of terrestrial vertebrates from the continent. In this way, they recognised four land connections, exactly corresponding to the four biozones. Recently, however, the theory of overseas sweepstake dispersal is becoming more established for the Ryukyu Islands, e.g. by Ota (2003), and not all four land connections are taken for granted anymore, as explained above in the paragraph on palaeogeography.

The endemic taxa found on the continental shelf islands presumably represent a mixture of vicariant taxa from before the opening of the East China Sea and new arrivals through sweepstake dispersal. This is indicated by the continuation of some taxa into the next biozone and the composition of the unbalanced faunas (for an overview of time ranges of the various taxa, see Jan van der Made, 2005).

Late Miocene or Early Pliocene

The geologically oldest mammalian fossil so far from the Ryukyu Islands is a molar of a gomphothere (*Gomphotherium* sp.). This fossil was found on Ogami, an islet of Miyako, not far from Ishigaki and Iriomote. It is reasonable to assume that *Gomphotherium* had reached these southern islands from Taiwan. Unfortunately, the fossil has not been studied in much detail, and nothing is known about its specific affinities. No further fossils belonging to this biozone have been found on the Ryukyu Islands.

Early Pleistocene

The earliest Pleistocene fauna of the Ryukyu Islands appears to have been impoverished and unbalanced, as far as can be concluded from the scarce material. The fauna is characterised by two deer (*Muntiacus* sp., a large *Cervus* sp.), a rat (*Diplothrix* sp.), and possibly a small mammoth (*Mammuthus* sp.). No further determination could be done other than on a genus level. The unbalanced and impoverished character of the fauna is best explained by overseas sweepstake dispersal, but due to the scarcity of exposed fossiliferous strata of this age, not much can be said with certainty. Fossils belonging to this biozone have been found only in Okinawa of the central group and are known as the Imadomari-Akagimata fossil assemblage. The bone bed was fission-track dated to 1.5 ± 0.3 million years ago, as given by Otsuka and Takahashi (2000).

The remains of the rat are limited to one specimen only, being a first upper molar. Otsuka (2002) reported it as belonging to the genus *Leopoldamys* of extant long-tailed giant rats, but, later, the same author considered it to be closer in morphology to *Diplothrix legata*, the extant Ryukyu long-tailed giant rat, as communicated to Ota (2003).

Deer fossils, including an antler, excavated from a mud stone bed called the Haneji Formation in Okinawa in the late 1990s, have been tentatively attributed to *Cervus astylodon*, the deer of the Late Pleistocene (see below). Its size is, however, much larger, and the relation between this form and that of the Late Pleistocene is by no means sure. Apart from the Imadomari-Akagimata findings, a few deer molars have also been found in shallow marine deposits on Kume, tentatively dated to the Middle Pleistocene. The molars were attributed to the genus *Dicroceros*, which implies that this is the first recorded occurrence of this deer on the islands. It is not clear whether the molars indeed belong to this biozone or to the next. In all likelihood, however, the three mentioned deer genera all belong to one, larger, ancestral form of the Late Pleistocene deer.

The mammoth molar, found at Cape Kyan (Kyan-Misaki), Okinawa, is very small and worn out, according to the original description by Nohara and Hasegawa (1973), indicating either that this mammoth was a dwarf or that the molar is a milk molar of a normal-sized, juvenile individual. The fossil probably originates from the Naha limestone stratum, corresponding to an age of 1.0–1.2 million years ago.

Otsuka and Takahashi (2000) suggested that the provenance of this Early Pleistocene fauna is either the Renzidong fauna (Anhui Province) or the Longupo (earlier Wushan) fauna of Central China, coming overland to the central Ryukyu Islands. The evidence for this is, however, limited. Ota (1998) concluded that during the Early Pleistocene the southern islands formed a peninsula with Taiwan, whereas the central islands formed a separate land mass that has never been connected to the other island groups at any time nor to the mainland, based on a detailed study of the endemic herpetofauna, living as well as extinct. This implies that Okinawa and nearby Kume harboured an endemic insular fauna comprising deer, rats, and mammoths which had arrived by swimming or floating, either from Taiwan or mainland Japan. Probably, this fauna survived and eventually gave rise to the Late Pleistocene fauna (see below).

Middle Pleistocene

Fossils belonging to this biozone are extremely rare and limited to two or three localities of the central group and one in the southern group. The composition of the central group indicates a long-term isolation, likely consisting of vicariant elements from the previous period. Species that have been recovered are deer (*Cervus astylodon* and a muntjac), a rabbit (*Pentalagus*), a spiny rat (*Tokudaia* sp.), a giant rat (*Diplothrix* sp.), and perhaps a bandicoot rat (*Bandicota* sp.?). The rats (*Tokudaia* and *Diplothrix*) are likely ancestral to the extant forms (*Tokudaia muenninki* and *Diplothrix legata*), while the other species are now extinct on Okinawa. The fossils of the southern group include a mammoth (*Mammuthus trogontherii*) and a roe deer (*Capreolus tokunagai*).

Only two sites on Okinawa (mud and gravel beds at Sonan and a limestone quarry at Nagahama) have been assigned to a Middle Pleistocene age with confidence, as summarised by Ai Kawamura and colleagues in their updated overview of 2016. The fossiliferous deposits at these two sites are considered to have been deposited during the early Middle Pleistocene. Fossils of the rat, the spiny rat, the rabbit, and the deer from Tokuno of the central group may date back to the late Middle Pleistocene, as given by Iryu and Matsuda (2010). The fossiliferous sediments with a mammoth in Tanabaru cave in Miyako of the southern group were deposited during the late Middle Pleistocene after 450,000 years ago, when the cave was formed. The taxonomic status of the mammoth – represented by two molars only – is uncertain. Otsuka (1978) suggested that they are closely related to the Early Pleistocene mammoths of Japan (Kyushu and Honshu) and Taiwan, both *trogontherii* stage mammoths. A relation with Taiwan is much more likely, as the mammoth is found only on the southern islands.

Fossils of the extant tortoises (*Manouria oyamai* and *Geoemyda japonica*) from Middle and Early Pleistocene fissure fillings on Okinawa have been preliminary reported from Early and Middle Pleistocene fissure fillings on Okinawa at a conference by Hirayama and Ozawa (2009). In earlier Japanese papers, the tortoises are supposed to have come overland through the 'Ryukyu Peninsula', a long and narrow peninsula extending between Taiwan and Okinawa due to regression and uplifting of the Ryukyu Coral Sea during the Middle Pleistocene, but sound geological, phylogeographical, and fossil evidence for this is lacking. The land bridge dispersal was in part inspired by the existence of limestone erosions found in the central parts of the Kerama and Tokara Gaps and the occurrence of a mainland mammoth on Miyako. On the other hand, mammoths and elephants are apt swimmers, and the composition of the faunas favours an overseas dispersal.

Late Pleistocene

The next biozone is largely a continuation of the Early and Middle Pleistocene faunas. The Late Pleistocene fossil record of the central group is characterised by the Amami rabbit (*Pentalagus furnessi*), the long-tailed giant rat (*Diplothrix legata*), the Ryukyu shrew and the lesser Ryukyu shrew (*Crocidura orii* and *Crocidura watasei*), the Ryukyu spiny rats (*Tokudaia osimensis* and *Tokudaia muennicki*), and several endemic deer (*Cervus astylodon*, *Dicroceros* sp., and two species of *Muntiacus*). The latter three deer are tentatively considered to constitute a single, new species of muntjac by Kawamura and colleagues (2016). Restricted to the southern islands are roe deer (*Capreolus tokunagai*) and a deer close to Japanese sika deer (*Cervus* sp.).

The small mammals – *Pentalagus*, *Diplothrix*, *Crocidura*, and *Tokudaia* – represent vicariant species since the earliest Pleistocene (see also below). Also, the tortoises (*Manouria oyamai* and *Geoemyda japonica*) are likely survivors from an earlier period. Fossils have been recovered from several Late Pleistocene fissure fillings and cave deposits on the islands Tokuno, Ie, and Okinawa of the Central Ryukyus and Miyako and Yonaguni of the Southern Ryukyus, as reported by Takahashi and colleagues (2003, 2018), as well as Ishigaki, listed by Kawamura and colleagues (2016).

Late Pleistocene fossils are abundant and have been retrieved from more than 100 localities on various islands ranging from Tokuno in the north to Yonaguni close to Taiwan.

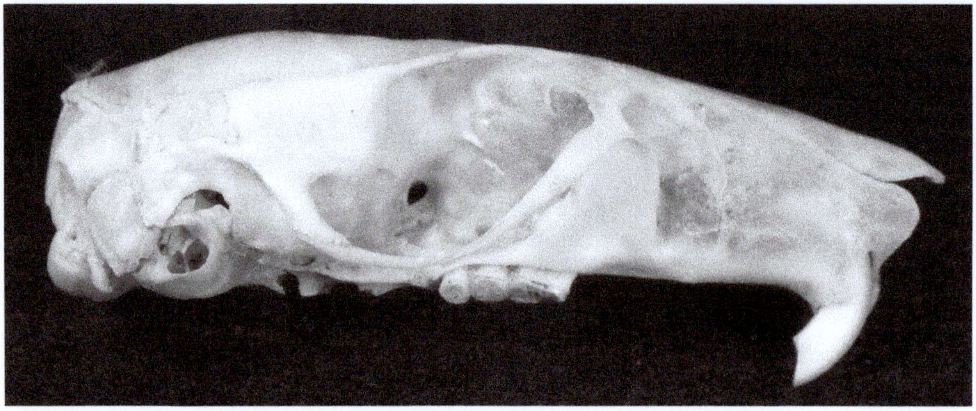

Figure 17.3 Skull of *Tokudaia osimensis* from Amami Island. Kagoshima University, Kyushu; lateral view. Skull length about 3.5 cm. (Photograph: Jan van der Made.)

All species of the central group – the deer (*Cervus astylodon*, muntjac), the Amami rabbit (*Pentalagus furnessi*), the spiny rats (*Tokudaia*) (Figure 17.3), the long-tailed giant rat (*Diplothrix legata*) (Figure 17.4), and the Ryukyu shrews (*Crocidura orii* and *Crocidura watasei*) – are best explained by vicariance of Late Pliocene or Early Pleistocene ancestors dating from the time before the isolation of the central Ryukyu Islands. This is confirmed by their distribution, which is restricted to the central islands.

Latest Pleistocene

Towards the end of this period, coinciding with the Last Glacial Maximum, the sea level lowered further, resulting in a colonisation by some more taxa to the southern islands. These new elements consist of a subspecies of the Iriomote cat (*Prionailurus bengalensis iriomotensis*), a reed vole (*Microtus fortis*), and a rat (*Rattus miyakoensis*). Only Palaeolithic humans (*Homo sapiens*) and a small wild boar (*Sus scrofa riukiuanus*, likely a human introduction) are new arrivals to the central Ryukyu Islands.

The fauna is still very unbalanced, with the wild cat as the only carnivore. The newcomers indicate, again, sweepstake dispersal. The Amami rabbit, the rat, and the spiny rat on the central islands survived the impact of pigs and humans and still live today, albeit with a very limited distribution on the central islands. The other endemics, including several tortoise species, gradually went extinct. On the central islands, there is an overlap of Palaeolithic humans and *Cervus astylodon* of the previous biozone, because, on Okinawa, the human fossils are found in assemblage with deer fossils. Several archaeological

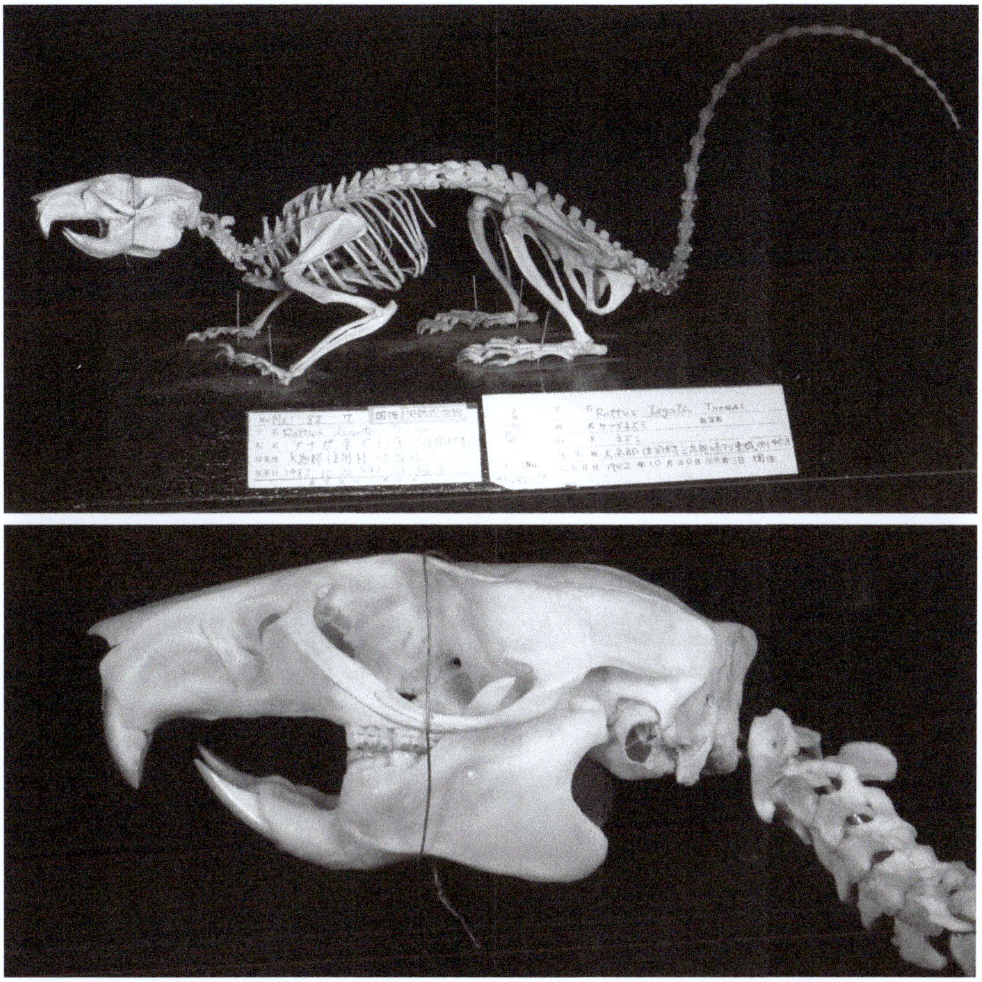

Figure 17.4 Skeleton of a female *Diplothrix legata* (above) and detail of the head (below); lateral view. Kagoshima University, Kyushu. Head and body length is 27 cm. (Photograph Jan van der Made.)

sites have been reported on Okinawa – aged between 13,000 and 37,000 years ago – of which Minatogawa Fissure is the most important because of the discovery here of Minatogawa Man, dated to about 20,000 years ago (Figure 17.5). The Sakitari-do cave on southern Okinawa yielded an almost continuing accumulation of marine shell tools and accessories with hardly any lithic artefact from 29,000 to 13,000 years ago, as reported at a conference by Masaki Fujita and colleagues (2015). This indicates that Palaeolithic people on Okinawa used shells as tools instead of lithic material. The earliest Palaeolithic occupation is that of Yamashita-cho fissure, radiocarbon dated to about 32,100 years ago by Kobayashi and colleagues (1971).

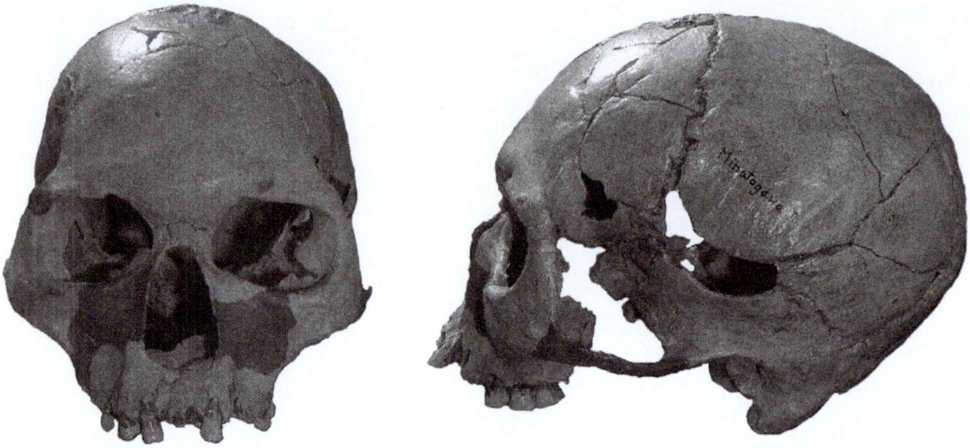

Figure 17.5 Human skull (cast) from Minatogawa, Okinawa; frontal and lateral views. National Museum of Natural History, Leiden. (Photograph: George Lyras.)

The wild cat, restricted to the southern islands, was once believed to have retained primitive features, on the basis of which it was considered to represent an ancestral felid stock, surviving as a relict on the island. Molecular phylogenetic studies of the 1990s, however, indicated its very close affinity with the fishing cat (*Prionailurus bengalensis*), from which it may have diverged at least 170,000 years ago.

Fossils of roe deer (*Capreolus tokunagai*) have been found only on Miyako (Pinza-Abu and Tana-Baru caves) of the southern group. Originally, rat fossils from the former cave (Pinza-Abu cave) were conferred by Hasegawa (1985) to the long-tailed giant rat (*Diplothrix legata*) of the central islands. Recently, the abundant remains from this cave were re-examined and renamed to *Rattus miyakoensis* by Satoshi Kawaguchi and colleagues (2009). Fossils of voles were found in the same cave as well, in association with those of the rat. Amongst the abundant material, Kaneko (1985) tentatively identified three species: the tundra or northern vole (*Microtus oeconomus*), the Chinese or reed vole (*Microtus fortis*), and an unnamed species. At present, there are no common voles anymore on the Ryukyu Islands. Charcoal from the deposits of Pinza-Abu cave was radiocarbon dated to 25,800 ± 900 and 26,800 ± 1,300 years ago by Hamada (1985).

Holocene

The arrival of mainland species continued into the Holocene, for the greater part facilitated by human activities and interferences. These are, for example, Japanese sika deer

(*Cervus nippon*), commensal rats and mice (*Rattus norvegicus*, *Rattus tanezumi*, and *Mus musculus*), and common musk shrews (*Suncus murinus*). The Palaeolithic Minatogawa people were replaced by the Neolithic Jōmon people (13,000–2,300 years ago). Some of the Holocene species managed to develop endemic characters in a remarkably short time, such as sika deer, which was introduced in the middle of the seventeenth century from Kyushu for velvet antler trade for its supposed medicinal value. It is at present recognised as an endangered subspecies, *Cervus nippon keramae*, characterised by a smaller body size, a dark brown coat, and reduced degree of sexual dimorphism.

Peculiarities and Evolution of Endemic Mammals

The Ryukyu Deer and Roe Deer

The Ryukyu deer (*Cervus astylodon*) of the Late Pleistocene is a typical island ruminant in the sense that it shows a considerable size variation together with a shortening of the lower limbs (Figure 17.6) and a high percentage (45%) of fusion of the tarsal bones (cubonavicular bone with the metatarsal bone; Figure 17.7). The antlers as well as the pedicles are in some cases very small to exceedingly small in size, and the brow-tine branches off from the very base of the antler. The

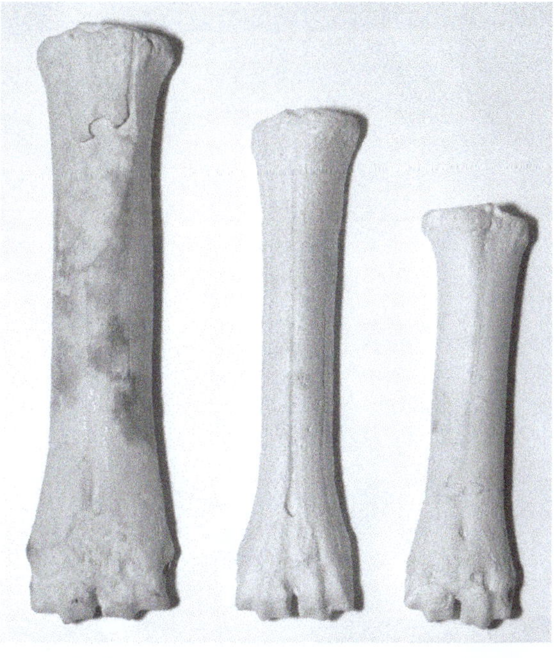

Figure 17.6 Three metacarpals (casts) of *Cervus astylodon* showing an unusual size variation and shortening; anterior (dorsal) view. National Museum of Natural History, Leiden. (Photograph: Eelco Kruidenier.)

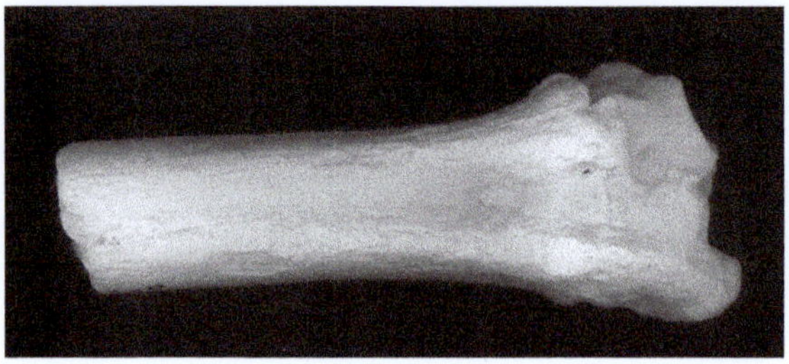

Figure 17.7 Proximal metatarsal (cast) of *Cervus astylodon* showing fusion with the tarsal (cubonavicular) bone; posterior (plantar) view. National Museum of Natural History, Leiden. (Photograph: John de Vos.)

beam is strongly curved, and the antler is more or less flattened and bears a strongly grooved surface. Four different body sizes can be recognised. Apart from the size variation, there is also a considerable variation in morphology. For example, the metacarpals from Kume of similar size can be divided into four different morphotypes (Figure 17.8). Yukihide Matsumoto and Otsuka (2000) state that each island had its own morphotype. The remarkable variation is in our view best explained as adaptive radiation. Fossils of this deer were found on the southern (Ishigaki and Yonaguni) and the central islands (Okinawa, Kume, Ie, and Tokuno) as well. Most fossils originate from Late Pleistocene cave and fissure deposits; on Okinawa, they are usually mixed with those of *Dicroceros*. A few sites on Okinawa, however, are reported to have yielded deer fossils from Early Pleistocene strata (1.3–1.7 million years ago), which would push the geological range for both *Cervus astylodon* and *Dicroceros* considerably back. The Ryukyu deer was originally thought to resemble the Chinese *Metacervulus capreolinus* most by Tokunaga and Takai (1939) and is hence sometimes referred to that genus.

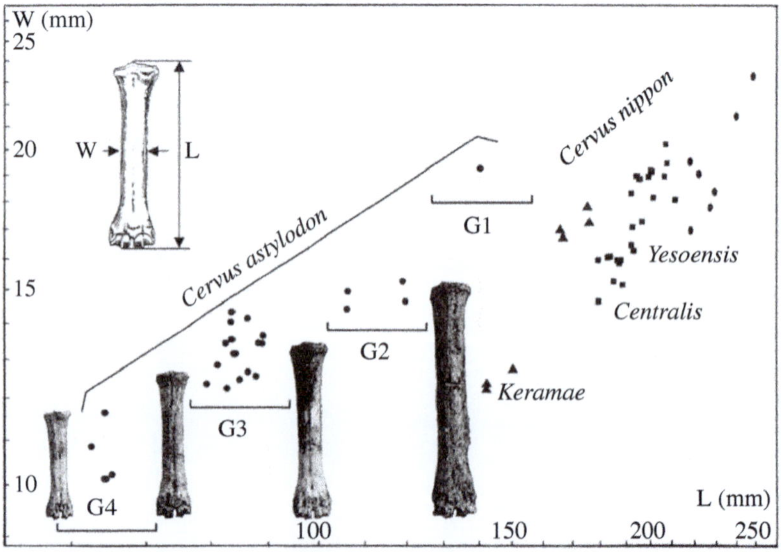

Figure 17.8 Four morphotypes of metacarpal bones in *Cervus astylodon* from fissure fillings on Kume Island; anterior view. (Adapted from Matsumoto and Otsuka, 2000.)

Figure 17.9 Skull of the roe deer of Miyako (*Capreolus tokunagai* = *C. miyakoensis*); latero-frontal view. Prefectural Museum, Okinawa. (Photograph: Jan van der Made.)

Miyako roe deer (*Capreolus tokunagai*) of the latest Pleistocene shares distinct but short blunt projections on the antler surfaces (Figure 17.9) with common roe deer (*Capreolus capreolus*), to which it is closely related. However, Miyako roe deer is much larger, which is in contrast to the general idea of size reduction under conditions of insularity in deer. On the other hand, Gargano (*Hoplitomeryx kriegsmani*) and Crete (*Candiacervus major*) were inhabited by species of deer that were larger than any known mainland deer, living contemporaneous with the dwarfed species. Another explanation for the unusually large size is that it evolved from a large-bodied mainland roe deer, typical for cold epochs.

The Amami Rabbit

The Amami rabbit (*Pentalagus furnessi*) is small and short-legged and retained some primitive characters. In the pattern of the lower third premolar, for example, it closely resembles the primitive *Alilepus*, the first leporine of Eurasia. Today, *Pentalagus* lives only on Amami and Tokuno of the central group (Box 17.1). Its fossils were reported from Late Pleistocene deposits at Tokuno by Tomida and Otsuka (1993).

Otsuka suggested *Pliopentalagus* from Late Pliocene deposits around the downstream of the Yang-Tsu River in eastern continental China as its possible ancestor. Fumio Yamada and colleagues (2002), on the other hand, concluded that it had split from other lagomorph taxa already in the Middle Miocene, based on mtDNA analysis. This might very well be an overestimation, considering the fact that *Alilepus* first appeared in the Late Miocene and was the first genus to display the characteristic upper third premolar with additional fold. Other living hares that retained the *Alilepus* pattern, apart from *Pentalagus*, are the Mexican volcano rabbit (*Romerolagus*) and the South African red rock hare (*Pronolagus*). Like the Amami rabbit, they have small, rounded ears, short legs, and short, thick fur. They may be leftovers of a primitive stock because their distribution is peripheral to that of the advanced leporines. *Pentalagus* then may have survived thanks to its isolation.

BOX 17.1

The Endangered Amami Rabbit

Due to being intensively hunted for food in combination with its supposedly medicinal value, *Pentalagus furnessi* (Figure Box 17.1) was on the brink of extinction in the early twentieth century. In 1921, the species was officially elected as a national symbol and protected since then. Since the 1980s, however, their numbers decreased again, this time because the rabbits were hunted by dogs, mongooses, and cats, all imported as domestic pets. In 1979, several individuals of the small mongoose (*Herpestes javanicus*) had been released to control the snakes and rats of Naze City on Amami. Since then, the mongoose has been expanding its range, covering almost the complete island 20 years later. In fact, the Amami rabbit is a (still) living example of the fatal results of exotic predators, often imported by humans, on a pre-existing endemic fauna.

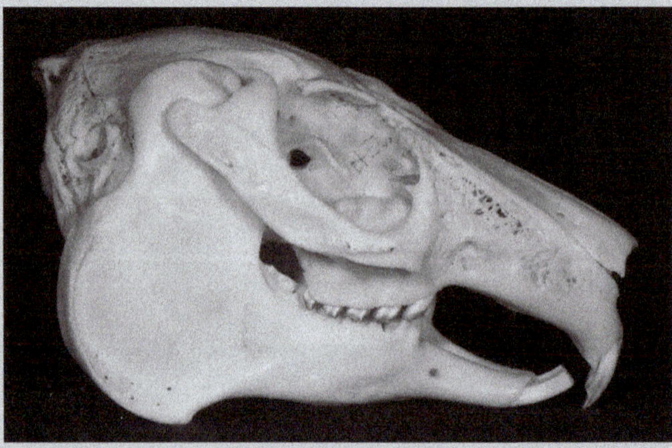

Figure Box 17.1 Skull of a female Amami rabbit, *Pentalagus furnessi*; lateral view. Kagoshima University, Kyushu. (Photograph: Jan van der Made.)

The Ryukyu Wild Pig

The Ryukyu dwarf wild boar (*Sus scrofa riukiuanus*) is at present jeopardised because of excessive hunting for local consumption, interbreeding with domestic pigs and diseases transmitted by the latter. The Ryukyu wild boar seems not to be an introduced feral population, as is the case on many other islands, but a true endemic wild pig, according to Oliver (1984). Mitochondrial DNA analysis by Watanobe and colleagues (1999) further suggests that the Ryukyu wild pig is not directly related to the Japanese wild boar (*Sus scrofa leucomystax*).

The Ryukyu Giant Rat

The Ryukyu long-tailed giant rat (*Diplothrix legata*; Figure 17.4) is a survivor since the Early Pleistocene. Fossils have been found only on the central islands (Tokuno, Kume, and Okinawa), where it still lives. The Ryukyu giant rat differs from common rats (*Rattus*) by its very thick and long fur. It is also quite large, with a head and body length ranging between 220 and 330 mm and a skull length of about 65 mm.

The Miyako rat (*Rattus miyakoensi*) of the latest Pleistocene is about 20% smaller than *Diplothrix legata* but is still a large rat. The total upper molar length is one of the largest amongst living species of the genus *Rattus*.

The Spiny Rats

The spiny rats (*Tokudaia*; Figure 17.3) resemble large voles but are covered with a dense coat consisting of fine hairs and coarse, grooved spines, hence their common name. They live at present only in the northern part of Okinawa, Tokuno, and Amami. Fossils, though, were recovered from the southern part of Okinawa, Tokuno, and Ie in Middle Pleistocene along with younger deposits. The populations from the three islands today are much diverged and were given a specific name – *muenninki* for Okinawa, *tokunoshimensis* for Tokuno, and *osimensis* for Amami – partly based on differences in chromosome number, being, respectively, 44, 45, and 25 (the latter two species have XO-type sex chromosomes). The fossils, however, are all described as *osimensis*, but these descriptions predated the new analyses of the extant spiny rats (e.g. Kaneko, 2001). Suzuki and colleagues (1999) suggested that the time of divergence of the Okinawa and Amami species is several million years, which coincides with the time when the central islands were still part of the continental shelf. Another possible explanation is a fast adaptive radiation, a not uncommon feature in island rodent communities. Evolution within island rodent communities is likely to proceed at a much faster speed than within comparable communities on the mainland due to the absence of mammalian predators. At present, the genus *Tokudaia* is considered closely related to Eurasian field mice (*Apodemus* and *Rhagamys*).

The Ryukyu Shrews

The lesser Ryukyu shrew (*Crocidura watasei*) is endemic to Amami and Okinawa of the central islands. It was initially considered a subspecies of Horsfield's Indian shrew, *Crocidura horsfieldii*. However, the chromosomal number of the latter is much higher than that of the Okinawa shrew. The lesser Ryukyu shrew seems more closely related to the musk shrew (*Suncus murinus*) than to other Asian *Crocidura*, based on mtDNA analysis by Masaharu Motokawa and colleagues (2004), but could also have derived through Robertsonian fusions from a form similar to the Japanese shrew *Crocidura dsinezumi*.

The larger Ryukyu shrew (*Crocidura orii*) is restricted to Amami and Tokuno. It was earlier considered a subspecies of the Japanese shrew based on morphological traits. At present, however, it is warranted specific status. The Pliocene, or even the Late Miocene, has been suggested as time for its arrival on the central islands.

Minatogawa Man

Human skeletal remains (Figure 17.5) (*Homo sapiens*), consisting of a complete skeleton and several incomplete skeletons, were excavated in 1970 from fissure deposits at the Minatogawa quarry in south-east Okinawa and are generally referred to as the Minatogawa people after the quarry. Charcoal from the fissure deposits was radiocarbon dated to about 18,250–16,000 years ago by Kobayashi and colleagues (1974) and repeated in the Minatogawa volume edited by Suzuki and Hanihara (1982). The faunal assemblage found in the same fissure is typical of the latest Pleistocene, mainly a mixture of deer and wild boar. Analysis by Matsu'ura (1982) of the fluorine content of the human and the animal bones from the fissure confirms that they are indeed contemporaneous. This, together with the absolute date, strongly suggests that the Minatogawa fossils are amongst the earliest modern human fossils in East Asia found to date. The isolation of the humans resulted in a certain degree of endemism. The Minatogawa people had stronger masticatory muscles in comparison to the average mainland humans of that time as described by Baba (2000). The skeletons of Minatogawa Man have been morphologically re-evaluated by Hisao Baba and Shuichiro Narasaki (1991).

References

Baba H. 2000. Physical adaptations of the Minatogawa people to island environments. *Tropics*, 10 (1): 231–241.

Baba H., Endo B. 1982. Postcranial skeleton of the Minatogawa Man. In: Suzuki H., Hanihara K. (eds), *The Minatogawa Man: The Upper Pleistocene Man from the Island of Okinawa. University Museum Bulletin*, 19. The University of Tokyo, pp. 61–159.

Baba H., Narasaki S. 1991. Minatogawa Man, the oldest type of modern *Homo sapiens* in East Asia. *The Quaternary Research*, 30 (2): 221–230.

Brauns D. 1884. Bemerkungen über die geographische Verbreitung der Säugethiere Japans. *Mitteilungen des Vereins fur Erdkunde zu Leipzig 2, Halle*, pp. 85–114.

Fujita M., Yamasaki S., Katagiri C., Kaifu Y. 2015. Late Pleistocene lifestyle adaptive to the island in Okinawa. *SEASIA Conference, Kyoto, 12-13 décembre* 2015: 92.

Hamada T. 1985. ^{14}C age of charcoal from Pinza-Abu Cave deposits, Miyako Island, Okinawa, Japan. *Excavation Report Pinza-Abu Cave*: 180. Department of Education, Okinawa Prefectural Government. [in Japanese.]

Hasegawa Y. 1985. Note on Carnivora, Chiroptera and larger rat (*Diplothrix legata*) from Pinza-Abu Cave, Miyako Island, Okinawa, Japan. In: *Department of Education, Okinawa Prefecture Government, Reports on Excavation of the Pinza-Abu Cave*, pp. 83–91. [in Japanese.]

Hasegawa Y., Otsuka H., Nohara T. 1973. Fossil vertebrates from the Mioko Island (studies of the Paleovertebrates Fana of Ryukyu Islands, Japan). Part I. *Memoirs of the National Science Museum*, 6: 40–52 [in Japanese with English abstract.]

Hirayama R., Ozawa T. 2009. Land turtles from the Early and Middle Pleistocene deposits of Okinawa Island, southeastern Japan. In: *Abstracts with Programs of the 158th Regular Meeting of the Palaeontological Society of Japan: 7*. The Palaeontological Society of Japan, Tokyo, Japan. [In Japanese]

Honda M., Okamoto T., Hikida T., Ota H. 2008. Molecular Phylogeography of the Endemic Five-lined Skink (*Plestiodon marginatus*) (Reptilia: Scincidae) of the Ryukyu Archipelago, Japan, with special reference to the relationship of a Northern Tokara Population. *Pacific Science*, 62 (3): 351–362.

Huxley T.H. 1868. *An Introduction to the Classification of Animals*. Churchill, London.

Iryu Y., Matsuda H. 2010. Neogene and Quaternary systems. In: *Nihon chiho chishitsushi 8 Kyushu Okinawa chiho (= Regional Geology of Japan 8, Kyushu and Okinawa Districts)*. Asakura-shoten, Tokyo, pp. 149–154. [in Japanese.]

Kamei T. 1970. Horizon of the elephant fossil in Miyoko Island. Tsukumo-Chigaku, 5: 1–8. [in Japanese.]

Kaneko Y. 1985. On some fossils of Microtinae Rodents from Pinza-Abu Cave, Miyako Island, Okinawa, Japan. In: *Department of Education, Okinawa Prefecture Government, Reports on Excavation of the Pinza-Abu Cave*, pp. 93–113. [in Japanese.]

Kaneko Y. 2001. Morphological discrimination of the Ryukyu spiny rat (genus Tokudaia) between the islands of Okinawa and Amami Oshima, in the Ryukyu Islands, southern Japan. *Mammal Study*, 26: 17–33.

Kawaguchi S., Kaneko Y., Hasegawa Y. 2009. A new species of the fossil murine rodent from the Pinza-Abu Cave, the Miyako Island of the Ryukyu Archipelago, Japan. *Bulletin of the Gunma Museum of Natural History*, 13: 15–28.

Kawamura A., Chang Ch.-H., Kawamura Y. 2016. Middle Pleistocene to Holocene mammal faunas of the Ryukyu Islands and Taiwan: An updated review incorporating results of recent research. *Quaternary International*, 397: 117–135.

Kizaki K., Oshiro I. 1977. Paleogeography of the Rytkyu [sic] Islands. *Marine Science Monthly*, 9: 542–549. [in Japanese with English abstract.]

Kobayashi H., Hirose Y., Sugino M., Watanabe N. 1974. TK-99 Minatogawa. *Radiocarbon*, 16: 100.

Kobayashi H., Matsui Y., Suzuki H. 1971. TK-78: Yamashita-cho Cave 1. *Radiocarbon*, 13: 100.

Matsu'ura S., 1982. Relative dating of the Minatogawa Man by fluorine analysis. In: Suzuki H., Hanihara K. (eds), *The Minatogawa Man: The Upper Pleistocene Man from the Island of Okinawa. University Museum Bulletin*, 19. The University of Tokyo, pp. 205–208.

Matsui M., Shimada T., Ota H., Tanaka-Ueno T. 2005. Multiple invasions of the Ryukyu Archipelago by Oriental frogs of the subgenus *Odorrana* with phylogenetic reassessment of the related subgenera of the genus *Rana*. *Molecular Phylogenetics and Evolution*, 37 (3): 733–742.

Matsumoto H. 1926. On some new fossil Cervicorns from Kazusa and Liukiu. *Science Reports of the Tôhoku Imperial University (2nd Series)*, 10: 21–23.

Matsumoto Y., Otsuka H. 2000. Morphotypes of fossil deer (*Cervus astylodon*) and its miniaturization process. *Tropics*, 10 (1): 145–154.

Motokawa M., Haradab M., Lin L.-K. 2004. Variation in the Y chromosome of *Crocidura tadae kurodai* (Insectivora, Soricidae). *Mammalian Biology – Zeitschrift für Säugetierkunde*, 69 (4): 273–276.

Nohara T., Hasegawa Y. 1973. An elephant tooth from Kyan, Okinawa-jima (Studies of the Paleovertbrates fauna of Ryukyu Islands, Japan). Part III. *Memoirs of the National Science Museum, Tokyo*, 6: 599–63. [in Japanese with English abstract.]

Okada Y. 1927. A study on the distribution of tailless batrachians of Japan. *Annotated Zoology of Japan*, 11: 137–143.

Oliver W.L.R. 1984. Introduced feral pigs. In: Papers presented at the Workshop *on Feral Mammals, International Union of the Conservation of Nature, Species Survival Commission, Caprinae Specialist Group, Helsinki*, pp. 85–126.

Ota H. 1998. Geographic patterns of endemism and speciation in amphibians and reptiles of the Ryukyu Archipelago, Japan, with special reference to their paleogeographic implications. *Researches on Population Ecology*, 40 (2): 189–204.

Ota H. 2000. The current geographic faunal pattern of reptiles and amphibians of the Ryukyu Archipelago and adjacent regions. *Tropics*, 10 (1): 51–62.

Ota H. 2003. Toward a synthesis of paleontological and neontological information on the terrestrial vertebrates of the Ryukyu Archipelago. I. Systematic and biogeographic review. *Journal of Fossil Research*, 36 (2): 43–59.

Ota H., Sakaguchi N., Ikehara S., Hikida T. 1993. The herpetofauna of the Senkaku Group, Ryukyu Archipelago. *Pacific Science*, 47: 248–255.

Otsuka H. 1978. On the fossil elephant dredged from East China Sea. In: Kizaki K. (ed.), *Geological Studies of the Ryukyu Islands*, 3: 149–156. [in Japanese.]

Otsuka H. 1996. On the fossil elephant from Miyako Island of the Nansei Islands. *Proceedings of the Nishinihon Branch, Geological Society of Japan*, 108: 26.

Otsuka H. 2002. Paleovertebrate fauna of the Ryukyu Archipelago and its origin. In: Kimura M. (ed.), *Establishment of the Ryukyu Arc and Colonization of Biota*. Okinawa Times, Naha, pp. 111–127. [in Japanese.]

Otsuka H., Takahashi A. 2000. Pleistocene vertebrate fauna in the Ryukyu Islands: Their migration and extinction. *Tropics*, 10 (1): 25–40.

Shuji M. 1999. A chronological review of Pleistocene human remains from the Japanese Archipelago. Interdisciplinary perspectives on the origins of the Japanese. International Symposium 1996. *International Symposium*, 11–B: 181–197.

Suzuki H. 1982. Skulls of Minatogawa Man. In: Suzuki H., Hanihara K. (eds), *The Minatogawa Man: The Upper Pleistocene Man from the Island of Okinawa. University Museum Bulletin of the University of Tokyo*, 19: 7–49.

Suzuki H. 1983. The Yamashita-cho Man. A late Pleistocene infantile skeleton from the Yamashita-cho Cave (Okinawa). *Bulletins et Mémoires de la Société d'Anthropologie de Paris*, 10 (1): 81–87.

Suzuki H., Hanihara K. (eds) 1982. The Minatogawa Man: The Upper Pleistocene Man from the Island of Okinawa. *University Museum Bulletin of the University of Tokyo*, 19.

Suzuki H., Isawa M.A., Ishii N., Nagaoka H., Tsuchiya K. 1999. The genetic status of two insular populations of the endemic spiny rat *Tokudaia osimensis* (Rodentia, Muridae) of the Ryukyu Islands, Japan. *Mammal Study*, 24: 43–50.

Takahashi A., Hirayama R., Otsuka, H. 2018. Systematic revision of *Manouria oyamai* (Testudines, Testudinidae), based on new material from the Upper Pleistocene of Okinawajima Island, the Ryukyu Archipelago, Japan, and its paleogeographic implications. *Journal of Vertebrate Paleontology*: e1427594.

Takahashi A., Otsuka H., Hirayama R. 2003. A new species of the genus *Manouria* (Testudines: Testudinidae) from the Upper Pleistocene of the Ryukyu Islands, Japan. *Paleontological Research*, 7: 195–217.

Takahashi A., Otsuka H., Ota H., 2008. Systematic review of late Pleistocene Turtles (Reptilia: Chelonii) from the Ryukyu Archipelago,

Japan, with special reference to paleogeographical implications. *Pacific Science*, 62 (3): 395–402.

Takamiya H. 1968. Archaeological findings of Naha City. *Source Materials for the History of Naha City*, 1: 247–255.

Tokunaga S. 1940. A fossil elephant tooth discovered in Miyakozima, an island of the Ryukyu Archipelago. *Proceedings of the Imperial Academy (Tokyo)*, 13 (2): 43–47.

Tokunaga S., Takai F. 1939. A study of *Metacervulus astylodon* (Matsumoto) from the Ryukyu Islands, Japan. *Transactions of the Biogeographical Society of Japan*, 3: 221–248.

Tomida Y., Otsuka H. 1993. First discovery of fossil Amami rabbit (*Pentalagus furnessi*) from Tokunoshima, southwestern Japan. *Bulletin of the National Science Museum (Series C, Geology & Paleontology)*, 19 (2): 93–97.

van der Made J. 2005. La fauna. In: Carbonell E. (ed.), *Hominidos: las primeras ocupaciones de los continentes*. Ariel, Barcelona, pp. 270–306.

Wallace A.R. 1876. *The Geographical Distribution of Animals; With A Study of the Relations of Living and Extinct Faunas as Elucidating the Past Changes of the Earth's Surface*, 2 vols. Macmillan & Co., London and Harper & Brothers, New York.

Watanobe T., Okumura N., Ishiguro N., Nakano M., Matsui A., Sahara M., Komatsu M. 1999. Genetic relationship and distribution of the Japanese wild boar (*Sus scrofa leucomystax*) and Ryukyu wild boar (*Sus scrofa riukiuanus*) analysed by mitochondrial DNA. *Molecular Ecology*, 8: 1509–1512.

Watase S. 1912. The boundary line between Palaearctic and Oriental regions. In: Holmgren N. (ed.), *Die Termiten Japans. Annotationes Zoologicae Japonenses*, 8(2): 108–111.

Yamada, F., Takaki, M., Suzuki, H. 2002. Molecular phylogeny of Japanese Leporidae, the Amami rabbit *Pentalagus furnessi*, the Japanese hare *Lepus brachyurus*, and the mountain hare *Lepus timidus*, inferred from mitochondrial DNA sequences. *Genes and Genetic Systems*, 77 (2): 107–116.

CHAPTER EIGHTEEN

The Californian Channel Islands

Evolution of Island Mammals: Adaptation and Extinction of Placental Mammals on Islands, Second Edition.
Alexandra van der Geer, George Lyras and John de Vos.

© 2021 John Wiley & Sons Ltd. Published 2021 by John Wiley & Sons Ltd.

The Channel Islands off the coast of southern California consist of two groups, a northern and a southern group. The northern group formed one large island – Santarosae – during the Pleistocene that was inhabited by dwarf mammoths, island foxes, and large deer mice. Only the fox survived until today. The dwarf mammoth is mainly known by an exceptionally complete skeleton of an adult male. The northern islands are famous for the discovery of the 'Arlington Springs Man', a Palaeolithic population that arrived at the end of the Last Glacial Maximum. The southern group was never connected to the northern group and yielded hardly any fossils.

Geology and Palaeogeography

The Channel Islands form an archipelago along the coast of southern California in the Pacific Ocean (Figure 18.1). The four northern islands – Santa Rosa, San Miguel, Santa Cruz, and Las Anacapas – formed one large island during the Late Pleistocene, dubbed Santarosae by Phil Orr (1968), derived from the name of the largest island today. Between about 24,000 and 19,000 years ago during the Last Glacial Maximum, this large island reached its maximum size and minimal distance to the mainland, ranging between 6.5 and 8 km as given by Louise Roth (1996). The southern islands – San Nicolas, Santa Catalina, Santa Barbara, and San Clemente – were never part of Santarosae. With the warming of the climate, the sea level rose again and about three quarter of the island submerged. Only the higher areas remained dry and eventually constituted the present four northern Channel Islands.

Earlier researchers, e.g. Fairbanks (1897) and Chester Stock and Eustace Furlong (1928), thought that Santarosae was once connected by land bridge to the mainland. This land bridge was

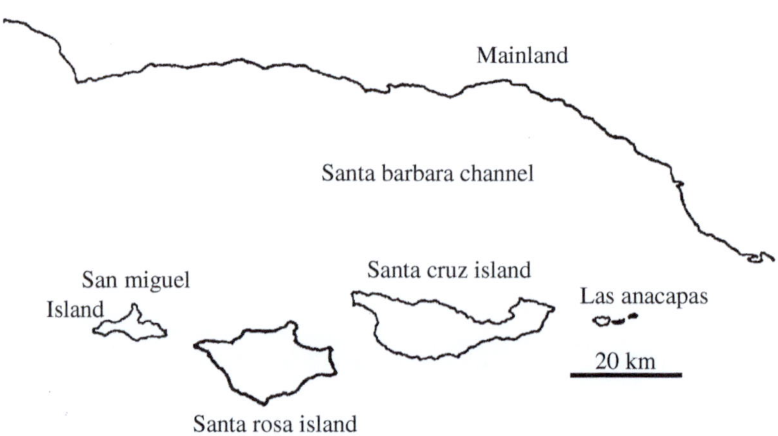

Figure 18.1 Map of the California Channel Islands.

considered an extension of the Santa Monica Mountains into the Pacific Ocean. This hypothesis persisted into the 1970s and was followed by Hooijer (1976) and Cary Madden (1977). The ancestors of the dwarf mammoths had crossed this bridge and got subsequently isolated when the rising sea level flooded the connection. After the evidence of the natatorial abilities of elephants by Johnson (1980), the need to explain the presence of mammoths on Santarosae by land bridge evaporated. Furthermore, the persistence of a deep-water strait of about 4–6 km width during the entire Pleistocene has been demonstrated by Johnson (1978) and Wenner and Johnson (1980).

Historical Palaeontology

Fossils from the Californian Channel Islands are known since 1856 when they were discovered during a Coast and Geodetic Survey of Santa Rosa. The remains were briefly reported by Robert Stearns (1873) in a meeting of the Academy of Sciences. Half a century later, Chester Stock and Eustace Furlong of the technical university of California went to Santa Rosa's north-western coast to excavate, hoping to find more material of the dwarf mammoth. Their excursion was not in vain, for they returned with fossils on which Stock and Furlong (1928) based their new species *Elephas* (=*Mammuthus*) *exilis*. In the late 1950s and the 1960s, archaeologist Phil Orr of the University of Santa Barbara – the mainland town, not the island – continued the excavations. Orr's primary goal was to reconstruct the life of the early island people. It was confirmed, in his opinion, that humans had led the mammoth to extinction through hunting.

The large collection of mammoth bones of Orr at the Santa Barbara Natural History Museum and the collections of Stock, Furlong, and others at the Los Angeles County Museum were studied in much detail by Roth (1982, 1990, 1992, 1993, 1996). Further excavations could not take place, hindered by the Vail and Vickers Cattle Company. New fossil specimens were only collected by Boris Woolley in the 1970s, who kept his collection at his ranch. In 1995, his widow donated the fossils to the museum at Santa Barbara. Meanwhile, the National Park Service had acquired San Miguel, Santa Rosa, Las Anacapas, and a portion of Santa Cruz in 1987, and, thus, the Channel Islands National Park was established.

In June 1994, geologist Tom Rockwell and his graduate student Kevin Colson discovered large fossil bones during a geological survey at Carrington Point, Santa Rosa. The bones protruded from a steep and sandy slope. Archaeologist Don Morris asked palaeontologist Larry Agenbroad for his opinion.

The latter, together with Jim Mead, visited the site and confirmed the earlier diagnosis: these were indeed the bones of *Mammuthus exilis*. A prompt excavation was decided because the skeleton was partly exposed and ran the risk of being lost due to erosion by rain and wind. More than 90% of the skeleton could be recovered. Even the tongue bone and the sternal bones were found in anatomical position.

Fossils of the dwarf mammoth were further found at several sites on San Miguel. A tusk and two limb bone portions were reported from the western coast of Santa Cruz Island. These have a large size and are attributed to the Columbian or American mammoth (*Mammuthus columbi*). In 2009, a complete tusk was discovered at the northern shore of Santa Cruz by archaeology graduate student Kristina Gill. Several ribs and a larger bone, possibly a femur, were found nearby, perhaps belonging to the same individual. The size seems to indicate a dwarf mammoth. From the southern islands, a single proboscidean tooth from San Nicolas has been described by Vedder and Robert Norris (1963).

Biozones and Faunal Units

The mammalian fossils from the northern Channel Islands (Santarosae) all belong to a single faunal unit and give no indication of any faunal turnover. Only one fossil biozone can thus be recognised for the Californian Channel Islands, characterised by a dwarf mammoth. The present-day fauna is in part a continuation of the Pleistocene fauna, added with human-introduced species (e.g. pigs) and culture-followers (e.g. skunks), but devoid of mammoths. The single proboscidean fossil from the southern Channel Islands (San Nicolas) cannot be reliably correlated to the fauna of Santarosae.

Late Pleistocene–Holocene

The fossil fauna of the northern Channel Islands consisted of a dwarf mammoth (*Mammuthus exilis*), the island fox (*Urocyon littoralis*), a sea otter (*Enhydra lutris*), a large deer mouse (*Peromyscus nesodytes*), and a small deer mouse (*Peromyscus anyapahensis*). Apart from the dwarf mammoth, a mainland American mammoth (*Mammuthus columbi*) also occurred on the island, in a ratio of about 1 : 10, as shown by Agenbroad (1998). Six fossils of a further undescribed sea otter, labelled as *Enhydra lutris*, were found in Indian kitchen middens at Santa Barbara by King A. Rickey in 1941 (according to the museum label) indicating that they had been eaten. Apart from mammals, the island was also inhabited by a flightless sea duck

(*Chendytes lawi*), but this species had, despite the fact that it was flightless, a wider distribution, including the Californian coast and possibly southern Oregon.

The fauna likely arrived during a sea level low stand, probably during the glacial periods around 150,000 or 250,000 years ago (Marine Isotope Stages 6 and 8, respectively). The oldest deposit – the lowest marine terrace of Santa Rosa – has yielded uranium-series ages between 78,600 and 83,800 years ago and contains fossils of the larger deer mouse and a tusk of the dwarf mammoth. The majority of deposits with the dwarf mammoth date (radiocarbon ages) to the last glacial period, between about 12,000 and 22,000 year ago, and only a few localities on the Santa Rosa Island have ages of over 41,000 years ago, as summarised by Agenbroad (2012).

At the end of the Pleistocene or in the Early Holocene, the ancestor of the island spotted skunk (*Spilogale gracilis amphiala*) colonised Santarosae. Today, it inhabits Santa Rosa and Santa Cruz, but analysis of genetic data – microsatellite loci – by Chris Floyd and colleagues (2011) reveals that it already arrived before the rising sea level separated the two islands some 11,500 years ago. The spotted skunks differ only marginally – by a shorter tail and wider-set eyes – from the mainland populations, reflecting a short time in isolation, in contrast to that of the fox and the deer mice. The western harvest mouse (*Reithrodontomys megalotis*), on the other hand, arrived much later, likely coinciding with and as a result of the arrival of Native Americans to the islands.

Peculiarities and Evolution of Endemic Mammals

Dwarf Mammoths: Mammuthus exilis

The Channel Islands dwarf mammoth was about half the size of its presumed mainland ancestor – the Columbian or American mammoth, *Mammuthus columbi* – which had a shoulder height of about 3.5–4.0 m (Plate 1). The size variation within the mammoth material, retrieved from the two islands Santa Rosa and San Miguel, is, however, rather large, and shoulder height estimations range from 1 m to almost 2.5 m according to Roth (1993). It is not clear whether this is to be explained by sexual dimorphism, by morphotypes, or by chronotypes. This makes body size estimations difficult, and the estimation of 1,150 kg as given by Burness and colleagues (2001) then needs to be regarded with caution. Agenbroad (2009) calculated the shoulder height as 1.37–1.93 m with an average of 1.72 m, based on 12 specimens. The average body mass of these 12 specimens is 759 kg, based on calculations using the femora. Agenbroad

(1998) challenged the earlier cited 2.5 m estimate as an error. The 1994 skeleton stands about 1.8 m at the shoulder, which falls within his new range.

The skeleton discovered at Carrington Point, Santa Rosa, in 1994, shows signs of osteoarthritis in several joints of the feet. This indicates a high individual age, which agrees with the heavily worn molars. Its age has been calculated at 56 African elephant years based on Laws' criteria using molar wear stages. Naturally, this is merely estimation because every change in diet or way of life influences the result. The skeleton itself has a geological age of about 12,840 years based on datings of collagen extracted from the interior of the right thigh bone, as reported by Agenbroad (2003).

The Californian dwarf mammoth was more gracile and short-legged compared to the Columbian mammoth and likely was a better climber or walker over rugose terrain. The dwarf mammoth was primarily feeding on leaves and twigs, as inferred from microwear analyses of the enamel surface of its teeth by Gina Semprebon and colleagues (2016). This means a distinct narrowing in dietary niche compared to its mainland ancestor. This ancestor probably arrived at Santarosae during the lowest sea level stand, when the island was at most 8 km away from the Californian coast. Elephants swim this distance with ease, and there is no reason to assume that mammoths could not do the same. Geological evidence disproves the existence of any land connection in the area.

The extinction of the dwarf mammoth is generally attributed to a climate change, resulting in a diminishing of the conifer forests in which it found its food and the subsequent drowning of major part of the island. The original habitat of the mammoth decreased substantially in size and apparently surpassed the critical size. However, this scenario has been challenged by Daniel Muhs and colleagues (2015), arguing that the mammoths had experienced a similar change, or even more dramatic, in the warm period after their arrival during a cold stage (Marine Isotope Stage 6 or 8). Orr (1968) was the first who postulated that humans had caused the extinction of the dwarf mammoths, since humans – the Arlington Springs Man – arrived around the extinction event. Archaeological excavations indicate that humans were already on the island about 13,000 years ago, as shown by Jon Erlandson and colleagues (1996, 1997). This early date – among the earliest human remains in the Americas – is confirmed by Mark Hardiman and colleagues (2016) on the basis of a sedimentary charcoal record from Arlington Canyon, which shows a significant period of charcoal deposition between about 12,500 and 14,000 years ago. Agenbroad (2003) presented Accelerator Mass Spectrometry (AMS) dates on bone collagen for the Arlington Springs Man of about 10,960 years

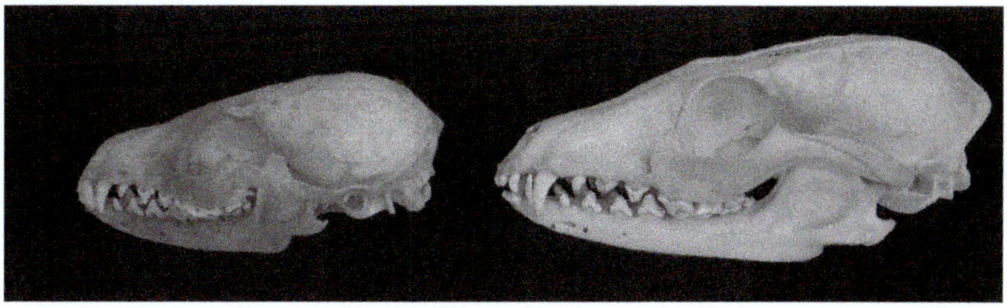

Figure 18.2 Skull of the Channel Island fox (*Urocyon littoralis*, left) compared to that of its mainland relative, the grey fox (*Urocyon cinereoargenteus*, right); lateral view. Field Museum of Natural History, Chicago. (Photograph: George Lyras.)

ago and the dwarf mammoth of about 11,010 years ago. It seems likely that a combination of factors, including human pressure, eventually led to the extinction of the mammoth, as suggested by Agenbroad and colleagues (2005).

Channel Island Fox

The island fox (*Urocyon littoralis*) is smaller than its ancestor, the grey fox (*Urocyon cinereoargenteus*) (Figure 18.2), and is the smallest living North American canid today. Linear measurements are 25% smaller than those of the grey fox. The most important differences with its ancestor are the pronounced muscle scars on the skeleton and the smaller body size. Furthermore, it has proportionally short limbs (Young, 2020). The dwarfism likely is the result of a combination of interspecific competition and the lack of larger prey. The grey fox probably arrived during the Late Pleistocene low sea level stand of about 18,000 years ago. Sometime between 5,000 and 2,000 years ago, the Amerindians brought the island fox to the southern Channel Islands. Today, the island fox is found on almost all Channel Islands. The differences between the foxes of the different islands are reflected at subspecific level: *littoralis* for San Miguel, *santarosae* for Santa Rosa, *santacruzae* for Santa Cruz, *dickeyi* for San Nicolas, *catalinae* for Santa Catalina, and *clementae* for San Clemente. The Santa Cruz subspecies is the smallest and the one of Santa Catalina the largest. The most remarkable difference between the subspecies is found in the number of the tail vertebrae. Other differences are mainly seen in coat pattern and colour.

Based on molecular data, the time of arrival per island was calculated by Robert Wayne and colleagues (1991) to 16,000–10,400 years ago for the northern Channel Islands, to 4,300–3,400 years ago for San Clemente, to 3,800–800 years ago for Santa Catalina, and to 2,200 years ago for San Nicolas. The fact that the smallest fox is found in the northern islands and the largest on Santa Catalina, which probably was not colonised before 800 years ago, indicates that the different subspecies simply reflect time of isolation. At the time of their study, the island fox was extinct on San Miguel, so this subspecies could not be analysed. The fox has been reintroduced into the wild on San Miguel after a programme of captive breeding.

The success of the island fox is partly thanks to its uncritical way of life. It eats practically everything and is found in several habitats. Nonetheless, at present, the species is endangered despite this dietary success. The reason for this is found in a shift in predatory species. For centuries, bald eagles (*Haliaeetus leucocephalus*) settled on the islands. They constituted no danger to the foxes, as they are primarily fish eaters. The bald eagles though were decimated in the 1950s caused by the accumulation of the insecticide DDT (dichlorodiphenyltrichloroethane) in fish, after which golden eagles (*Aquila chrysaetos*) had the chance to populate the island as explained by Gary Roemer and colleagues (2002). The golden eagles were attracted by the imported domestic pigs, which were herded like goats and sheep. When the pig population had reached high numbers in the 1990s, the number of golden eagles increased simultaneously. However, the eagles hunted the foxes as well, and the latter's numbers dropped drastically as a result. Population dynamics of islanders is as a rule more vulnerable to changes, and an additional cause of death is generally not easily balanced, but leads to an exponential decrease in numbers. The populations of San Miguel and Santa Rosa were eventually extirpated, but later re-established from captive breeding colonies.

Deer Mice

Fossils of the large deer mouse (*Peromyscus nesodytes*) are restricted to Late Pleistocene deposits on Santa Rosa and archaeological sites on San Miguel. From the same sediments, a dwarf mammoth tusk was retrieved. Hardly anything is known of this species, except for its large size. The large Pleistocene deer mouse was probably the largest species of the genus, except maybe for the species of southern Mexico and Central America.

The remains from San Miguel are dated to about 2,000 years ago as given by Phillip Walker (1980), while the oldest remains date back to about 80,000 years ago from sediments exposed in Garañon Canyon of Santa Rosa, as reported by Agenbroad and colleagues (2014).

On Las Anacapas, a small deer mouse (*Peromyscus anyapahensis*) is known from Late Pleistocene deposits. Mice of this genus live at present on most uninhabited islets in the Gulf of California. Today, the closely related *Peromyscus maniculatus* lives today on the islands, with a distinct subspecies on each of the eight islands it inhabits.

References

Agenbroad L.D. 1998. New pygmy mammoth (*Mammuthus exilis*) localities and radiocarbon dates from San Miguel, Santa Rosa, and Santa Cruz Islands, California. In: Weigand P. (ed.), *Contributions to the Geology of the Northern Channel Islands, Southern California*. Pacific Section of the American Association of Petroleum Geologists, Bakersfield, California, pp. 169–175.

Agenbroad L.D. 2003. New absolute dates and comparisons for California's *Mammuthus exilis*. *Deinsea*, 9: 1–16.

Agenbroad L.D. 2009. *Mammuthus exilis* from the California Channel Islands: Height, mass, and geological age. In: Damiani C.C., Garcelon D.K. (eds), *Proceedings of the 7th California Islands Symposium*. Institute for Wildlife Studies, Arcata, CA, pp. 15–19.

Agenbroad L.D. 2012. Giants and pygmies: Mammoths of Santa Rosa Island, California (USA). *Quaternary International*, 255: 2–8.

Agenbroad L.D., Bugbee M.M., Morris D.P., Wilkins W.J. 2014. Preserving the pygmy mammoth: Twenty years of collaboration between Channel Islands National Park and the mammoth site of Hot Springs, SD, Inc. *Proceedings of the 10th Conference on Fossil Resources, Rapid City, SD May 2015. Dakoterra*, 6: 119–123.

Agenbroad L.D., Johnson J., Morris D., Stafford T., Jr. 2005. Mammoths and humans as late Pleistocene contemporaries on Santa Rosa Island. In: Garcelon D., Schwemm C. (eds), *Proceedings of the Sixth California Islands Symposium. Institute for Wildlife Studies.* Arcata, California, pp. 3–7.

Burness G.P., Diamond J., Flannery T. 2001. Dinosaurs, dragons, and dwarfs: the evolution of maximal body size. *Proceedings of the National Academy of Sciences of the USA*, 98: 14518–14523.

Erlandson J.M., Kennett D.J., Ingram B.L., Guthrie D.A., Morris D.P., Tveskov M.A., West G.J., Walker P.L. 1996. An archaeological and paleontological chronology for Daisy Cave (CA-SMI261), San Miguel Island, California. *Radiocarbon*, 38: 1–29.

Erlandson J.M., Tveskov M.A., Kennett D.J., Ingram B.L. 1997. Further evidence for a terminal Pleistocene occupation of Daisy Cave, San Miguel Island, *California. Current Research in the Pleistocene*, 13: 161–168.

Fairbanks H.W. 1897. Oscillations of the coast of California during the Pliocene and Pleistocene. *American Geologist*, 20: 213–245.

Floyd C.H., Van Vuren D.H., Crooks K.R., Jones K.L., Garcelon D.K., Belfiore N.M., Dragoo J.W., May B. 2011. Genetic differentiation of island spotted skunks, *Spilogale gracilis amphiala. Journal of Mammalogy*, 92 (1): 148–158.

Hardiman M., Scott A.C., Pinter N., Anderson R.S., Ejarque A., Carter-Champion A., Staff R.A. 2016. Fire history on the California Channel Islands spanning human arrival in the *Americas. Philosophical Transactions of the Royal Society B*, 371: 20150167.

Hooijer D.A. 1976. Observations on the pygmy mammoths of the Channel Islands, California. In: Churcher C.S. (ed.), *Essays on Palaeontology in Honor of Loris Shano Russell*. Royal Ontario Museum, Ontario, Canada, pp. 220–225.

Johnson D. L. 1978. The origin of island mammoths and the Quaternary land bridge history of the Northern Channel Islands, California. *Quaternary Research*, 10: 204–225.

Johnson D.L. 1980. Problems in the land vertebrate zoogeography of certain islands and the swimming powers of elephants. *Journal of Biogeography*, 7: 383–398.

Madden C. T. 1977. Elephants of the Santa Barbara Channel Islands, southern California. *Geological Society of America. Abstracts with Programs*, 9: 458–459.

Muhs D.R., Simmons K.R., Groves L.T., McGeehin J.P., Schumann R.R., Agenbroad L.D. 2015. Late Quaternary sea-level history and the antiquity of mammoths (*Mammuthus exilis* and *Mammuthus columbi*), Channel Islands National Park, California, USA. *Quaternary Research*, 83: 502–521.

Orr P.C. 1968. *Prehistory of Santa Rosa Island. Santa Barbara Museum of Natural History*. Santa Barbara, California.

Roemer G.W., Donlan C.J., Courchamp F. 2002. Golden eagles, feral pigs and insular carnivores: How exotic species turn native predators into prey. *Proceedings of the National Academy of Sciences of the United States of America*, 99: 791–796.

Roth V.L. 1982. *Dwarf Mammoths from the Santa Barbara, California Channel Islands: Size, Shape, Development, and Evolution.* Unpublished PhD thesis. University Microfilms. Yale University, Ann Arbor.

Roth V.L. 1990. Insular dwarf elephants: A case study in body mass estimation and ecological inference. In: Damuth J., MacFadden B.J. (eds), *Body Size in Mammalian Paleobiology: Estimation and Biological Implications.* Cambridge University Press, Cambridge, pp. 151–179.

Roth V.L. 1992. Inferences from allometry and fossils: Dwarfing of elephants on islands. *In*: Futuyma D., Antonovics J. (eds), *Oxford Surveys in Evolutionary Biology, 8: 259–288.* Oxford University Press, New York.

Roth V.L. 1993. Dwarfism and variability in the Santa Rosa island mammoth: An interspecific comparison of limb-bone sizes and shapes in elephants. In: Hochberg F.G. (ed.), *Third California Islands Symposium: Recent Advances in Research on the California Islands.* Santa Barbara Museum of Natural History, pp. 433–442.

Roth V.L. 1996. Pleistocene dwarf elephants from the California Islands. In: Shoshani J.H., Tassy P. (eds), *The Proboscidea.* Oxford University Press, Oxford, pp. 249–253.

Semprebon G.M., Rivals F., Fahlke J.M., Sanders W.J., Lister A.M., Göhlich U.B. 2016. Dietary reconstruction of pygmy mammoths from Santa Rosa Island of California. *Quaternary International*, 406: 123–136.

Stearns R.E.C. 1873. [no title: paraphrase of comments attributed to Stearns in minutes of the Academy's regular meeting.] *Proceedings of the California Academy of Sciences*, 5: 152.

Stock C., Furlong E.L. 1928. The Pleistocene elephants of Santa Rosa Island. *Science*, 68: 140–141.

Vedder J.G., Norris R.M. 1963. Geology of San Nicholas Island California. *Geological Survey Professional Paper*, 369 (1–6): 1–65.

Walker P.L. 1980 Archaeological evidence for the extinction of three terrestrial mammals on San Miguel. Island. In: Power D.M. (ed.), *The California Islands: Proceedings of a Multidisciplinary Symposium*. Santa Barbara Museum of Natural History, Santa Barbara, California, pp. 703–717.
Wayne R., George S., Gilbert D., Collins P., Girman D. 1991. A morphologic and genetic study of the Island Fox *Urocyon littoralis*. *Evolution*, 8: 1849–1868.
Wenner A.M., Johnson D.L. 1980. Land vertebrates on the California Channel Islands: Sweepstakes or bridges? In: Power D.M. (ed.), *The California Islands: Proceedings of a Multidisciplinary Symposium*. Santa Barbara Museum of Natural History, Santa Barbara, California, pp. 497–530.
Young C.B. 2020. Static allometry of a small-bodied omnivore: body size and limb scaling of an island fox and inferences for *Homo floresiensis*. *Journal of Human Evolution*, 149: 102899.

CHAPTER NINETEEN

The Greater Antilles

Evolution of Island Mammals: Adaptation and Extinction of Placental Mammals on Islands, Second Edition.
Alexandra van der Geer, George Lyras and John de Vos.
© 2021 John Wiley & Sons Ltd. Published 2021 by John Wiley & Sons Ltd.

The Greater Antilles are known for their Pleistocene fossils of endemic sloths, hutias, coneys, spiny rats, rice rats, bats, monkeys, and shrew-like insectivores. The earliest fossils of terrestrial mammals date back to the Eocene and belong to a rhinocerotid and a terrestrial sirenian, both likely of a continental stock. There are also a few fossils of endemic sloths, monkeys, hutias, and insectivores known from the Oligocene and the Miocene. The fossils show that the Greater Antilles has been a region with a high degree of endemism and high species diversity since at least the Early Miocene. Despite the fact that even today the West Indies constitutes a hotspot of biodiversity, the present mammalian fauna is only a poor representation of its extinct Late Quaternary fauna. Among the about 70 species of endemic non-volant mammals known from the Late Quaternary, only about 15 species survived into the present time.

Geology and Palaeogeography

The Greater Antilles form part of the West Indies, a region that consists of more than 7,000 islands, islets, reefs, and cays in the Caribbean Sea and adjacent part of the Atlantic Ocean north of Cuba (Figure 19.1). The island group is named as such because Christopher Columbus, who landed there in 1492, believed that he was in the 'Indies', the collective name for the European colonies in the Malayan Archipelago of that time. Apart from the Greater Antilles, the region consists of the Lesser Antilles, including the Leeward, the Bahamas, and the Turks and Caicos Islands. The geological history and palaeogeography of this very large and varied island group is exceedingly complex and has been explained differently by different authors. Subduction of the North American plate beneath the Caribbean plate resulted in regional volcanism, leading to the formation of the proto-Antillean island arc. This arc more or less connected North and South America during the Late Cretaceous (c. 100–70 million years ago). The proto-Antilles was fragmented by plate tectonic movement, and in due time, the current islands were formed through collisions, movements, and fusions. The taxa present on the islands got isolated, and their evolutionary pathways took their own directions. This model is referred to as the *vicariance model* starting with Donn Rosen (1975) and is sometimes applied to the earliest inhabitants of the Greater Antilles as an island group. The impact of the asteroid that struck the Earth in the Caribbean region 65 million years ago, as shown by Allen Hildebrand and William Boynton (1990), almost certainly included massive waves (tsunamis) in the order of a kilometre

Figure 19.1 Map of the Greater Antilles (Cuba, Jamaica, Hispaniola, and Puerto Rico) with their numerous satellite islands (e.g. Isla de Pinos and Gonave), the Bahamas archipelago (Bahamas, Turks, and Caicos Islands) and Cayman Islands.

or more in height, as calculated by Florentin Maurrasse (1991), and gigantic hurricanes (hypercanes), as briefly mentioned by Kerry Emanuel and colleagues (1995). Likely, these massive events destroyed most if not all life on the proto-Antilles at that time and with it the vicariance model.

Later, at the Eocene–Oligocene transition, a land span was formed, reconnecting South America with the landmasses of the developing Greater Antilles (Figure 19.2), according to Manuel Iturralde-Vinent and Ross MacPhee (1999), elaborating on their earlier proposal (MacPhee and Iturralde-Vinent, 1994, 1995). They refer to their model as the land span hypothesis,

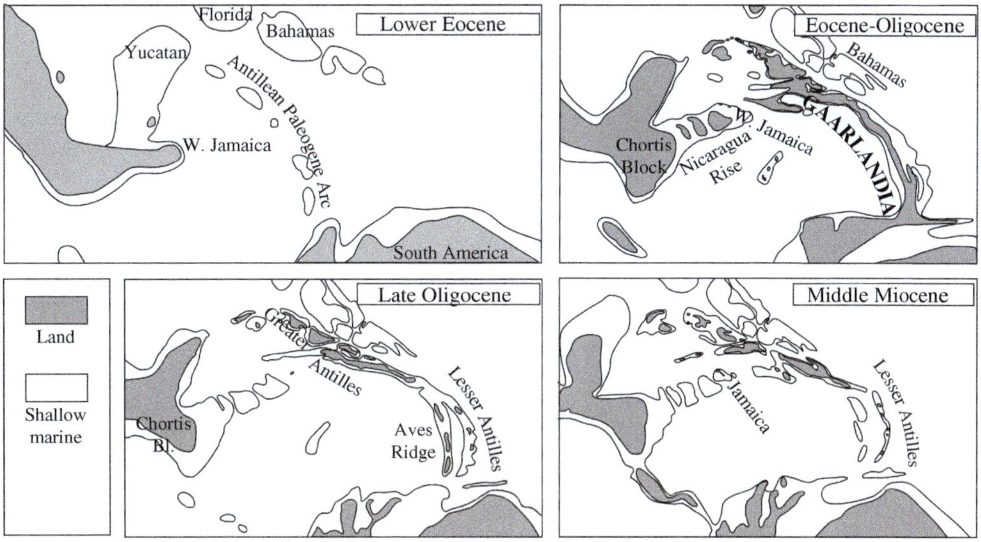

Figure 19.2 Palaeogeography of the Caribbean region from the Eocene till the Middle Miocene, as proposed by Iturralde-Vinent and MacPhee since 1999. Early Eocene (around 55 Ma): With the exception of Jamaica, which was connected to the mainland, the only subaerial parts of the Caribbean region are the small islands of the Antillean Palaeogene Arc. Note that the exact location of these islands is unknown. Late Eocene–Early Oligocene (33–35 Ma): During this time interval, a general tectonic uplift coincided with a major eustatic sea level drop. As a result, a land span was formed along the Aves Ridge, connecting South America with the proto-Antilles. Late Oligocene (25–27 Ma): During this period, there were extensive marine ingressions, probably due to a combination of tectonic subsidence and sea level rise. The presence of rocks that were formed in non-marine environments and of plants and terrestrial sediments in marine basins suggests that some higher parts remained persistently emerged along the axis of GAARlandia. Early Middle Miocene (14–16 Ma): Neogene tectonic movements and deformations along the margins of the Caribbean Plate subdivided GAARlandia and created isolated tectonic blocks and terrains separated by deep-water gaps. This process isolated the faunas of the former GAARlandia, resulting in, for example, independent sloth clades in Cuba and Hispaniola.

making a distinction between a land bridge – a connection between two continental landmasses – and a land span – a connection between a continent and an off-shelf island. MacPhee and Iturralde-Vinent name this land span GAARlandia because it combined the Greater Antilles and the Aves Ridge (Box 19.1). GAARlandia was formed 35–34 million years ago as a result of a major sea level drop of about 160 m, as calculated by Kenneth Miller and colleagues (1996), in combination with a substantial tectonic uplift along the Greater Antilles and Aves Ridge (see Figure 19.2). Although the land span lasted for a relatively short period of geologic time, just 1 or 2 million years, it may have created the conditions for overland dispersal of several mammalian taxa to what we call now the Greater Antilles.

BOX 19.1

Sweepstake Tickets to the West Indies

The GAARlandia land span model of MacPhee and Iturralde-Vinent was heavily criticised by Stephen Blair Hedges (2001). Hedges supports the dispersal model and the over water dispersal of terrestrial vertebrates by floating on natural rafts. He presented data provided by molecular clock studies and taxonomic composition of the reptilian fauna, which support neither an arrival during the Late Cretaceous (the vicariance model) nor at the onset of the Oligocene (the land span model). The calculated times of origin for West Indian reptilian lineages appear to be scattered throughout the Cainozoic and are not nicely clustered during one time period. Another point of evidence is, in Hedges' view, the fact that adaptive radiation had taken place within most lineages studied. This suggests that niches were left vacant by groups absent from the Antilles, which would not have been the case if the landmasses with their original mainland fauna would have been gradually disconnected. Furthermore, the major source area for the reptiles is South America, which is in agreement with the water currents in the Caribbean region, flowing from south-east to north-west, though this does not necessarily hold for the situation in the past. Salamanders, finally, are lacking altogether, indicating the absence of a continuous land bridge. Hedges does not dismiss the vicariance model entirely, but accepts it for the tiny frogs of the genus *Eleutherodactylus*, which in his view colonised the West Indies close to the Cretaceous–Tertiary boundary. Many of Hedges' arguments were contested in a reply by MacPhee and Iturralde-Vinent (2005).

Historical Palaeontology

Cuba

The first mammalian fossil of the Greater Indies was discovered in 1860 in Cuba. It was a mandible of a mammal found in a warm spring at Ciego Montero near Cienfuegos (Figure 19.3) by Jose de Figueroa, as is written on the label. The mandible was given to the naturalist Felipe Poey, who presented it in 1861 before the Academy of Sciences of Havana as a giant rodent. The Spanish geologist Manuel Fernández de Castro (1864) realised that the specimen belonged to a sloth and depicted it in a paper about the extinct animals of Cuba. De Castro

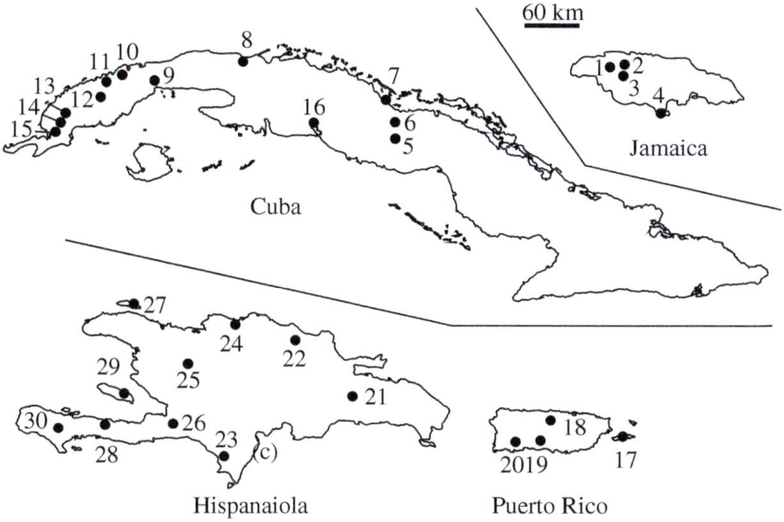

Figure 19.3 The most important localities of the Greater Antilles. (1) Seven Rivers, (2) Sheep Pen, (3) Wallingford Roadside cave, (4) Jackson's Bay, (5) Domo de Zaza, (6) Casimba (Sierra de Jatibonico), (7) Cayons Salinas y Lucas, (8) Cueva de Bellamar, (9) San Antonio de los Baños, (10) Santa Fé, (11) Anafe, (12) Cueva Alta, (13) Quemado de Pineda, (14) Sumidero, (15) Guane, (16) Ciego Montero, (17) Vieques Island, (18) Barahona caves, (19) Juana Díaz, (20) Yauco, (21) Bayaquana, (22) La Toca, (23) Cabo Rojo, (24) Cap Haitien, (25) Trou Manman Jumu (St Michael de l'Atalaye), (26) Morne la Visite, (27) Trou Gallery (Île de la Tortue), (28) Trou Wòch Dadier (Étang Miragoâne), (29) Trou Zombie (Île de la Gonave), and (30) Plain Formon.

sent the specimen to Paris for the 1867 exposition, where it was studied by Alphonse Pomel. Meanwhile in Philadelphia, De Castro's paper came to the attention of Joseph Leidy, who being unaware of the intentions of Pomel, described the mandible as well. As a result, two publications appeared in 1868 about the same specimen reaching similar conclusions. Both Leidy and Pomel recognised that the sloth was related to the mainland genus *Megalonyx* but differed enough to establish a new name. Because the incisors had an unusual rodent-like position, which had led Poey to his misidentification, Leidy named it *Megalocnus rodens* (Plate 34). Though Pomels' name (*Myomorphus cubensis*) was described in a more extensive paper, Leidy's name has a priority of a few months.

The discovery was later discredited by many scientists, amongst others Wayland Vaughan (1902), who thought that the mandible had been brought to Cuba by humans. The reason underlying this misconception was the original publication of De Castro (1865) where he described the Ciego Montero sloth together with fossils of a horse and a peccary, which latter he regarded as a hippo. However, both the peccary and the horse turned out to be recent and had been brought to Cuba from the mainland by the Spanish. In analogy with this, many scientists of that time assumed that the sloth fossil was not coming from Cuba either, further confirmed by the similarity of *Megalocnus* with the mainland *Megalonyx*. In addition, a series of wars between Cuban pre-independency forces and the Spanish army prevented any extensive exploration of the island for decades, and thus, no further evidence was provided.

Figure 19.4 Barnum Brown (left) and presumably Charles Falkenbach (right) draining the Chapapote Spring at Baños de Ciego Montero, Cuba, in 1918. (Image number 103891, American Museum of Natural History Library.)

At the beginning of the twentieth century, Carlos de la Torre, professor at the University of Havana, collected a considerable amount of sloth fossils from the Ciego Montero spring but also at the fissure-spring deposits of Cashimba in the Sierra de Jatibonico. He gave the material for further study at the American Museum of Natural History in New York. In 1911 and 1918, the Museum sent Barnum Brown to collect more fossils from these two Cuban sites, together with De la Torre (Figure 19.4). William Diller Matthew noted about the excavation of the Ciego Montero spring,

> The exploration of the deposit was a rather difficult matter as the spring is a powerful one and the water had to be pumped out and drained away by means of a gasoline pump, and the spring openings cemented up, before the deposit around it could be thoroughly explored. (Matthew, 1919: 163)

The fossils from the two spring deposits consist mainly of ground sloths (Plate 34), crocodiles, and giant tortoises. Brown also collected fossils from a number of caves. The majority of these fossils belonged to small mammals.

Brown's Cuban collection was studied by Matthew. With De la Torre (1915) as first author, he published a short paper describing three new sloth genera and species, *Microcnus gliriformes*, *Mesocnus browni*, and *Miocnus antillensis*. However, they did not assign holotypes, thus leaving the new names as *nomina nuda*, empty names without meaning. Four years later, Matthew published more data and wrote an extensive manuscript about the fossil sloths from Cuba. Unfortunately, his paper was never completed and remained a manuscript in the archives of the American Museum. Some parts were published a year after his death (as Matthew, 1931), and a more complete and an edited version was published many years later (as Matthew and De Paula Couto, 1959). At present, Matthew's species *Mesocnus browni* is assigned to *Parocnus, Microcnus gliriformes* to *Neocnus*, and *Miocnus antillensis* to *Acratocnus*.

In early 1991, Oscar Arredondo (as reported by Morgan and Ottenwalder, 1993) found an unusual partial skull of the insectivore *Solenodon* in the collections of the National Natural History Museum of Cuba. He noticed that this skull was larger than that of the living *Solenodon cubanus* of Cuba. About 20 years earlier, he (Arredondo, 1970) had mentioned a large femur from a late Quaternary fossil site in western Cuba, and 10 years later, he illustrated and described it together with Gary Morgan and Clayton Ray (Morgan and colleagues, 1980), plus two additional large fossil femurs from western Cuba. The finding of the skull confirmed that all these specimens belong to a much larger extinct Cuban *Solenodon*. The species was described by Morgan and Jose Ottenwalder (1993) as the giant extinct insectivore *Solenodon arredondoi* in honour of its discoverer.

Manuel Rivero and Arredondo (1991) described a new Late Pleistocene Cuban howler monkey species as *Paralouatta varonai*, based on a nearly complete skull, a number of isolated teeth, and some postcranial bones. The holotype was discovered in 1987 by Rolando Crespo Diaz of the Cuban speleological society at Cueva del Mono Fósil on the south slope of the Sierra de Galeras near Viñales, Pinar del Río province, western Cuba. A decade later, a second new species of the same genus was described as *Paralouatta marianae* by MacPhee, Iturralde-Vinent and Gaffney (2003). They based the description on an astragalus recovered at Domo de Zaza, an Early Miocene locality from south-central Cuba (Figure 19.5) and suggested that it could be ancestral to *Paralouatta varonai*. From the same locality, MacPhee and Manuel Iturralde-Vinent (1994) had described a new sloth species (*Imagocnus zazae*) and a new species of isolobodontine hutia (*Zazamys veronicae*).

Puerto Rico

The first fossils in Puerto Rico were collected in 1915 in an ash deposit in Cueva de la Ceiba – a cave between Utuado and Areeibo – during an excavation led by the anthropologist Franz Boas. The material – remains of rodents and a sloth – was given to Joel A. Allen, who described the rodent material a year later

Figure 19.5 Excavation detail of the Early Miocene site Domo de Zaza, Cuba, with Ross MacPhee. (Photograph: Ross MacPhee.)

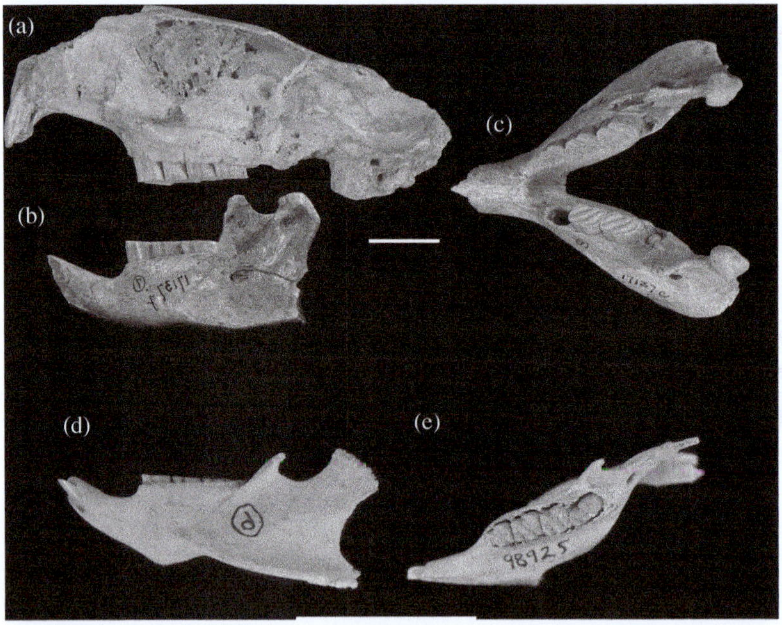

Figure 19.6 Extinct caviomorph rodents from Puerto Rico. *Elasmodontomys obliquus*: (a) lateral view of skull and (b) lateral and (c) occlusal views of mandible. *Isolobodon portoricensis*: (d) lateral and (e) occlusal views of semi-mandible. Scale bar: 2 cm. Views of (b) and (d) mirrored from the right side. American Museum of Natural History. (Photographs: Bartholomeus van der Geer.)

(Allen, 1916) as *Isolobodon portoricensis*, a recently extinct hutia (Figure 19.6). The fieldwork in Puerto Rico was continued by Harold Anthony, who in the summer of 1916 visited several caves accompanied by his wife and collected a large number of fossils of bats, rodents, and sloths. These fossils, together with many recent mammal specimens, were deposited at the American Museum of Natural History. In his excavation report (Anthony, 1917), he described several endemic genera, published in the years thereafter in a series of short papers

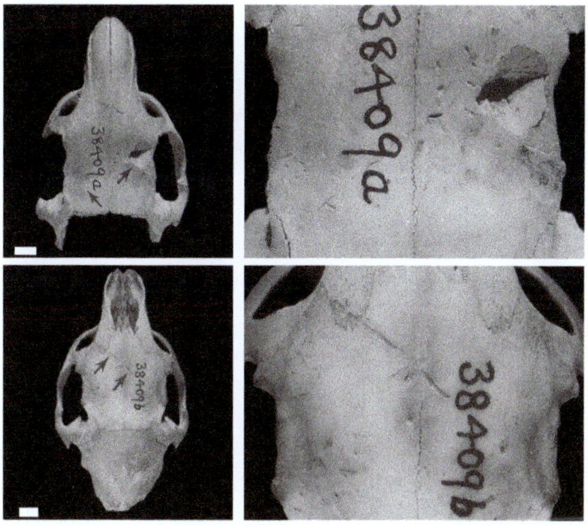

Figure 19.7 Holotype (top) and paratype (below) of *Isolobodon portoricensis*. Left: dorsal view showing premortem damage (arrows). Right: close-up of frontal bones of the same specimens. The paratype has been consistently misidentified as the holotype since Anthony's (1918) publication. Scale bar: 5 mm. (From Flemming and MacPhee, 1999; Photograph: Ross Mac Phee.)

(Anthony, 1919, 1920, 1926) in which he described the material more extensively. In the report, he named a new insectivore genus and species *Nesophontes edithae*, a new genus and species of spiny rats *Heteropsomys insulans*, a new genus and species of giant hutia *Elasmodontomys obliquus* (Figure 19.6), and a new sloth genus and species *Acratocnus odontrigonus*. In the same paper, he depicted *Isolobodon portoricensis*, but with the paratype as holotype due to a mistake in numbering. This error remained unnoticed till Clare Flemming and MacPhee (1999) repaired it (Figure 19.7). Anthony (1918) described a new genus and species of cave rat as *Homopsomys antillensis*, but this species is now considered a junior synonym of *Heteropsomys insulans*.

In March 1994, MacPhee and Iturralde-Vinent found a fossil fragment of another sloth near the town of Yauco in the south-western part of Puerto Rico. The specimen was nothing more than the proximal part of a small femur, but there was something amazing about it. It was found within the deposits of a near-shore environment of an Early Oligocene age, which implies that it is the earliest sloth fossil ever found in the West Indies. The discovery provides evidence that sloths had been present in the Greater Antilles already from the Early Oligocene.

Jamaica

Anthony, after excavating in Puerto Rico, expected to find fossil sloths in Jamaica as well, and from November 1919 till March 1920, he conducted the first palaeontological investigation on this island together with Charles Falkenbach. Although Anthony failed to find any sloths – and so did everyone after him – he did discover fossils of rodents. To accommodate them, he described four new genera (Anthony, 1920), of which only one, *Clidomys*, is widely accepted. The other two (*Spiridontomys* and *Speoxenus*) are now considered synonyms of *Clidomys* and the same probably is valid for his *Alterodon*, which is based on a broken tooth only. He recognised two different *Clidomys* species, the smaller *Clidomys parvus* and the larger *Clidomys osborni*. *Spiridontomys jamaicensis* is now included in the former and *Speoxenus cundalli* and probably *Alterodon major* in the latter.

During this field trip, Anthony discovered some monkey postcranial material at Long Mile cave and New cave and a monkey mandible at Long Mile cave but did not describe them. The material remained instead stored for decades in the drawers of the American Museum of Natural History till it attracted the attention of two graduate students, Ernest Williams and Karl Koopman (1952). They described the mandible under the name of *Xenothrix mcgregori*, a new taxon of platyrrhine monkey. They also pointed out that several unusual mammalian postcranial elements in Anthony's Jamaican faunal collections possibly could be attributed to this monkey as well (later described by MacPhee and Fleagle, 1991). The Long Mile assemblage, however, is mixed; it includes material that is plausibly regarded as primate as well as other elements that are not.

In the 1990s, an enigmatic perissodactyl fossil was found in lagoon deposits at Seven Rivers in Jamaica consisting of a right mandible with teeth. Daryl Domning and colleagues (1997) described it as a species of the rhinocerotid *Hyrachyus*, comparable to the North American *Hyrachyus affinis*. The site is rich in fossils and yielded hundreds of bones of fishes, crocodilians, turtles, sirenians, bats, and possibly a primate. Domning (2001) described the walking sirenian from this site as *Pezosiren portelli*, based on a nearly complete skeleton and named it in honour of Roger W. Portell, the discoverer of the site. The Seven Rivers site is the oldest site discovered so far, with an age of late Early or early Middle Eocene, and is one of the very rare examples of a non-cave deposit in the West Indies.

Hispaniola

In 1921, two geologists discovered some fossil bones in two caves in the north-eastern part of the Dominican Republic, Hispaniola. They sent the fossils to Gerrit Miller at the Smithsonian Institution in Washington, who described them a year later as rodents and a ground sloth, the latter tentatively attributed to *Megalocnus*. In the years thereafter, he excavated fossils of insectivores, bats, rodents, and sloths at a locality near St Michael, Haiti, and additional materials in the area around the bay of Samaná, Dominican Republic. He published his results (Miller, 1929) and erected the new sloth genus and species *Parocnus serus* and the new sloth species, *Acratocnus comes*, now transferred to *Neocnus*. Miller also reported a tibia of a monkey, which he had found in an Indian kitchen midden at Samaná Bay. He thought it belonged to an extant monkey, but unable to tell which species, he assumed that it was an Old World monkey that was imported from Africa. Today, this tibia is considered the first finding of *Antillothrix bernensis*, an extinct platyrrhine monkey once endemic to Hispaniola, described as *Saimiri* by Rímoli (1977) and transferred to the new genus *Antillothrix* by Ross MacPhee and colleagues (1995).

Cayman Islands

Walter Guinness, Lord Moyne, collected fossils from a cave near Stake Bay, Cayman Brac in 1937. The following year, the entomologist C. Bernard Lewis, a member of the Oxford University Cayman Islands Biological Expedition, collected several skulls and mandibles of *Capromys* from a cave near Stake Bay. In 1965, Thomas H. Patton excavated two cave deposits in the Bluff of Cayman Brac. According to Gary S. Morgan and colleagues (2019), Cayman Brac was systematically explored between 1976 and 1986 by various researchers, and several sites were excavated. Exploration on Little Cayman led to the discovery of only two small fossil deposits. The fossil materials from the Cayman Islands are described by Morgan and colleagues (2019).

Peculiarities and Evolution of Endemic Mammals

A clear biozonation of the West Indies is lacking altogether due to the paucity of pre-Pleistocene fossils and the unclear relation between the various islands. Furthermore, faunal turnovers have not been detected so far, and evidence from Dominican amber indicates a long-term stasis in faunal composition. These amber deposits are generally dated to the Early Miocene (about 23–15 million years ago) and might contain representatives of extant Greater Antillean vertebrate groups such as perhaps an early hutia and a nesophontid insectivore, as shown by George Poinar and Roberta Poinar (1999). The presence of a small carnivore was suggested by them on the basis of fossil hair retrieved from the same Dominican amber, but was not further verified. In addition, a detailed microscopical study by Enrique Peñalver and David Grimaldi (2006) of other strains of hair in Dominican amber shows that the hairs belong to a small insectivore instead and resembles the hair of *Solenodon* most.

It appears that Tertiary vertebrates of the West Indies fit the same taxonomic pattern seen in Quaternary and extant vertebrates of the same region (overview by Borroto-Páez, 2012), with the exception of a single Eocene rhinocerotid finding from Jamaica.

The Rhino and the Sea Cow

The oldest fossil of a fully terrestrial mammal collected in the West Indies belongs to a juvenile specimen of the early rhinocerotid genus *Hyrachyus*. The fossil was found near Seven Rivers, parish of St James, as reported by Domning and colleagues (1997). The formation was attributed to the late Early or early Middle Eocene by Robinson (1988). The presence of this odd-toed ungulate in Jamaica can be explained as the result of sweepstake dispersal over water or, more likely, as the evidence of an Eocene land bridge connecting North America, the Mexican Arc, the Chortis Block, the Nicaragua Rise, and Jamaica. Whatever might be the origin of this rhinocerotid, it had no further impact in the history of the Antillean mammals because Jamaica submerged in the late Middle Eocene and with it any terrestrial form. This submergence persisted until the middle Late Miocene.

The ancient sirenian (sea cow) *Pezosiren portelli*, also from Seven Rivers, Jamaica, illustrates the evolutionary transition between terrestrial and aquatic life because it had four limbs perfectly adapted to walking but shared at the same time the typical skull, teeth, and horizontal tooth replacement with the fully aquatic, 'normal' sea cows. Its heavy, pachyosteosclerotic ribs indicate that it probably spent most of its time in the water, perhaps like a hippo, and its tail vertebrae indicate that it swam like an otter, not like modern sea cows and whales. *Pezosiren* had an estimated body length of 2.1 m, short limbs, barrel-shaped trunk, and a relatively short neck. *Pezosiren* was described by Richard Owen (1855) based on remains found between the parishes of St Elisabeth and Trelawney, Jamaica – some 40 km east–south-east of Seven Rivers – in earliest Middle Eocene beds. It is placed in the basal family Prorastomidae, together with the only other member *Prorastomus sirenoides* (Figure 19.8), also found in Jamaica. The phylogenetic relation between these two earliest members of the family is unresolved. Submergence of Jamaica had no fatal impact on this family as Late Eocene prorastomid fossils were found in Florida.

Figure 19.8 Skull of *Prorastomus sirenoides* (holotype). Scale bar: 5 cm. Natural History Museum, London. (Photograph: George Lyras.)

The Monkeys

The Greater Antillean primate fossil record is represented by platyrrhine monkeys (literally, flat-nosed monkeys), or New World monkeys. These are arboreal monkeys, living in forests, with a small to medium size. New World monkeys are all diurnal, except for the Central and South American owl monkeys (*Aotus*), so the Caribbean species likely were diurnal as well, especially in the absence of terrestrial predators. An easy diagnostic feature of New World monkeys is the retention of an upper second premolar, which is lost in the Old World monkeys and apes. The five Antillean species presently recognised are *Paralouatta marianae* (Figure 19.9) and *Paralouatta varonai* (Figure 19.10) from Cuba, *Xenothrix mcgregori* from Jamaica, and *Antillothrix bernensis* and *Insulacebus toussaintiana* from Hispaniola. They are attributed to a clade on their own, that of the Xenotrichini, endemic to the Greater Antilles, and considered most closely related to the titi monkeys (*Callicebus*), according to MacPhee and Inés Horovitz (2004). There are also a number of primate skeletal elements, which are not yet assigned to any species and, therefore, could represent additional taxa.

Initially, an endemic spider monkey (*Ateles anthropomorphus*) was reported from Cuba. The species is known only from the holotype, a few isolated teeth found in the cave Boca del Purial in the Cordillera del Escambray by Luis Montané in 1888 during an archaeological excavation. He passed the

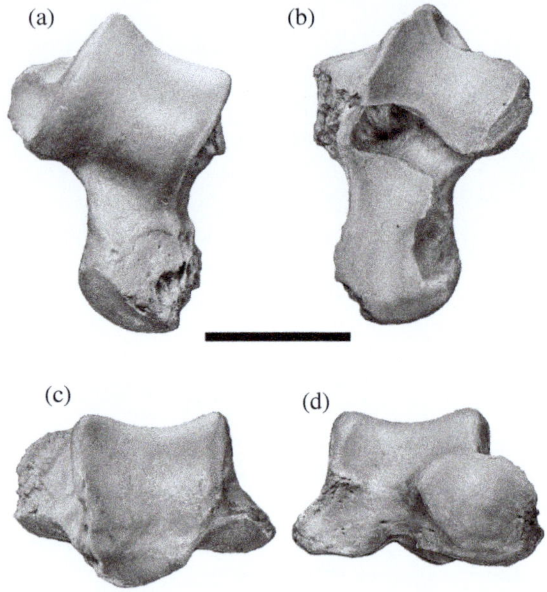

Figure 19.9 Astragal of *Paralouatta marianae* (holotype) from Domo de Zaza, Cuba. (a) Dorsal, (b) plantar, (c) proximal, and (d) distal views. Scale bar: 1 cm. (From MacPhee et al., 2003; Photograph: Ross MacPhee.)

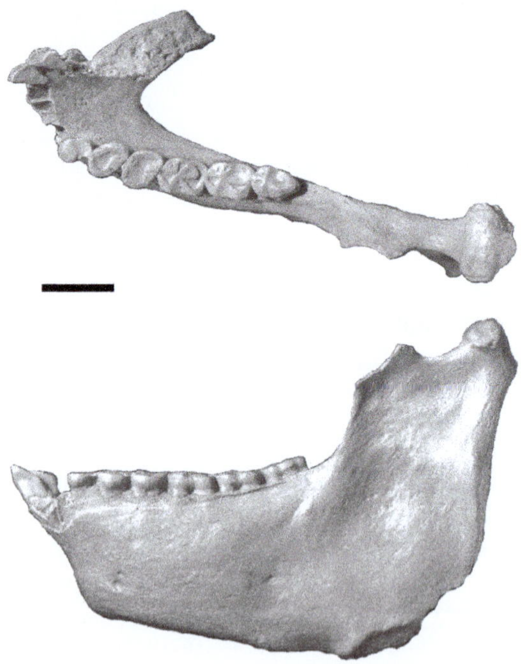

Figure 19.10 Lower jaw of the monkey *Paralouatta varonai* from Cuba. (a) Occlusal and (b) lateral views. Scale bar: 1 cm. (From MacPhee et al., 1995; Photograph: Ross MacPhee.)

teeth to Florentino Ameghino who in 1910 named the new species *Montaneia anthropomorpha*, the first native monkey from the Caribbean. Miller (1916) was of the opinion that the teeth were virtually indistinguishable from those of the extant brown-headed spider monkey *Ateles fusciceps* from Central and northern South America. Radiocarbon age determination by MacPhee and Manuel Rivero de la Calle (1996) confirmed a young age for the specimen – at most 300 years – and concluded that it represented nothing else but a recent import of a brown-headed spider monkey. Finally, from the Eocene Seven Rivers locality of Jamaica, an incomplete right petrosal bone of a primate of unknown affinity was reported by MacPhee and colleagues (1999). Later, MacPhee (2005) refers to this petrosal as belonging to yet unidentified mammal.

The species *Paralouatta marianae* is so far the oldest known primate from the West Indies. The single fossil retrieved is a talus of Early Miocene age. MacPhee and colleagues (2003) noted that it represents a form closely allied with or perhaps even ancestral to *Paralouatta varonai* of the Late Pleistocene. The body mass of *Paralouatta varonai* is estimated to be about 9–10 kg, which means that it is one of the largest known platyrrhine monkeys. The body mass of extant Platyrrhini ranges from 13.5 kg for the woolly spider monkey (*Brachyteles arachnoides*) to 125 g for the pygmy marmoset (*Callithrix* (= *Cebuella*) *pygmaeus*), as given by Karen Sears and colleagues (2008). The postcranial skeleton of *Paralouatta* shows a puzzling combination of anatomical features: it is principally a New World monkey with some skeletal features of the semi-terrestrial Old World cercopithecines like the mangabeys (*Cercopithecus*). This let MacPhee and Jeff Meldrum to suggest that it spend a significant time on the ground. At the same time, *Paralouatta* appears to have given up the suspensory activities so typical for New World monkeys. *Paralouatta* has a large face with well-developed supra-orbital tori and a low cranial vault. It differs from most other New World monkeys in possessing three-rooted third and fourth premolars. Rivero and Arredondo (1991) suggested that this might be a retained primitive feature, which sporadically occurs in a few living New World monkeys and in the Miocene *Stirtonia*. *Paralouatta* is unique in having exceptionally large eyes relative to its skull size. This is seen only in the night monkeys of the genus *Aotus* and seems to indicate a nocturnal lifestyle, which sounds bizarre for a monkey living in a predator-free forest, unless it took this evolutionary pathway to escape avian predators.

The Jamaican monkey, *Xenothrix mcgregori*, was a small, rather short-limbed, slow-moving arboreal quadruped with a body mass of an estimated 2–5 kg. It lost the last molar, in parallel with the marmosets, but it differed from the latter by a unique molar structure. It probably was a fruit-eater. *Xenothrix* fossils are found in latest Pleistocene to Late Holocene cave deposits, e.g. at the Long Mile cave near Windsor – the type locality, the Jacksons Bay caves, and New cave in the Portland Ridge, Clarendon. According to the radiometric dates reported by MacPhee (1984) on the Long Mile cave, *Xenothrix* must have been present on the north coast of Jamaica until late in the Holocene. Siobhán Cooke and colleagues (2017) calculated the last occurrence of *Xenothrix* to be about 900 years before present.

Antillothrix bernensis was initially described as a new species of the South American genus *Saimiri* by Rímoli (1977). MacPhee and colleagues (1995) erected the new genus *Antillothrix* to accommodate it; they considered it more closely related to the Cuban *Paralouatta* than to any mainland species. Its fossils are known from teeth, partial skulls, and a distal tibia from three localities in Hispaniola: Cueva de Berne (eastern Dominican Republic), Trou Woch Sa Wo (south-western Haiti), and Rio Naranjo Abajo (western end of Bahia de Samaná, Dominican Republic). The latter is the locality where Miller discovered the tibia of what he thought to be an imported Old World monkey, which MacPhee and colleagues have now attributed to *Antillothrix*. Richard Kay and colleagues (2011) reported a very well-preserved skull of *Antillothrix bernensis* found in the Padre Nuestro underwater cavern complex of Hispaniola. In 2012, another complete skull of this species was discovered in the same cavern complex, described by Lauren Halenar and colleagues (2017). The body mass of *Antillothrix* was tentatively estimated to 2–5 kg by MacPhee and Jeff Meldrum (2006) and thus falls in the middle range of body sizes found in extant New World monkeys. Its locomotor and postural behaviour is unresolved due to the paucity of postcranial remains. Tallman and Cooke (2016) reconstructed *Antillothrix* as an arboreal quadruped monkey. Cooke and colleagues (2011) described a fossil maxillary and mandibular fragment with almost complete dentition found in Haiti under the new species *Insulacebus toussaintiana*.

The undescribed primate material includes three enigmatic fossil femora, two from Jamaica (one from Coco Ree cave and one from Sheep Pen) and one from Hispaniola (Trou Woch Sa Wo). These specimens, although possessing primate-like features, differ from the known Antillean forms. Their specific attribution thus remains uncertain.

Earlier works considered that the Antillean primates were very recent derivatives of separate mainland groups of platyrrhines. Iturralde-Vinent and MacPhee – in accordance with their model of GAARlandia – suggested that New World monkeys entered the landmasses of the Caribbean at the Eocene to Oligocene transition, about 33–34 millions of years ago. Other researches – e.g. Liliana Dávalos (2004) – doubted such an early colonisation. However, the presence of *Paralouatta marianae* already during the Early Miocene in Cuba supports an early arrival. In addition, the anatomy of Antillean monkeys is very specialised, indicating long-term isolation.

The Sloths

Fossils of sloths have been found on several Greater Antilles islands (Table 19.1). Their earliest remains might date back to the Early Oligocene (see below), but the majority of the fossil remains is not older than the latest Pleistocene. The sloths underwent significant evolutionary radiation in the Greater Antilles. During the Pleistocene, the sloths were represented by a variety of species ranging from gigantic ground-dwellers (e.g. *Megalocnus*) (Figure 19.11) to small arborealists (e.g. *Neocnus*). Most Antillean sloths went extinct before the Holocene, and, today, there are no sloths living in the West Indies.

According to work on preserved ancient proteins by Samantha Presslee and colleagues (2019), the Antillean sloths all belong to the family Megalocnidae. There is a considerable disagreement among experts concerning the family content and interrelationships. As a result, there are various phylogenetic and taxonomic schemes published on the Antillean sloths, such as those of De Paula Couto (1967), Karl-Heinz Fischer (1971), Luis Varona (1974), Nestor Mayo (1980), and Jennifer White and MacPhee (2001). Some works place the Antillean sloths in two subfamilies, suggesting a diphyletic origin. White and MacPhee (2001) conducted a cladistic analysis using osteological and dental characters. They too came up with a diphyletic origin of the Antillean sloths. According to their analysis, the Antillean sloths can be divided into two subfamilies, the Megalocninae (giant ground sloths) and Choloepodinae (small tree sloths), both with a Quaternary distribution. However, the ancient proteins analysis by Presslee and colleagues (2019) demonstrated that diphyly is unlikely and that the Antillean sloths have a monophyletic origin. Therefore, they removed the Megalocninae from the Megalonychidae where they were traditionally placed and raised the subfamily to family level.

The genera *Megalocnus* and *Parocnus* are restricted to Cuba (*Megalocnus rodens*, *Parocnus browni*) and Hispaniola (*Megalocnus zile* and *Parocnus serus*). They are the only large-sized sloths of the West Indies. De Paula Couto (1979) estimated that *Megalocnus rodens* (Plate 34) probably weighed more than 270 kg. This might be an overestimation, as inferred from MacPhee's estimate of 150 kg for *Megalocnus rodens* and *M. zile*, as communicated to Burness and colleagues (2001). *Megalocnus* is characterised by its specialised maxillary teeth, which are pseudo-rodentiform or incisiform and are easily distinguished from the caniniform teeth of *Parocnus*. There

Table 19.1 List of the sloths of the West Indies and their synonyms as they were revised by White and MacPhee (2001). Two taxa, *Galerocnus jaimezi* (named by Arredondo and Rivero in 1997) and *Paramiocnus riveroi* (named by Arredondo and Arredondo, 2000), are referred to by White and MacPhee (2001) as Species B and C, because these two species are based on extremely scanty remains that they did not have the chance to study.

***Imagocnus* (MacPhee and Iturralde-Vinent,** 1994)
Imagocus zazae (MacPhee and Iturralde-Vinent, 1994); Early Miocene of Cuba
Gen. et sp. indet
Species A; Early Oligocene of Puerto Rico
Species B; Quaternary of Cuba
Species C; Quaternary of Cuba
Species D; Late Pliocene and/or Quaternary of Grenada

***Paulocnus* (Hooijer,** 1962)
Paulocnus petrifactus (Hooijer, 1962); Quaternary of Curacao

***Acratocnus* (Anthony,** 1916) **[=*Miocnus*; Matthew,** 1931; ***Habanocnus*; Mayo,** 1978.**]**
Acratocnus odontrigonus (Anthony, 1916) [=*Acratocnus major*, Anthony, 1918]; Quaternary of Puerto Rico
Acratocnus antillensis (Matthew, 1931) [=*Miocnus antillensis* Matthew, 1931 (introduced as nomen nudum by de la Torre and Matthew, 1915); *Habanocnus hoffstetteri* (Mayo, 1978); *Habanocnus paulacoutoi* (Mayo, 1978)]; Quaternary of Cuba
Acratocnus ye (MacPhee and colleagues, 2000); Quaternary of Hispaniola

***Neocnus* (Arredondo,** 1961) **[=Cubanocnus (Kretzoi,** 1968)**]**
Neocnus gliriformis (Matthew, 1931) [=*Microcnus gliriformis* (Matthew, 1931); *Cubanocnus gliriformis* (Kretzoi, 1968)]; Quaternary of Cuba
Neocnus major (Arredondo, 1961) [=*Neocnus minor* (Arredondo, 1961); *Neocnus baireiensis* (Mayo, 1980)]; Quaternary of Cuba
Neocnus comes (Miller, 1929) [='*Acratocnus* (?)' *comes* (Miller, 1929); *Synocnus comes* (De Paula Couto, 1967)]; Quaternary; Hispaniola
Neocnus dousman (MacPhee and colleagues, 2000); Quaternary of Hispaniola
Neocnus toupiti (MacPhee and colleagues, 2000); Quaternary of Hispaniola

***Megalocnus* (Leidy,** 1868) **[= *Myomorphus* (Pomel,** 1868); ***Neomesocnus* (Arredondo,** 1961)**]**
Megalocnus rodens (Leidy, 1868) [=*Myomorphus cubensis* (Pomel, 1868); *Megalocnus rodens rodens* (Leidy, 1868); *Megalocnus rodens casimbae* (Matthew, 1959 in Matthew and De Paula Couto, 1959); *Megalocnus ursulus* (Matthew, 1959 in Matthew and De Paula Couto, 1959); *Megalocnus junius* (Matthew, 1959 in Matthew and De Paula Couto, 1959); *Megalocnus intermedius* (Mayo, 1969); *Neomesocnus brevirrostris* (Arredondo, 1961)]; Quaternary of Cuba
Megalocnus zile (MacPhee and colleagues, 2000) [= *Megalocuus*? (lapsus calami) sp.? in parte, Miller, 1922]; Quaternary of Hispaniola

***Parocnus* (Miller,** 1929); **[='*Megalocuus*?' [lapsus calami], in parte Miller,** 1922; ***Mesocnus* (Matthew,** 1931)**]**
Parocnus serus (Miller, 1929) [='*Megalocuus*? [lapsus calami] sp?', in parte, Miller, 1922]; Quaternary of Hispaniola.
Parocnus browni (Matthew, 1931:2) [= *Mesocnus browni* (Matthew, 1931:2); *Mesocnus torrei* (Matthew, 1931); *Mesocnus herrerai* (Arredondo, 1977:2)]; Quaternary of Cuba.

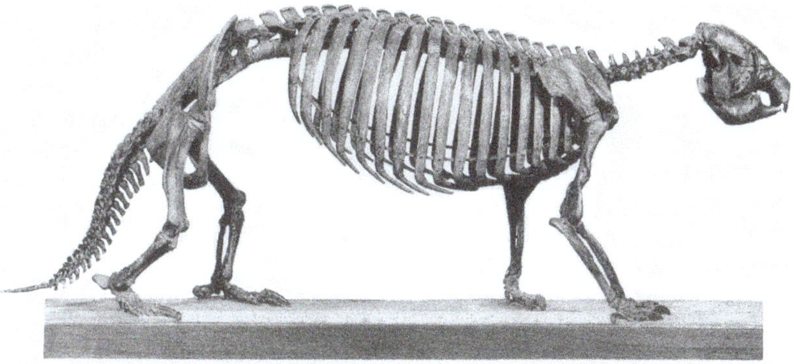

Figure 19.11 Composite skeleton of the giant Cuban ground sloth *Megalocnus rodens*. American Museum of Natural History, New York. Plate 26 from Matthew, W. D., Couto, C. P., and Simpson, G. G. (1959) The Cuban Edentates. (Bulletin of the AMNH, 117(1); New York: American Museum of Natural History.)

are some postcranial differences as well, e.g. the femoral and humeral head of *Megalocnus* are more convex and there are some differences in their scapular morphology. In general, the Antillean sloths present a large morphological variation, as actually many fossil insular species do. Because of this large intraspecific variation, a separate genus (*Neomesocnus*), three additional species, and two subspecies of *Megalocnus* were used in the past. Today, these names are considered junior synonyms of *Megalocnus rodens*. *Mesocnus* – a name originally used by De la Torre and Matthew (1915) – is now synonymous with *Parocnus*.

The genera *Neocnus* and *Acratocnus* are restricted to Cuba (*Neocnus gliriformis*, *Neocnus major*, and *Acratocnus antillensis*), Hispaniola (*Acratocnus ye*, possibly *Acratocnus simorhyncus*, *Neocnus comes*, *Neocnus dousman*, and *Neocnus toupiti*) (Figure 19.12), and Puerto Rico (*Acratocnus odontrigonus*).

The most outstanding features of the *Neocnus* sloth species are their small body sizes and skeletal indications of great arboreal agility. In addition, it seems that the jugal bones of *Neocnus* were highly reduced, something that might be an indication of weakened masticatory apparatus. Of all the *Neocnus* species, *Neocnus toupiti* is the smallest. Actually, it is one of the smallest, if not the smallest, sloths known to science. In the past, more *Neocnus* species were recognised, but Wood and MacPhee synonymised them into the present arrangement. It is important to note that the distinction between the various *Neocnus* species is not based on morphological traits which are associated with sexual dimorphism in other mammalian groups, and, therefore, the possibility of making the mistake of describing male and female individuals to different species can be ruled out.

Apart from the well-represented Quaternary taxa, there are some very rare sloth findings known from the Miocene and Oligocene. These findings, although limited, are of particular importance, since they provide important information on the natural history of Antillean sloths. The oldest Antillean

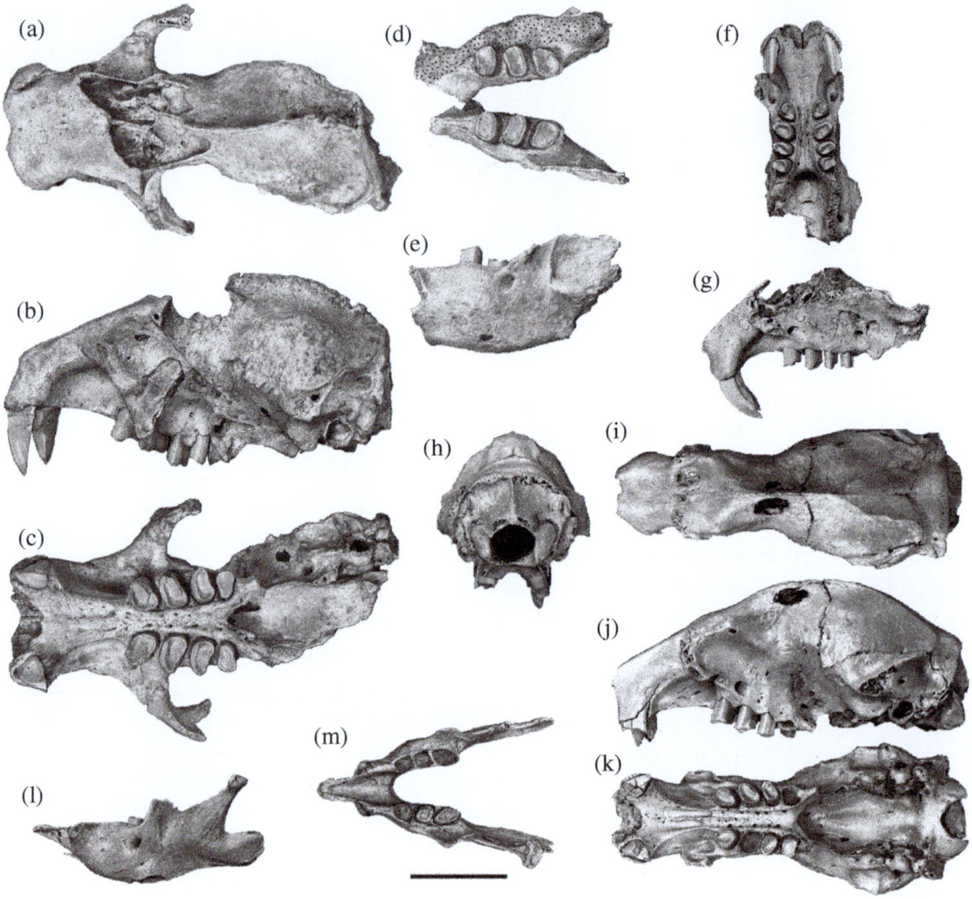

Figure 19.12 Skulls and jaws of sloths from the Quaternary of Hispaniola. (a) Dorsal, (b) lateral, and (c) ventral views of skull; (d) occlusal and (e) lateral views of mandible of *Acratocnus ye*, holotype individual from Trouing Vapè Deron, Haiti. Striped area is a calcite matrix that has not been removed. (f) Ventral and (g) lateral views of skull of the holotype partial skeleton of *Neocnus toupiti*, Trouing Jérémie 5, Haiti. Upper caniniform teeth of this specimen were partially dislodged after death into unnatural positions. (h) Occipital, (i) dorsal, (j) lateral, and (k) ventral views of holotype skull of *Neocnus dousman*, Trouing de la Scierie, Haiti. (l) Lateral and (m) occlusal views of referred mandible of *Neocnus dousman*. Scale bar: 3 cm. (From MacPhee et al., 2000; Photograph: Ross MacPhee.)

sloth specimen is known from Yauco, an Early Oligocene site in south-western Puerto Rico. It is a proximal femur fragment which MacPhee and Iturralde-Vinent attributed to a xenarthean – the order of sloths, anteaters, and armadillos –and tentatively to a megalonychid. This animal was small, as small as, or smaller than the living three-toed tree sloth *Bradypus*, which has a body weight of 2.25–6.20 kg. The next sloth on the geological time ladder is *Imagocnus zazae*. The species is based on a palate found in the Early Miocene site Domo de Zaza in Cuba (Figure 19.13). This palate differs from that of

Figure 19.13 Palate of *Imagocnus zazae* (holotype). (a) Ventral and (b) lateral views. Scale bar: 1 cm. T: torus on intermaxillary suture; P: palatine canal; F: fossa in advance of alveolus of first molariform. (From MacPhee and Iturralde-Vinent, 1994; Photograph: Ross MacPhee.)

the other Antillean sloths with the exclusion of *Acratocnus* and *Parocnus* in lacking large foramina at the height of the first molariform teeth. The palate differs from that of *Acratocnus* and *Parocnus* in other anatomical features. From the same site, a partial pelvis is known, which is larger than that of any known Antillean sloth, even larger than that of the large ground sloth *Megalocnus rodens*. Because the size of the pelvis does not match the size of the palate and due to the limited comparative material, the pelvis could not be attributed to a particular (new) species or genus.

The Rodents

The rodents of the Greater Antilles belong to two main groups: the caviomorph rodents and the sigmodontine murids. The diversity of Antillean caviomorph rodents is overwhelming (Table 19.2). As MacPhee and Clare Flemming (2003) noted, they are 'the most diversified of the non-volant mammal groups of the West Indies, whether this is measured by number of species, body size range, presumed ecological specialisations, or any other appropriate index'. In total, 59 species of Antillean caviomorphs are recognised (Table 19.2). Unfortunately, the majority of these species are extinct today. The earliest caviomorph species is *Borikenomys praecursor* of the late Early Oligocene of Puerto Rico, dated to about 29.5 million years ago. Laurent Marivaux and colleagues (2020) resolved its phylogenetic status as a chinchilloid. This find demonstrates an early dispersal of chinchilloids from South America to the West Indies, perhaps via the emerged Aves Ridge (c. 33–35 million years ago), or shortly after by rafting.

The many species can be grouped into three main groups. These are the hutias and coneys (Capromyidae), the Antillean spiny rats (Heteropsomyinae), and the Antillean platetooths

Table 19.2 Classification of the Antillean rodents and their geographic distribution, mainly based on Woods and colleagues (2001), as corrected by MacPhee (2009) and new discoveries.

Capromyinae		
Capromys gundlachianus	Cuba	Living
Capromys latus	Cuba	Extinct
Capromys pilorides	Cuba	Living
Capromys sp.	Cayman Islands	Extinct
Mesocapromys angelcabrerai	Off Cuba (Cayo Salinas in C. de Ana María)	Living
Mesocapromys auritus	Off Cuba (Cayo Fragoso)	Living
Mesocapromys barbouri	Cuba	Extinct
Mesocapromys beatrizae	Cuba	Extinct
Mesocapromys gracilis	Cuba	Extinct
Mesocapromys kraglievichi	Cuba	Extinct
Mesocapromys minimus	Cuba	Extinct
Mesocapromys nanus	Cuba, Isla de Pinos	Living
Mesocapromys sanfelipensis	Off Cuba (Cayo Juan Garcia)	Living
Mysateles garrilordi	Cuba	Living
Mesocapromys melanurus	Cuba	Living
Mysateles meridionalis	SW Isla de Pinos	Living
Mysateles prehensilis	Cuba	Living
Geocapromys columbianus	Cuba	Extinct
Geocapromys brownii	Jamaica	Living
Geocapromys pleistocenicus	Jamaica	Extinct
Geocapromys ingrahami	Bahamas	Living
Geocapromys thoracatus	Little Swan	Extinct
Geocapromys spp. (2 species)	Cayman Islands	Extinct
Plagiodontinae		
Plagiodontia aedium	Hispaniola	Living
Plagiodontia araeum	Hispaniola	Extinct
Plagiodontia ipnaeum	Hispaniola	Extinct
Rhizoplagiodontia lemkei	Hispaniola	Extinct
Isolobodontinae		
Isolobodon portoricensis	Hispaniola	Extinct
Isolobodon montanus	Hispaniola	Extinct
Zazamys veronicae	Cuba	Extinct
Hexolobodontinae		
Hexolobodon phenax	Hispaniola	Extinct
Hexolobodon sp.	Hispaniola	Extinct
Heptaxodontinae		
Quemisia gravis	Hispaniola	Extinct
Elasmodontomys obliquus	Puerto Rico	Extinct
Amblyrhiza inundata	Anguilla and St Martin	Extinct
Borikenomys praecursor	Puerto Rico	Extinct
Clidomys osborni	Jamaica	Extinct
Xaymaca fulvopulvis	Jamaica	Extinct
Heteropsomyinae		
Boromys offella	Cuba, Isla de Pinos, off Cuba	Extinct
Boromys torrei	Cuba, Isla de Pinos	Extinct

Table 19.2 (*Continued*)

Brotomys contractus	Hispaniola	Extinct
Brotomys voratus	Hispaniola	Extinct
Heteropsomys antillensis	Puerto Rico	Extinct
Heteropsomys insulans	Puerto Rico	Extinct
Puertoricomys corozalus	Puerto Rico	Extinct
Sigmodontinae		
Oryzomys antillarum	Jamaica	Extinct
Oryzomys curasoae	Curacao	Extinct
Oligoryzomys victus	St Vincent	Extinct
Megalomys desmarestii	Martinique	Extinct
Megalomys luciae	St Lucia	Extinct
Megalomys audreyae	Barbuda	Extinct
Megalomys georgina	Barbados	Extinct
'Megalomys' curazensis	Curacao	Extinct
Pennatomys nivalis	Nevis, St Eustatius and St Kitts	Extinct
Agathaeromys praeuniversialis	Bonaire	Extinct
Agathaeromys donovani	Bonaire	Extinct
Dushimys larsi	Curacao	Extinct
Hydrochoerinae		
Hydrochoerus gaylordi	Grenada	Extinct
Unassigned		
Tainotherium valei	Puerto Rico	Extinct

or giant hutias (Heptaxodontidae). The extant agoutis – e.g. *Dasyprocta mexicana* in Cuba – are all descendants of common agouti introduced by Amerindian settlers for the purpose of supplying food.

The family of the hutias and coneys (Capromyidae) includes nine genera (*Zazamys*, *Capromys*, *Mesocapromys*, *Mysateles*, *Geocapromys* (the only terrestrial mammal of the Bahamas; Box 19.2), *Plagiodontia*, *Rhizoplagiodontia*, *Isolobodon*, and *Hexolobon*). In total, 44 species are classified in this family, the majority of which are today extinct. All of them have an exclusively West Indian history, which means that must have evolved wholly within this region. The earliest known hutia is *Zazamys veronicae* (Figure 19.14) from Domo de Zaza, an Early Miocene site in Cuba (Figure 19.5). *Zazamys* resembles *Isolobodon*, a genus described from the Quaternary of Hispaniola, Puerto Rico, Mona Island, and several of the Virgin Islands, though not from Cuba. However, this poses no problem, as according to MacPhee and Iturralde-Vinent (1995), the two genera are at the base of the radiation of hutias. According to them, the caviomorph rodents reached the Antilles at the time of the existence of the GAARlandia land span, some 34–33 millions of years ago. Subsequently, they underwent a significant evolutionary radiation, which resulted in their Late Quaternary diversity. This hypothesis was further supported by the molecular analysis of Charles Woods and colleagues

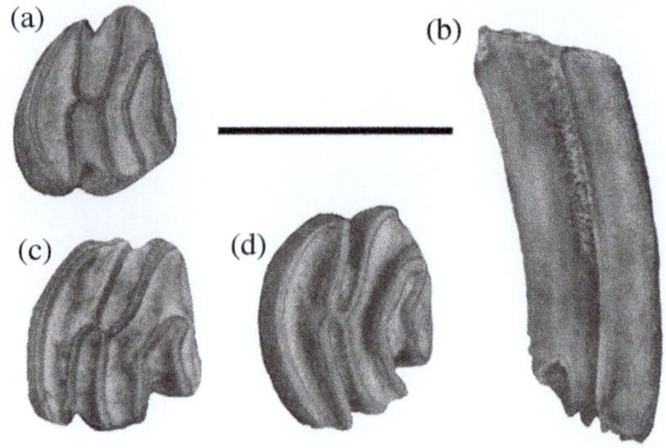

Figure 19.14 Molars of *Zazamys veronicae* from Domo de Zaza, Cuba. (a) Occlusal and (b) lateral views of left lower M3 (holotype), (c) right lower M1 or M2 (referred specimen), and (d) left lower M1 or M2 (referred specimen). Scale bar: 5 mm. (From MacPhee and Iturralde-Vinent, 1995; Photograph: Ross MacPhee.)

(2001). They show that the extant hutias of Hispaniola, which is situated at the centre of the West Indies and close to the now submerged Aves Ridge, are more basal than the rest in terms of adaptive radiation.

Isolobodon portoricensis (Figure 19.7) likely went extinct less than 500 years ago. Its year of extinction is estimated at 1525 on Mona Island by Angel M. Nieves-Rivera and Donald McFarlane (2001), following European settlement. Reports suggesting a survival to approximately 1800 were not confirmed.

The Antillean spiny rats (Heteropsomyinae) constitute a subfamily of the Echimyidae. Four Antillean genera are recognised: *Boromys* from Cuba and Isla de Pinos, *Brotomys* from Hispaniola, *Heteropsomys* (Figure 19.6), and *Puertoricomys* from Puerto Rico, which are all extinct. This subfamily does not have an exclusively West Indian distribution, but occurs also on the mainland of South America. It is widely agreed upon that heteropsomyins are closely related to capromyids. The Puerto Rico spiny rat (*Puertoricomys corozalus*), described by Woods (1989), may date back to the Pliocene, according to MacPhee and Wyss (1990), because it differs from the Late Quaternary cave rats (*Heteropsomys*).

The phylogeny of the Antillean platetooths or giant hutias is still unresolved. These animals have multilamellar teeth – hence their first common name – and some were truly large – hence their second common name. In the past, they were placed in the family Heptaxodontidae together with the Late Miocene South American mainland genera *Pentastylomys* and *Tetrastylomys*, but this arrangement was questioned by Clayton E. Ray (1965). Today, this family is considered as a paraphyletic assemblage consisting of *Elasmodontomys obliquus* (Puerto Rico; Figure 19.6), *Quemisia gravis* (Hispaniola), *Clidomys osborni*, *Clidomys parvus*, *Xaymaca fulvopulvis*, and perhaps *Alterodon major* (Jamaica) and *Amblyrhiza inundata* of the Lesser Antilles. Their exact phylogenetic position and interrelations are not yet solved, apart from the fact that they are all caviomorphs. Marivaux and colleagues (2020) resolved *Elasmodontomys* and *Amblyrhiza* as chinchilloids and placed it tentatively close to the Oligocene *Borikenomys* of Puerto Rico. *Alterodon*, based on a single broken cheek tooth, might be a synonym of *Clidomys osborni* as suggested by MacPhee, Woods and Morgan (1983). A fifth giant rodent from Jamaica, *Tainotherium valei*, has been described by Turvey and colleagues (2005) and might be related to the other Jamaican giant hutias. Unfortunately, *Tainotherium* is known only from an incomplete right femur, and, therefore, no secure phylogenetic scheme can be drawn. The anatomy of the femur indicates an arboreal way of life.

The giant hutias *Elasmodontomys*, *Quemisia*, and *Tainotherium* are giants indeed, at least for rodent standards. The body mass of *Elasmodontomys obliquus* is estimated at about 50 kg and that of *Tainotherium valei* (unnamed sp. A) at about 100 kg as communicated by MacPhee to Burness and colleagues (2001).

The single murid species of the Greater Antilles, the Antilles rice rat of Jamaica (*Oryzomys antillarum*), is part of the subfamily Sigmodontinae. The genus is widespread in North and South America.

BOX 19.2

The Bahamas

The Bahamas form an archipelago of more than 700 low limestone islands, clays, and islets. The islands lie on shallow carbonate banks separated by deep water. According to Storrs Olson and Gregory Pregill (1982), these banks define the extent and configuration of the palaeo-islands of the Bahamas during the last glaciation. The fossil mammals from the Bahamas consist of rodents and bats. The Bahaman hutia (*Geocapromys ingrahami*) is the only non-volant native mammal of the archipelago, where it still lives today. They were targeted and consumed as food by the indigenous people and translocated beyond their native range within the Bahamas to the Turks and Caicos islands. According to Michelle LeFebvre and colleagues (2019), the archaeological record of the Bahaman hutia extends to several islands of the Bahamas (Eleuthera, San Salvador, Samana Cays, and Crooked Island) and Caicos (Middle Caicos and Providenciales). Barbara Lawrence (1934) named the subspecies *Geocapromys ingrahami irrectus* on the basis of the archaeological material from Crooked Island and the subspecies *Geocapromys ingrahami abaconis* for a few specimens from Great Abaco. Remains of the Bahaman hutia recovered from Banana Hole on New Province island have been identified as the result of predation by the extinct giant Bahaman barn owl (*Tyto pollens*). Fossils of the Bahaman hutia were collected from Great Exuma and Little Exuma, and a single specimen from Sawmill Sink of Greater Abaco. Despite preserving only one hutia specimen, the latter deposit is the richest locality of the Bahamas and according to David Steadman and his colleagues (2015), more than 5,000 vertebrate fossils of bats, birds, reptiles, and fishes have been collected there.

According to Armand Rodríquez-Durán and Thomas Kunz (2001), locally extinct populations of Antillean ghost-faced bat (*Mormoops blainvillei*) are known from Abaco, Exuma, and New Providence. Remains of the common ghost-faced bat (*Mormoops megalophylla*) are known from Abaco and Andros and Parnell's moustached bat (*Pteronotus parnellii*) from Abaco and New Providence. Both species are locally extinct from those islands. Steadman and colleagues in 2015 also reported the presence of the now extirpated southeastern myotis (*Myotis austroriparius*) in Abaco.

The Insectivores

There are two genera of insectivores in West Indies, the still living solenodons (*Solenodon*) and the extinct so-called West Indian shrews (*Nesophontes*). The former genus is known only from Cuba (the living *Solenodon cubanus* and the fossil *Solenodon arredondoi*) and Hispaniola (the living *Solenodon paradoxus* and the fossil *Solenodon marcanoi*). The extinct Hispaniola solenodon is referred to *Antillogale* by Patterson (1962), but the generic validity of the latter was questioned by Van Valen (1967) and relegated to the level of subgenus by Varona (1974), though it was again resurrected by Alfred Roca and colleagues (2004) as possibly valid after all.

The genus *Nesophontes* is known from skulls and fragmentary skeletal material found in owl pellets from several Antillean islands. They have been described from Cuba (*Nesophontes micrus, Nesophontes longirostris, Nesophontes major, Nesophontessuperstes*, and *Nesophontes submicrus*) and its satellite Isla de Pinos (*Nesophontes micrus*), Hispaniola (*Nesophontes paramicrus, Nesophontes Hypomicrus*, and *Nesophontes zamicrus*), Puerto Rico and its satellite Isla de Vieques (*Nesophontes edithae*), and from Cayman Islands (*Nesophontes hemicingulus*). Although there have been suggestions that *Nesophontes* survived into the twentieth century, MacPhee and colleagues (1999) showed that the genus probably became extinct shortly after European arrival about 500 years ago.

The extant solenodons resemble large, robust shrews and have a body mass of about 1 kg. The extinct *Solenodon arredondoi* from Cuba was even larger, and its body mass has been estimated at 1.5–2 kg by Morgan and Ottenwalder (1993). The second extinct species, *Solenodon marcanoi* from Hispaniola, on the other hand, was smaller than the living species. A very typical feature of *Solenodon paradoxus* is the presence of a submaxillary venomous gland. The duct of this gland ends at the base of the second lower incisor, which is deeply grooved to guide the toxic saliva. Much work has been carried out on the origin of *Solenodon*. However, these studies – e.g. by Michael Stanhope and colleagues (1998), Robert Asher (1999), and Ginny Emerson and colleagues (1999) – present major conflicts in their conclusions. As a result, the answer to the bio-geographical origin of the Antillean insectivores remains elusive. The molecular study performed by Roca and colleagues (2004) suggests that solenodons are extremely ancient mammals as they formed a clade of their own since the Late Cretaceous. Already, Matthew (1918) had remarked that the West Indies insectivores apparently had arrived very early, probably in the early Tertiary.

The extinct *Nesophontes* probably ranged in size from about the size of a common house mouse to that of a brown rat, thus between about 10 and 125 g. The skull of the largest species, *Nesophontes edithae*, is 52 mm long in males. Possibly the Puerto Rican species attained its large size because of lack of competition, as there was no *Solenodon* on the island. A peculiar feature of *Nesophontes* is the lack of ossified auditory bullae, the jugal bones, and a zygomatic arch. The thorax was narrow instead of widened as in the solenodons. Its anatomy suggests a burrowing lifestyle, perhaps converging to that of moles.

Anthony (1916), who originally named the genus, considered *Nesophontes* to have 'no evident close relationships with any known genus'. However, based on the considerable osteological similarity between *Solenodon* and *Nesophontes*, Glover Allen (1918) and McDowell (1958) suggested that the two genera are closely related and on their turn to the Soricidae. Van Valen (1967), however, maintained the Nesophontidae as a family separate from the Solenodontidae and even placed each in a separate order, based on differences in molar cusp patterns. Terry Yates (1984) followed the earlier arrangement and thus considered them closely related. Selina Brace and her colleagues (2016) performed a phylogenetic analysis on the basis of ancient DNA evidence collected from an about 750-year-old

Nesophontes paramicrus specimen from Hispaniola and concluded that the closest living relatives of Nesophontidae are indeed the Solenodontidae. They further estimated the split between *Nesophontes* and *Solenodon* to have occurred at least 43.6 million years ago, to be more precise, between 43.6 and 68.6 million years ago. They further placed Nesophontidae and Solenodontidae in the newly erected suborder Solenodonota.

The West Indian insectivores constitute a problem for both the land span model and the rafting model. MacPhee and Grimaldi (1996) described a partial axial skeleton of a small mammal embedded in a piece of Late Oligocene–Early Miocene Dominican amber. According to them, it was possibly the fossil of an insectivore because it was very small and not a bat, and, therefore, this group could have used the Early Oligocene land span to colonise the Caribbean islands. This poses, however, a problem, because there are no insectivore fossils from South America prior to the establishment of the Panama Isthmus. An alternative hypothesis is that the Antillean insectivores originate from North American taxa instead. Evidence supporting this model came from the molecular studies carried out by Roca and colleagues (2004). They suggested that the West Indian insectivore *Solenodon* is extremely ancient and that its ancestors appeared in the Late Cretaceous. The phylogenetic analysis of Brace and colleagues (2016) confirms an early origin of Solenodonota. This happens to be the time that the proto-Antillean arc moved north-eastward relative to the North American mainland, reducing the distance between the proto-Antilles and North America. Whether the insectivores came over land or by rafting (or drifting islands, see MacFadden, 1980) is unclear, but in any case, if their dispersal to the proto-Antillean arc indeed took place soon after their appearance, they seem holdovers of a primitive stock and may represent vicariance.

The Bats

Some species of bats, known from the fossil record, are still living in the Greater Antilles (e.g. *Noctilio leporinus*, *Mormoops blainvillii*, *Phyllonycteris aphylla*, *Pteronotus macleayii*, and *Molossus molossus*). Five species are extinct (*Artibeus anthonyi*, *Mormoops magna*, *Phyllops vetus*, *Pteronotus pristinus* [Cuba], and *Tonatia saurophila* [Jamaica]), and two species are expatriated from the Greater Antilles, but continue to live elsewhere: locally extinct populations of *Mormoops megalophylla* are known from Cuba, Jamaica, and Hispaniola, and *Desmodus rotundus* fossils have been described from Cuba. All bat fossils have a Late Pleistocene or Holocene age.

The majority of bat species that went extinct or have undergone local extinction or extirpation are specialised or obligate cave-dwelling bats that roost in hot, humid chambers deep within large caves. It is possible that their extinction took place when many caves were flooded due to the sea level rise at the end of the Pleistocene. Vampire bats are unknown from the recent fauna of the West Indies, but fossils of the extant vampire, *Desmodus rotundus*, have been described from two Late Quaternary localities in Cuba. Karl Koopman (1989) suggested that the vampire bat fed on the blood of the small ground sloths inhabiting Cuba in the Late Quaternary and that the extinction of the sloths consequently led to their disappearance.

Liliane Dávalos (2007) explains the presence of short-faced frugivorous bats on the continent by dispersal from the Caribbean islands. In her view, the ancestor of all short-faced bats (Stenodermatina) reached the islands somewhere during the Miocene. After a long-term isolation, they underwent adaptive radiation on the islands. One of these new lineages dispersed back to the continent where it expanded its range and radiated into the four extant genera. Independently, similar re-colonisations of the continent are reported for aerial insectivores – natalids and probably *Mormoops* – and nectarivores (*Glossophaga* and *Leptonycteris*), according to her (Dávalos, 2005, 2006) for the former and according to Genoways and colleagues (2005) for the latter. Her main argument is that the continental short-faced bats share a most recent common ancestor, and not the Caribbean species.

References

Allen G.M. 1918. Fossil mammals from Cuba. *Bulletin of the Museum of Comparative Zoology*, 62 (4): 133–148.

Allen J.A. 1916. An extinct octodont from the island of Porto Rico, West Indies. *Annals of the New York Academy of Sciences*, 27: 17–22.

Ameghino F. 1910. *Montaneia anthropomorpha.* Un género de monos hoy extinguido de la isla de Cuba. Nota preliminar. *Anales de Museo Nacional de Buenos Aires* (Serie 3), 13: 317–318.

Anthony H.E. 1916. *New York Academy of Sciences Expedition, Porto Rico, 1916, 31 May to 30 July*. Unpublished typescript, Mammology Departmental Library and Archives, American Museum of Natural History.

Anthony H.E. 1917. New fossil rodents from Porto Rico. *Bulletin of the American Museum of Natural History*, 37: 185–186.

Anthony H.E. 1918. The indigenous land mammals of Porto Rico, living and extinct. *Memoirs of the American Museum of Natural History (New Series)*, 2 (2): 331–435.

Anthony H.E. 1919. Mammals collected in eastern Cuba in 1917, with descriptions of two new species. *Bulletin of the American Museum of Natural History*, 41: 625–643.

Anthony H.E. 1920. New mammals from Jamaica. *Bulletin of the American Museum of Natural History*, 42: 469–475.

Anthony H.E. 1926. Mammals of Porto Rico, living and extinct – Rodentia and Edentata. *Scientific Survey of Porto Rico and Virgin Island*, 9 (2): 97–238.

Arredondo C., Arredondo O. 2000. Nuevo género y especie de perezoso (Edentata: egalonychidae) del Pleistoceno de Cuba. *Revista Biología*, 14: 66–72.

Arredondo O. 1961. Descripciones preliminares de dos nuevos géneros y especies de edentados del Pleistoceno Cubano. *Boletin del Grupo de Exploraciones Científicas*, 1: 19–40.

Arredondo O. 1970. Dos nuevas especies subfosiles de mamiferos (Insectivora: Nesophontidae) del Holoceno Precolombino de Cuba. *Memoria de la Sociedad de Ciencias Naturales La Salle*, 30 (86): 122–152.

Arredondo O. 1977. Nueva especie de *Mesocnus* (Edentata: Megalonychidae) del Pleistoceno de Cuba. *Poeyana*, 172: 1–10.

Arredondo O., Rivero M. 1997. Nuevo género y especie de Megalonychidae del Cuaternario Cubano. *Revista Biología*, 11: 105–112.

Asher R.J. 1999. A morphological basis for assessing the phylogeny of the "Tenrecoidea" (Mammalia, Lipotyphla). *Cladistics*, 15: 231–252.

Borroto-Páez R. Woods C. A. Sergile F. E. (eds.). 2012. *Terrestrial Mammals of the West Indies*. Florida Museum of Natural History and Wacahoota Press, Gainesville, Florida.

Brace S., Thomas J.A., Dalén L., Burger J., MacPhee R.D.E., Barnes I., Turvey S.T. 2016. Evolutionary history of the Nesophontidae, the last unplaced recent mammal family. *Molecular Biology and Evolution*, 33 (12): 3095–3103.

Burness G.P., Diamond J., Flannery T. 2001. Dinosaurs, dragons, and dwarfs: the evolution of maximal body size. *Proceedings of the National Academy of Sciences of the USA*, 98: 14518–14523.

Cooke S.B., Mychajliw A.M., Southon J., MacPhee R.D.E. 2017. The extinction of *Xenothrix mcgregori*, Jamaica's last monkey. *Journal of Mammalogy*, 98 (4): 937–949.

Cooke S.B., Rosenberger A.L., Turvey S. 2011. An extinct monkey from Haiti and the origins of the Greater Antillean primates. *Proceedings of the National Academy of Sciences*, 108 (7): 2699–2704.

Dávalos L.M. 2004. Phylogeny and biogeography of Caribbean mammals. *Biological Journal of the Linnean Society*, 81 (3): 373–394.

Dávalos L.M. 2005. Molecular phylogeny of funnel-eared bats (Chiroptera: Natalidae), with notes on biogeography and conservation. *Molecular Phylogenetics and Evolution*, 37: 91–103.

Dávalos L.M. 2006. The geography of diversification in the mormoopids (Chiroptera: Mormoopidae). *Biological Journal of the Linnean Society*, 88: 101–118.

Dávalos L.M. 2007. Short-faced bats (Phyllostomidae: Stenodermatina): A Caribbean radiation of strict frugivores. *Journal of Biogeography*, 34 (2): 364–375.

de Castro M.F. 1864. De la existencia de grandes mamíferos fósiles en la Isla de Cuba. *Anales de la Real Academia de Ciencias Médicas, Físicas y Naturales de la Habana*, 1: 17–21, 54–60, 96–107. [Published as a pamphlet in 1865.]

de la Torre C., Matthew W.D. 1915. *Megalocnus* and other Cuban ground sloths. *Bulletin of the Geological Society of America*, 26: 152.

De Paula Couto C. 1967. Pleistocene edentates of the West Indies. *American Museum Novitates*, 2304: 1–55.

De Paula Couto C. 1979. *Tratado de paleomastozoologia*. Academia Brasileira de Ciencias, Rio de Janeiro.

Domning D.P. 2001. The earliest known fully quadrupedal sirenians. *Nature*, 413: 625–627.

Domning D.P., Emry R.J., Portell R.W., Donovan S.K., Schindler K.S. 1997. Oldest West Indian land mammal: Rhinocerotoid Ungulate from the Eocene of Jamaica. *Journal of Vertebrate Paleontology*, 17 (4): 638–641.

Emanuel K.A., Speer K., Rotunno R., Srivastava R., Molina M., 1995. Hypercanes: A possible link in global extinction scenarios. *Journal of Geophysics Research*, 100 (D7): 13755–13765.

Emerson G.L., Kilpatrick C.W., NcNiff B.E., Ottenwalder J.A., Allard M.W. 1999. Phylogenetic relationships of the order Insectivora based on complete 12S rRNA sequences from mitochondria. *Cladistics*, 15: 221–230.

Fischer K. 1971. Riesenfaultiere (Megalonychidae, Edentata, Mammalia) aus dem Pleistozän der Pio-Domingo-Höhle in Kuba. *Wissenschaftliche Zeitschrift der Humboldt-Universität zu Berlin Mathematisch-Naturwissenschaftliche Reihe*, 20 (4/5): 609–673.

Flemming C., MacPhee R.D.E. 1999. Redetermination of holotype of *Isolobodon portoricensis* (Rodentia, Capromyidae), with notes on recent mammalian extinctions in Puerto Rico. *American Museum Novitates*, 3278: 1–11.

Genoways H.H., Baker R.J., Bickham J.W., Philips C.J. 2005. Bats of Jamaica. *Special Publications of the Museum of Texas Technical University*, 48: 1–155.

Halenar L.B., Cooke S.B., Rosenberger A.L., Rímoli R. 2017. New cranium of the endemic Caribbean platyrrhine, *Antillothrix bernensis*, from La Altagracia Province, Dominican Republic. *Journal of Human Evolution*, 106: 133–153.

Hedges S.B. 2001. Biogeography of the West Indies: An Overview. In: Woods C.A., Sergile F.E. (eds), Biogeography of the West Indies. Patterns and Perspectives, 2nd edn. CRC Press, Boca Raton, Florida, pp. 15–33.

Hildebrand A.R., Boynton W.V. 1990. Proximal Cretaceous–Tertiary boundary impact deposits in the Caribbean. *Science*, 248: 843–847.

Hooijer D.A. 1962. A fossil ground sloth from Curaçao, Netherlands Antilles. *Proceedings of the Koninklijke Akademie van Wetenschappen (Series B)*, 65 (1): 46–60.

Iturralde-Vinent M.A., MacPhee R.D.E., 1999. Paleogeography of the Caribbean Region: implications for Cenozoic biogeography. *Bulletin of the American Museum of Natural History*, 238: 1–95.

Kay R.F., Hunt K.D., Beeker C.D., Conrad G.W., Johnson C.C., Keller J. 2011. Preliminary notes on a newly discovered skull of the extinct monkey *Antillothrix* from Hispaniola and the origin of the Greater Antillean monkeys. *Journal of Human Evolution*, 60 (1): 124–128.

Koopman K.F. 1989. A review and analysis of the bats of the West Indies. In: Woods C.A. (ed.), *Biogeography of the West Indies, Past, Present and Future.* Sandhill Crane Press, Gainesville, Florida, pp. 635–644.

Kretzoi M. 1968. New generic names for homonyms. *Vertebrata Hungarica*, 10 (1–2): 163–166.

Lawrence B. 1934. New *Geocapromys* from the Bahamas. *Occasional Papers of the Boston Society of Natural History*, 8: 189–196.

LeFebvre M.J., deFrance S.D., Kamenov G.D., Keegan W.F., Krigbaum J. 2019. The zooarchaeology and isotopic ecology of the Bahamian hutia (*Geocapromys ingrahami*): Evidence for pre-Columbian anthropogenic management. *PLoS ONE*, 14 (9): e0220284.

Leidy J. 1868. Notice of some vertebrate remains from the West Indian Islands. *Proceedings of the Academy of Natural Sciences of Philadelphia*, 20: 178–180.

MacFadden B. 1980. Rafting mammals or drifting islands? Biogeography of the Greater Antillean insectivores *Nesophontes* and *Solenodon. Journal of Biogeography*, 7: 11–22.

MacPhee R.D.E. 1984. Quaternary mammal localities and heptaxodontid rodents of Jamaica. *American Museum Novitates,* 2803:1–34.

MacPhee R.D.E. 2005. 'First' appearances in the Cenozoic land-mammal record of the Greater Antilles: significance and comparison with South American and Antarctic records. *Journal of Biogeography*, 32: 551–564.

MacPhee R.D.E. 2009. *Insulae infortunatae*: Establishing a chronology for Late Quaternary mammal extinctions in the West Indies. In: Haynes, G. (ed.), *American Megafaunal Extinctions at the End of the Pleistocene.* Springer Netherlands, Dordrecht, pp. 169–193.

MacPhee R.D.E., Fleagle J.G. 1991. Postcranial remains of *Xenothrix mcgregori* (Primates, Xenothrichidae) and other late Quaternary mammals from Long Mile Cave, Jamaica. *Bulletin of the American Museum of Natural History*, 206: 287–321.

MacPhee R.D.E., Flemming C. 2003. A possible heptaxodontine and other caviidan rodents from the Quaternary of Jamaica. *American Museum Novitates*, 3422: 1–42.

MacPhee R.D.E., Flemming C., Domning D.P., Portell R.W., Beatty B. 1999. Eocene? primate petrosal from Jamaica: morphology and biogeographical implications. *Journal of Vertebrate Paleontology*, 19 (supplement to 3): 61A.

MacPhee R.D.E., Grimaldi D.A. 1996. Mammal bones in Dominican amber. *Nature*, 380: 489–490.

MacPhee R.D.E., Horovitz I. 2004. New craniodental remains of the Quaternary Jamaican monkey *Xenothrix mcgregori* (Xenotrichini, Callicebinae, Pitheciidae), with a reconsideration of the *Aotus* hypothesis. *American Museum Novitates*, 3434: 1–51.

MacPhee R.D.E., Horovitz I., Arredondo O., Vasquez O.J. 1995. A new genus for the extinct Hispaniolan monkey *Saimiri bernensis* Rimoli,

1977, with notes on its systematic position. *American Museum Novitates*, 3134: 1–21.
MacPhee R.D.E., Iturralde-Vinent M. 1994. First Tertiary land mammal from Greater Antilles: An early Miocene sloth (Xenartha, Megalonychidae) from Cuba. *American Museum Novitates*, 3094: 1–13.
MacPhee R.D.E., Iturralde-Vinent M. 1995. Origin of the Greater Antillean land mammal fauna. 1. New Tertiary fossils from Cuba and Puerto Rico. *American Museum Novitates*, 3141: 1–31.
MacPhee R.D.E., Iturralde-Vinent M. 2005. The interpretation of Caribbean paleogeography: Reply to Hedges. *Monogafies de la Societat d Historia Natural de les Balears*, 12: 175–184.
MacPhee R.D.E., Iturralde-Vinent M., Gaffney E.S. 2003. Domo de Zaza, an early Miocene vertebrate locality in south-central Cuba: With notes on the tectonic evolution of Puerto Rico and the Mona Passage. *American Museum Novitates*, 3394: 1–42.
MacPhee R.D.E., Meldrum J. 2006. Postcranial remains of the extinct monkeys of the Greater Antilles, with evidence for semiterrestriality in *Paralouatta*. *American Museum Novitates*, 3516: 1–65.
MacPhee R.D.E., Rivero de la Calle M. 1996. Accelerator mass spectrometry 14C age determination for the alleged "Cuban spider monkey," *Ateles* (=*Montaneia*) *anthropomorphus*. *Journal of Human Evolution*, 30: 89–94.
MacPhee R.D.E., White J., Woods C.A. 2000. New megalonychid sloths (Phyllophaga, Xenarthra) from the Quaternary of Hispaniola. *American Museum Novitates*, 3303: 1–32.
MacPhee R.D.E., Woods C.A., Morgan G.S. 1983. The Pleistocene rodent *Alterodon major* and the mammalian biogeography of Jamaica. *Palaeontology*, 26: 831–837.
MacPhee R.D.E., Wyss A.R. 1990. Oligo-Miocene vertebrates from Puerto Rico, with a catalog of localities. *American Museum Novitates*, 2965: 1–45.
Marivaux L., Vélez-Juarbe J., Merzeraud G., Pujos F., Viñola López L.W., Boivin M., Santos-Mercado H., Cruz E.J., Grajales A., Padilla J., Vélez-Rosado K.I., Philippon M., Léticée J.-L., Münch P., Antoine P.O. 2020. Early Oligocene chinchilloid caviomorphs from Puerto Rico and the initial rodent colonization of the West Indies. *Proceedings of the Royal Society B*, 287: 20192806
Matthew W.D. 1918. Affinities and origin of the Antillean mammals. *Bulletin of the Geological Society of America*, 29: 657–666.
Matthew W.D. 1919. Recent discoveries of fossil vertebrates in the West Indies and their bearing on the origin of the Antillean fauna. *Proceedings of the American Philosophical Society*, 58: 161–181.
Matthew W.D. 1931. Genera and new species of ground sloths from the Pleistocene of Cuba. *American Museum Novitates*, 511: 1–5.
Matthew W.D., de Paula Couto C. 1959. The Cuban edentates. *Bulletin of the American Museum of Natural History*, 117: 1–56.
Maurrasse F.J.-M.R. 1991. Impacts, tsunamis and the Haitian Cretaceous-Tertiary boundary layer. *Science*, 252: 1690–1693.
Mayo N.A. 1969. Nueva especie de Megalonychidae y descripción de los depósitos cuaternarios de la Cueva del Vaho, Boca de Jaruco, La Habana. *Memorias de la Facultad de Ciencias de la Universidad de La Habana*, 3: 1–58.

Mayo N.A. 1978. Ein neues Genus und zwei neue Arten fossiler Riesenfaultiere von der Familie der Megalonychidae (Edentata: Mammalia) aus dem Pleistozän Kubas. *Eclogae Geologicae Helvetiae*, 71 (3): 687–697.

Mayo N.A. 1980. Nueva especie de *Neocnus* (Edentata: Megalonychidae de Cuba) y consideraciones sobre la evolución, edad, y paleoecología de las especies de este género. *Actas 2 Congreso Argentino de Paleontología y Bioestratigrafía y I Congreso Latinoamericano de Paleontología, Buenos Aires*, T. 3: 223–236.

McDowell S.B. 1958. The Greater Antillean insectivores. *Bulletin of the American Museum of Natural History*, 115 (3): 113–214.

Miller G.S. 1916. Bones of mammals from Indian sites in Cuba and Santo Domingo. *Smithsonian Miscellaneous Collection*, 66 (12): 1–11.

Miller G.S. 1922. Remains of mammals from caves in the Republic of Haiti. *Smithsonian Miscellaneous Collections*, 74 (3): 1–8.

Miller G.S. 1929. A second collection of mammals from caves near St. Michel, Haiti. *Smithsonian Miscellaneous Collections*, 81 (9): 1–30.

Miller K.G., Mountain G.S., the Leg 150 Shipboard Party, and Members of the New Jersey Coastal Plain Project. 1996. Drilling and dating New Jersey Oligocene-Miocene sequences: Ice-volume, global sea level, and Exxon records. *Science*, 271: 1092–1095.

Morgan G.S. Ottenwalder J.A., 1993. A new extinct species of *Solenodon* (Mammalia: Insectivora: Solenodontidae) from the Late Quaternary of Cuba. *Annals of the Carnegie Museum*, 62: 151–164.

Morgan G.S., MacPhee R.D.E., Woods R. Turvey S. 2019. Late Quaternary fossil mammals from the Cayman Islands, West Indies. *Bulletin of the American Museum of Natural History*, 428: 1–82.

Morgan G.S., Ray C.E., Arrendondo O. 1980. A giant insectivore from Cuba (Mammalia: Insectivora: Solenodontidae). *Proceedings of the Biological Society of Washington*, 93: 597–608.

Nieves-Rivera A., McFarlane D.A. 2001. In search of the extinct hutia in cave deposits of Isla de Mona. *National Speleological Society News*, 59 (4): 92–95.

Olson S.L., Pregill G.K. 1982. Fossil vertebrates from the Bahamas: Introduction to the palaeontology of Bahaman vertebrates. *Smithsonian Contributions to Paleobiology*, 48: 1–21.

Owen R. 1855. On the fossil skull of a mammal (*Prorastomus sirenoides*, Owen), from the island of Jamaica. *Quarterly Journal of the Geological Society*, 11: 541–543.

Patterson B. 1962. An extinct solenodontid insectivore from Hispaniola. *Breviora* 165: 1–11.

Peñalver E., Grimaldi D. 2006. Assemblages of mammalian hair and blood-feeding midges (Insecta: Diptera: Psychodidae: Phlebotominae) in Miocene amber. *Transactions of the Royal Society of Edinburgh (Earth Sciences)*, 96: 177–195.

Poinar G.O. Jr., Poinar R. 1999. *The Amber Forest: Reconstruction of a Vanished World*. Princeton University Press, Princeton.

Pomel A. 1868. Sur le *Myomorphus cubensis*, sous-genre nouveau du *Megalonyx*. *Comptes Rendus de l'Académie des Sciences de Paris*, 67: 665–668, 850.

Presslee S., Slater G.J., Pujos F., Forasiepi A.M., Fischer R., Molloy K., Mackie M., Olsen J.V., Kramarz A., Taglioretti M., Scaglia F., Lezcano, M., Lanata J.L., Southon J., Feranec R., Bloch J., Hajduk A., Martin F.M., Gismondi R.S., Reguero M., de Muizon C., Greenwood A., Chait B.T., Penkman K., Collins M., MacPhee R.D.E. 2019. Palaeoproteomics resolves sloth relationships. *Nature Ecology and Evolution*, 3 (7): 1121–1130.

Ray C.E. 1965. The relationships of *Quemisia gravis* (Rodentia: ?Heptaxodontidae). *Smithsonian Miscellaneous Collections*, 149(3): 1–12.

Rímoli R.O. 1977. Nuevas citas para mamiferos precolombinos en la Hispaniola. *Universidad Autonoma de Santo Domingo, Cuadernos del CENDIA*, 259 (5): 1–15.

Rivero M., Arredondo O. 1991. *Paralouatta varonai*, a new Quaternary platyrrhine from Cuba. *Journal of Human Evolution* 21: 1–11.

Robinson E. 1988. Late Cretaceous and early Tertiary sedimentary rocks of the Central Inlier, Jamaica. *Journal of the Geological Society of Jamaica*, 24: 49–67.

Roca A.L., Bar-Gal G.K., Eizirik E., Helgen K.M., Maria R., Springer M.S., O'Brien S.J., Murphy W.J. 2004. Mesozoic origin for West Indian insectivores. *Nature*, 429: 649–651.

Rodríquez-Durán A., Kunz T.H. 2001. Biogeography of West Indian bats: an ecological perspective. In: Woods C.A., Sergile F.E. (eds), *Biogeography of the West Indies: patterns and perspectives*. CRC Press, Boca Raton, pp. 333–368.

Rosen D.E. 1975. A vicariance model of Caribbean biogeography. *Systematic Zoology*, 24: 431–464.

Sears K.E., Finarelli J.A., Flynn J.J., Wyss A.R. 2008. Estimating body mass in New World "monkeys" (Platyrrhini, Primates), with a consideration of the Miocene platyrrhine, *Chilecebus carrascoensis*. *American Museum Novitates*, 3617: 1–29.

Stanhope M.J., Waddell V.G., Madsen O., de Jong W.W., Hedges S.B., Cleven G.C., Kao D., Springer M.S. 1998. Molecular evidence for multiple origins of Insectivora and for a new order of endemic African mammals. *Proceedings National Academy of Sciences of the United States of America*, 95: 9967–9972.

Steadman D.W., Albury N.A., Kakuk B., Mead J.I., Soto-Centeno J.A., Singleton H.M., Franklin J. 2015. Vertebrate community on an ice-age Caribbean island. *Proceedings of the National Academy of Sciences of the United States of America*, 112 (44): E5963–E5971.

Tallman M., Cooke S.B. 2016. New endemic platyrrhine humerus from Haiti and the evolution of the Greater Antillean platyrrhines. *Journal of Human Evolution*, 91: 144–166.

Turvey S.T., Grady F.V., Rye P. 2005. A new genus and species of "giant hutia" (*Tainotherium valei*) from the Quaternary of Puerto Rico: An extinct arboreal quadruped? *Journal of Zoology*, 270 (4): 585–594.

Van Valen L. 1967. New Paleocene insectivores and insectivore classification. *Bulletin of the American Museum of Natural History*, 135 (5): 217–284.

Varona L.S. 1974. *Catálogo de los mamíferos vivientes y extinguidos de las Antillas*. Editorial de la academia de ciencias de Cuba, Habana.

Vaughan T.W., 1902. Notes on Cuban fossil mammals. *Science*, 15: 148–149.

White J.L., MacPhee R.D.E. 2001. The sloths of the West Indies: A systematic and phylogenetic review. In: Woods, C.A., Sergile F.G. (eds), *Biogeography of the West Indies: Patterns and Perspectives*, 2nd edition. CRC Press, Boca Raton, London, New York, Washington, D.C., pp. 201–234.

Williams E.E., Koopman K.F. 1952. West Indian Fossil Monkeys. *American Museum Novitates*, 1546: 1–16.

Woods C.A. 1989. The biogeography of West Indian rodents. In: Woods C.A. (ed.), *Biogeography of the West Indies: Past, Present, and Future*. Sandhill Crane Press, Gainesville, Florida, pp. 741–798.

Woods C.A., Borroto Paez R., Kilpatrick C.W. 2001. Insular Patterns and Radiations of West Indian Rodents. In: Woods C.A., Sergile F.G. (eds), *Biogeography of the West Indies: Patterns and Perspectives*, 2nd edn. CRC Press, Boca Raton, London, New York, Washington, D.C., pp. 335–368.

Yates T.L. 1984. Insectivores, elephant shrews, tree shrews, and dermopterans. In: Anderson S., Jones J.K. (eds), *Orders and Families of Recent Mammals of the World*. John Wiley, New York, pp. 117–144.

CHAPTER TWENTY

The Lesser Antilles

Evolution of Island Mammals: Adaptation and Extinction of Placental Mammals on Islands, Second Edition.
Alexandra van der Geer, George Lyras and John de Vos.
© 2021 John Wiley & Sons Ltd. Published 2021 by John Wiley & Sons Ltd.

The Lesser Antilles are known for their Quaternary remains of endemic hutias, rice rats, capybaras, sloths, and bats. The earliest fossils of terrestrial mammals date back to the Pleistocene. The present mammalian fauna is only a poor representation of its extinct Late Quaternary fauna. Many of them went extinct after the European arrival in the region.

Geology and Palaeogeography

The Lesser Antilles, including the Leeward Antilles, form part of the West Indies and constitute a vast archipelago in the Caribbean Sea (Figure 20.1). For the late Cretaceous (c. 100–70 million years ago) history, see the Greater Antilles. According to Robert Allen and colleagues (2019), the history of Eastern Caribbean is characterised by the presence of three separate subduction systems that have been active at different times (Figure 19.2). The oldest is a volcanic system that was active during the Cretaceous-Paleogene. The Aves Ridge, a mostly underwater ridge east of Lesser Antilles, is a remaining expression of that system (see the Greater Antilles for the importance of this ridge to mammalian dispersals). The second system was active during the Eocene and Oligocene. Arc magmatism from 40 Ma is found on the islands of the Limestone Caribbees. The Limestone Caribbees became extinct in the Early Miocene, and the volcanic arc migrated to the modern Lesser Antilles Arc.

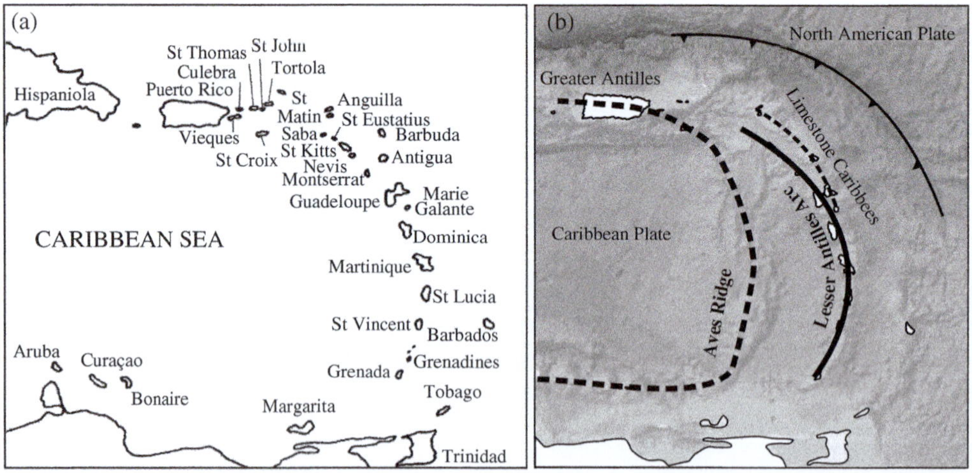

Figure 20.1 Map of the Lesser Antilles. (a) The Lesser Antilles consist of the Leeward and Windward Islands (the many islands extending from the Virgin Islands in the north to Grenada and the Grenadines in the south) and the Leeward Antilles (the Dutch ABC islands and a group of Venezuelan islands). (b) Location of key topographic and tectonic features of the Lesser Antilles.

Historical Palaeontology

The first fossil discovery in the Lesser Antilles dates back to 1868. That year, a quantity of earth, breccia, and limestone fragments collected at a cave, possibly Cavannagh Cave, on Anguilla was brought to the port of Philadelphia. The material had been shipped to Henry Waters & Brothers, manufacturers of phosphatic fertilisers, and was meant to be used as land fertiliser. Henry Waters, however, noticed the presence of bones in it and informed the American palaeontologist Edward Drinker Cope. Cope examined the material, verified the presence of fossils, and collected some teeth and bone fragments. Hendrik van Rijgersma, a Dutch physician who was also a keen amateur naturalist, visited the locality and collected more teeth and bones for Cope. In a meeting at the Philadelphia Academy of Sciences, Cope (1868) announced the discovery of the fossiliferous site and showed the audience the fossils of a previously unknown rodent of phenomenal size, which he named *Amblyrhiza inundata* (Figure 20.2, Plate 35). In the three following years, he announced three more species – *latidens, longidens,* and *quadrans* – for which Cope (1869) erected the new genus *Loxomylus*. Cope's three additional species have been regarded junior synonyms of *inundata* since the study of Antje Schreuder (1933). Cope (1883) finally published an extensive description of the skull and postcranial of his new, enigmatic genus after a long delay:

The remains [of *Amblyrhiza*] were first obtained in 1868, and brief notices of them have been made at various times since, but the publication of the full account was delayed in the hope that other objects might be added to the collection. The death of the gentleman who procured the specimens [i.e., Hendrik

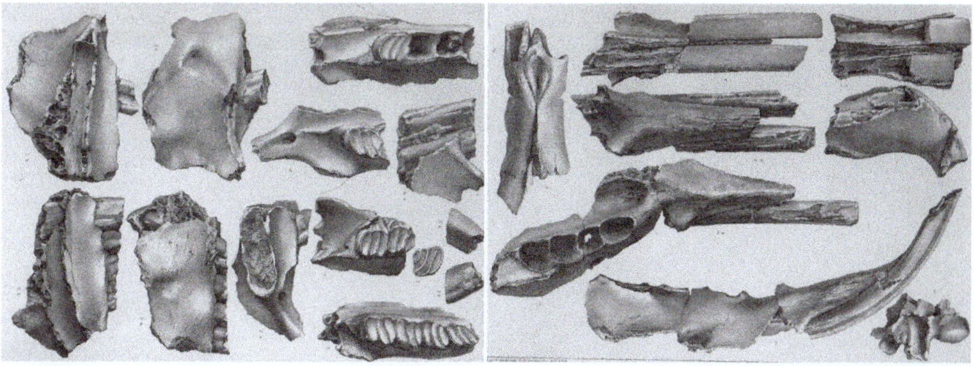

Figure 20.2 First drawing of skull fragments of *Amblyrhiza*, amongst others, the nasals and upper and lower jaw; various views. (From Cope, 1868.)

van Rijgersma], and other causes having shown that no further exploration was practicable, the memoir was prepared and submitted to the Smithsonian Institution, for publication, in 1878. The other works in progress prevented the publication until the present time, but the interval has been taken advantage of by the author to revise the manuscript and superintend the preparation of the plates. (Cope, 1883: p. iii by Spencer Baird, secretary of the Smithsonian Institution.)

Cope realised in this work that his two genera were identical and kept *Amblyrhiza* and reduced *Loxomylus* to junior synonym. He could not explain the occurrence of this gigantic rodent on such a remote island, and thus presumed the existence of a now submerged or 'inundated' land bridge to the mainland, hence the specific name *inundata*. Geological evidence indeed suggests that during the Late Pleistocene, the Lesser Antilles formed a larger island, mainly due to a low sea level (Figure 20.3).

The earliest discoveries of fossil mammals from the Netherlands Antilles were a few capybara teeth from Curacao that were first reported and described by Hooijer (1959) as *Hydrochoerus hydrochaeris*. Hooijer could not believe that such a water-loving animal was able to survive on this arid island and, thus, must have been introduced by early settlers. The second fossil species is a West Indian giant rice rat, which he described in the same paper as *Megalomys curazensis*. Among the material, Hooijer (1962) also discovered a small ground sloth,

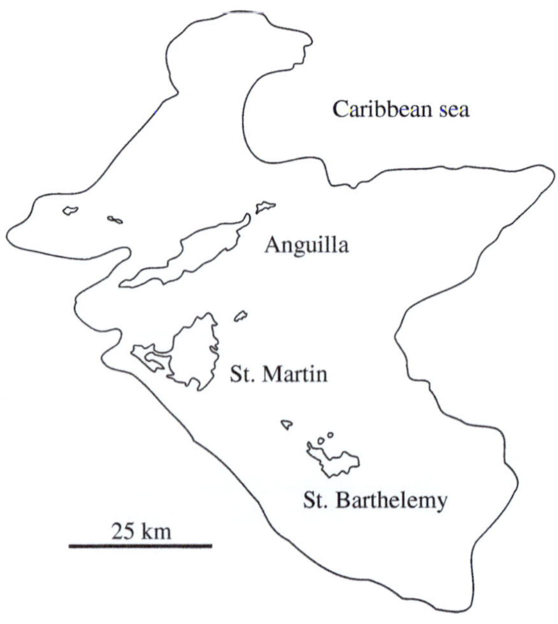

Figure 20.3 Map of the Anguilla Bank. The isobath is drawn at 100 m, which is approximately the sea level drop at 20,000 years before present. (Redrawn from McFarlane and colleagues, 1998.)

which he named *Paulocnus petrifactus*, based on a partial skull and some limb bones. The bone-bearing deposit from which it came was discovered in 1961 by Paul Stuiver at the Tafelberg Santa Barbara, eastern Curacao. Paul de Buisonjé treated the blocks with acetic acid, after which a large skull portion with one tooth *in situ*, a rostrum fragment, and several postcranial materials appeared, all belonging to one individual (Plate 36).

In 1980 and 1983, David Steadman and his collaborators excavated late Quaternary sediments from a fissure filling (Burma Quarry) in Northern Antigua. They found two, still unnamed, species of rice rats, as they reported in 1984.

Joseph Gaylord discovered capybara fossils in a Plio–Pleistocene lahar deposit near 12 Degrees North Hotel in Grenada in 1982. These specimens were described by MacPhee and colleagues (2000) as *Hydrochaeris* (=*Hydrochoerus*) *gaylordi* in honour of its discoverer. The age of the specimens was dated as possibly older than 2 million years, clearly predating any human influence. Hooijer's earlier conclusion that the capybara on Curacao had been brought there by humans could thus be dismissed. Some further undetermined sloth teeth are reported from the same locality as well. Their size fits well with the sloth from Curacao (*Paulocnus petrifactus*), but there are too few distinctive features preserved to draw any conclusion.

Peculiarities and Evolution of Endemic Mammals

The Lesser Antilles lack a clear biozonation, as is also the case for the Greater Antilles. The known fossils are all Pleistocene or early Holocene. The Old World monkey living today in the Lesser Antilles is the mona monkey (*Cercopithecus mona*), which was introduced to Grenada in the 1600s and is not endemic to the region.

The Sloths

Fossils of sloths, belonging to the family Megalocnidae (Table 19.1), have been found only on Curacao and Grenada in Pleistocene deposits. The sloth from Curacao (*Paulocnus petrifactus*) is tentatively assigned to the family of fossil sloths of the Greater Antilles. The sloth from Curacao (Plate 36) is known only by the type specimen, being a skull, and the original hypodigm material, consisting of some postcranial elements. The relationship of *Paulocnus* within the Megalocnidae is not yet fully understood, likely due to the limited and badly preserved available material. As a result, it is not clear whether this genus is part of the West Indian sloth radiation, or an overseas immigrant from South America as Curacao lies at a reasonable distance from the Venezuelan coast. *Paulocnus* is larger than *Acratocnus* and *Neocnus* of the Greater Antilles. Although Hooijer, who described the species, claimed that it is morphologically different from the other Antillean sloths, no re-evaluation of the material has been done so far. According to White and MacPhee, it seems to have some similarities with *Acratocnus*.

The sloth teeth from a Plio-Pleistocene lahar deposit in Grenada remain undetermined but have the same size as those of the species from Curacao.

Giant Hutias and Capybaras

The Lesser Antilles have a much less diverse range of caviomorph species with representatives of only two main groups: the Antillean platetooths or giant hutias (Heptaxodontidae; Table 19.2) and the capybaras (*Hydrochoerus*). The extant agoutis – e.g. *Dasyprocta leporina* in the Lesser Antilles and *Dasyprocta punctata* in Cayman Islands – of the various Lesser Antilles are all descendants of common agouti introduced by Amerindian settlers as food supplies. The agoutis are not to be considered endemic to the region, though several subspecies of *Dasyprocta leporina* have been mentioned in the literature, such as *Dasyprocta leporina antillensis* on St Lucia and Martinique. At present, these subspecies are no longer widely accepted.

Figure 20.4 Isolated incisor (top; lateral view) and maxillary fragment with three molars (bottom; occlusal view) of *Amblyrhiza inundata*. Scale bar: 1 cm. National Museum of Natural History, Leiden. (Photograph: Eelco Kruidenier.)

The giant hutia (*Amblyrhiza inundata*; Plate 35, Figure 20.4) was originally placed in the same caviomorph family Heptaxodontidae as the Greater Antilles giant hutias *Elasmodontomys* (Puerto Rico), *Quemisia* (Hispaniola), *Clidomys*, *Xaymaca*, and perhaps *Alterodon* (Jamaica). Today, however, the family is considered as a paraphyletic assemblage and their exact phylogenetic interrelations are not yet solved. *Amblyrhiza inundata* is the largest rodent that ever lived on an island. Just how large has been a matter of speculation for more than a century, till Audrone Biknevicius and colleagues (1993) calculated on the basis of limb bones that its body mass probably ranged between slightly less than 50 kg to over 200 kg – depending on method of analysis and specimen used.

The capybaras are represented by a genus of the extant subfamily Hydrochoerini of the Caviidae. Capybara fossils are known only from two islands, Curacao (*Hydrochoerus hydrochaeris*) and Grenada (*Hydrochoerus gaylordi*; Figure 20.5). On the contrary to the other caviomorphs mentioned above, the capybaras may have reached the West Indian islands via overseas dispersal. The distance between these islands and the mainland is relatively short, so such dispersal could take place. In addition, extant capybaras are excellent swimmers.

(Giant) Rice Rats

Rice rats, according to Turvey and colleagues (2010), probably occurred on all of the Lesser Antillean islands from Grenada to Anguilla, and the lack of records from a few islands in this chain represents a sampling artefact. Six genera of rice rats have been described so far, *Oryzomys*, *Oligoryzomys*, *Megalomys*, *Pennatomys*, *Agathaeromys*, and *Dushimys*. Furthermore, on several islands (Gouadeloupe, La Desirade, Marie Galante, Montserrat, Sab, St Martin, Grenada, Antigua, and Anguilla), there are remains of still undescribed taxa. *Oryzomys* is a widespread genus in North and South America but is restricted to Jamaica (*Oryzomys antillarum*) and Curacao (*Oryzomys curasoae*) in the West Indies. The St Vincent rice rat is attributed to *Oligoryzomys victus*. The genus *Megalomys* on the other hand is endemic to the Lesser Antilles and once occurred on Martinique (*Megalomys desmarestii*), St Lucia (*Megalomys luciae*), Barbuda (*Megalomys audreyae* and *Megalomys* sp.), and Barbados (*Megalomys georgina*). *Pennatomys nivalis* is known from Nevis, St Eustatius, and St Kitts, which formed a single larger island during the Pleistocene low sea-level stands. The body mass of the Barbuda rice rat has been estimated at 0.79 kg by Burness and colleagues (2001). The genus *Megalomys* is also reported from Curacao (*M. curazensis*).

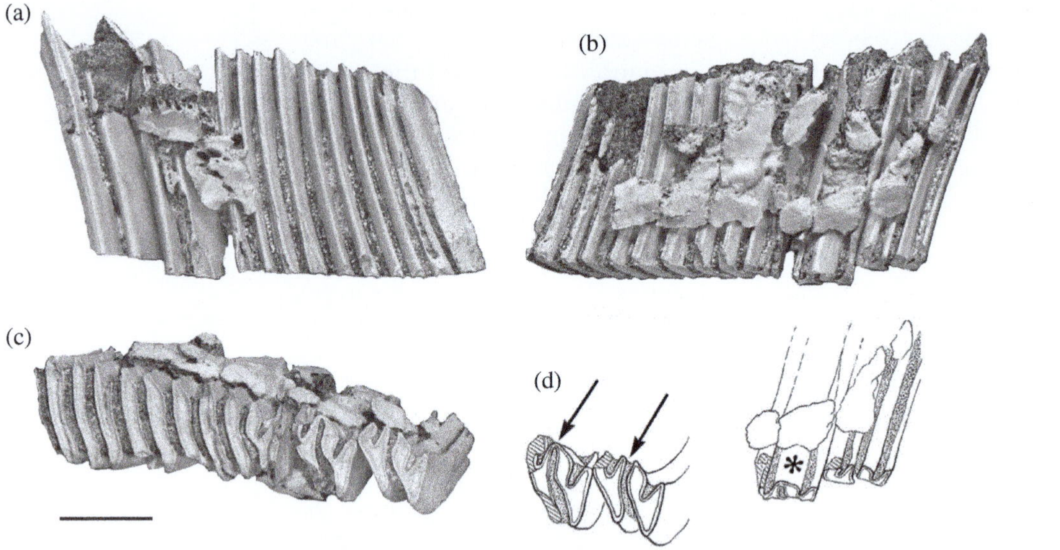

Figure 20.5 Partial maxilla retaining M1–M3 of the Grenada capybara (*Hydrochoerus gaylordi*), holotype. (a) Lingual, (b) buccal, and (c) occlusal views; (d) drawing of occlusal and buccal aspects of M1–M2. The asterisk on M2 identifies buccally conjoined lamellae. The left arrow indicates the buccally conjoined lamellae in M2, which are absent in M1 (right arrow). On the contrary to *Hydrochoerus gaylordi*, in extant capybaras, both the M1 and the M2 consist of two independent lamellae. Scale bar: 1 cm. From MacPhee and colleagues, 1998. (Photograph: Ross MacPhee.)

However, Turvey and colleagues (2010, 2012) suggested that it may not be related to *Megalomys* in the strict sense. Two species of *Agathaeromys*, *Agathaeromys praeuniversitatis* and *Agathaeromys donovani,* were described by Jelle Zijlstra and colleagues (2010). The genus is known only from Bonaire. Its age is probably Pleistocene, approximately 900,000–230,000 years ago. The genus *Dushimys*, named by Zijlstra (2012), was endemic to the Middle Pleistocene of Curacao. Only one species is known thus far, *Dushimys larsi*, was named by Zijlstra (2012).

All five Antillean murid genera are today extinct from the Lesser Antilles. *Megalomys curazensis* and *Dushimys* of Curacao and *Agathaeromys* of Bonaire are known only from fossils. Other taxa, however, survived within the period of human colonisation. According to Turvey and colleagues (2010), although rice rats were heavily exploited by prehistoric Amerindians, many taxa survived until close to the time of the first European arrival. A factor in the extinction of rice rats probably was the predation by the introduced mongooses and black rats. The idea behind the introduction of the mongoose was to control the introduced rats. From 1870 till 1900, mongooses were introduced in 23 West Indian islands. That turned out to be a bad plan, as the mongoose wiped out the native animals instead of the rats. The introduction of alien species seems to be the reason behind the extinction of *Megalomys desmarestii*. The latter had the size of a cat, and in addition to the mongoose, the species was hunted by humans to safeguard their coconut crops but also for food. The final blow for its extinction might have come in 1902, during the great volcanic eruption of Mount Pelée, as there are no reports of the species anymore after that catastrophic year.

The Bats

Mormoops blainvillii is known from Anguilla, Antigua, and Barbuda. The species is today extinct from those islands but continues to live on the Greater Antilles. Fossils of *Pteronotus parnellii* and *Phyllonycteris major* are known from Antigua. The former species still lives in the Greater Antilles. *Phyllonycteris major* is an extinct bat known only from Antigua and Puerto Rico. Locally extinct populations of *Macrotus waterhousii* are known from Anguilla, St Martin, and Barbuda in the northern Lesser Antilles.

Endemic extant bats from the Windward Islands of the Lesser Antilles are *Monophyllus plethodon*, *Ardops nichollsi*, *Brachyphylla cavernarum*, and *Myotis martiniquensis*. All bat fossils are estimated to be of a Late Pleistocene or Holocene age.

The majority of bat species that went extinct or have undergone local extinction or extirpation are specialised or obligate cave-dwelling bats that roost in hot, humid chambers deep within large caves. It is possible that their extinction took place when many caves were flooded due to the sea level rise at the end of the Pleistocene.

References

Allen R.W., Collier J.S., Stewart A.G., Henstock T., Goes S., Rietbrock A., the VoiLA Team. 2019. The role of arc migration in the development of the Lesser Antilles: A new tectonic model for the Cenozoic evolution of the eastern Caribbean. *Geology*, 47: 891–895.

Biknevicius A.R., McFarlane D.A., McPhee R.D.E. 1993. Body size in *Amblyrhiza inundata* (Rodentia: Caviomorpha), an extinct megafaunal rodent from the Anguilla Bank, West Indies: Estimates and implications. *American Museum Novitates*, 3079: 1–25.

Burness G.P., Diamond J., Flannery T. 2001. Dinosaurs, dragons, and dwarfs: the evolution of maximal body size. *Proceedings of the National Academy of Sciences of the USA*, 98: 14518–14523.

Cope E.D. 1868. Exhibition of bones and teeth of a large rodent from the cave deposits of Anguilla, one of the Virgin West India Islands. *Proceedings of the Academy of Natural Sciences of Philadelphia*, 1868: 313.

Cope E.D. 1869. Synopsis of the extinct Mammalia of the cave formations in the United States with observations on some Myriapoda found in and near the same. *Proceedings of the American Philosophical Society*, 11: 171–192.

Cope E.D. 1883. On the contents of a bone cave in the island of Anguilla (West Indies). *Smithsonian Contributions to Knowledge*, 25: 1–30.

Hooijer D.A. 1959. *Fossil rodents from Curaçao and Bonaire. Studies Fauna Curaçao & Other Caribbean Islands*, 9 (35): 1–27.

Hooijer D.A. 1962. A fossil ground sloth from Curaçao, Netherlands Antilles. *Proceedings of the Koninklijke Akademie van Wetenschappen (Series B)*, 65 (1): 46–60.

MacPhee R.D.E., Singer R, Diamond M. 2000. Late Cenozoic Land Mammals from Grenada, Lesser Antilles Island-Arc. American Museum Novitates 3302: 1–20.

McFarlane D.A., MacPhee R.D.E., Ford D.C. 1998. Body size variability and a Sangamonian extinction model for Amblyrhiza, a West Indian megafaunal rodent. *Quaternary Research*, 50: 80–89.

Schreuder A. 1933. Skull remains of *Amblyrhiza. Tijdschrift der Nederlandse Dierkundige Vereeniging*, 3: 242–266.

Steadman D.W., Pregill G.K., Olson S.L. 1984. Fossil vertebrates from Antigua, Lesser Antilles: Evidence for late Holocene human-caused extinctions in the West Indies. *Proceedings of the National Academy of Science of the United States of America*, 81: 4448–4451.

Turvey S.T., Brace S., Weksler M. 2012. A new species of recently extinct rice rat (*Megalomys*) from Barbados. *Mammalian Biology*, 77: 404–413.

Turvey S.T., Weksler M., Morris E.L., Nokkert M. 2010. Taxonomy, phylogeny, and diversity of the extinct Lesser Antillean rice rats (Sigmodontinae: Oryzomyini), with description of a new genus and species. *Zoological Journal of the Linnean Society*, 160: 748–772.

Zijlstra J.S. 2012. A new oryzomyine (Rodentia: Sigmodontinae) from the Quaternary of Curaçao (West Indies). *Zootaxa*, 3534: 61–68.

Zijlstra J.S., Madern P.A., van den Hoek Ostende L.W. 2010. New genus and two new species of Pleistocene oryzomyines (Cricetidae: Sigmodontinae) from Bonaire, Netherlands Antilles. *Journal of Mammalogy*, 91 (4): 860–873.

PART III

Species and Processes

in a taxonomic arrangement. These are, in this order, proboscideans, primates, sloths, lagomorphs, rodents, insectivores and bats, cervids and bovids, hippos and pigs, and carnivores. Each paragraph starts with geographic distribution and geological range, followed by a section on common morphological trends and sometimes by a remark on taxonomy. Detailed description of the taxa is found under the respective island in the previous part. Primates are only briefly described, as the extensive body of literature on fossil insular primates is not easily summarised in a comprehensive chapter in a general reference book without doing injustice to the various specialists on this topic.

After the separate treatment of the endemic taxa, the evolutionary processes that are observed in insular lineages are presented. Observed anatomical and morphological patterns or trends are discussed, such as body mass changes, changes in dentition, and morphological changes in the limbs. The drive behind these patterns is explained where possible, including the island rule, and the processes of speciation on islands summarised. Finally, the last chapter deals with extinction of island endemics.

CHAPTER TWENTY ONE

An Overview of Endemic Species

Evolution of Island Mammals: Adaptation and Extinction of Placental Mammals on Islands, Second Edition.
Alexandra van der Geer, George Lyras and John de Vos.

© 2021 John Wiley & Sons Ltd. Published 2021 by John Wiley & Sons Ltd.

This chapter lists the most important fossil island species worldwide, with notes on their distribution, range, and dispersal. Species without a fossil record, such as the vast majority of murids of the Philippines, Sulawesi, New Guinea, and so on, most lemurs and tenrecs of Madagascar, and many more are here excluded. For details, see the respective island Chapters.

Proboscidea: Mammoths, Elephants, and Stegodons

Apparently, proboscideans were the most successful lineage of large-sized island colonisers, including stegodons, gomphotheres, mammoths, and elephants. Endemic proboscideans have been reported from islands all over the Northern Hemisphere. Everywhere, they developed a smaller size, eventually reaching dwarf or even pygmy proportions compared to their mainland ancestors (Plate 1, Box 5.1). It is hard to think of an island with a rich fossil record lacking any proboscidean remains. Notable Exceptions include the Balearics, Gargano, the central Ryukyu Islands, and the West Indies.

Distribution and Range

Endemic stegodons, elephants with low-crowned teeth with ridges, were common elements in insular South East and East Asia. They are known from the following islands, often with two species from subsequent geological periods, and all belonging to the genus *Stegodon*: Java, Flores (Plate 26), Sumba, Timor (Plate 28), Sulawesi, Luzon (Plate 32), Mindanao, and Japan (Plate 33). All species lived during the Pleistocene.

Endemic primitive proboscideans, on the other hand, are rare and are known only from East Java and Sulawesi (*Stegoloxodon*, a four-tusked elephant) (Plate 29) and Honshu, Japan (*Stegolophodon*). The latter represents the oldest records of insular dwarfed proboscideans, dating to the Middle Miocene. Endemic gomphotheres, elephants with primitive teeth with cusps, are even rarer and reported only from East and Central Java (*Sinomastodon*).

Endemic mammoths (*Mammuthus*) occurred on Crete, Sardinia, Japan, the northern Californian Channel Islands (Plate 1), and an isolated population of the mainland species on Wrangel Island (Lister, 1993). The Cretan mammoth is the oldest (Early Pleistocene), while the dwarfed population of Wrangel were the last surviving woolly mammoths, dated to about 4,000–5,000 years ago.

Endemic elephants (*Palaeoloxodon* and *Elephas*) finally are reported from the Middle and Late Pleistocene of Cyprus, Crete, Kasos, Tilos (Plate 7 (a–d)), Rhodes, Naxos and Delos (palaeo-Cyclades) (Plate 7 (e)), Astypalaia, Sicily (Plate 12), Malta, and Japan with species of *Palaeoloxodon* and Java with *Elephas*. There are also anecdotal references for the presence of dwarf elephants on the Greek islands Paros, Milos, Sériphos and Kýthnos. Elephants from other Greek islands are considered mainland forms according to Athanassios Athanassiou and colleagues (2019). The elephants from the Philippines (Luzon, Visayas) can belong to either genus as their taxonomy is unresolved.

Dispersal

During the Early Pleistocene, dispersal to various islands of *Mammuthus meridionalis* (in Europe) and *Mammuthus trogontherii* (in Asia) can be observed, evolving into dwarf species. Examples are *Mammuthus creticus* of Crete and *Mammuthus protomammonteus* of Japan. During the Late Pleistocene, a dispersal of woolly mammoths (*Mammuthus primigenius*) to Wrangel Island took place and of American mammoths (*Mammuthus columbi*) to the Californian Channel Islands, where they evolved into the dwarf species *Mammuthus exilis*.

The dispersal history of *Stegoloxodon celebensis* from Sulawesi and *Stegoloxodon indonesicus* from Java is less clear. Originally, the first insular species was supposed to be derived from *Archidiskodon planifrons* by Hooijer in 1949. Van den Bergh (1999) considered the Sulawesi *'Elephas'* (renamed as *Stegoloxodon* by Markov and Saegusa, 2008) congeneric with the Javan species and agreed with an early stock of *'Elephas' planifrons* as the most likely ancestor. The most striking feature is the occasional presence of lower tusks. This, however, is also seen in the earliest forms of *'Elephas' planifrons*.

During the Middle and Late Pleistocene, a widespread dispersal of the huge straight-tusked elephants – *Palaeoloxodon antiquus* in Europe and *Palaeoloxodon namadicus* in Asia – took place, again resulting in dwarf forms. The Mediterranean islands each harboured a different species of dwarf elephant, all descendant from the same mainland ancestor. Insular Southeast Asia had only one dwarf palaeoloxodontine elephants (Japan). The Japanese species (*Palaeoloxodon naumanni*) is the type species of the genus.

Primates

Endemic extinct primates are extremely rare in the fossil record, partly due to the paucity of primate fossils in general. They include two adaptive radiations – the lemurs of Madagascar and the New World monkeys of the Greater Antilles – and four isolated occurrences of three hominids and one monkey.

Distribution and Range

Endemic New World monkeys (Platyrrhini) are known by five extinct species of the Greater Antilles and form a clade on their own (Xenotrichini). These are *Paralouatta marianae* (Early Miocene; Figure 19.9) and *Paralouatta varonai* (Late Pleistocene; Figure 19.10) from Cuba, *Xenothrix mcgregori* (latest Pleistocene-Late Holocene) from Jamaica, and *Antillothrix bernensis* and *Insulacebus toussaintiana* from Hispaniola. In addition, there is still undescribed primate material from Jamaica and Hispaniola, unlike that of known species, which may thus belong to new species.

Endemic fossil lemurs (Plate 20) are restricted to Madagascar (Chapter 11) and include monkey lemurs (*Archaeolemur* and *Hadropithecus*), the sloth lemurs (*Mesopropithecus*, *Babakotia*, *Archaeoindris*, and *Palaeopropithecus*, Plate 23), the koala lemurs (*Megaladapis*), the aye-ayes (*Daubentonia*), and 'true' lemurs (*Pachylemur*).

Endemic fossil hominids include the Late Miocene *Oreopithecus bambolii* of Tuscany and the two Late Pleistocene hominins of insular South East Asia, being *Homo floresiensis* (Flores; Figure 13.8) and *Homo luzonensis* (Luzon). *Homo erectus* of Java (Early to late Middle Pleistocene) is not considered an endemic species here.

The single endemic Old World monkey is a macaque from the Early Pleistocene of Sardinia-Corsica (*Macaca majori*).

Dispersal

Regarding the arrival of the New World monkeys to the Greater Antilles, two dispersals must have taken place. First, the trans-Atlantic voyage of the ancestral Platyrrhini from Africa to South America. Seiffert and colleagues (2020) estimate the time of this travel at the Eocene-Oligocene boundary around 34 million years ago when the sea level lowered (see further Box 3.3). They base this estimation on their discovery of the very small monkey *Ucayalipithecus perdita*, which is closely related to Oligocene parapithecid primates of Africa. Second, the voyage from South America to Cuba, Hispaniola, and Jamaica. This included at least a partial land span, referred to as GAARlandia by MacPhee and Iturralde-Vinent in 1994, the extent of which is unclear (see Box 3.2).

Sloths

Endemic fossil sloths are restricted to the Greater Antilles, Curacao (Table 19.1; Plates 34 and 36) and Grenada ranging in time between the Early Oligocene and the latest Pleistocene. They all belong to the endemic family Megalocnidae and likely represent an adaptive radiation within the West Indies. They have been described for Cuba (*Megalocnus*, *Parocnus*, two species of *Neocnus*, and *Acratocnus*), Hispaniola (*Megalocnus*, *Parocnus*, *Acratocnus*, and three species of *Neocnus*), Puerto Rico (*Acratocnus*), Curacao (*Paulocnus*; Plate 36), and Grenada (undescribed). Except for the exceptionally large *Megalocnus* (about 150 kg) and *Parocnus*, they are medium to small sized.

Lagomorpha: Rabbits, Hares, and Pikas

Endemic lagomorphs have been reported from the fossil record of several islands worldwide. A number of these developed a larger or even gigantic size – for small mammal species standards – whereas others retained a more modest size. Apart from size changes and independent thereof, morphological changes are observed as well, e.g. in the dentition in some lineages.

Distribution and Range

Fossil insular pikas (Ochotonidae) are restricted to Mediterranean islands and known from Gargano (Plate 10), Sardinia, and Corsica with *Prolagus* (Plate 13), Tuscany with *Paludotona*, and Mallorca and Minorca with *Gymnesicolagus* (overview by Blanca Moncunill-Solé in 2021). Pikas apparently were very successful on these islands, as can be inferred from their often abundant remains in fossiliferous deposits, especially of Sardinia. The earliest forms are the species from Tuscany and Gargano (Late Miocene) and Mallorca-Minorca (Middle–Late Miocene). The Sardinian species survived well into the Holocene.

Fossil insular hares and rabbits (Leporidae) are much rarer and often known from a few remains only. They have been described from Sicily, Ibiza, and Mallorca (*Hypolagus*), Sardinia (*Sardolagus*), Minorca (*Nuralagus*), and the central Ryukyu Islands (*Pentalagus*; Box 17.1).

The endemic rabbit of the Ryukyu Islands, the Amami rabbit (*Pentalagus furnessi*), is still extant, though its numbers have decreased drastically due to anthropogenic factors, such as hunting and the introduction of mongooses as pest controllers. The endemic rabbit of Sumatra (*Nesolagus netscheri*) is believed to have lived on the island already since the Late Miocene, but no fossils or subfossils have been found so far. This species has short, blackish brown ears and a striking pattern of dark brown stripes and patches. Although the Sumatran rabbit is indeed rabbit-like in appearance, it is not closely related to other hares and rabbits but represents a group on its own. Patnaik (2002) suggested that the *Nesolagus* lineage diverged from its nearest relatives (*Alilepus*-like hares) during the Late Miocene. The closest living relative is the Annamite rabbit (*Nesolagus timminsi*) from Laos and Vietnam. Surridge and colleagues (1999) estimated that the Sumatran and the Annamite rabbit diverged from each other around 8 million years ago, that is, very close to the phylogenetic split of the genus from the other hares and rabbits. This would indicate that vicariance of the Sumatran rabbit started towards the end of the Miocene. However, no fossil findings corroborate this hypothesis. In addition, at present, no consistent robust phylogeny of lagomorphs has been established, and the true relationships within the Leporidae remain unknown.

Dispersal

Lagomorphs are considered bad overseas dispersers, and in this regard, it is interesting to note that they are entirely lacking on remote islands such as Flores, the Philippines, and Cyprus. Angelone (2007) remarked that lagomorphs also lack on islands relatively close to the mainland – unless purposely introduced by humans – until a land connection is established. The presence of a lagomorph in an insular fauna therefore suggests, but does not prove, the existence in the past of some sort of land connection. The Balearics provide such an example. Here, the older faunas (Late Miocene–Early Pliocene) contain endemic pikas and hares, whereas the younger faunas lack lagomorphs. The latest Miocene (Messinian) filter connection between the Balearics and the mainland enabled the dispersal of the lagomorphs, but, in due time, they went extinct. The absence of fossil lagomorphs from the mainland of Japan indicates that the islands probably were always disconnected from the mainland, or at least that the connection was too difficult to cross. On the central Ryukyu Islands, however, an endemic lagomorph, the Amami rabbit (*Pentalagus*), survived for perhaps millions of years. The divergence time of this rabbit has been estimated at about 9.5 million years ago, coinciding with the isolation of the islands in the Late Miocene–Early Pliocene.

The present-day occurrence of feral rabbits (*Oryctolagus cuniculus*) on remote islands is entirely thanks to human agency. Already, the Romans brought rabbits to small islands because their notorious burrowing and gnawing habits make them difficult to keep. On islands, they are controlled much more easily. They were, amongst others, brought to the Balearics. By the time of emperor August, their number had increased so much that they constituted a plague. Ferrets or mongooses, referred to as 'viverrae' by Pliny the Elder in the early first century (Plinius, ca. 77–79), had to be introduced to control the prolific rabbits. Later, seafarers brought rabbits to oceanic islands to guarantee a constant food supply for later visits. This is the reason why today rabbits are found on practically all islands, even including the Falkland Islands. Early colonists took rabbits to Australia and New Zealand for the same reason, with well-known disastrous impact. In parallel with the Romans, almost two millennia before, the colonists introduced ferrets to hunt the rabbits, but the eventual outcome was the extinction of the endemic birds.

Rodentia: Rats, Dormice, Hamsters, and Caviomorphs

Fossil and subfossil insular rodents are found on most islands, continental as well as oceanic, and are represented by several families, such as murine rats and mice (Muridae), dormice (Gliridae), gundis or comb rats (Ctenodactylidae), voles, hamsters, rice rats (Cricetidae), coneys, hutias (Capromyidae), spiny rats (Echimyidae), and capybaras (Caviidae). The rice rats, coneys, hutias, spiny rats, and capybaras are restricted to the New World and the dormice and gundis to the Old World.

Distribution and Range

Fossil or subfossil rats and mice (Murinae) are perhaps the commonest among insular rodents. They have a worldwide distribution and time range spanning from the Late Miocene to the Holocene. They have been reported from the following islands, in a number of cases represented with several anagenetic species or different genera for each biozone. In alphabetical order: Californian Channel Islands (*Peromyscus*), Canary Islands (*Canariomys*), Crete (*Kritimys*, *Mus*), Flores (*Hooijeromys*, *Papagomys*, *Spelaeomys;* Plate 27), Gargano (*Mikrotia*, *Apatomys*) (Plate 10), Ilin (*Crateromys*), Japan (*Apodemus*), Madagascar (*Macrotarsomys*), Masbate, Philippines (*Rattus*; Plate 30), New Guinea (*Mallomys*), Ryukyu Islands (*Diplothrix*, *Tokudaia*), Sardinia (*Apodemus*, *Rhagapodemus*, *Rhagamys*), Sicily (*Apodemus*), Timor (*Coryphomys*; Box 13.3), Sulawesi (*Lenomys*, *Paruromys*), and Tuscany (*Parapodemus*, *Anthracomys*, *Huerzelerimys*).

A few of these are still extant, such as *Paruromys dominator* and *Lenomys meyeri* of Sulawesi, *Papagomys armandvillei* of Flores, *Rattus everetti* of the Philippines, and *Macrotarsomys petteri* of Madagascar. Timothy Flannery (1995) suggested that arboreal species are more prone to survive than are terrestrial species, based on a study of the three species of *Uromys* on Guadalcanal, Solomon Islands. However, on Flores, the probably arboreal *Spelaeomys* and the terrestrial *Papagomys theodorverhoeveni* both went extinct, while another terrestrial species, *Papagomys armandvillei*, survived. Furthermore, the perhaps extinct Ilin cloud rat of the Philippines, *Crateromys paulus* (Ilin Island, Greater Mindoro; listed as data deficient by the International Union for Conservation of Nature [IUCN Red List]) was arboreal, however, whereas the terrestrial *Rattus everetti* survived.

Fossil endemic dormice (Gliridae) are restricted to the central and western Mediterranean region and have been reported only from Gargano (*Stertomys*), Sicily, and Malta (*Leithia* and *Maltamys*; Plate 37), Sardinia (*Peridyromys*, *Microdyromys*, and *Tyrrhenoglis*), Tuscany (*Anthracoglis*), Mallorca and Minorca (*Moissenetia*, *Bransatoglis*, *Carbomys*, *Margaritamys*, *Peridyromys*, *Muscardinus*, and *Hypnomys*), Ibiza (*Eivissia* and *Hypnomys*), Las Murchas, and Spain (*Pseudodryomys*).

Fossil endemic voles (Arvicolinae) are even more restricted and are known only from Sardinia (*Microtus* [*Tyrrhenicola*]) and Malta (*Microtus* [*Pitimys*]), both belonging to the extant genus of small-eared meadow voles.

Fossil hamsters (Cricetinae) are known only from Gargano (*Hattomys* and *Mystemys*), Tuscany (*Kowalskia*), and Mallorca (*Tragomys*).

Fossil giant rice rats (Sigmodontinae) are described from Santa Cruz, Galápagos (*Megaoryzomys*), and the Antilles, from where they are only known from Curacao (*Megalomys*). Their distribution, though, was likely much larger, because giant rice rats are known from historical times from the Lesser Antilles (Martinique, Barbuda, Santa Lucia, Montserrat, Anguilla, St Eustatius, St Kitts, Antigua, Guadeloupe, and Marie Galante). Fossils of lesser rice rats (*Oryzomys*) have been reported only from Jamaica and St Vincent.

Fossil and subfossil hutias and coneys (Capromyidae) are reported from Cuba, Puerto Rico, Hispaniola, Virgin Islands, and a few small adjacent islands (*Zazamys*, *Capromys*, *Mesocapromys*, *Mysateles*, *Geocapromys*, *Plagiodontia*, *Rhizoplagiodontia*, *Isolobodon*, and *Hexolobon*) in great specific diversity (see Table 19.2).

Several fossil and subfossil Antillean spiny rats (Echimyidae) have been reported from Cuba, Isla de Piños, Hispaniola, and Puerto Rico (*Puertoricomys*, *Boromys*, *Brotomys*, and *Heteropsomys*).

Fossil and subfossil giant hutias (Heptaxodontidae) are known from the Lesser Antilles (*Amblyrhiza*) (Plate 35), Puerto Rico (*Elasmodontomys* and *Tainotherium*), Hispaniola (*Quemisia*), and Jamaica (*Clidomys*, *Xaymaca*, and perhaps *Alterodon*).

Fossil capybaras are restricted to Curacao and Grenada (*Hydrochaeris*).

Finally, fossil endemic gundis or comb rats (Ctenodactylidae) are extremely rare elements in island faunas and are thus far restricted to Sicily (*Pellegrinia*) (Figure 7.4), Sardinia (*Sardomys* and *Pireddamys*), and perhaps Mallorca (tentatively ascribed to *Tataromys*). The attribution of the latter Oligocene remains to *Tataromys* needs confirmation.

Dispersal

How gundis reached the Italian islands Sicily and Sardinia is an unsolved issue. Gundis evolved in Asia during the Oligocene and reached Africa during the Middle Miocene, from where they seem to have colonised the two Italian islands, as Thaler (1972) recognised African characteristics in the Sicilian form. A problem with this scenario is that the Sicilian gundi is strongly endemic, which fact obscures phylogenetic relationships. Regarding the Sardinian form, probably they dispersed via the Tuscan Archipelago to which Sardinia and Corsica were connected during the Miocene.

Thick Layers of Small Mammal Fossils

Accumulation of small mammal remains is generally thanks to the food habits of birds of prey, especially so those of owls (Strigiformes). The hunting, ingestive, and digestive processes of these owls cause hardly any modification to skeletal elements found in their pellets, as demonstrated by Andrews (1990). The least destructive owls are barn owls (*Tyto alba*), snowy owls (*Bubo scandiacus*), and long-eared and short-eared owls (*Asio otus* and *Asio flammeus*). Somewhat more destructive in their habits are eagle owls (*Bubo bubo*) and tawny owls (*Strix aluco*). Insular owls have been reported, amongst others, from the Pleistocene of the West Indies, Sicily, and Crete and the Late Miocene of Gargano. Here, thick deposits entirely consisting of murid remains are found.

Insect-eaters: Shrews, Moonrats, Solenodons, Tenrecs, and Allies

Fossil endemic shrews, shrew-like insectivores, moonrats, moles (order Eulipotyphla), and tenrecs (order Afrosoricidae) have been reported from several islands worldwide. The fossil record of bats (order Chiroptera, Box 21.1), however, is extremely fragmentary and incomplete.

Distribution and Range

Fossils of red-toothed shrews (Soricinae) are known from Gargano (*Lartetium*), Sicily, Sardinia and Corsica (*Asoriculus*), Tuscany (cf. *Crocidosorex*), Mallorca and Minorca (*Nesiotites*), and perhaps Ibiza (undescribed insectivore). The Gargano and Tuscany species are the earliest (Late Miocene), while the last survivors are the species from Sardinia, Corsica, Mallorca, and Minorca (early Holocene).

Fossils of white-toothed shrews and musk shrews (Crocidurinae) have a wider distribution and have been described from the Middle and Late Pleistocene of Crete, Sicily, Malta, possibly Flores, the central Ryukyu Islands (all species of *Crocidura*), and Flores (*Suncus*). At present, many endemic *Crocidura* species still live on islands worldwide. Several of these, such as the extant species of the Canary Islands (*Crocidura canariensis*, *Crocidura osorio*), Sicily (*Crocidura sicula*), Crete (*Crocidura zimmermanni*), have colonised the islands well before humans and underwent a long-term evolution *in situ*, exceeding a million years.

Fossil from West Indies shrew-like insectivores (*Nesophontes*) have been described from the Late Pleistocene to Holocene of Cuba, Isla de Piños, Hispaniola, Puerto Rico, and its satellite Isla de Vieques and Cayman Islands. On the largest islands (Cuba and Hispaniola), several species were likely contemporaneous (adaptive radiation).

Fossil endemic moonrats or gymnures (Galericinae) are restricted to the Late Miocene of Gargano (*Deinogalerix*, Figure 6.7, Plate 37) and the early Oligocene of Mallorca (*Tetracus*). The Gargano lineage constitutes an adaptive radiation of several species of different body sizes.

Fossil endemic moles (Talpidae) are restricted to Sardinia (*Geotrypus*, *Nuragha*, and *Talpa*), with each geological epoch having its own mole genus ranging from the Early Miocene up to the Early Holocene.

Fossil solenodons (*Solenodon*) are restricted to Cuba and Hispaniola. From Madagascar, only one subfossil tenrec is known (*Microgale macpheei*), which is considered most closely related to the extant short-tailed shrew tenrec (*Microgale brevicaudata*).

Dispersal

Shrews and shrew-like insect-eaters are claimed to be amongst the most unlikely candidates for overseas dispersal because of their extremely high metabolism and tiny body size. Yet, several species made it to the islands and established successful colonies. In this respect, it is perhaps illuminating to remark that Georges Cuvier, the father of comparative anatomy, observed similarities between the Sardinian extinct shrew (*Asoriculus similis*) and the Eurasian water shrew (*Sorex fodiens*, now *Neomys fodiens*). The water shrew is a large shrew with a head–body length of about 100 mm and a thick subcutaneous layer in order not to cool down in the water. Its buoyancy is amazing due to its dense fur that traps air bubbles. The chance that a water shrew finds itself trapped on a flotsam floating on a river heading towards the open sea and survives subsequent transport – presumably facilitated by its ability to enter stupor but also by its relatively large fat reserves for a shrew – is certainly higher than it is for a terrestrial shrew. *Asoriculus* and *Nesiotites* and perhaps some other insular red-toothed shrews as well are phylogenetically related to water shrews.

BOX 21.1

Bats on Islands

Though, at present, many island bat species, often insect-eaters as well, are endemic to their island, hardly anything is known about fossil insular bats. In most cases, bat remains have been rarely identified or described, and many bat fossils remain largely unidentified.

Subfossil endemic bats have been reported only from Madagascar (Late Pleistocene: *Triaenops goodmani* and *Hipposideros besaoka*) and the Greater Antilles (Late Pleistocene: *Artibeus anthonyi*, *Mormoops magna*, *Pteronotus pristinus*, *Phyllonycteris major*, and *Phyllops vetus*; Late Pleistocene–Recent: *Mormoops blainvillii*, *Phyllonycteris aphylla*, and *Pteronotus macleayii*). Fossil possibly endemic bats are known only from Minorca (Pliocene: *Rhinolophus* cf. *grivensis*).

Fossil bats have further been reported from amongst others Cyprus (Late Pleistocene), Sicily (middle Middle Pleistocene), Ibiza (Late Pliocene), Flores (Late Pleistocene and Early Holocene), and the West Indies (Late Pleistocene). Many fossil bats, however, are not endemic to the island, for example, the fruit bat *Rousettus aegyptiacus* which has an extensive geographical range. The same holds for most subfossil bats of the West Indies (e.g. *Molossus molossus*, *Mormoops megalophylla*, *Antrozous pallidus*, *Monophyllus plethodon*, *Desmodus rotundus*, *Tonatia saurophila*, and *Noctilio leporinus*), some of which are locally extinct.

Ruminantia: Deer and Bovids

Most palaeo-islands harboured at least one species of insular ruminant, evidenced by the fossil record. They are, together with proboscideans and rodents, the most successful insular taxa regarding distribution in time and space. Generally, where insular deer are present, bovids are absent and vice versa. Only in the case of a relative rich island fauna, both taxa occur simultaneously, as is the case with Sicily during the Late Pleistocene and Java during the Middle Pleistocene.

Distribution and Range

Fossils of endemic deer (Cervidae) have been reported from Corsica (*Cervus*), Crete (*Candiacervus*; Plates 5 and 6), Japan (*Elaphurus* and *Cervus*), Java (*Axis lydekkeri* and *Rusa*), Jersey (*Cervus*), Karpathos and Kassos ('*Candiacervus*'), Malta (*Cervus*), the Philippines (*Rusa*), Pianosa (*Cervus*), the central Ryukyu Islands (*Cervus*, *Muntiacus*, and *Dicroceros*), the southern Ryukyu Islands (*Capreolus*), Sardinia-Corsica (*Praemegaceros*), and Sicily (*Dama* and *Cervus*). No deer fossils have been reported from the New World, though today endemic white-tailed deer (*Odocoileus*) live on islands of South Carolina (see below). The earliest island deer are from the Early Pleistocene of Japan, while the majority is of a Late Pleistocene age.

Fossil endemic bovids (Bovidae) have been reported from Ibiza (two undescribed bovids), Japan (possibly *Bubalus*), Java (*Duboisia*, *Bubalus*, *Bibos*, *Epileptobos*), Mallorca and Minorca (*Myotragus*; Plate 19), Pianosa (*Bos*), Sardinia-Tuscany (*Maremmia*, Plate 16; *Tyrrhenotragus*, *Etruria*, and *Turritragus*), Sardinia-Corsica (*Nesogoral* and *Asoletragus*), Sicily (*Bos* and *Bison*), Sulawesi (*Bubalus* (*Anoa*)), and the Philippines and Taiwan (*Bubalus*).

The earliest species are from the Late Miocene of Sardinia and Tuscany, represented by four non-related species. The *Myotragus* lineage from Mallorca and Minorca has the longest temporal range starting from the Early Pliocene into the Holocene.

Fossils of less derived ruminants are restricted to the Late Oligocene (tentatively ascribed to *Pomelomeryx*) and Early Miocene ('*Amphitragulus*', *Sardomeryx*) of Sardinia and the Late Miocene of Gargano (*Hoplitomeryx*; Plate 9), and Scontrone (*Scontromeryx*).

Dispersal

The presence of endemic bovids apart from water buffaloes indicates a land bridge or at most a filtered connection in the past, whereas the ancestors of endemic deer can have reached the island by sweepstake dispersal as well. The absence of ruminants from Madagascar and remote oceanic islands such as Cyprus (mouflons are an introduced species) is entirely related to their large distance.

Hippopotamidae: Hippos

Hippopotamuses are common elements in several fossil insular faunas. Faunas with endemic hippos generally contain also dwarf elephants, but lack deer, except for Java. On Madagascar, they occur simultaneously with elephant birds and leaf-eating lemurs.

Distribution and Range

Fossils of endemic hippos are known from Cyprus (Plate 2), Crete (Plate 4), Sicily (Plate 11), Malta, Madagascar (Plates 21 and 22) (all *Hippopotamus*), and Java (*Hexaprotodon*). The earliest species are those from Crete and Java (Early Pleistocene), while the Madagascar dwarf hippos survived into the Holocene. For their long-distance overseas dispersal to Madagascar, see Box 11.1.

Suidae: Pigs

Fossils of endemic pigs have been reported from Luzon (*Celebochoerus*); Sardinia (*Eumaiochoerus*, *Sus sondaari,* and perhaps *Hyotherium*); Sulawesi (*Celebochoerus*, *Babyrousa*, and *Sus*); Tuscany (*Eumaiochoerus*); and Java, the central Ryukyu Islands and Taiwan (*Sus*). The earliest species is from Sardinia (Early Miocene), while native pigs are still present in insular Southeast Asia, e.g. babirusa and the warty pigs of the Philippines and Sulawesi. Pigs on islands with several species always belong to different geological periods.

Carnivora: Dogs, Hyenas, Otters, and Martens

Vertebrate carnivores other than mustelids are generally lacking from endemic insular faunas due to their poor overseas dispersal abilities combined with their dependency on a minimal availability of prey animals and often solitary lifestyle. Non-mustelid carnivore fossils are thus among the rarest fossils on earth. Nevertheless, a few exceptions do exist. Moreover, up to a very recent past endemic carnivores lived on a few islands, until humans caused their untimely extinction. Otters, on the other hand, regularly form part of endemic insular faunas. They are excellent swimmers and mainly depend on fish and crustaceans, which are widely available along the coasts of islands.

Distribution and Range

Endemic fossil canids have been reported only from the Middle and Late Pleistocene of Sardinia and Corsica (*Cynotherium*; Plate 18), the Early and early Middle Pleistocene of Java (*Xenocyon*; Plate 24), and the Late Pleistocene of the northern Channel Islands of California (*Urocyon*). Recently, extinct insular canids are the Falkland wolf or fox (*Dusicyon australis*) and the Japanese wolf (*Canis hodophilax*), whereas the Channel Island fox (*Urocyon littoralis*) is still living.

Endemic fossil hyenas are restricted to the Early Pleistocene of Sardinia (*Chasmaporthetes*) and the middle Middle Pleistocene of Java (*Pachycrocuta*). The latter species was element of an impoverished but balanced mainland-like fauna (Kedung Brubus).

Fossil island otters are known from Crete (*Lutrogale*; Plate 3), Gargano (*Paralutra*), Sicily (*Lutraeximia*), Malta (*Lutra*), Sardinia (*Sardolutra*, *Algarolutra*, and *Megalenhydris*), Corsica (*Lutra* and *Algarolutra*), Tuscany (*Tyrrhenolutra* and *Paludolutra*), Java (*Lutrogale*), and Japan (*Lutra*, see also below), Californian Channel Islands (*Enhydra*). The species from Gargano and Tuscany date back to the Late Miocene, whereas the remainder lived during the Middle and/or Late Pleistocene.

Endemic fossil mustelids other than otters reported so far belong to the weasel-like Galictinae. These are from the Early Pleistocene of Sicily (*Mustelercta*) and the Middle Pleistocene of Sardinia (*Enhydrictis*). The Middle Pleistocene Japanese otter (*Lutra nipponica*) was attributed to the new polecat genus *Oriensictis* by Ogino and Otsuka (2008) based on tooth morphology.

Endemic fossil viverrids finally are known only from the terminal Pleistocene of Cyprus (*Genetta*).

Dispersal

The otters from Malta, Sardinia, and Corsica (including *Megalenhydris* and *Sardolutra*) represent a radiation from the widespread Pleistocene European mainland species *Lutra simplicidens*, as suggested by Willemsen (2006). The ancestral form presumably reached the islands during the early Middle Pleistocene or even before. The radiation gave rise to species adapted to different ecological niches. Whether also *Algarolutra* formed part of this radiation is unclear because its remains are too sparse.

Taphonomy

Otter fossils are rarely found, and the endemic island otters are known by only one individual or a few fragments only. The good news, however, is that, in some cases, the only recovered remains consists of a nearly complete skeleton. This is the case with the otters from Crete (*Lutrogale cretensis*), Sardinia (*Megalenhydris barbaricina, Sardolutra ichnusae*), and Sicily (*Lutraeximia trinacriae*).

Carnivora: Felids

Fossil felids have been reported from Java (*Panthera* and *Prionailurus*), Japan (*Panthera*), and Iriomote in the central Ryukyu Islands (*Prionailurus*), but the degree of their endemism is unclear. Subfossil tigers have been reported from Borneo, Sumatra, and the Philippines. The recently extinct tigers of Bali and Java (*Panthera tigris balica* and *Panthera tigris sondaica*) and the endangered tiger of Sumatra (*Panthera tigris sumatrae*) differ on the subspecies level, indicating that some degree of endemism might have been present in the past as well. The native wild cat of Iriomote, the central Ryukyu Islands (*Prionailurus iriomotensis*), is still extant.

Island Tigers: Endemic Subspecies or Size Clines?

Abundant tiger fossils (*Panthera tigris*) are found in Java and Sumatra in early Middle–Late Pleistocene beds. The Javanese tiger is found partly in association with the Javanese Siwalik open woodland fauna, or *Homo erectus* – *Stegodon* fauna of the early Middle to Late Pleistocene. The Sumatran tiger (dated 73–63 kyr ago in Westaway and colleagues, 2017), on the other hand, and the Javanese tiger from the Late Pleistocene Punung fauna (dated 128 kyr ago in Westaway and colleagues, 2007) are found in caves, and in association with the tropical rain forest fauna from South China. The earlier fauna from the South Asian Siwaliks had migrated into Java through some sort of filter, because the fauna is impoverished and shows some degree of endemism. The tiger of the youngest site (Ngandong) of the Javanese *Homo erectus* – *Stegodon* fauna – is larger than that of the oldest site (Ci Saat). The tigers from the Punung fauna and from Sumatra, on the other hand, are part of a mainland fauna that arrived during a low sea level from South China via the Sunda Shelf in Java, Sumatra, and Borneo. The Punung tiger might be ancestral to the recently extinct Javan tiger (*Panthera tigris sondaica*).

According to Christine Hertler and Rebekka Volmer (2008), the body mass reduction of the tigers of the middle Middle Pleistocene Kedung Brubus faunal unit of Java is the result of island dwarfing. They also speculated that the delay in appearance of dwarfing (the two previous faunal units have forms with body mass comparable to that seen in mainland extant tigers) might be due to a gene flow from the mainland. Tigers are good swimmers; therefore, the possibility of continuous colonisation of Java by tigers during the preceding Early and early Middle Pleistocene faunal units cannot be ruled out. Tigers are highly mobile and may disperse for more than 1,000 km and swim for up to 29 km across rivers or 15 km across the sea, as reported by D. P. Garga (1948). The distances between the various Sunda Islands are only between 12.5 and 20 km at most, taking into account the small islets in between. During periods of low sea level, Sumatra, Java, Bali, and Borneo even formed a single land mass (van den Bergh and colleagues, 2001). Hertler and Volmer explained the unusually large size of the late Middle Pleistocene Ngandong tiger as the result of competition among the felids, canids, and hyenas that inhabited Java at that time.

On Bali, no mammalian fossils were found to date, but tigers (*Panthera tigris balica*) lived there until 1937, when the last individual was shot at Sumbar Kima in west Bali. A direct link between the Balinese and Javan subspecies is likely but unproven, as is the link of these recent tigers with the Pleistocene forms. The only surviving Indonesian tiger is the Sumatran tiger (*Panthera tigris sumatrae*), but also this subspecies is heading towards extinction, as it is critically endangered.

DNA analysis by Joel Cracraft and colleagues (1998) indicated that the Sumatran tigers have been isolated since the end of the last glacial period, coinciding with the transition from the Punung to the Wajak fauna on Java. It is the smallest living tiger subspecies, but still somewhat larger than the extinct Bali tiger, which had an adult male body mass of about 90–100 kg.

The progressively decreasing size towards the equator might be explained by Bergmann's rule – size increases proportionally with latitude. The Bali tiger is the smallest tiger subspecies known so far. All three Indonesian tigers are smaller than the Malayan tiger (*Panthera tigris jacksoni*). Size clines are seen today in the jaguar (*Panthera onca*) and the puma (*Felis concolor*) and are most probably linked to a combination of climatic and habitat factors, as explained by Alan Turner and Mauricio Antón (1997). The smallest individuals occur in equatorial regions. An alternative explanation is prey size, but this can be dismissed because the herbivorous megafauna of Java included large banteng and water buffaloes.

References

Andrews P. 1990. *Owls, Caves and Fossils; Predation, Preservation and Accumulation of Small Mammal Bones in Caves, with an Analysis of the Pleistocene Cave Faunas from Westbury-Sub-Mendip, Somerset, U.K.* The University of Chicago Press, Chicago.

Angelone C. 2007. Messinian *Prolagus* (Ochotonidae, Lagomorpha) of Italy. *Geobios*, 40: 407–421.

Athanassiou A., van der Geer A.A.E., Lyras G.A. 2019. Pleistocene insular Proboscidea of the Eastern Mediterranean: A review and update. *Quaternary Science Reviews*, 218: 306–321

Cracraft J., Feinstein J., Vaughn J., Helm-Bychowski K. 1998. Sorting out tigers (*Panthera tigris*): Mitochondrial sequences, nuclear inserts, systematics, and conservation genetics. *Animal Conservation*, 1 (2): 139–150.

Flannery T.F. 1995. *Mammals of New Guinea*. Robert Brown and Associates, Carina, Queensland.

Garga D. P. 1948. How far can a tiger swim? *Journal of the Bombay Natural History Society*, 47 (3): 545–546.

Hertler C., Volmer R. 2008. Assessing prey competition in fossil carnivore communities – a scenario for prey competition and its evolutionary consequences for tigers in Pleistocene Java. *Palaeogeography, Palaeoclimatology, Palaeoecology*, 257: 67–80.

Hooijer D.A. 1949 (a). Pleistocene vertebrates from Celebes. IV. *Archidiskodon celebensis* nov. spec. *Zoologische Mededelingen*, 30: 205–226.

Lister A.M. 1993. Mammoths in miniature. *Nature*, 362: 288–289.

MacPhee R.D.E., Iturralde-Vinent M. 1994. First Tertiary land mammal from Greater Antilles: An early Miocene sloth (Xenartha, Megalonychidae) from Cuba. *American Museum Novitates*, 3094: 1–13.

Markov G.N., Saegusa H., 2008. On the validity of *Stegoloxodon* Kretzoi, 1950 (Mammalia: Proboscidea). *Zootaxa*, 1861: 55–56.

Moncunill-Solé B. 2021. Eco-evolutionary adaptations of ochotonids (Mammalia: Lagomorpha) to islands: new insights into Late Miocene pikas from the Gargano palaeo-archipelago (Italy). *Biological Journal of the Linnean Society*, 132: 400–413.

Ogino S., Otsuka H. 2008. New middle Pleistocene Galictini (Mustelidae, Carnivora) from the Matsugae cave deposits, northern Kyushu, West Japan. *Paleontological Research*, 12 (2): 159–166.

Patnaik R. 2002. Pliocene Leporidae (Lagomorpha, Mammalia) from the Upper Siwaliks of India: Implications for phylogenetic relationships. *Journal of Vertebrate Paleontology*, 22 (2): 443–452.

Plinius G. (the Elder). ca. 77–79. Naturalis Historia. Text: ed. Gronovius, Leyden, 1669. Trans.: Philemon Holland, London 1634–1635. Bostock and Riley, Bohn, London, 1855.

Seiffert E.R., Tejedor M.F., Fleagle J.G., Novo N.M., Cornejo F.M., Bond M., de Vries D., Campbell K.E. Jr., 2020. A parapithecid stem anthropoid of African origin in the Paleogene of South America. *Science*, 368: 194–197.

Surridge A.K., Timmins R.J., Hewitt G.M., Bell D.J. 1999. Striped rabbits in Southeast Asia. *Nature*, 400: 726.

Thaler L. 1972. Les rongeurs (Rodentia et Lagomorpha) du Monte Pellegrino et la question des anciens isthmes de la Sicilie. *Comptes Rendus de l'Académie des Sciences de Paris (D)*, 274: 188–190.

Turner A., Antón M., 1997. *The Big Cats and Their Fossil Relatives. An Illustrated Guide to Their Evolution and Natural History*. Columbia University Press, New York.

van den Bergh G.D. 1999. The Late Neogene elephantoid-bearing faunas of Indonesia and their palaeozoogeographic implications; A study of the terrestrial faunal succession of Sulawesi, Flores and Java, including evidence for early hominid dispersal east of Wallace's line. *Scripta Geologica*, 117: 1–419.

van den Bergh G.D., de Vos J., Sondaar P.Y. 2001. The late quaternary palaeogeography of mammal evolution in the Indonesian Archipelago. *Palaeontology, Palaeoclimatology, Palaeoecology*, 171: 385–408.

Westaway K.E., Louys J., Due Awe R., Morwood M.J., Price G.J., Zhao J.X., Aubert M., Joannes-Boyau R., Smith T.M., Skinner M.M., Compton T., Bailey R.M., van den Bergh G.D., de Vos J., Pike A.G.W., Stringer C., Saptomo E.W., Rizal Y., Zaim J., Santoso W.D., Trihascaryo A., Kinsley L., Sulistyanto B. 2017. An early modern human presence in Sumatra 73,000–63,000 years ago. *Nature*, 548: 322–325.

Westaway K.E., Morwood M.J., Roberts R.G., Rokus A.D., Zhao J.X., Storm P., Aziz F., van den Bergh G., Hadi P., Jatmiko, de Vos J. 2007. Age and biostratigraphic significance of the Punung rainforest fauna, East Java, Indonesia, and implications for *Pongo* and *Homo*. *Journal of Human Evolution*, 53 (6): 709–717.

Willemsen G.F. 2006. *Megalenhydris* and its relationship to *Lutra* reconsidered. *Hellenic Journal of Geosciences*, 41: 83–87.

CHAPTER TWENTY TWO

Speciation Processes in Island Environments

Evolution of Island Mammals: Adaptation and Extinction of Placental Mammals on Islands, Second Edition.
Alexandra van der Geer, George Lyras and John de Vos.
© 2021 John Wiley & Sons Ltd. Published 2021 by John Wiley & Sons Ltd.

Since Charles Darwin's publication on the origin of species (1859), it is widely accepted that evolution through natural selection underlies current biodiversity. This selection is for the greater part initiated and directed by the nature and availability of ecological niches. Islands are not different in this respect. Once the founder taxon has colonised an island successfully, that is, survives long enough and reproduces in sufficient numbers not to get extinct, the possibility for niche expansion is open. The main difference between the speciation processes on islands versus those on the mainland is that the evolving taxa are more closely related to each other because of the extremely limited number of founder taxa. For example, where on the mainland the different deer species in a large ecosystem may belong to different genera, the different deer on an island belong to the same genus, as we will see below.

What Influences Speciation?

New insular species, evolved *in situ* on the island, should be better adapted to their new environment than their ancestors were. The major drive behind this speciation is explained differently by different authors, for example, lack of predation pressure (Sondaar, 1977; Case, 1978; Heaney, 1978; Lomolino, 1985), restricted area (the latter two authors), physiological optimal body size (Brown and colleagues, 1993; Damuth, 1993; Lomolino, 2005), resource limitation (McNab, 2001, 2002), decreased interspecific competition (Lomolino, 1985), increased intraspecific competition (Box 22.1), or, in a more complex unifying model (Lomolino, 2016), a combination of fundamental processes – immigration, extinction, and evolution – taken together, influencing each in combination with ecological feedbacks in the form of interspecific interactions. Naturally, the circumstances differ from island to island, and from species to species, as we shall see below.

Ecological Release from Competitors

Initially, islands harbour only a few vertebrate species, and as a result, ecologically relevant competing species are mostly absent. The evolution of a primary coloniser will thus be released from the constraining forces that were active on the mainland. This allows the successful coloniser to lose particular behavioural and morphological features that were necessary on the mainland but useless once on the island. Furthermore, the nature of the competitive taxa may have changed, for example, mammals are absent in New Zealand, so competition from passerine birds is much more important for large birds than in species-rich systems. As a result, species on islands evolve towards new forms that have lost particular specialisations or that are adapted to lifestyles or habitats unknown to their mainland relatives.

The most spectacular, and certainly the best known, effect of ecological release is body mass change. Many large mammals, like elephants and hippos, evolved towards miniature forms, while many small mammals, like rodents, evolved towards giant forms (see Chapter 23). Size evolution is, however, not the only result of ecological release. Also, an increase in other morphological variation may evolve, generally towards niche expansion, to include a greater variety of food. As we will see in Chapter 24, the majority of island mammals evolved peculiar adaptations in their skulls, teeth, and limb bones to expand their niche to increase survival chances.

Release from Predators

In mammals, the total lack of predators often leads to an increased 'tameness', sometimes referred to as ecological naiveté. In many historical reports of first encounters with island animals, this absence of fear is noted. Darwin (1839) observed that the now-extinct wolf-like fox of the Falkland Islands (*Dusicyon australis*) could be easily lured with meat and then stabbed to death. It was, in fact, so naive that Darwin correctly predicted that 'in all probability, this fox will be classed with the dodo, as an animal which has perished from the face of the earth'. An equally naive fox was the native fox or zorro (*Lycalopex fulvipes*) of San Pedro Island, Chiloé archipelago of Chile, where he, in 1834, collected the type specimen simply by hitting it over the head with his geological hammer. This release from predator pressure opens the way for the evolution of functional traits into a different direction to meet the requirements of new priorities. These may include the evolution towards an optimal body size, loss of long limbs, and adaptation to a different reproductive strategy and different habitats, with different trade-offs.

In parallel, many birds lose their capacity to fly once they are settled on an island. Flying is a costly way of transportation, more so in combination with increased body mass, and is less useful on predator-free islands. There are many examples of flightless island birds, including the flightless crane (*Grus cubensis*) and flightless giant owls (*Ornimegalonyx*) of Pleistocene Cuba, the flightless ibis (*Xenicibis xympithecus*) of Pleistocene Jamaica, the extinct flightless ibises of Hawaii (*Apteribis*), the walking owl of Pleistocene Crete (*Athene cretensis*), the giant flightless swan of Late Miocene Gargano (*Garganornis ballmanni*), the kiwis (*Apteryx*) of New Zealand, the dodo (*Raphus cucullatus*) on Mauritius, the solitaire (*Pezophaps solitaria*) on Rodriquez, the flightless cormorant (*Phalacrocorax harrisii*) on the Galápagos, the giant flightless pigeon (*Natunaornis gigoura*) on Fiji Island, and several others, most of them extinct. Especially rails (Rallidae) are prone to give up their airborne skills, as shown by David W. Steadman (1995), while this seems to apply to swans and owls for the fossil record.

BOX 22.1

Increased Intraspecific Competition

The absence of ecologically relevant competitors does not necessarily mean that on islands there are no competitors at all. The observation that several Pleistocene insular genera contain dwarf, medium and sometimes even giant species at the same time (adaptive radiation), can be explained as driven by intraspecific competition for food and area. Such processes are described not only for some mainland rodent communities by Tamar Dayan and Daniel Simberloff (1994) and Virginie Parra and colleagues (1999) but also for a Late Miocene island rodent, the burrowing mouse *Mikrotia* from Gargano by Virginie Millien-Parra 2000. Difference with the mainland is that now the various morphotypes or species within one lineage compete with each other, whereas on the mainland this happens between different genera. For the burrowing mouse *Mikrotia*, this resulted in three different lineages, first proposed by Freudenthal (1976) and followed by Virginie Millien and Jean-Jacques Jaeger (2001). This may also apply to the cervoid *Hoplitomeryx* from the same fauna. The initial overgrazing and the resulting resource limitation may have pushed the survivors to enter new niches or broaden existing niches to avoid intraspecific competition as much as possible. Eventually, this may have led to a radiation into contemporaneous different size classes and morphotypes, each adapted to its own ecological niche. Increased intraspecific competition likely also played a major role in the adaptive radiation of the Pleistocene deer of Crete, each with a different type of antler, skull, and body size, as proposed by van der Geer (2018a). Apart from adaptive radiation, intraspecific competition may also drive the accelerated evolution towards dwarfism, as is suggested in the case of the mouse-goat *Myotragus* of Mallorca by Palombo and colleagues (2006) and the Sicilian dwarf elephant *Palaeoloxodon falconeri* by Raia and colleagues (2003).

Island Isolation

Islands that are close to the mainland have low extinction rates because additional individuals of the same species regularly arrive from the mainland and thus prevent a population decline. This process – termed *the rescue effect* by James H. Brown and Astrid Kodric-Brown (1977) – is characterised by high rates of colonisation and low rates of extinction. In addition, due to a constant gene flow from the mainland, new species do not evolve and endemic forms are not found. In the fossil record, there are several cases of mainland megafauna on islands. For example, on the Greek island Kefallonia, Late Pleistocene hippo fossils were found that are identical to those of *Hippopotamus amphibius* of the mainland. These hippos seem not to have evolved into an insular species because in all likelihood genetic contact with the mainland populations remained unchanged. During glacial periods, this type of island was very close to the mainland.

Naturally, the maximal distance for this effect to take place depends on the dispersal abilities of the taxon. The limits to long-distance dispersal over seawater gaps vary greatly among mammals. Bats have greater dispersal abilities than non-volant mammals, and as such, they have colonised distant islands such as the Hawaiian islands 3,500 km away from the mainland. Non-volant mammals have lesser dispersal abilities. Nevertheless, rice rats managed to cross a distance of roughly 800 km and successfully colonise the Galápagos islands (*Megaoryzomys*), where they evolved a body mass of more than 500 g (see Chapter 23). Although rats have been recorded to actively swim both in the wild and in experimental conditions, their dispersal over such long distances can only be explained by rafting. Large mammals such as elephants on the other hand are capable of actively crossing sea gaps of over 50 km. The number of species that lose contact with their mainland relatives is thus correlated with distance and species composition of the mainland pool. The larger the distance, the more difficult to cross the sea barrier, and as a result, less and less species manage to colonise the island. Species with better dispersal abilities require a higher isolation than less vagile species. This is reflected in the occurrence of mainland bat species on islands with endemic rodents.

On islands on which the gene flow with the mainland is small or absent altogether, the composition of the fauna is largely the result of *in situ* speciation. Heaney (2000) presented a conceptual diagram illustrating the importance of distance on colonisation and speciation (Figure 22.1). Species richness in isolated archipelagos appears to rise through time as a product of adaptive radiation. Van der Geer and colleagues (2017)

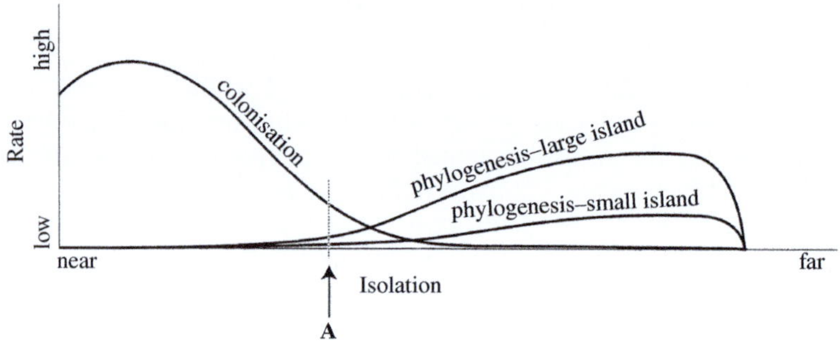

Figure 22.1 The effect of distance on colonisation success and speciation. A model developed by Heaney (2000) based on an extensive study on the modern fauna of the Philippines. (From Heaney, 2000).

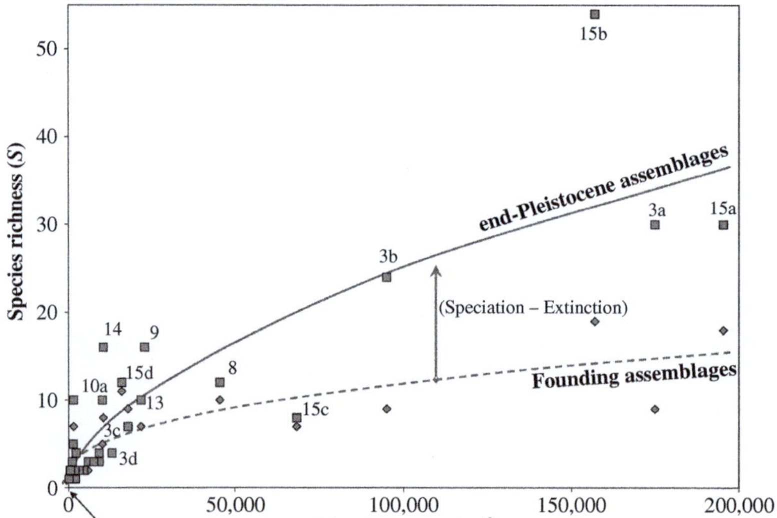

Figure 22.2 Species–area relationships for founding assemblages and latest Pleistocene–early Holocene assemblages of mammals on 35 palaeo-islands during the Pleistocene (without Madagascar). Black arrow and text describe thresholds for *in situ* speciation, here corresponding to the smallest islands with endemic mammals. The largest islands are indicated as follows, (3a) Cuba, (3b) Hispaniola, (3c) Puerto Rico, (3d) Jamaica, (8) Sardinia and Corsica, (9) Sicily, (10a) Crete, (13) Flores, (14) Timor, (15a) Greater Luzon, (15b) Greater Mindanao, (15c) Greater Negros-Panay, and (15d) Mindoro. (From Van der Geer, Lomolino and Lyras, 2017.)

show for 35 palaeo-islands worldwide that indeed although the number of founding species decreases with increasing distance, the number of Late Pleistocene resulting species is independent of isolation, and entirely the result of *in situ* speciation minus extinction (Figure 22.2).

Island Area

Island size plays perhaps an even more important role in speciation than isolation. The smaller the island, the earlier the vacant niches are filled by colonisers, thus reducing the opportunities for speciation, as already observed by Heaney (1978). Island size in itself is an inadequate measure for speciation, because important parameters, such as habitat diversity and local isolation, are not necessarily directly correlated with island size. In practice, however, as Robert Whittaker and José Maria Fernández-Palacios (2007) noticed, island size is the

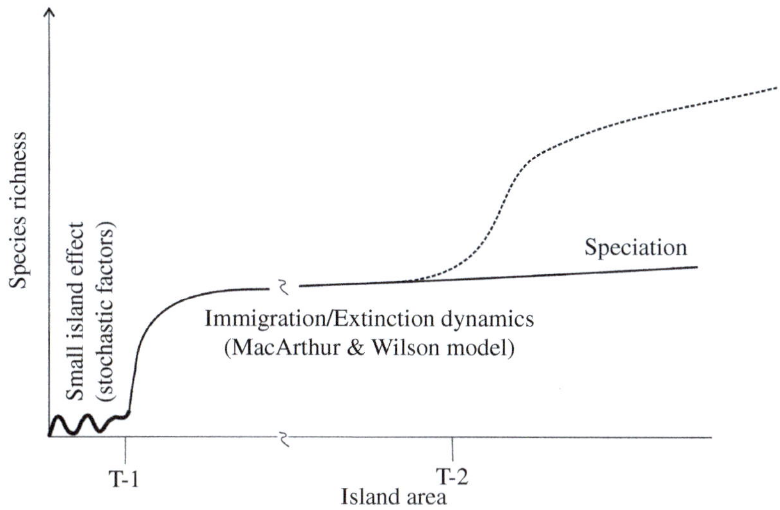

Figure 22.3 Species–area relationship as suggested by Lomolino (2000). For very small islands (smaller than t-1), the number of species is unpredictable due to stochastic factors. On larger islands, there is an equilibrium between immigration and extinction (MacArthur and Wilson, 1963), resulting in an almost constant number of species, which only very slowly increases with increasing area. For islands larger than t-2, Lomolino expected an explosive increase in species richness, due to *in situ* speciation. (From Lomolino, 2000.)

most important and comprehensive variable in terms of its correlation with species numbers.

Lomolino (2000) proposed a model describing the so-called species–area relation (Figure 22.3) according to which species richness increases at varying rates with the area of the island. Three area-related main factors that influence species richness are the total of available resources, the level of habitat diversity, and the degree of potential gene flow. With increasing island area and thus carrying capacity, population sizes and densities may increase, allowing these populations to survive long enough for speciation to take place. Increasing habitat diversity on larger islands provides more potential niches, thus enabling the evolution of new species. Finally, the larger the island, the more potential barriers, such as rivers and mountains, exist, preventing gene flow between the populations. In the Philippines, for example, many murid species are endemic to a single mountain or mountain chain, with valleys acting as a barrier.

It appears that on very small islands, species richness is greatly affected by stochastic factors, such as episodic disturbances, habitat characteristics, and interspecific interactions, and, therefore, no particular trend can be inferred for them, known as *the small-island effect*. Its rarity might be a sampling artefact, as most studies tend to avoid or ignore data from small islands in order to investigate a more general pattern. On larger islands, the factors that are responsible for the small island pattern are dampened. Another deviation from the progressive trend is predicted for the very large islands. The number of species here increases faster due to the processes of *in situ* speciation. Indeed, data from 35 palaeo-islands confirm this pattern. Van der Geer and colleagues (2017) found that although the number of founding species followed the predictions for the species–area relationship, the patterns for the resulting or 'terminal' species at the end of the Late Pleistocene were anomalous, with steep log-log slopes as a result of the explosion in species numbers relative to area on very large islands (Figure 22.2). The present-day island biodiversity (excluding introduced fauna) is more in line with the earlier models, but this is an artefact of extinctions since the onset of the Holocene and, therefore, not natural.

Island Area Plus Isolation

Island isolation and island size together influence the fundamental processes of immigration and extinction. As Lawrence Heaney (2000) noted, speciation is a process that interacts in a complex manner with both immigration and extinction. Therefore, speciation should be modelled as a separate process. Lomolino (2000) integrated these three processes of island biogeography in a single graphical model (Figure 22.4a). This tripartite model provides a summary of immigration, extinction, and speciation in relation to the island isolation and area. Immigration rates should increase with proximity to the source region (e.g. the mainland) and should decrease with isolation. Extinction rates should decrease as island area increases. Speciation should be most important where

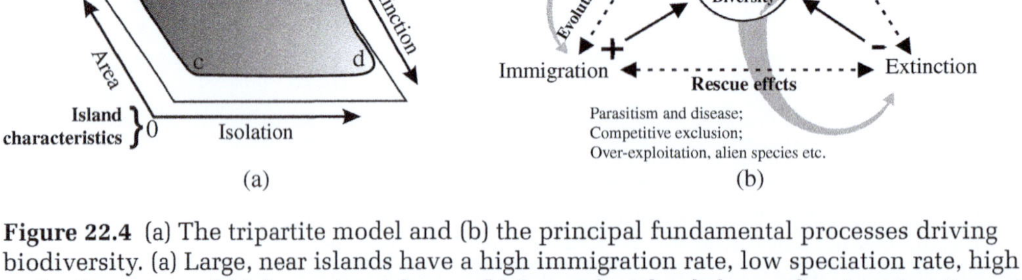

Figure 22.4 (a) The tripartite model and (b) the principal fundamental processes driving biodiversity. (a) Large, near islands have a high immigration rate, low speciation rate, high extinction rate, and high species richness. (b) Large, far islands have a low immigration rate, high speciation rate, low extinction rate, and high species richness. (c) Small, near islands have a high immigration rate, no speciation rate, high extinction rate, and moderate to low species richness. (d) Small, far islands have a low immigration rate, low speciation rate, high extinction rate, and low species richness. Examples from the Late Pleistocene fossil record are Sicily (a), Madagascar (b), Tilos (c), and Grand Cayman (d). (Adapted from Lomolino, 2016).

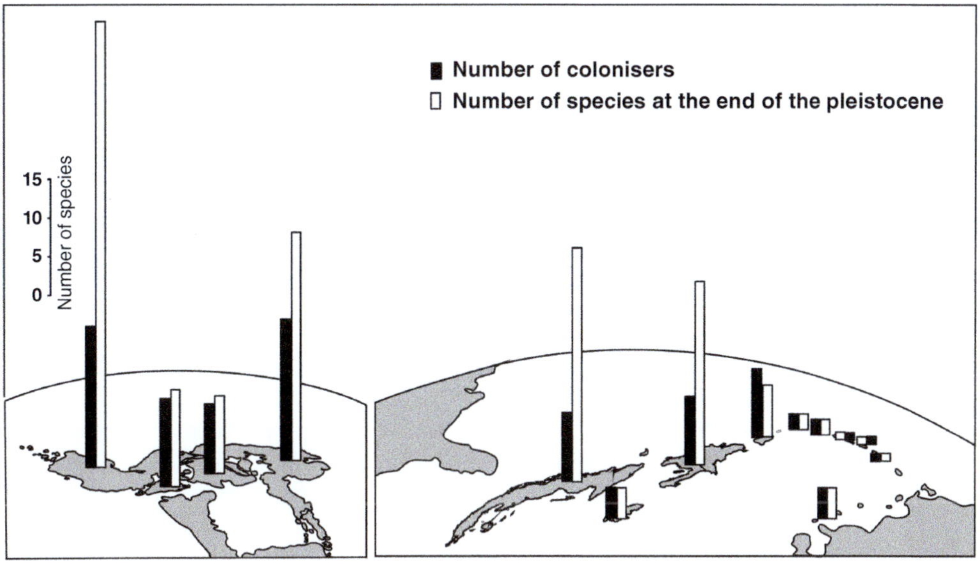

Figure 22.5 Graphic illustration of the effect of area and isolation for two archipelagos considered separately – to the left, the Philippines; to the right, the West Indies. (From Van der Geer, Lomolino, and Lyras, 2017)

extinction and immigration are lowest and, therefore, will be greater on very large and remote islands. This is illustrated with the Late Pleistocene biodiversity of the West Indies and the Philippines (Figure 22.5), where the number of founding species (colonisers) increases with area, but decreases with isolation, and the number of resulting species increases with area. An elaboration of the tripartite model by Lomolino (2016) includes the interdependence of the three fundamental biogeographic processes – immigration, extinction, and speciation – as well as ecological feedback at all points in the form of ecologically relevant competition and predation (Figure 22.4b). This feedback is an essential addition to the model as it influences the possibilities of other species to colonise the island successfully and survive long enough for speciation to take place as well as the degree to which these other species may evolve (ecological release). The fossil record shows several examples of the importance of this feedback system, not only on speciation, as we shall see below, but also on the morphological divergence that may evolve, as we shall see in the next two chapters.

Dietary Niche Width

The ability to explore different ecological niches or broaden an ancestral ecological niche appears only possible for clades that are not too specialised. In other words, generalists cope better with the array of food items available on the colonised island than specialists, as first pointed out by Lawlor (1982). This is why deer can adapt to a new diet, if necessary, whereas a panda cannot. The plasticity of deer is illustrated by brow-antlered deer (*Cervus eldi*) from Manipur, India. Here, it lives in very wet swamps and on floating grass islands, whereas the same species from Thailand lives in higher, dry areas, as reported by van Bemmel (1973). The same plasticity accounts for the easy fusion of deer species, for example, Schomburgk's deer (*Cervus duvauceli schomburgki*) dissolved into herds of brow-antlered deer (*Cervus eldi*) when their swampy habitats disappeared in the 1920s and they had to withdraw to the higher areas. The versatile character of deer likely explains their great success on palaeo-islands, in combination with their good dispersal capacities. Pleistocene insular deer evolved high-crowned teeth, relative to their ancestors, indicating a wider dietary niche.

The same holds for elephants, which are perhaps the most effective feeding generalists. Elephants can thrive on a wide variety of food items, even of very low quality, because of their hind-gut-fermenting digestive strategy, as shown by Montague Demment and Peter Van Soest (1985). In addition, elephants are able to switch from browsing to grazing or mixed feeding, despite nearly unchanged morphological features in their dentition, as pointed out by Paul Koch and colleagues (1998). Dwarf elephants appear indeed to be the most common faunal elements of palaeo-islands, represented by several genera: *Palaeoloxodon*, *Elephas*, *Mammuthus*, *Stegodon*, *Stegoloxodon*, and *Stegolophodon*.

The Speed of Evolution

Dwarfism in large mammals might require only a few thousand years of evolution to occur, as first indicated by Adrian Lister (1989). He showed that in the Last Interglacial, red deer (*Cervus elaphus*) on Jersey, one of the British Channel Islands, evolved a sixfold weight reduction in less than 6,000 years. At the other side of the scale, Heaney (1978) had observed that red-backed voles (*Myodes*, earlier named *Clethrionomys*) and field mice (*Apodemus*) need just a few thousand years and common mice (*Mus*) even only 70 years to change substantially in size. Van der Geer (2018b) showed a twofold size increase in less than 3,500 years in Polynesian rats (*Rattus exulans*) on Pacific islands. Millien (2006) demonstrated for a larger dataset that rates of morphological evolution are indeed significantly greater – up to a factor of 3.1 – for island mammal populations than for mainland mammal populations, based on a literature study mainly for rodents. On the other hand, the size decrease within the Miocene early stegodon *Stegolophodon pseudolatidens* lineage of Japan seems to have required between 2.3 and 0.6 million years of isolation and the Sardinian dwarf deer (*Praemegaceros*) decreased about 30–40% in size in about 1.7 million years. The Balearic mouse-goat (*Myotragus*) needed even more, as about 5 million years passed between the arrival of the earliest form (Messinian) and the extinction of the youngest form (Early Holocene). The Sardinian dwarf dog (*Cynotherium sardous*) evolved to half its ancestral size in about 800,000 years (Figure 22.6), as shown by Lyras and colleagues (2010). To sum up, evolution of dwarfism or gigantism may be rapid, but this is not always the case.

Types of Speciation on Islands

Anagenetic Evolution

The term anagenetic evolution refers to the gradual evolution of an ancestral species into a single descendent species without a branching event. A major part of the fossil insular mammals evolved through this way of speciation. In most cases, all we have in the fossil record is one insular species, from which the mainland ancestor often can be inferred. Such is the case with all dwarf elephants worldwide. In some lineages, however, one or more intermediate forms are preserved, forming a lineage that shows progressively changes over time. These forms then represent chronospecies or chronomorphs, often partially overlapping in size and morphology. In ungulates, the most obvious change is towards a smaller body size, while in small mammals, the opposite is often observed.

A progressive decrease in size is observed in several Pleistocene insular ruminants, e.g. the deer *Praemegaceros* lineage of Sardinia and Corsica (Chapter 9) and the bovid *Myotragus* lineage from Mallorca and Minorca (Chapter 10). The subsequent stages of the Corso-Sardinian *Praemegaceros* are from large to small and ranging from the early Middle Pleistocene to the Late Pleistocene, *Praemegaceros* sp., *Praemegaceros sardus*, and, eventually, *Praemegaceros cazioti*. Apart from progressive dwarfing, this megacerine deer lineage underwent a slight proportional reduction in metapodial length combined with a very slight increase in robusticity of these metapodials, though not to the degree as seen in typical insular ruminants.

The anagenetic lineage of the Balearic mouse-goat *Myotragus* spans an even longer time span, ranging from the Early Pliocene to the Latest Pleistocene. The subsequent evolutionary

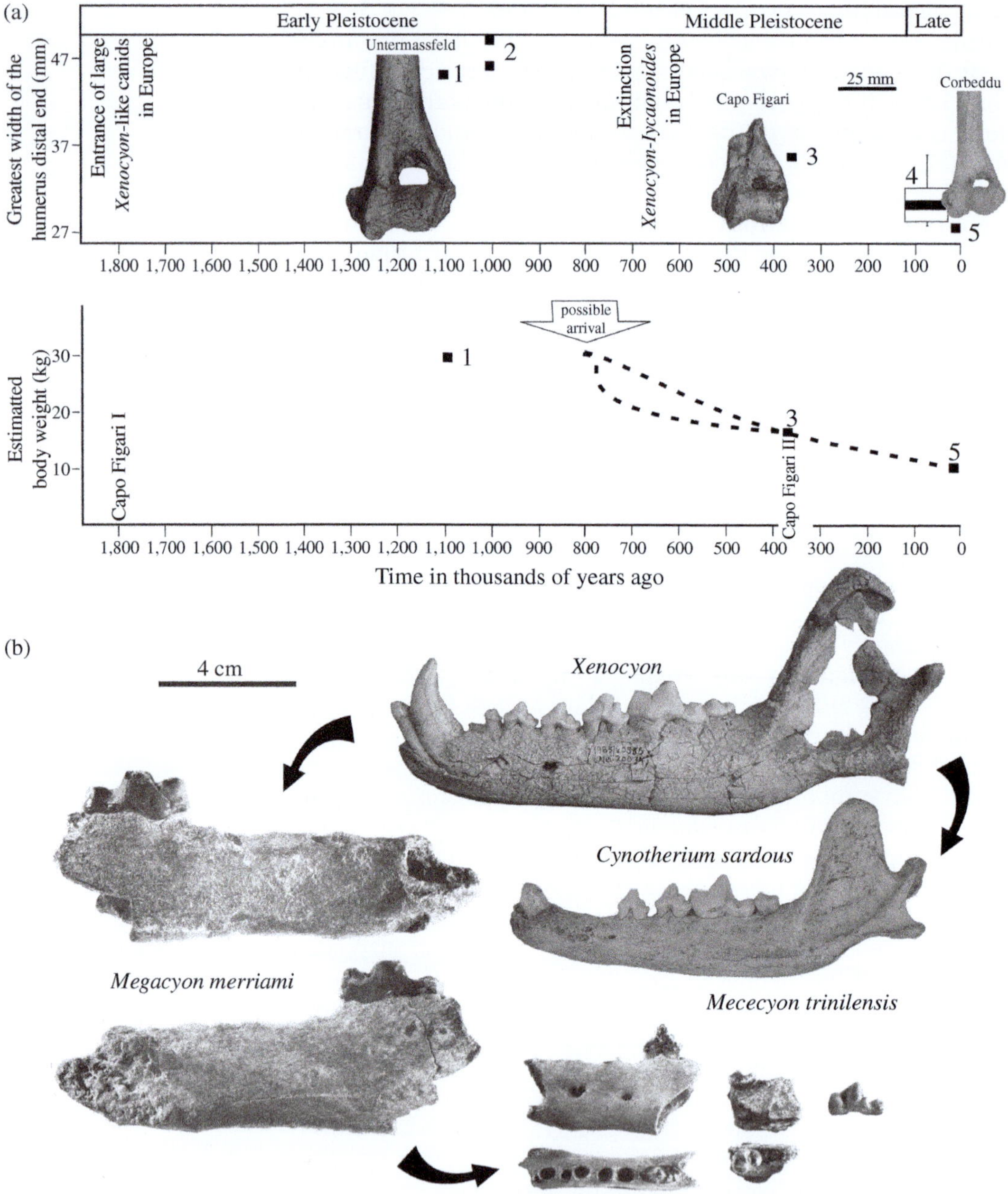

Figure 22.6 (a) Speed of size reduction in the *Cynotherium* lineage. The upper diagram shows the width of the humerus at the distal end versus the geochronological time of the specimens. The lower diagram shows the estimated body mass as calculated from the circumference of the humerus trochlea. 1, *Xenocyon lycaonoides* (Untermassfeld, Germany); 2, *Xenocyon lycaonoides* (Stránská Skála, Czech Republic); 3, *Cynotherium* sp. (Capo Figari, Sardinia); 4, average and range of *Cynotherium sardous* (Dragonara Cave, Sardinia); 5, *Cynotherium sardous* (Corbeddu Cave, Sardinia). (From Lyras and colleagues, 2010.) (b) Mandibles of *Xenocyon* and of insular canids that descended from it. The Javanese genera *Megacyon* and *Mececyon* are considered junior synonyms of *Xenocyon* by van der Geer and colleagues (2018). (Photographs of *Xenocyon* (Germany) and *Cynotherium* (Sardinia) from Lyras and colleagues (2006), of *Megacyon* (Java) and *Mececyon* (Java) from Schütt (1973).)

stages, from large to small, are *Myotragus pepgonellae*, *Myotragus antiquus*, *Myotragus kopperi*, *Myotragus batei*, and, finally, *Myotragus balearicus*. Apart from the progressive dwarfing and a relative shortening of the metapodials, a remarkable change in dentition occurs, with, as most bizarre feature, the acquirement of an ever-growing (monophyodont) dentition (Chapter 24), in convergence with vicuñas, rodents, lagomorphs, and the fossil antelope *Maremmia* from Tuscany.

Another example of a series of chronospecies is provided by the Sardinian dog lineage (*Cynotherium*). The oldest species (*Cynotherium* sp.; late Early Pleistocene) is significantly larger, followed by the early Middle Pleistocene *Cynotherium malatestai*, and finally the youngest and smallest species Late Pleistocene *Cynotherium sardous*. Within the latter, a continuation of the gradual size decrease is observed. Likely, a convergent evolution took place in Java in the hypercarnivorous dog lineage *Xenocyon*, phylogenetically related to *Cynotherium*. Here, the Early Pleistocene large dog *Xenocyon merriami* evolved anagenetically into the early Middle Pleistocene small dog *Xenocyon trinilensis*. An example within the stegodons is provided by late Early Miocene *Stegolophodon pseudolatidens* of Japan. Three developmental stages of gradual size decrease are recognised in this lineage. Another is the Pleistocene dwarf stegodon of Flores (*Stegodon florensis*) with two distinct subspecies (Middle Pleistocene: *florensis* and Late Pleistocene: *insularis*), again of decreasing size.

The opposite, a progressive size increase, is observed in the lineage of giant dormice of Mallorca (*Hypnomys*). The subsequent stages of gigantism, ranging from the Early Pliocene to the Early Holocene, are *Hypnomys waldreni*, *Hypnomys intermedius*, and *Hypnomys morpheus*. Although the latter species was a descendent of the common garden dormouse *Eliomys*, it had the size of the extant edible dormouse (*Myoxus glis*), as did the closely related Sicilian and Maltese dormice (*Leithia*). Progressive gigantism is also observed in the Sardinian giant pika *Prolagus sardus* with the earlier chronospecies *Prolagus figaro*.

Adaptive Radiation

The term adaptive, or evolutionary, radiation is applied when an ancestral species rapidly evolves in several species in a branching way to occupy new environments, or adopt new feeding strategies, as reflected in new phenotypes. Classic examples from the islands are Darwin's finches of the Galapagos (tribe Geospizini) and the honeycreepers of Hawaii (tribe Drepanidini). When a species extends its range onto several islands, then, on each island, a different species might evolve simply as a result of genetic drift. This type is sometimes termed non-adaptive radiation.

Adaptive radiation is recognisable on all taxonomic levels. The processes behind all these adaptive radiations appear to be the same at all levels: innovation – or simply a mere survival after a mass extinction of the others –followed by a rapid, monophyletic radiation to occupy as many available ecological niches or to expand the intrinsic ecological niche as far as possible. A comparable situation is found when a taxon colonises an empty region: an island for terrestrial and flying taxa, an isolated lake for aquatic taxa. The end result, after an innovation, a mass extinction, or isolation, is a radiation in order to adapt to new strategies.

On islands, primary colonisers have no competitors (ecological release) and so have the unique possibility to evolve beyond the degree seen on the mainland. Clear examples are the island continents: the carnivorous marsupials and herbivorous mammals of South America during the Tertiary – a mega-island in those days – and the marsupials of Australia during the Cainozoic – coming from South America through Antarctica in the Early Eocene – were able to radiate into a huge variety of taxa with a large degree of convergent evolution with placental mammals of Africa, North America, and Eurasia. But, also, much smaller islands provide examples of evolutionary radiation of extinct mammals. These include Pleistocene Hispaniola with four or five sloth species (*Neocnus* and *Acratocnus*), Pleistocene Crete with its eight species (*Candiacervus*), and Late Miocene Gargano with four five-horned deer species (*Hoplitomeryx*) and at any time slice two related species of gymnures (*Deinogalerix*) and the mid-Eocene of the Pontide terrane of Anatolia with a radiation of archaic ungulates (pleuraspidotheriids). Today, spectacular adaptive mammalian radiations are observed in, for example, murids (e.g., Luzon, Sulawesi, New Guinea, and Australia), lemurs, and tenrecs (Madagascar).

Speciation Within an Archipelago

The adaptive radiation as observed in endemic insular taxa is often explained in terms of allopatric speciation. In actual fact, this implies the existence of an archipelago or group of islands. In the first instance, the archipelago-driven speciation model was developed for birds, which are considerably more vagile than mammals. A classic example is provided by Darwin's finches from the Galapagos. The 6 genera and about 14 species are remarkably similar in their dull-coloured plumage and body size, but differ in beak shape and size as already observed by David Lack (1947) in response to differences in habitat as shown by Peter B. Grant (1986). Colin Patterson (1978) had already recognised several trophic groups, ranging from seed- and insect-eating ground (coastal zone, lowlands) and tree finches (forest zone); the cactus flower eating ground finch (lowlands); the insect-eating warbler finch; the fruit-, bud- and soft seed-eating tree finch (forest zone); and the insect-eating woodpecker-like finches (forest zone, including mangroves). Congeneric species each use a different part of the habitat, as observed by E. White Gifford (1919). On the other hand, continental passerine birds that are closely related to each other tend to differ from each other in plumage, whereas they are usually similar in beak and other structural characters. On the mainland, differences in beak more usually characterise broader units or genera. In Darwin's finches, the opposite seems to be the case.

There are more such cases of passerine birds displaying extreme radiation, for example, the honeycreeper finches of Hawaii (Drepanididae) and the birds of paradise of New Guinea (Paradisaeidae). The Hawaiian finches show even more differentiation in morphology (9 genera and 22 species) than Darwin's Finches. The diets range from honey and small seeds (e.g. *Palmeria*), hard seeds, and nuts (e.g. *Psittirostra*, *Chloridopa*, and *Pseudonestor*) to insects (e.g. *Hemignathus*) and the types of beak accordingly range from short and slender to the robust parrot-like beak of *Pseudonestor*. However, although these differences in diet are huge, other differences are much less. For example, they are all medium sized (10–20 cm) and have a monochrome plumage (olive green, yellow, red, or black), without distinct pattern, except for the tufted honeycreeper *Palmeria dolei*, which is grey with red spots and stripes and has a short tuft above the nasal openings. The extreme adaptive radiation as seen in the passerine island birds is driven by the total lack of competition by other land birds and led to the occupation of all possible trophic niches.

Regarding the fossil record of island mammals, this archipelago theory has been brought to the fore to explain the adaptive radiation as observed in several fossil taxa. An example is the Late Miocene island Gargano, which supposedly formed part of a larger archipelago (Apulo-Dalmatic Realm). The presence of several rapidly evolving sister taxa in the murid *Mikrotia* and the pika *Prolagus* was explained in the light of an archipelago by Claudio De Giuli in the 1980s. The various taxa that appear in chronological order in the fossil record of Gargano are in this view considered to represent successive invasions from another island in the island arc.

A seemingly related case might be provided by the Pleistocene deer of Crete and Karpathos. During the Late Pleistocene, both islands were inhabited by the insular deer genus *Candiacervus*. There were at least eight species on Crete and two species on Karpathos. However, there is a disagreement concerning the origin (and therefore the taxonomy) of these species. If *Candiacervus* is a multiple-island endemic, then it must have originated on one island and have subsequently colonised the other one. In that case, the attribution of both Karpathos and Cretan deer to a single endemic genus is correct. More plausible, though, is that *Candiacervus* is a single island endemic of Crete, phylogenetically unrelated to the Karpathos deer. The similarity between the Karpathos deer and the Cretan deer, which formed the basis for the taxonomy, is then the result of convergent evolution. The similar environment of the islands drove the same evolutionary changes, resulting in two morphologically similar but phylogenetically unrelated taxa. This option is the most likely, based on antler morphology. If so, the name *Candiacervus* should be restricted to the Cretan deer and a new name should be applied to the Karpathos deer (Figure 22.7).

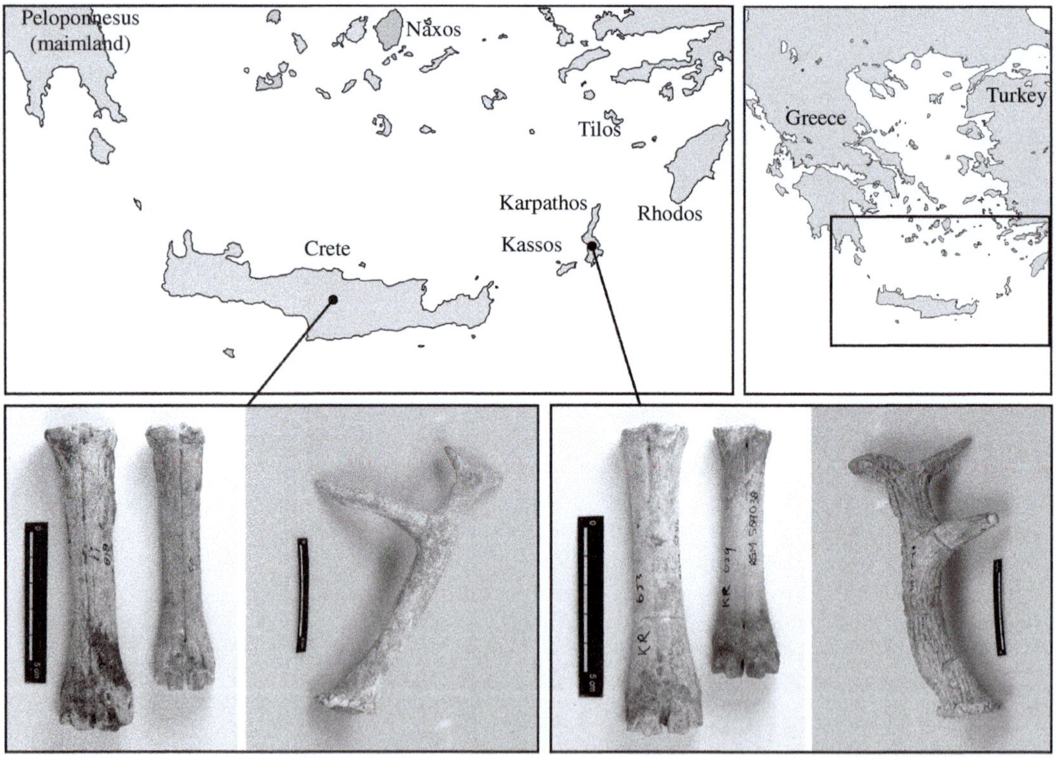

Figure 22.7 Map of the southern Aegean Sea and metapodials and antlers of the deer *Candiacervus* from Crete (right antler) and '*Candiacervus*' from Kassos and Karpathos (left antler); all in anterior view. Although morphological differences exist between deer of these three islands, no detailed study has been published so far on deer from Kassos and Karpathos. The latter appear to have more robust metapodials and different antler morphology and aspect, including a grooved surface.

Speciation Within One Island

A counter-theory to the archipelago speciation is that of speciation *in situ* on the island itself. Although allopatric speciation may be a major type of speciation in island species, a number of endemic species evolved within the same island, as already indicated by John D. Sauer (1990). The rodents of the Philippines provide an interesting example of diversification both within and between islands. The Philippines Archipelago consists of about 7,000 islands. Some islands merged during the low sea-level stands of the Pleistocene but many remained isolated by deep sea channels (Figure 15.1). Most native rodents of the Philippines are members of just two endemic clades. According to Heaney and colleagues (2013), diversification within these two clades has involved both colonisation between islands and speciation within each island. The largest Philippine island, Luzon, has the greatest concentration of endemic mammalian species in the world, as observed by Heaney, Balete, and Rickart (2016) in their book on Luzon. According to Heaney and colleagues (2013), despite this island being part of an archipelago, a significant percentage of the species at any given locality of Luzon are members of clades that have co-occurred for million years on that island, and only a few species are members of clades that have arrived more recently. Furthermore, Heaney and colleagues (2018) document for Mindoro the presence of an endemic clade (*Apomys*) consisting of probably four species. With an area of only 9,735 km^2, Mindoro is the smallest oceanic island supporting a living mammalian clade of within-island diversification.

Within-island speciation is the only plausible explanation for the extreme variation observed in the Pleistocene deer *Candiacervus* of Crete in the Mediterranean, as this large island is not part of an archipelago (see also above). Here, six size groups, and in one of them three types of skulls and antlers, can be distinguished based on biometrics of cranial and postcranial materials from different sites. These eight species clearly differ too much from each other to assume a single ecological niche, based on the spectacular differences in body size between the individual species. The largest, *Candiacervus major*, is even the tallest deer that ever lived. It stood 1.65 m at the shoulder and was thus taller than the elk (*Alces alces*) of today. Speciation here is best explained by a gradual evolution into different ecological niches, initiated by differential habitat preferences, a theory postulated by Scott Diehl and Guy Bush (1989). On the ground of body proportions, molar morphology, and wear pattern, the specialist trophic niches occupied by the eight taxa might, tentatively, can be summarised as follows: grassy food or prickly bushes on a rocky hill (*Candiacervus ropalophorus* and *Candiacervus* sizes 2), grasses on a steppe-like plain (*Candiacervus cretensis*), leaves and branches in a forest, like red deer (*Candiacervus rethymnensis*), and leaf-like food and soft bushes in a forested terrain with many obstacles (*Candiacervus dorothensis* and *Candiacervus major*).

Similar cases are provided by other island deer. In the Late Miocene, five-horned deer (*Hoplitomeryx)* of Gargano four adult size groups and one very large juvenile size group can be recognised, of which at least two occur in each fissure, representing a distinct unit in time. The tallest species, *Hoplitomeryx kriegsmani*, was comparable to *Candiacervus major* with its extremely long, slender giraffe-like limbs. A third case is the Pleistocene deer from the Japanese Ryukyu Islands (*Cervus astylodon*). Four size groups can be distinguished among the metacarpal bones from Kume island. Presumably, the different morphotypes belong to different populations, perhaps occupying different habitats. Again, the same is observed in the dwarf deer (*Cervus*) of Malta. The variation in length between the metatarsals is unusually large, and three size groups are recognised. Only two size classes are seen in the Late Pleistocene dwarf deer (*Cervus* sp.) of Masbate (Philippines) and Karpathos (*Candiacervus*), both deer of very small islands.

A significant step towards the formation of new species is the limitation of hybridisation during the early stages of divergence. As Peter Grant (1994) noted, species that are resource specialists are less likely to hybridise than resource generalists. Therefore, factors such as environmental heterogeneity and absence of competitive species play an important role to this type of speciation.

Species Flock

The term species flock is sometimes applied to a group of closely related species within a restricted area. The morphological differences between these species are small, and it is often not easy to distinguish them from each other, particularly when dealing with fossil forms. A species flock likely represents an adaptive radiation which is still at its initial stage. As a result, the separate species did not yet reach the morphological diversity seen in later stages.

In several fossil island mammals, an unusually large variation in body size is present. Fossils of smaller and larger specimens are found together in the same layer of the same locality, and proper size groups are often not easily made. For example, a very large size variation or unusual range is seen in the Late Miocene gymnure *Deinogalerix* of Gargano. At present, five species within two lineages have been formally described, but many specimens are intermediate in size and cannot be assigned to a species. The large variation is not easily explained by clear-cut species.

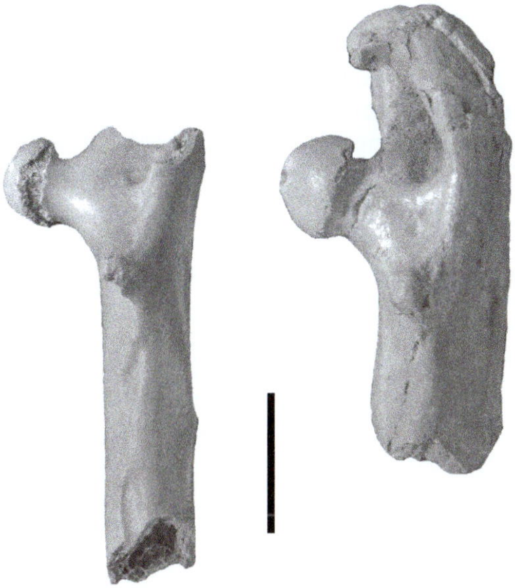

Figure 22.8 'Small' and 'large' proximal femora of the giant hutia of Anguilla Bank (*Amblyrhiza inundata*); posterior view. The individual represented by the small femur is estimated to have weighed about 74 kg, whereas the individual represented by the large femur was more than twice as heavy, with an estimated body mass of about 168 kg. Scale bar: 5 cm. (Photograph: Ross MacPhee.)

A rather similar situation is perhaps met on Mallorca. The Late Pleistocene endemic mouse-goat *Myotragus balearicus*, the last species of a lineage progressively evolving smaller body size, exhibits a great size variation, far exceeding that of mainland relatives. The largest metatarsal is three times the smallest from the same cave (e.g. Muleta) and the same level, and all intermediate sizes are also found. Grouping of bones into different size ranges or morphotypes is not possible, as first reported by Sondaar (1977). It seems that this terminal species of its lineage had expanded its niche maximally compared to earlier species.

Other examples of an unusually large variation in size are provided by the giant hutia (*Amblyrhiza inundata*, Figure 22.8) of the West Indies, the Cretan dwarf hippo (*Hippopotamus creutzburgi*), the Cretan elephant (*Elephas antiquus creutzburgi*), the dwarf mammoth (*Mammuthus exilis*) of the northern Californian Channel Islands, and the small Japanese mammoth (*Mammuthus protomammonteus* or *Mammuthus trogontherii*).

The evolution of a species flock was first observed in the East African Great Lakes. Here, the haplochromine cichlid fishes underwent a rapid speciation (adaptive radiation), resulting in a wide range of trophotypes as first observed by Geoffrey Fryer and T.D. Iles (1972). Initially, they differed little from their immediate ancestors. The subsequent rapid differentiation resulted mainly in differences in the mouth, which gradually became adapted to different types of food: detritus, fishes, shells, crabs, insects, phytoplankton, and zooplankton. The driver behind this species flock was a sudden availability of new niches, when the Holocene cichlids entered a new and still unoccupied lake. Lake Victoria was filled only about 14,000 years ago, due to the creation of the Rift Valley, which started to arise from 750,000 years ago. Immediately after the formation of the lake, a host of adaptive zones became available which gave the cichlids the possibility to radiate beyond the degree seen in related cichlids. The fossil cases mentioned above, showing a large variation in size and shape, may represent such a type of species flock as well, which is, in turn, best explained as the initial phase of an adaptive radiation in different species.

References

Brown J.H., Kodric-Brown A. 1977. Turnover rates in insular biogeography: Effect of immigration on extinction. *Ecology*, 58: 445–449.

Brown J.H., Marquet P.A., Taper M.L. 1993. Evolution of body size: Consequences of an energetic definition of fitness. *American Naturalist*, 142: 573–584.

Case T.J. 1978. A general explanation for insular body size trends in terrestrial vertebrates. *Ecology*, 59: 1–18.

Damuth J. 1993. Cope's rule, the island rule and the scaling of mammalian population density. *Nature* 365: 748–750.

Darwin Ch. 1839. *Narrative of the Surveying Voyages of His Majesty's Ships Adventure and Beagle between the years 1826 and 1836, Describing Their Examination of the Southern Shores of South America, and the Beagle's Circumnavigation of the Globe. Journal and Remarks 1832–1836.* London: Henry Colburn.

Darwin Ch. 1859. *On the Origin of Species by Means of Natural Selection, or the Preservation of Favoured Races in the Struggle for Life.* John Murray, London.

Dayan T., Simberloff D. 1994. Morphological relationships among coexisting heteromyids: an incisive dental character. *American Naturalist*, 143: 462–477.

Demment M.W., Van Soest P.J. 1985. A nutritional explanation for body-size patterns of ruminant and nonruminant herbivores. *American Naturalist*, 125: 641–672.

Diehl S., Bush G.L. 1989. The role of habitat preference in adaptation and speciation. In: Otte D., Endler J.A. (eds), *Speciation and Its Consequences*. Sinauer, Sunderland, MA, pp. 345–365.

Freudenthal M. 1976. Rodent stratigraphy of some Miocene fissure fillings in Gargano (prov. Foggia, Italy). *Scripta Geologica*, 37: 1–23.

Fryer G., Iles T.D. 1972. *The Cichlid Fishes of the Great Lakes of Africa: Their Biology and Evolution*. Oliver and Boyd, Edinburgh.

Gifford E.W. 1919. Field notes on the land birds of the Galápagos Islands and of Cocos Island, Costa Rica. *Proceedings of the California Academy of Sciences (Series 4)*, 2: 189–258.

Grant P.R. 1986. *Ecology and Evolution of Darwin's Finches*. Princeton University Press, Princeton, New Jersey.

Grant P.R. 1994. Population variation and hybridization: Comparison of finches from two archipelagos. *Evolutionary Ecology*, 8: 598–617.

Heaney L.R. 1978. Island area and body size of insular mammals: Evidence from the tri-colored squirrel (*Callosciurus prevosti*) of Southeast Asia. *Evolution*, 32 (1): 29–44.
Heaney L.R. 2000. Dynamic disequilibrium: A long-term, large-scale perspective on the equilibrium model of island biogeography. *Global Ecology and Biogeography*, 9: 59–74.
Heaney L.R., Balete D.S., Rickart E.A. 2013. Models of oceanic island biogeography: Changing perspectives on biodiversity dynamics in archipelagoes. *Frontiers of Biogeography*, 5 (4): 249–257.
Heaney L.R., Balete D.S., Rickart E.A. 2016. *The Mammals of Luzon Island: Biogeography and Natural History of a Philippine Fauna*. Johns Hopkins University Press.
Heaney L.R., Kyriazis C.C., Balete D.S., Rickart E.A., Steppan S.J. 2018. How small an island? Speciation by endemic mammals (*Apomys*, Muridae) on an oceanic Philippine island. *Journal of Biogeography*, 45 (7): 1675–1687.
Koch P.L., Hoppe K.A., Webb D.S. 1998. The isotopic ecology of late Pleistocene mammals in North America Part 1. Florida. *Chemical Geology*, 152: 119–138.
Lack D. 1947. *Darwin's Finches*. Cambridge University Press, Cambridge.
Lawlor T.E. 1986. Comparative biogeography of mammals on islands. *Biological Journal of the Linnean Society*, 28: 99–125.
Lister A.M. 1989. Rapid dwarfing of red deer on Jersey in the Last Interglacial. *Nature*, 342: 539–542.
Lomolino M.V. 1985. Body size of mammals on islands: The island rule reexamined. *American Naturalist*, 125: 310–316.
Lomolino M.V. 2000. Ecology's most general, yet protean pattern: The species-area relationship. *Journal of Biogeography*, 27: 17–26.
Lomolino M.V. 2005. Body size evolution in insular vertebrates: Generality of the island rule. *Journal of Biogeography*, 32: 1683–1699.
Lomolino M.V. 2016. The unifying, fundamental principles of biogeography: understanding *Island Life*. *Frontiers of Biogeography*, 8 (2): fb_29920.
Lyras G.A., vVan der Geer A.A.E., Dermitzakis M.D., de Vos J. 2006. *Cynotherium sardous*, an insular canid (Mammalia: Carnivora) from the Pleistocene of Sardinia (Italy), and its origin. *Journal of Vertebrate Paleontology*, 26 (3): 735–745.
Lyras G.A., van der Geer A., Rook L. 2010. Body size of insular carnivores: Evidence from the fossil record. *Journal of Biogeography*, 37: 1007–1021.
MacArthur R.H., Wilson E.O. 1963. An equilibrium theory of insular zoogeography. *Evolution*, 17: 373–387.
Mansino S., Crespo V.D., Montoya P., Ruiz-Sánchez F.J. 2017. Muridae from the late Miocene site of Venta del Moro (Eastern Spain). *Historical Biology*, 29 (5): 677–691.
McNab B.K. 2001. Functional adaptations to island life in the West Indies. In: Woods C.A., Sergile F.E. (eds), *Biogeography of the West Indies. Patterns and perspectives*, 2nd edn CRC Press, Boca Raton, Florida, pp. 191–210.
McNab B.K. 2002. Minimizing energy expenditure facilitates vertebrate persistence on oceanic islands. *Ecology Letters*, 5: 693–704.

Millien V. 2006. Morphological evolution is accelerated among island mammals. *PLoS Biology*, 4 (10): e321.

Millien V., Jaeger, J.-J. 2001. Size evolution of the lower incisor of *Microtia*, a genus of endemic murine rodents from the late Neogene of Gargano, southern Italy. *Paleobiology*, 27 (2): 379–391.

Millien-Parra V. 2000 (a). Species differentiation among muroid rodents on the basis of their lower incisor size and shape: Ecological and taxonomical implications. *Mammalia*, 64 (2): 221–239.

Millien-Parra V. 2000 (b). The evolution of *Microtia* Freudenthal, 1976 (Mammalia, Rodentia), an endemic genus from the Neogene of the Gargano island, southern Italy. In: Rheinwald G. (ed.), *Isolated Vertebrate Communities in the Tropics. Proceedings 4th International Symposium, Bonn. Bonner Zoologische Monografien*, 46: 381–389.

Palombo M.R., Bover P., Valli A.M.F., Alcover J.A. 2006. The Plio-Pleistocene endemic bovids from the Western Mediterranean Islands: Knowledge, problems and perspectives. *Hellenic Journal of Geosciences*, 41 (1): 153–162.

Parra V., Loreau M., Jaeger J.-J. 1999. Incisor size and community structure in rodents: Two tests of the role of competition. *Acta Oecologica*, 20: 93–101.

Patterson C. 1978. *Evolution*. British Museum (Natural History), London.

Raia P., Barbera C., Conte M. 2003. The fast life of a dwarfed giant. *Evolutionary Ecology* 17: 293–312.

Sauer J.D. 1990. Allopatric speciation: Deduced but not detected. *Journal of Biogeography*, 17: 1–3.

Schütt G. 1973. Pleistozäne Caniden (Carnivora, Mammalia) aus Java. *Proceedings of the Koninklijke Academie van Wetenschappen* B, 76: 446–471.

Sondaar P.Y. 1977. Insularity and its effect on mammal evolution. In: Hecht M.N., Goody P.L., Hecht B.M. (eds), *Major Patterns in Vertebrate Evolution*. Plenum, New York, pp. 671–707.

Steadman D.W. 1995. Prehistoric extinctions of Pacific Island birds: Biodiversity meets zooarchaeology. *Science*, 267: 1123–1131.

van Bemmel A.C.V. 1973. The concept of superspecies applied to Eurasiatic Cervidae. *Zeitschrift für Säugetierkunde*, 38 (5): 295–302.

van der Geer A.A.E. 2018(a). Uniformity in variety: Antler morphology and evolution in a predator-free environment. *Palaeontologia Electronica*, 21.1: 9A.

van der Geer A.A.E. 2018(b). Changing Invaders: Trends of gigantism in insular introduced rats. *Environmental Conservation*, 45 (3): 203–211

van der Geer A.A.E., Lomolino M., Lyras G.A. 2017. Islands before man: The species-area relationship during the late Pleistocene. *Journal of Biogeography*, 44, 995–1006.

van der Geer A.A.E., Lyras G.A., Volmer R. 2018. Insular dwarfism in canids on Java (Indonesia) and its implication for the environment of *Homo erectus* during the Early and earliest Middle Pleistocene. *Palaeogeography, Palaeoclimatology, Palaeoecology*, 507: 168–179.

Whittaker R.J., Fernández-Palacios J.M. 2007. *Island Biogeography: Ecology, Evolution, and Conservation*, 2nd edn. Oxford University Press, Oxford.

CHAPTER TWENTY THREE

The Island Rule: Dwarfism and Gigantism

Evolution of Island Mammals: Adaptation and Extinction of Placental Mammals on Islands, Second Edition.
Alexandra van der Geer, George Lyras and John de Vos.
© 2021 John Wiley & Sons Ltd. Published 2021 by John Wiley & Sons Ltd.

> *.....wherever we find pygmy elephants we have also giant rodents; in these presumably small isolated populations gigantism in small rodents appears to have been as advantageous to the animals as is dwarfing in elephants. (Hooijer 1967, 143)*

Dramatic size changes (Figure 23.1), observed in many insular mammalian taxa, are certainly the best-known adaptations to insular conditions.

The Island Rule: A Graded Trend

On islands, small mammal species tend to evolve larger body size, whereas large mammal species tend to evolve smaller body size. The existence of this pattern was first observed by J. Bristol Foster (1964) based on a large dataset containing 116 insular species or subspecies from Western North American and European islands. He found a clear tendency towards gigantism in rodents but towards dwarfism in lagomorphs, carnivores, and artiodactyls. The trend is so common in island

Figure 23.1 Artistic reconstruction of the dwarf elephant of Sicily (*Palaeoloxodon falconeri*) and the dwarf hippo of Cyprus (*Hippopotamus minor*) against outlines of their respective ancestors, *Palaeoloxodon antiquus* and *Hippopotamus amphibius*. (Drawing George Lyras.)

species (Figure 23.2) that it has been dubbed the island rule since Van Valen (1973) coined the term, or, alternatively, Foster's rule. Lomolino (2005) and Lomolino and colleagues (2012) elaborated on the generality of the pattern by emphasising the importance of the combination of various selective forces whose importance varies with body size of the species. Together these forces, or variables, contribute to the overall, emergent pattern among across species, which is the graded trend on islands from gigantism in small species to dwarfism in large species. On the mainland, ecological interactions and displacement drive diversification in all traits, including body size, whereas on species-poor islands, these selective pressures are reversed and ecological release results in insular species converging on intermediate body sizes. As we shall see in Chapter 24, this ecological release also drives convergent evolution of functional traits, often resulting in bizarre morphologies, such as ever-growing incisors and disproportionate short limbs.

This trend is particularly evident in the fossil record, which provides some of the best examples as shown by Lomolino and colleagues in 2013. Several insular elephant and hippo species evolved a body mass reduction of at least 90% (Figure 23.1), while several rodent species were over five times larger, compared to their mainland ancestor. The trend seems to apply to fossil hominids as well, taken the diminutive species *Homo floresiensis* (Chapter 13, Figure 13.8) and *Homo luzonensis* (Chapter 15) into account.

The Dutch palaeontologists Dirk Albert Hooijer (1967) and Paul Yves Sondaar (1977) were among the first to recognise the generality of the trend in palaeo-insular species of the Indonesian archipelago, the Netherlands Antilles, and the Mediterranean. Lomolino and colleagues (2013) – based on an extensive dataset containing 63 fossil island species worldwide – confirmed the antiquity of the island rule and its ecological basis. The island rule is an antique and pervasive trend, yet with much scatter about this general trend, especially for carnivore species. This variation is sometimes studied in isolation and considered as exceptions to the rule (e.g. Meiri and colleagues, 2006, 2008). However, much of the residual variation is consistent with an ecological explanation of the island rule, where not only ancestral body mass is taken into account but also variables like environmental characteristics, species bauplan, island size and isolation, type of prey, and above all, ecological interactions with mammalian predators and competitors. In other words, populations of exactly the same species but on very different islands are unlikely to evolve exactly the same degree and direction of body size change. For example, Lyras and colleagues (2010) found that

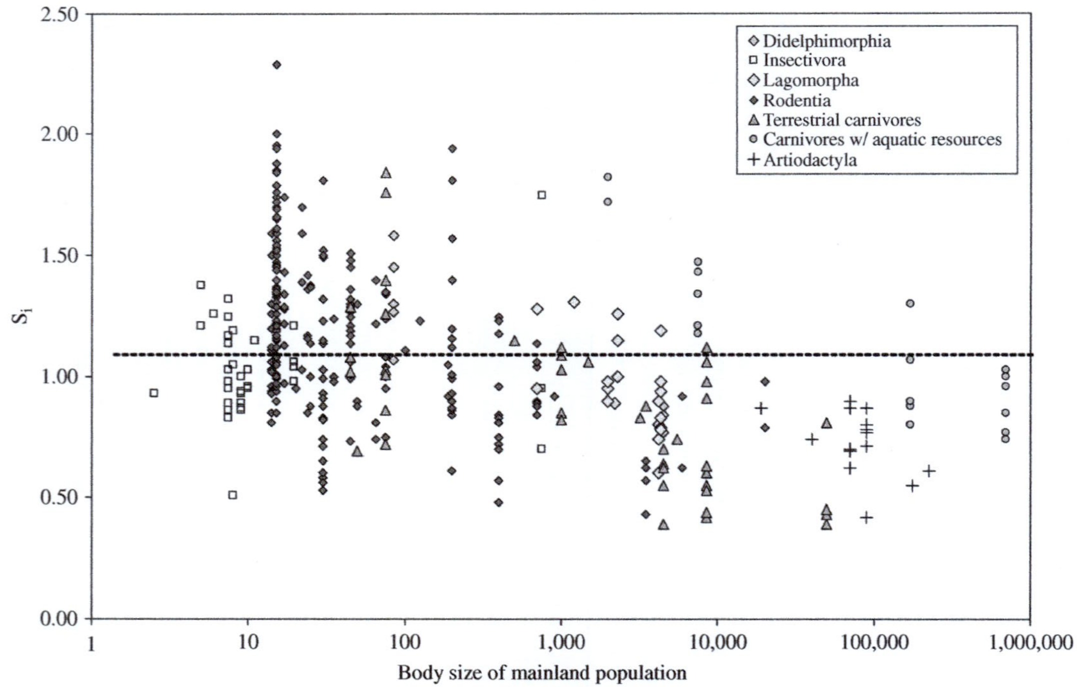

Figure 23.2 Body mass trends for populations of insular mammals. Si = relative mass of insular forms expressed as a proportion of body mass of their mainland relative. (From Lomolino, 2005.)

Pleistocene island carnivores either evolved smaller or larger body size, depending on their prey (see below) rather than on their ancestral size. On the other hand, populations of different species but on the same island are equally unlikely to evolve towards the same end result.

As we will see below, amongst the most spectacular examples are a dwarf elephant of Sicily (*Palaeoloxodon falconeri*; early Middle Pleistocene) with an estimated body mass reduction of over 98%, the dwarf hippo of Cyprus (Pleistocene) with a body mass reduction of 96%, and the giant gymnure or moon-rat (*Deinogalerix koenigswaldi*; Late Miocene) of Gargano, a palaeo-island of Italy (Plate 37), which was about 200 times the size of its mainland shrew-sized ancestor.

Island Isolation

Natural selection already operates, not just during residence on the island but during immigration as well. As is clear from the fossil record, which only preserves long-time survival successes, the founding populations are not a random subset of the nearest mainland faunas. Typically, remote palaeo-islands harbour at most elephants, ruminants, rodents, and shrews, while less remote islands and continental shelf islands are richer in species and more representative of mainland assemblages. Extremely remote islands, such as those of Polynesia and Micronesia, have no native terrestrial mammals at all, except for bats in some cases. Lomolino (1984, 2005) predicted that 'immigrant selection' (natural selection for the better dispersers) should favour larger individuals in the founding population. Foster (1964) had indeed noted a positive correlation between island isolation and body size in insular populations of deer mouse (*Peromyscus*). This was confirmed by later authors for nine more rodent species (summarised by Lomolino, 2005). Based on a large global dataset of extant island species, Lomolino and colleagues (2012) found that isolation indeed influences the evolution of small mammals – those lighter than 2,692 g – while body size evolution in larger mammals was independent of island isolation. This was confirmed for island proboscideans by Athanassiou and colleagues (2019) and van der Geer and colleagues (2016) who found no correlation between the degree of dwarfism and island isolation.

However, geographic distance is not a good estimator for accessibility and thus isolation. For example, the Indonesian island of Sulawesi has been colonised by murine rodents, squirrels, water buffaloes, pigs, macaques, and hominins prior to the Late Pleistocene (van den Bergh and colleagues, 2016), despite the great width of the Makassar Strait. At the other end of the spectrum is the island of Flores, which is relatively near the mainland, especially during low sea-level stands, but has been colonised only by murids, stegodons, and hominids. Flores appears to be practically unreachable due to the strong Indonesian through-flow currents that pass through the straits. Presumably, prevailing oceanic currents play a major role in determining island accessibility and isolation, as shown by Jason Ali and Matthew Huber (2010) for Madagascar.

Island Area and Resource Limitation

Island size not only influences mammalian diversity on islands – as predicted by the theory of island biography – but body mass evolution as well. Large mammals have higher resource requirements, while smaller islands have a lower total productivity, which, in combination with the often higher population densities and ecological release from smaller competitors, should favour smaller individuals, and thus drive evolution towards dwarfism. Island area is in this view mainly a proxy for resource availability and diversity and competition from large species. However, analyses by Lomolino and colleagues (2012, 2013) have shown that for large mammal species, body size reduction seems not correlated to island area per se, counter to predictions based on resource limitation. The only exception seems to be exhibited by dwarf proboscideans on predator-free islands, which are smaller on the smaller islands (see below), especially stegodons. Potentially, resource limitation may only play a significant role in truly large mammals.

For small mammals, island area is not postulated as a driving factor behind body mass changes; yet, there are indications that this might be the case. Heaney (1978), based on a large dataset of 39 subspecies of the Asian tri-coloured squirrel (*Callosciurus prevostii*), found that squirrels on very small islands were smaller than on larger islands, which he attributed to resource limitation. Furthermore, these squirrels were much smaller on very large islands, likely due to more intense competition. Heaney (1978) postulated a hump-shaped or parabolic pattern, where body size increased with island area to a point, and then decreased on larger islands (perhaps as diversity of other mammals increased beyond some threshold of ecological tolerance). Heaney (1978) predicted that the smaller the species, the smaller the threshold or optimal island size conferring peak/maximal size of insular populations. Although Adrian Lister and Charlotte Hall (2014) did find that body size of mice (*Mus musculus*) was positively correlated to island size on the Scottish and Faroe islands, this trend included the largest island (900 km^2) they studied, thus failing to exhibit any decrease phase as predicted by Heaney's hypothesis. Yom-Tov and colleagues (1999) reported that body size of introduced Polynesian rats (*Rattus exulans*) not only fails to exhibit the predicted, parabolic pattern in body size with island size but also actually exhibits a negative trend – i.e. body size of these Pacific island rats *decreases* as island size increases. This apparently anomalous pattern likely reflects the more direct effects of species richness and competition from other vertebrates, which increase with island size and likely dominates over any influence of resource limitation (van der Geer 2018).

For large mammals, resource limitation was widely considered the drive behind the evolution of dwarfism in larger species, as integrated in the ecological interpretation of the island rule first proposed by Lomolino (1985), followed by, amongst others, Roth (1990, 1992) to explain dwarfism in fossil insular elephants. However, Lomolino and colleagues (2012), based on a much larger dataset than the original one (1985), unexpectedly failed to find evidence for resource limitation as driver.

Sondaar (1977) was one of the first to reject resource limitation as extrinsic factor. He explained dwarfism in elephants, hippos, and deer as adaptation to facilitate locomotion in rugged terrains, which characterised islands inhabited by dwarfed forms of these large mammals (e.g. islands of the Mediterranean Sea). This is evidenced, amongst others, by the relatively shorter limbs, in combination with a higher degree of bone fusions in limb and foot bones of these grazing and browsing mammals.

Resource limitation may, however, play a role in the first phase of evolutionary changes in island populations of large mammals before an equilibrium between herbivore and plant populations has been established. Large herbivores on islands often overgraze and then experience a bottleneck phenomenon at some point. Today, this can be observed in isolated areas, such as predator-free forest fragments in Venezuela. After collapse of herbivore populations, plant populations recover and rebound, eventually triggering another eruption in herbivore populations, albeit with a time lag that varies with generation time of the herbivores. Such boom-and-bust cycles of herbivore populations have been reported for caribou on Arctic islands and white-tailed deer in predator-free suburbs of the eastern United States. The fossil record notoriously lacks such evidence, likely because this very early phase is too short to be detected on a geological timescale. One instance was reported in the literature as putative example – the Late Pleistocene fossil assemblage of dwarf deer (*Candiacervus ropalophorus*) in the cave Gerani 4 on Crete. Although originally presumed to represent a complete herd, the taphonomy is best explained as an attritional profile, instead of a catastrophic profile (van der Geer and colleagues, 2014).

Evidence for resource limitation in the form of decline in nutritive quality (in contrast to plant biomass and carrying capacity, per se) is provided by the remains retrieved from another cave on Crete (Mavro Muri). Here, the deer bones are brittle and thin, suffering from some sort of osteopenia. Likely, this pathology was caused by a mineral deficiency, as suggested by Dermitzakis and colleagues (2006), and explained as a nutritional imbalance leading to a disturbed bone metabolism by Lyras and colleagues (2019) (but see Reumer and Robert, 2005).

Aquatic Prey

The generality of the island rule has sometimes been questioned in the case of a few larger mammals. Meiri and colleagues (2006), for example, present exceptions among carnivores in case of prey abundance. Indeed, resource limitation is certainly not the case for those fortunate island inhabitants that find unusually large resource abundance, as is the case with the Kodiak bear (*Ursus arctos middendorffi*) on Kodiak Island, Alaska. Here, ample populations of salmon along with carcasses of whales and other marine wildlife washed ashore provide rich resources for these opportunistic carnivores. As a result, the bears here are on average larger than their mainland relatives. As Lomolino (1985) observed, large size in these already large carnivores is not unexpected because they are essentially marine, not insular – at least in terms of the bulk of their diets. Especially mustelids with aquatic prey evolve much larger body masses. This is confirmed by the fossil record, where almost all island otters are larger than their ancestors, except for one Sardinian otter (*Algarolutra majori*, about 10% body mass reduction). The largest otter (*Megalenhydris barbaricina*) was even about 2.8 times the size of its presumed ancestor (*Lutra simplicidens*).

Ecologically Relevant Competition

As emphasised earlier, the presence or absence of competing taxa – ecological release from competition – is likely the most important factor in body size evolution in island mammals. In the total absence of other taxa, a coloniser may evolve the optimal body mass for its bauplan and ecological niche, whereas ecological interactions in species-rich ecosystems result in divergent selection and displacement from the hypothetical (species-neutral) optimal size. The effects of ecological release are most likely to be observed in small mammals because they are, by definition, likely to be dominated by larger species, at least on the mainland (Lomolino, 1985). This is best observed in the fossil record when small mammals establish populations on species-poor islands. As van der Geer and colleagues (2013) reconstructed, the resident giant species, having evolved for hundreds of thousands of years in isolation free of competition, undergoes reduction in body size following colonisation by ecologically similar mammals, eventually suffering extinction (a series of phenomena consistent with Edward O. Wilson's (1961) theory of taxon cycles).

The fossil record also provides evidence for the effect of ecological release from competition in large mammal as well. A much lower degree of body size reduction is exhibited by proboscidean populations with competitors (59% reduction on average) than without competitors (83% reduction on average). An example is provided by the large mammals of Crete, where the Early-early Middle Pleistocene is characterised by a dwarf mammoth (*Mammuthus creticus*; 96% reduction) without competitors and the late Middle-Late Pleistocene by a small elephant (*Palaeoloxodon creutzburgi*; 62% reduction) which co-occurred and presumably competed with several deer species (*Candiacervus*). On Flores, the Early and Late Pleistocene are characterised by very small stegodons (*Stegodon sondaari*, *Stegodon florensis insularis*, both about 85% reduction in mass) as the only large herbivores. Other clear examples of the effect of release from competition are found on Mallorca, where the total absence of any other herbivore gave the caprine *Myotragus* the possibility to reduce its body mass by about 80% – the smallest specimens were just 23 kg, as calculated by Bover (2004).

Similar bouts of extreme gigantism suggesting the interplay between competition and body size are seen in herbivorous, insular birds in the absence of grazing and browsing mammals, such as the takahe (*Porphyrio mantelli*) and the moas (Dinornithidae) of New Zealand. The same applies to the extinct elephant birds (Aepyornithidae) of Madagascar and the dodo (*Raphus cucullatus*) of Mauritius, which evolved extreme gigantism under ecological release from mammalian competitors. The largest species of elephant bird (*Vorombe titan*) reached a mass of 730 kg, as calculated by Hansford and Turvey (2018).

Temporal Variation

Along with competition and other ecological variables that contribute to the spread about the general trend line of the island rule, time is an important factor influencing body size evolution of insular mammals. In contrast to extant and recent mammals, palaeo-insular mammals exhibit much more extreme cases of body mass evolution – some dwarfing to less than 5% the ancestral body mass in otherwise large mammals, or increasing to over 100 times the ancestral body mass in initially small mammals. The most plausible explanation is that these different degrees between extant and fossil mammals are simply a function of the much more extended geological period during which the latter populations evolved in isolation. Extant, native island species and populations occurring on continental islands spent a comparatively short time span in isolation (typically < 15,000 years) since the last sea-level rise at the start of the Holocene (Lomolino and colleagues, 2013). This is just a small fraction of the hundreds of thousand or even millions of years of time in isolation of the palaeo-insular species.

The direction and degree of body size evolution is a result of the complex interaction of physiogeographic variables and ecological feedback; both of these subject to dramatic changes throughout the palaeo-record, thus triggering equally dramatic dynamics in body size of insular species when characteristics of the islands or of the insular communities change. As alluded to above, fluctuating body size trends have indeed been identified in the fossil record, including that for the three insular lineages of burrowing mouse of Gargano, *Mikrotia*, by Freudenthal (1976). Van der Geer and colleagues (2013) found similar trends also in the Gargano giant hamster (*Hattomys*), Cretan giant mice (*Kritimys*, *Mus*), the Cretan shrew (*Crocidura zimmermanni*), the Balearic giant dormouse (*Hypnomys*), the Balearic shrew (*Nesiotites*), and the Sardinian giant pika (*Prolagus sardus*). In particular, the arrival of a phylogenetically related competitor coincides with an abrupt and dramatic reversal of the initial body size trendallotropy (van der Geer and colleagues, 2013). This is not unique to islands but has been observed in mainland rodent lineages as well. For example, Samuel Mansino and colleagues (2017) noted that the *Paraethomys meini* tends to evolve larger body size in localities without species of the related lineage *Paraethomys abaigari* – *Paraethomys jaegeri*, while after the gradual appearance of *Paraethomys abaigari* in the early Pliocene, *Paraethomys meini* evolved smaller body size in sympatric conditions.

Optimal Body Mass

These transformations in body size may represent shifts towards an optimal body mass for a particular body plan (bauplan) and resource exploitation strategy in isolated, species-poor environments (Lomolino, 2005). Indeed, small mammals should have a certain level of optimal energetic trade-off between resource provisioning and offspring production abilities at a certain small body mass, according to Brown and colleagues (1993), Damuth (1993) and Douglas Kelt and Brown (1998). This theoretical optimal body mass for small mammals is supposed to range between 0.1 and 1 kg, depending on, amongst other things, the taxon being considered and its physiological and ecological traits. Thus, the fundamental body size differs among mammalian orders, ranging from approximately 0.3 kg in rodents to 6.9 kg in ungulates (Lomolino, 2005).

Examples of insular giant rodents with a mass between 100 g and 1 kg are numerous. These include (data from van den Hoek Ostende and colleagues, 2017), in order of increasing body size, the giant dormice *Muscardinus cyclopeus* (Pliocene, Minorca; 100 g) and *Maltamys wiedincitensis* (Middle Pleistocene, Sicily; 160 g), the giant rice rats *Megalomys luciae* (Late Pleistocene, Saint Lucia; 160 g) and *Megalomys audreyae* (Late Pleistocene, Barbuda; 180 g), the giant dormouse *Hypnomys morpheus* (Late Pleistocene, Mallorca; 230 g), *Megalomys desmarestii* (Holocene, Martinique; 288 g), the giant rats *Kritimys catreus* (Middle Pleistocene, Crete; 518 g) and *Megaoryzomys curioi* (Pleistocene, Galapagos; 570 g) (Figure 23.3), the giant hamster *Hattomys gargantua* (Late Miocene, Gargano; 874 g), the giant rat *Diplothrix legata* (Pleistocene-Holocene, Central Ryukyu Islands; 581 g), the giant dormouse *Leithia melitensis* (Middle Pleistocene, Sicily; 1 kg; Plate 37), and the giant *Canariomys tamarani* (Late Pleistocene, Gran Canaria; 1 kg).

Furthermore, there are many cases where small mammals evolved a body mass exceeding 1 kg. This is observed in, for example, the giant murids *Papagomys armandvillei* (Late Pleistocene-Holocene, Flores; 1.2 kg), *Canariomys bravoi* (Late Pleistocene, Tenerife; 1.5 kg), *Mikrotia magna* (Late Miocene, Gargano; 1.9 kg), *Coryphomys buehleri* (Late Pleistocene, Timor; over 2 kg), *Phloeomys pallidus* (Pleistocene-Holocene, Luzon; 2.25 kg), the giant gymnure *Deinogalerix koenigswaldi* (Late Miocene, Gargano; c. 3 kg; Plate 37), the giant pika *Gymnesicolagus gelaberti* (Middle-Late Miocene, Mallorca and Minorca; c. 5.4 kg), and the giant hare *Nuralagus rex* (Pliocene, Minorca; c. 14 kg).

In a minority of cases, in about 10 out of 60 insular species (excluding species that are part of island adaptive radiations), body mass of small mammals remained below 100 g. These include, in order of descending mass, *Rhagamys orthodon* (Late Pleistocene, Sardinia; 90 g), *Tokudaia osimensis* (Pleistocene-Holocene, Amami; 87 g), *Agathaeromys praeuniversitatis* (Middle Pleistocene, Bonaire; 78 g), *Oryzomys antillarum* (Holocene, Jamaica; 55 g), *Mus minotaurus* (Late Pleistocene, Crete; 54 g), *Apodemus mystacinus*

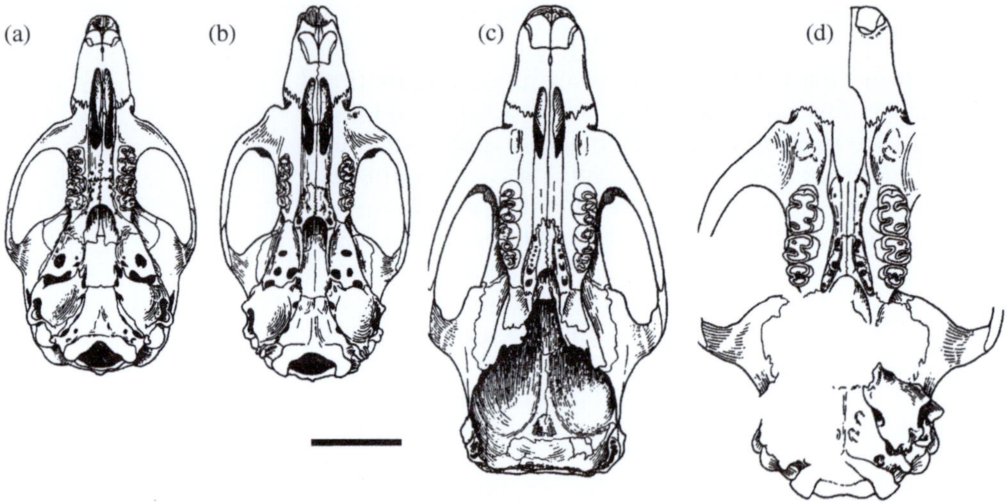

Figure 23.3 Skulls of rice rats in ventral aspect. (a) *Thomasomys aureus*, female, Peru; (b) *Nesoryzomys swarthi*, male, Isla Santiago, Galapagos; (c) *Megalomys desmarestii*, Martinique, West Indies; (d) *Megaoryzomys curioi*, Santa Cruz, Galapagos. Scale bar: 1 cm. The giant Galapagos rice rat (d) is considered most closely related to *Thomasomys* of South America. Initially, it was placed in the same genus with the West Indian giant rice rats (c), but it differs from the latter in a number of morphological characters. The extinct Galapagos rice rat of Santiago (b) did not develop a giant size as the fossil species (d) and likely represents a later colonisation. (Adapted from Steadman and Ray, 1982.)

(Late Pleistocene, Naxos; 46 g), *Malpaisomys insularis* (Late Pleistocene-Holocene, Fuerteventura-Lanzarote; 40 g), *Peromyscus nesodytes* (Late Pleistocene-Holocene, Santa Rosae; 40 g), and *Mus cypriacus* (Holocene; potentially Middle Pleistocene, Cyprus; 18 g).

Taking all instances together, it seems that an optimal body size for palaeo-insular species of rodents, if existing, would fall somewhere between 1 and 2 kg. That is significantly higher than around 300 g predicted by data only on modern island rodents but may possibly indicate the influence of cooler temperatures on mammalian body size (i.e. consistent with Bergmann's rule).

The Island Rule in Detail

Because different functional groups of mammals respond differently but predictably to the same selective forces, it is informative to consider the patterns exhibited by each group in more detail, beginning with the proboscideans, then on to consider the primates, lagomorphs, rodents, insectivores, ungulates, and carnivores. The pattern is not limited to mammals, but has also been demonstrated for other vertebrates, including dinosaurs (Box 23.1).

Elephants, Mammoths, and Stegodons

Proboscideans are the most exemplary island mammals in the fossil record, not only because of their wide occurrence in time and space but also because of their sometimes spectacular body size evolution. Almost all island Proboscidea are smaller than their mainland ancestors, none is larger. Their size reduction, however, varies greatly, and the island species have been reported as pygmies, dwarfs, small-sized, or medium-sized accordingly (Plate 1). The degree of evolutionary body size decrease is predictably correlated to similar variables as were shown to impact size evolution in extant species (see above), as is shown by van der Geer and colleagues (2016) for 22 species of island proboscideans, and Athanassiou and colleagues (2019) for additional Aegean populations. Especially the presence of megaherbivore competitors (deer and bovids) appears to prevent the evolution of extreme body size reduction ('pygmy size' in earlier literature) (see above).

An unexpected variable is island area. Lomolino and colleagues (2013) have shown that for large mammal species, body size does not increase with island area (see above), counter predictions based on resource limitation. On the other hand, a positive correlation appears to exist between island area and body mass for proboscideans, especially for species of *Stegodon* of maritime South East Asia (van der Geer and colleagues, 2016). Island isolation on the other hand fails to predict the degree of body size evolution in island proboscideans. There seems, however, to be threshold of between 6 and 10 km distance between the island and the mainland, below which no dwarfism evolved in proboscideans, as observed by Athanassiou and colleagues (2019).

BOX 23.1

Dwarf Dinosaurs of Europe

Dinosaurs were not exempt from the generality of the island rule. The presence of diminutive dinosaurs in some fossil localities of Romania and Germany is today widely accepted as the result of insular dwarfism. The earliest report of insular dwarfism in dinosaurs was by Baron Franz Nopcsa von Felső-Szilvásof (1912). Nopcsa noticed that the dinosaurs from the Late Cretaceous deposits (about 66–68 million years ago) of Transylvania (Romania) – then referred to as Siebenbürgen in German – were almost always smaller than their nearest relatives from elsewhere and concluded that the area was an island back then. This observation reminded

Othenio Abel, who attended Nopcsa's report, of the then recent discoveries of dwarf elephants in the Mediterranean. He explained the dwarfism of these island species as proof of degenerative inbreeding, to which Nopcsa replied that more likely, it was the effect of the fight for survival in a restricted area that caused the dwarfism of the elephants and hippos. Julius von Pia added to the discussion that the terms like 'large' and 'small' should always be considered in relation to the closest relatives.

Recent work on the dinosaurs from Haţeg Basin of Transylvania verified the early observations of Nopcsa. According to Michael Benton and colleagues (2010), there is a wealth of osteo-histological, phylogenetic, and tectonic data that support his original idea that the small size of some of them was indeed the result of phylogenetic dwarfing in an island setting. One of these dwarf dinosaurs is the titanosaurian sauropod *Magyarosaurus dacus*. Its length was estimated to about 6 m and its weight to less than a ton. Kohen Stein and colleagues (2010) provided histological evidence that the specimens they analysed belonged to fully mature individuals, despite their small size. They also noted their reduced growth rate.

Another example of insular dwarfing in dinosaurs is the macronarian sauropod *Europasaurus holgeri*, discovered within Jurassic marine sediments (about 154 million years ago) of Lower Saxony, Germany. Several individuals of this quadrupedal herbivorous species were discovered, the largest of which had an estimated length of only 6.2 m, a very small size compared to that of other macronarians, such as *Giraffatitan* and *Camarasaurus* that were 15–23 m long. Histological studies by Martin Sander and colleagues (2006) demonstrate that some individuals were adult and thus the small size of *Europasaurus* was the result of insular dwarfism. According to the same authors, the largest palaeo-islands of the region had areas of less than 200,000 km^2, which apparently was too small to support large-bodied sauropods.

Primates

The most important group of insular fossil primates are the Malagasy lemurs, and they evolved small as well as very large body sizes, being part of a spectacular adaptive radiation likely starting from a small-sized ancestral lemur. All 17 extinct lemur species are large or even very large (Table 11.1), larger than the surviving lemur species today. The heaviest species was the giant sloth lemur *Archaeoindris fontoynontii* with its weight of about 200 kg, the size of heavier male gorillas. Another noteworthy group, the Platyrrhini or New World monkeys of the Greater Antilles, consists of five fossil species, whose body sizes all fall in the middle or upper part of the range of extant Platyrrhini. No extreme body size seems to have evolved in this group, but the degree and direction of their body size evolution are unknown by lack of an ancestral species.

Insular hominids include the Late Miocene *Oreopithecus* of Tuscany, the Late Pleistocene *Homo floresiensis* and Middle Pleistocene small-sized *Homo* of Flores, and the late Pleistocene *Homo luzonensis* of Luzon. Their body size evolution has proved to be a challenging case study in insular evolution because of lack of an unequivocal ancestor (see discussions in the relevant island chapters). Both human species (*Homo*) are diminutive, while the Flores human also has primitive wrists and large feet.

Other fossil insular primates include the Early Pleistocene macaque of Sardinia (*Macaca majori*), which evolved a smaller body size relative to its ancestor. This is in line with what is observed in insular primates today, as shown by Lindell Bromham and Marcel Cardillo in 2007. Extant primates in general exhibit the island rule's graded trend (see chart in Lomolino and colleagues, 2017).

Pikas and Hares

Insular pikas appear to evolve a large size without exception. The largest pika known so far is *Gymnesicolagus gelaberti* of the Balearic Islands (Majorca and Minorca), weighing on average 5.4 kg, estimated from the length of its lower molar row. Next in line are *Prolagus imperialis* and *Prolagus apricenicus* of Gargano, followed by *Prolagus figaro* and *Prolagus sardus* (Figure 23.4) of Sardinia and Corsica and *Paludotona etruria* of Tuscany. The Sardinian pikas are chronospecies of a lineage that gradually evolved increasing body mass throughout the Pleistocene, also between successive populations within the species. The youngest populations of *Prolagus figaro* are about 10% heavier than the oldest populations and those of *Prolagus sardus* are about 32% heavier than the respective oldest populations, according to body mass estimations based on limb bones by Blanca Moncunill-Solé and colleagues (2016).

The giant hare of Minorca (*Nuralagus rex*) presumably is the prize winner and reached the amazing body mass of an estimated 14 kg, about twice that of the European hare (*Lepus europaeus*). Not all insular hares, though, developed a large

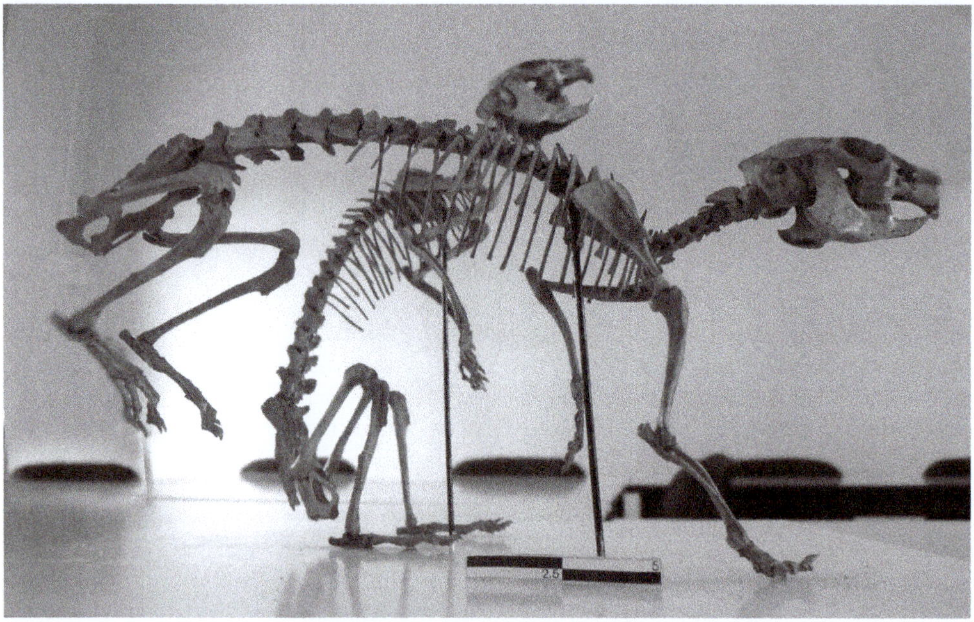

Figure 23.4 Two skeletons of Sardinian pika (*Prolagus sardus*). Palaeontological and Speleological Museum 'E.A. Martel', Carbonia, Sardinia. (Photograph: Alexandra van der Geer.)

size. The *Hypolagus balearicus* from the latest Miocene or Early Pliocene (Messinian) of Ibiza retained a small body size. It might be that it showed endemic features apart from size increase, but the fauna of Ibiza of this period is poorly understood, and endemic characters as well as potentially competing species are unknown so far.

Rodents

The majority of extinct insular rodents developed a larger body mass, compared to their mainland relatives (overview by van den Hoek Ostende and colleagues, 2017). The largest include the giant dormice of Malta and Sicily (*Leithia melitensis*; 1 kg, 12.5 times ancestral size; Plate 37) and Gargano (*Stertomys laticrestatus*; 458 g, 6.4 times size of an early chronospecies and *Mikrotia magna*; 1.9 kg, 9.5 times size of an early chronospecies), the giant rice rats of Martinique (*Megalomys desmaresti*; 288 g, 4.8 times ancestral size), and the Galapagos (*Megaoryzomys curioi*; 570 g, 5.9 times ancestral size) (Figure 23.3). Other fossil rodents are very large, but due to lack of a known ancestor, no data on relative body mass is available. These include the giant rats of Flores (*Papagomys armandvillei*; 1200 g), Timor (*Coryphomys buhleri*; over 2 kg), and the giant hutia of the Lesser Antilles (*Amblyrhiza inundata*; between 50 kg and 200 kg).

A number of exceptions to the general trend towards gigantism are known from the fossil record, other than species that are part of adaptive radiations. These include the Cretan Late Pleistocene mouse and the Naxos rock mouse. The Cretan species (*Mus minotaurus*) was hardly larger than the extant common house mouse (*Mus musculus*). The Naxos species (*Apodemus* cf. *mystacinus*) remained even unchanged.

The general trend for body size increase among small mammals like rodents was hypothetically associated with the feeding behaviour of birds of prey, especially owls. On many islands, birds are the only homoeothermic predators. Owls swallow their prey as a whole, which sets an upper limit to their prey size. Barn owls (*Tyto alba*), for example, cannot take prey with jaw lengths over 17 mm, as shown by Peter Andrews (1990). Larger prey, such as large water voles, rats, or rabbits are only caught as neonates or juveniles, but, even then, rarely. This implies that for island mice, it pays back when they reach at least the size of a large rat. As a response, the owls follow the progressive trend of size increase as well. An extreme outcome is seen in Gargano, where a giant barn owl (*Tyto gigantea*) was contemporaneous with a giant murid (*Mikrotia magna*).

An inconsistency with this raptor-gape size hypothesis, however, is that in the *Mikrotia* lineages, small species seem to coexist with giant species, as is also the case with *Stertomys* of the same island, and with the murids of Luzon, Sulawesi etcetera. More likely, therefore, it may be the other way around; instead of raptor predation driving increased body size in rodents, gigantism of rodents may have been driven by other ecological opportunities (e.g. release from competitors), and the raptors simply followed suit (Box 23.2).

BOX 23.2

Giant Birds as Top Predators and Megaherbivores

The relation between the evolution of gigantism of insular small-sized mammal species and simultaneous gigantism of birds of prey may be much more complicated than just a prey–predator size relation. In fact, gigantism of birds of prey is seen regularly where they constitute the top predators, not only on islands – e.g. New Zealand with the Holocene Haast's eagle (*Harpagornis moorei*) with a wingspan up to 3 m and Gargano with the Late Miocene eagle (*Garganoaetus freudenthali*). Another example of giant predatory birds is the South American phorusrhachids or 'terror birds' of the

Oligocene and Miocene – when South America constituted a continental island – which lived at a time when mammalian carnivores yet had to colonise the island. These predatory birds were too large to fly, with their stature of over 2 m. On islands, the trend is not restricted to predatory birds, since herbivorous birds may also develop gigantic size in the absence of herbivorous mammals; e.g. extreme cases include the moas (Dinornithiformes) of New Zealand, the elephant birds (Aepyornithiformes) of Madagascar, while less extreme but still substantial cases of gigantism include the dodo (*Raphus cucullatus*) on Mauritius, the solitaire (*Pezophaps solitaria*) on Rodriguez, a giant waterfowl (*Garganornis ballmanni*) on Gargano, and a giant flightless pigeon (*Natunaornis gigoura*) on Fiji. Gigantism of these birds apparently is not directly related to gigantism of small mammals, but is instead a result of lack of competition from mammals, perhaps compounded by intensified competition from other insular birds, which are generally small in body size. Apart from, but possibly co-evolving with, body size increase, are trends towards flightlessness in some fossil insular bird lineages, such as the dodo and the Fiji pigeon and the Cretan walking owl (*Athene cretensis*), as well as in the hundreds of recently extinct species of flightless rails and insular waterfowl of the tropical Pacific that David Steadman (1995) reported.

Shrews and Shrew-like Insect-eaters

A number of island shrews evolved a considerably larger body size. A progressive size increase is observed in the *Nesiotites* lineage of red-toothed shrews of Mallorca. The Pliocene species is considerably smaller than the Pleistocene species, and within the latter, a gradual size increase between specimens from older localities and specimens from younger localities is seen. The largest populations at the terminal Pleistocene were even 4.6 times the ancestral size. The closely related Sicilian red-toothed shrew (*Asoriculus burgioi*) of the Early Pleistocene Pellegrino fauna was about twice as large as its mainland relatives. Also, the extinct West Indies shrews (*Nesophontes*) were large, ranging in size from about the size of a common house mouse to that of a brown rat. The body mass of the extinct Cuban solenodon (*Solenodon cubanus*) has been estimated at 1.5–2.0 kg, though the extinct Hispaniola solenodon (*Solenodon marcanoi*) was smaller than the living species (*Solenodon paradoxus*).

By far, the most spectacular cases of island gigantism are provided by the Late Miocene gymnures (*Deinogalerix*) of Gargano. The largest form (*Deinogalerix koenigswaldi*) evolved a body mass increase of over 200 times that of its mainland ancestor, a form close to *Parasorex socialis* (Plate 37). This bizarre species probably survived on aquatic prey, a variable that is correlated to substantially increased body masses in extant island populations, as shown by Lomolino and colleagues (2012).

However, the trend for increasing size is certainly not universal for island shrews. For instance, the Cretan white-toothed shrew (*Crocidura zimmermanni*) retained its body size from the late Middle Pleistocene until today, with minor fluctuations, likely coinciding with climate fluctuations. Also, the Oligocene gymnure of Mallorca (*Tetracus daamsi*) did not increase in size, based on the size of its posterior teeth. Instead, it developed longer and wider anterior teeth than its mainland relative, probably in response to a change in diet.

Deer and Bovids

Dwarfing, together with proportional shortening of the limbs (Chapter 24), evolved in several species of island deer and bovids. These include, amongst others, the lineage of Late Pleistocene Cretan deer (*Candiacervus*), Late Pleistocene Karpathos and Kassos deer ('*Candiacervus' pygadiensis* and '*Candiacervus' cerigensis*), the lineage of Late Miocene five-horned deer of Gargano (*Hoplitomeryx*), the lineage of the Balearic mouse-goat (*Myotragus*), Pleistocene Ryukyu deer (*Cervus astylodon*), Late Pleistocene Malta deer (dwarf *Cervus*), late Middle Pleistocene red deer of Jersey (*Cervus elaphus*; Chapter 22), Late Pleistocene deer of Sardinia (*Praemegaceros cazioti*), the Late Pleistocene water buffaloes of Cebu, Philippines (*Bubalus cebuensis*), Mindoro (*Bubalus mindorensis*), Luzon (*Bubalus* sp.), and the anoas of Sulawesi (*Bubalus depressicornis*, *Bubalus quarlesi*, and *Bubalus grovesi*). Consistent with the ecological explanation for the island rule, such significant body size decrease is restricted to ruminants in unbalanced and depauperate faunas, characteristic for the true islands in the sense of Alcover and colleagues (1998).

An only moderate body size reduction is observed in insular deer and bovids in more balanced faunas. These include, amongst others, red deer of Sicily (*Cervus elaphus siciliae*), Pianosa (*Cervus elaphus*), Corsica (*Cervus elaphus rossii*), Japanese sika deer (*Cervus nippon*), roe deer of Miyako (*Capreolus miyakoensis*), and aurochs of Sicily (*Bos primigenius siciliae*) and Pianosa (*Bos primigenius bubaloides*).

A few examples of island populations of extant deer species show a moderate degree of dwarfing as well, though not to the extent seen in most extinct species. These include three subspecies of white-tailed deer (*Odocoileus virginianus*) of South Carolina (*hiltonensis* on Hilton Head Island, *taurinsulae* on Bulls Island, and *venatorius* on Hunting Island), one subspecies of white-tailed deer of Georgia (*Odocoileus v. nigribarbis* on Blackbeard Island), and a reindeer subspecies on Spitsbergen (*Rangifer tarandus platyrhynchus*).

Gigantism in insular deer occurs as well but is extremely rare. The three known cases are species that are part of an adaptive radiation, again consistent with the ecological explanation for body size evolution – in this case with increasing intensity of competition promoting body size divergence in a diversifying lineage. Such giant species with slender, extremely long limbs are found along with dwarf and middle-sized species on Crete (*Candiacervus major*), Gargano (*Hoplitomeryx kriegsmani*), and Scontrone (perhaps *Scontromeryx*). The metapodials of these deer are longer than those of any known continental deer, exceeding even those of the elk or moose (*Alces alces*).

Hippos and Pigs

All insular hippos are smaller than their mainland ancestors. The size of these endemic forms though differs from island to island. The smallest form is *Hippopotamus minor* from Cyprus (Figures 4.3 and 23.1), which evolved a body mass reduction of 96%, relative to its mainland ancestor. The largest insular forms are *Hippopotamus pentlandi* from Sicily (25% mass reduction) and *Hexaprotodon sivajavanicus* from Java. Other dwarf hippo species are intermediate in size – Crete, 87% reduction; Madagascar, 75% reduction; Malta, 40% reduction – apparently unrelated to island size, as Malta is the smallest island with a hippo. On Java, an island with a long history of gradually decreasing isolation, a trend towards larger size can be observed, as demonstrated by Johanna de Visser (2008), with the largest specimens in the Ngandong fauna at the end of the Middle Pleistocene.

Body mass changes in island pigs are invariably towards smaller size. The Sardinian *Eumaiochoerus etruscus* of the Late Miocene is about 15% smaller than the mainland forms. Also, Sondaar's pig (*Sus sondaari*) of the Pliocene, also from Sardinia, is about 15% smaller than its ancestor (*Sus arvernensis*). In this lineage, however, a secondary body mass increase has been observed by van der Made (1988) because the remains from an older site (Capo Mannu, western Sardinia) are smaller than those from a younger site (Capo Figari, north-eastern Sardinia), perhaps parallel to what is seen in the Javanese hippo. The Middle Pleistocene Sulawesi pig (*Celebochoerus* sp.) is smaller than its island ancestor, the giant Sulawesi pig (*Celebochoerus heekereni*) of the preceding biozone. Also, the closely related *Celebochoerus cagayanensis* of Luzon is small, as small as the smallest Sulawesi species, based on the size of its teeth, according to Thomas Ingicco and colleagues (2016). An only slightly smaller size is seen in the Ryukyu Islands wild boar (*Sus scrofa riukiuanus*) as compared to wild boar of the Asian mainland.

Dogs and Otters

Carnivore body size on islands is closely related to the relative abundance and size spectrum of available resources, as proposed by Shai Meiri and colleagues (2006). However, where small prey is easily and widely available, carnivore body size tends to decrease substantially, as evidenced by the fossil record of Sardinia and Java.

A progressive body mass decrease is observed in the lineages of the Sardinian dog (*Cynotherium sardous*), the Sardinian hunting hyena (*Chasmaporthetes melei*), the Javan dogs (*Xenocyon merriami–Xenocyon trinilensis*), and the Channel Islands fox (*Urocyon littoralis*). They all, except for the fox, evolved from large-sized carnivores that preyed on animals larger than themselves, yet evolved towards smaller body size. The Sardinian dog (Late Pleistocene) was half the size of its mainland ancestor (*Xenocyon*). The reason for this significant size reduction is that it shifted from hunting large prey to hunting small prey such as the abundantly present lagomorphs (Chapter 9). This latter feeding strategy requires relatively low hunting costs, but it cannot sustain a large body size because of energetic constraints as shown by Carbone and colleagues (1999, 2007). Therefore, the lineage evolved towards a smaller form (Figure 22.6). A similar size reduction within a *Xenocyon* lineage is observed on Java, where the large species *Xenocyon merriami* evolved into the 60% smaller species *Xenocyon trinilensis* (Chapter 12). On Java, the simultaneous presence of large cats may have caused it to shift prey and expand its niche to mainly rodents, as proposed by us (AG, GL) and Rebekka Volmer (2018). On Sardinia and Corsica, however, no such competitor was present, but shift in prey took place nonetheless, perhaps due to a very low prey diversity compared to Java. On both Sardinia and Java, such ecological release is possible, as foxes (small prey hunters) are missing in both faunas.

On the other hand, the Early Pleistocene Sardinian hunting hyena (*Chasmaporthetes melei*) is only moderately smaller (Figure 9.3) than its mainland relative (*Chasmaporthetes lunensis*). It was an active hunter that was able to bring down large prey and crush their bones, much like its mainland relative. Likely, the only moderate size decrease might thus be due to functional constraints because even the smallest available prey (goral-like bovids and pigs) were still medium-sized, strong, and massive. Presumably, this limited the hyena from further evolving reduced size.

Dwarfism is not observed in all insular canids. An exception is provided by the Falkland Islands fox (*Dusicyon australis*). Unfortunately, not much is known about this endemic canid as it became extinct in the nineteenth century. What is known, however, is that it was preying on birds, such as geese and penguins, and on seals, as summarised by Nowak (1999). The wolf was the only terrestrial mammal on these islands, contrary to the other islands with endemic canids. *Dusicyon australis* had about the same size as the South American fossil species *Dusicyon avus*, to which it is phylogenetically related. This implies that it was larger than any other closely related living species of the South American mainland. It seems that the total absence of small prey mammals from the Falklands hampered the size decrease of *Dusicyon*, simply because it had to remain large in order to hunt big prey, which, in addition, was abundantly available. Alternatively, this island wolf may be ecologically analogous to the grizzly bears of the Pacific Northwest, switching to a marine and therefore less limited diet in contrast to the more terrestrial diets of these carnivores on the mainland.

Island otters do not always evolve body mass shifts; actually, most species retained the ancestral body mass as far as can be concluded from the scattered remains, often limited to some teeth and at most one or two postcranial elements. Such is the case with the Late Miocene otter of Gargano (*Paralutra garganensis*) and one of the Late Pleistocene otters of Sardinia (*Algarolutra majori*). The Sicilian otter (*Lutraeximia trinacriae*), on the other hand, was slightly smaller than its ancestor, probably related to the common otter. In contrast, the Late Pleistocene Cretan otter (*Lutrogale cretensis*) was slightly larger than its mainland ancestor (*Lutrogale perspicillata*), the Javanese otter (*Lutrogale robusta*) was rather large, and one of the Sardinian otters (*Megalenhydris barbaricina*) was even gigantic with a head and body length of more than 1.4 m (Chapter 9). Increased body mass in otters, with only one exception, *Algarolutra majori* of Sardinia, is correlated with aquatic prey, in agreement with Lomolino's (2005) and Lomolino and colleagues' (2012) ecological explanation for the island rule – there may be an optimal size for mammals, but that optimum varies with ecological characteristics of the islands and of the species as well.

References

Alcover J.A., Campillo X., Macias M., Sans A. 1998. Mammal species of the world: Additional data on insular mammals. *American Museum Novitates*, 3248: 1–29.

Ali J.R., Huber M. 2010. Mammalian biodiversity on Madagascar controlled by ocean currents. *Nature*, 463: 653–656.

Andrews P. 1990. *Owls, Caves and Fossils; Predation, Preservation and Accumulation of Small Mammal Bones in Caves, with an Analysis of the Pleistocene Cave Faunas from Westbury-Sub-Mendip, Somerset, U.K.* The University of Chicago Press, Chicago.

Athanassiou A., van der Geer A.A.E., Lyras G.A. 2019. Pleistocene insular Proboscidea of the Eastern Mediterranean: A review and update. *Quaternary Science Reviews*, 218: 306–321.

Benton M.J., Csiki Z., Grigorescu D., Redelstorff R., Sander P.M., Stein K., Weishampel D.B. 2010. Dinosaurs and the island rule: The dwarfed dinosaurs from Haţeg Island. *Palaeogeography, Palaeoclimatology, Palaeoecology*, 293: 438–454.

Bover P. 2004. *Noves aportacions al coneixement del gènere Myotragus Bate, 1909 (Artiodactyla, Caprinae) de les illes Balears.* Unpublished PhD thesis, Universitat de les Illes Balears, Palma de Mallorca. (In Catalan.)

Bromham L., Cardillo M. 2007. Primates follow the "island rule": Implications for interpreting *Homo floresiensis*. *Biology Letters*, 3: 398–400.

Brown J.H., Marquet P.A., Taper M.L. 1993. Evolution of body size: Consequences of an energetic definition of fitness. *American Naturalist*, 142: 573–584.

Carbone C., Mace M., Craig Roberts S., Macdonald D.W. 1999. Energetic constraints on the diet of terrestrial carnivores. *Nature,* 408: 286–287.

Carbone C., Teacher A., Rowcliffe J.M. 2007. The costs of carnivory. *PLoS Biology*, 5: e22.

Damuth J. 1993. Cope's rule, the island rule and the scaling of mammalian population density. *Nature* 365: 748–750.

de Visser J.A. 2008. *The Extinct Genus Hexaprotodon Falconer & Cautley, 1836 (Mammalia, Artiodactyla, Hippopotamidae) in Asia: Paleoecology and Taxonomy*. Unpublished PhD thesis. University of Utrecht, the Netherlands.

Dermitzakis M., van der Geer A.A.E., Lyras G. 2006. Palaeopathological observations on a population of fossil deer from the Late Pleistocene of Crete. In: Kalofourtis A., Papadopoulos N., Spiliopoulou C., Marabellas K., Chatzioannou A. (eds), *Volume in Honor of Prof. A.S. Koutselinis*. Parisianos, Athens, pp. 43–51. [In Greek with English abstract.]

Foster J.B. 1964. Evolution of mammals on islands. *Nature*, 202: 234–235.

Freudenthal M. 1976. Rodent stratigraphy of some Miocene fissure fillings in Gargano (prov. Foggia, Italy). *Scripta Geologica*, 37: 1–23.

Hansford J.P., Turvey S.T., 2018. Unexpected diversity within the extinct elephant birds (Aves: Aepyornithidae) and a new identity for the world's largest bird. *Royal Society Open Science*, 5 (9): 181295.

Heaney L.R. 1978. Island area and body size of insular mammals: Evidence from the tri-colored squirrel (*Callosciurus prevosti*) of Southeast Asia. *Evolution*, 32 (1): 29–44.

Hooijer D.A. 1967. Indo-Australian insular elephants. *Genetica*, 38: 143–162.

Ingicco T., van den Bergh G.D., de Vos J., Castro A., Amano N., Bautista A., 2016. A new species of *Celebochoerus* (Suidae, Mammalia) from the Philippines and the paleobiogeography of the genus *Celebochoerus* Hooijer. 1948. *Geobios*, 49 (4): 285–291.

Kelt D.A., Brown J.H. 1998. Diversification of body sizes: Patterns and processes in the assembly of terrestrial mammal faunas. In: McKinney M., Drake J. (eds), *Biodiversity Dynamics*. Columbia University Press, New York, pp. 109–131.

Lister A.M., Hall C. 2014. Variation in body and tooth size with island area in small mammals: A study of Scottish and Faroese house mice (*Mus musculus*). *Annales Zoologici Fennici*, 51: 95–110.

Lomolino M.V. 1984. Mammalian island biogeography: effects of area, isolation and vagility. *Oecologia*, 61(3): 376–382.

Lomolino M.V. 1985. Body size of mammals on islands: The island rule reexamined. *American Naturalist*, 125: 310–316.

Lomolino M.V. 2005. Body size evolution in insular vertebrates: Generality of the island rule. *Journal of Biogeography*, 32: 1683–1699.

Lomolino M.V., Riddle B.R., Whittaker R.J., 2017. *Biogeography: Biological Diversity Across Space and Time*, 5th edn. Sinauer Associates, Sunderland, MA.

Lomolino M.V., Sax D.F., Palombo M.R., van der Geer A.A.E. 2012. Of mice and mammoths: Evaluations of causal explanations for body size evolution in insular mammals. *Journal of Biogeography*, 39: 842–854.

Lomolino M.V., van der Geer A.A.E., Lyras G.A., Palombo M.R., Sax D., Rozzi R. 2013. Of mice and mammoths: Generality and antiquity of the island rule. *Journal of Biogeography*, 40: 1427–1439.

Lyras G., Giannakopoulou G., Lillis T., van der Geer A.A.E. 2019. Paradise lost: Evidence for a devastating metabolic bone disease in an insular Pleistocene deer. *International Journal of Paleopathology*, 24: 213–226.

Lyras G.A., van der Geer A., Rook L. 2010. Body size of insular carnivores: Evidence from the fossil record. *Journal of Biogeography*, 37: 1007–1021.

Meiri S., Cooper N., Purvis A. 2008. The island rule: Made to be broken? *Proceedings of the Royal Society B*, 275: 141–148.

Meiri S., Dayan T., Simberloff D. 2006. The generality of the island rule reexamined. *Journal of Biogeography*, 33: 1571–1577.

Moncunill-Solé B., Tuveri C., Arca M., Angelone C. 2016. Comparing the body mass variations in endemic insular species of the genus *Prolagus* (Ochotonidae, Lagomorpha) in the Pleistocene of Sardinia (Italy). *Rivista Italiana di Paleontologia e Stratigrafia*, 122 (1): 25–36.

Nopcsa F. 1912. Über das Vorkommen der Dinosaurier in Siebenbürgen. *Verhandlungen der zoologische-botanischen Gesellschaft, Wien*, 54: 12–14. Published 1914.

Nowak R.M. 1999. *Walker's Mammals of the World*, 6th edn. The Johns Hopkins University Press, Baltimore and London.

Reumer J.W.F., Robert I. 2005. Two bony enigmas: Not everything is what it looks like. *Mayurqa*, 30: 33–43.

Roth V.L. 1990. Insular dwarf elephants: A case study in body mass estimation and ecological inference. In: Damuth J., MacFadden B.J. (eds), *Body Size in Mammalian Paleobiology: Estimation and Biological Implications.* Cambridge University Press, Cambridge, pp. 151–179.

Roth V.L. 1992. Inferences from allometry and fossils: Dwarfing of elephants on islands. *In*: Futuyma D., Antonovics J. (eds), *Oxford Surveys in Evolutionary Biology*, 8: 259–288. Oxford University Press, New York.

Sander P.M., Mateus O., Laven T., Knötschke N. 2006. Bone histology indicates insular dwarfism in a new Late Jurassic sauropod dinosaur. *Nature*, 441: 739–741.

Sondaar P.Y. 1977. Insularity and its effect on mammal evolution. In: Hecht M.N., Goody P.L., Hecht B.M. (eds), *Major Patterns in Vertebrate Evolution.* Plenum, New York, pp. 671–707.

Steadman D.W. 1995. Prehistoric extinctions of Pacific Island birds: Biodiversity meets zooarchaeology. *Science*, 267: 1123–1131.

Steadman D.W., Ray C.E. 1982. The relationships of *Megaoryzomys curioi*, an extinct Cricetine rodent (Muroidea: Muridae) from the Galápagos Islands, Ecuador. *Smithsonian Contributions to Paleobiology*, 51: 1–23.

Stein K., Csiki Z., Rogers K.C., Weishampel D.B., Redelstorff R., Carballido J.L., Sander P.M. 2010. Small body size and extreme cortical bone remodeling indicate phyletic dwarfism in *Magyarosaurus dacus* (Sauropoda: Titanosauria). *Proceedings of the National Academy of Sciences of the USA*, 107 (20): 9258–9263.

van den Bergh G.D., Li B., Brumm A., Grün R., Yurnaldi D., Moore M.W., Kurniawan I., Setiawan R., Aziz F., Roberts R.G., Suyono, Storey M., Setiabudi E., Morwood M.J. 2016. Earliest hominin occupation of Sulawesi, Indonesia. *Nature*, 529: 208–211.

van den Hoek Ostende L.W., van der Geer A.A.E., Wijngaarde, C. 2017. Why are there no giants at the dwarf's feet? Insular micromammals in the eastern Mediterranean. In: Mol D. et al. (eds), *Proceedings of the VIth International Conference on Mammoths and their Relatives, May 2014, Grevena-Siatista, Greece. Quaternary International*, 445: 269–279.

van der Geer A.A.E. 2018. Changing Invaders: Trends of gigantism in insular introduced rats. *Environmental Conservation*, 45 (3): 203–211.

van der Geer A.A.E., Lyras G.A., Lomolino M., Palombo M.R., Sax D. 2013. Body size evolution of palaeo-insular mammals: Temporal variations and interspecific interactions. *Journal of Biogeography*, 40: 1440–1450.

van der Geer A.A.E., Lyras G.A., MacPhee R.D.E., Lomolino M., Drinia H. 2014. Mortality in a predator-free insular environment: The extinct dwarf deer from Crete. *American Museum Novitates*, 3807, 1–29.

van der Geer A.A.E., Lyras G.A., Volmer R. 2018. Insular dwarfism in canids on Java (Indonesia) and its implication for the environment of *Homo erectus* during the Early and earliest Middle Pleistocene. *Palaeogeography, Palaeoclimatology, Palaeoecology*, 507: 168–179.

van der Geer A.A.E., van den Bergh G., Lyras G.A., Prasetyo U.W., Due R.A., Setiyabudi E., Drinia H. 2016. The effect of area and isolation on insular dwarf proboscideans. *Journal of Biogeography*, 43: 1656–1666.

van der Made J. 1988. *Sus nanus* nov. sp., a Pliocene dwarf pig from Capo Figari (Northeastern Sardinia). *Bolletino della Società Paleontologica Italiana*, 27: 367–378.

Van Valen L.M. 1973. Body size and numbers of plants and animals. *Evolution*, 27 (1): 27–35.

Wilson E.O. 1961. The nature of the taxon cycle in the Melanesian ant fauna. *American Naturalist*, 95: 169–193.

Yom-Tov Y. Yom-Tov S., Moller H. 1999. Competition, coexistence, and adaptation amongst rodent invaders to Pacific and New Zealand Islands. *Journal of Biogeography*, 26: 947–958.

CHAPTER TWENTY FOUR

Parallel Patterns and Trends

Evolution of Island Mammals: Adaptation and Extinction of Placental Mammals on Islands, Second Edition.
Alexandra van der Geer, George Lyras and John de Vos.
© 2021 John Wiley & Sons Ltd. Published 2021 by John Wiley & Sons Ltd.

Dramatic body mass changes, observed in many insular mammalian taxa, are certainly the best-known adaptations to insular conditions. However, these size changes, no matter how spectacular they often are, are not the only typical modifications of island lineages. Generally, adaptations of island species are reflected in their craniodental anatomy in response to changes in diet and defensive systems, and in their postcranial anatomy in response to changes in locomotion, as first pointed out by Sondaar (1977) (Box 24.1). In some cases, these morphological changes are so extensive that it is not easy to trace back their direct mainland ancestor with certainty.

Changes in body proportions and functional structures in island endemics tend to evolve in parallel ways, as we have seen in the island chapters. These parallel evolutionary changes are not restricted to certain genera but occur in a wide taxonomic range. The parallel patterns and trends are functionally best explained by shared adaptational needs, such as those to cope with rugose or mountainous terrain, abrasive or nutrient-low diets, and increased intraspecific competition. Every functional trait evolves as a trade-off, or compromise, between costs and gains. When the relative importance of the various selective pressures changes, as on islands, the new compromise drives evolution of the functional trait into a different direction. Generally, dietary niche expansion now gains in importance over anti-predator skills.

Some structures lose their so-called derived condition and return to a condition shared with less derived or primitive members of the taxon, as noticed by van der Geer (2014). This may complicate cladistic analysis, as islanders may be plotted closer to less-derived forms of the same taxon than to their own ancestors.

BOX 24.1

Paul Yves Sondaar (1934–2003)

The Dutch palaeontologist Paul Sondaar (Figure Box 24.1) is renowned for his research on fossil island faunas, first in the Mediterranean, later also in Indonesia and Japan. His participation in an international conference on *Major Patterns in Vertebrate Evolution* (1976) resulted a year later in perhaps his most influential publication on insularity and its effect on mammal evolution. Here, he elaborated on sweepstake routes and the typical composition of fossil island faunas, especially their often bizarre adaptations. Among his famous similes are 'elephants have a snorkel' and 'low-gear locomotion'.

Sondaar's legacy includes a wealth of unpublished autobiographic anecdotes. In one of these, he described his very first discovery of fossils of insular mammals, a tibia and an astragal from the dwarf hippo of Crete.

Figure Box 24.1 Paul Yves Sondaar (1934–2003) standing next to the fossiliferous layer of Mavro Muri cave, Crete. (Photograph: John de Vos, 1977.)

This discovery had a big impact on him, as from that moment on, he switched his scientific focus from horses to insular mammals. He wrote (in translation):

> 'It was more a professional routine than real interest which played a role, when I compared the two bones from the hind foot [of the Cretan dwarf hippo] with those of an extant hippo. They were much smaller, but what arouse my interest most was that they had a different shape. It was pure coincidence that I found just two bones from the ankle joint that fitted each other perfectly well. It appeared to me that the ankle joint of the dwarf hippo had been much more mobile than that of its large continental ancestor. They could walk much better and more tiptoed. Very different from all hippos from the past and present, in fact more like a goat. After having made the walk to the mountain of Katharo myself, I thought, of course, a normal hippo would never have made it. No degeneration but adaptation to the landscape of Crete! Besides, land bridges are also geologically unlikely, and offer no explanation for why only elephants, hippopotamuses and deer crossed them. The very same animals, elephants, hippos and deer, happen to be good swimmers, and because of their digestion system where gasses develop, they are at the same time good floaters. The trunk of an elephant is comparable to a snorkel. Predators do swim, but hardly ever in herds, besides, they are very bad floaters. Deer, elephants and hippos came to the

uninhabited islands and changed (evolved). The changes are to be considered as adaptations to a predator-free island environment, where it turned out that being small was advantageous. Deer had now short legs; being fast was not important, in contrast to increased stability in the mountains. Short legs as the first gear in a jeep'.

Evolutionary Changes in the Teeth

Several insular taxa evolved a dentition with higher molar crowns (hypsodonty) than their mainland ancestors. This is seen in elephants, deer, bovids, and pigs (see below). In an exceptional case – the mouse-goat *Myotragus balearicus* of Late Pleistocene Mallorca – the roots of the lower first and second molars even distort normal ossification of the base of the mandibular ramus (Plate 19). Apart from the extremely hypsodont molars, the latter lineage also evolved ever-growing incisors to cope with a higher degree of wear of the front teeth, in parallel with the unrelated antelope *Maremmia* (Plate 16) of the Late Miocene of Tuscany.

Many insular endemics also have proportionally shorter muzzles. In these short-snouted endemics, the maxillary tooth row is more arcuate and less complete. Loss of premolars is seen in several insular taxa, such as ruminants (*Hoplitomeryx matthei*, *Myotragus batei*, *Myotragus balearicus*, and *Maremmia lorenzi*), dogs (*Cynotherium sardous*), pigs (*Sus sondaari*), and hippos (*Hippopotamus minor*). Also the extant Shenkaku mole (*Nesoscaptor uchidai*) has a reduced number of teeth in combination with a shorter muzzle (Box 24.2). A relatively shorter muzzle is sometimes explained as a paedomorphic feature but is probably better explained as an adaptation to enable a more powerful mastication, as suggested by van der Geer (2005). Snout length is primarily controlled by the requirements of a mechanical nature, as remarked by Leonard Radinsky (1987) and not, for example, by requirements for the senses. With increasing snout length, the more muscle action is needed to retain the same level of grinding force at the molar part, needed for chewing. Shortening of the muzzle, therefore, results in a relative increase in power exerted at the molariforms, as in cats. In the case of island ruminants, the trade-off between fast and strong insicival cutting together with the possibility of processing large amounts of food at the time (typical for long snouts) and powerful molar grinding or shearing (typical for short snouts) is shifted in favour of the grinding, in a way similar to the situation seen in hypercarnivorous carnivores (the Cape hunting dog: *Lycaon pictus*), who tear their prey apart using their molars as pliers. The reduction in loss of elements further increases the pressure at the remaining ones.

The Shenkaku Mole

BOX 24.2

The Shenkaku mole (*Nesoscaptor uchidai*) is an extant endemic mole, restricted to the largest island of the Pinnacle Islands. These islands form a group of uninhabited rocks at the edge of the continental shelf of mainland Asia. They got disconnected towards the end of the Early Pleistocene when the East China Sea filled. The long-term isolation resulted in a distinct genus, different from all East Asian moles (*Mogera*) and South East Asian moles (*Euroscaptor*). Based on the fact that it resembles the Taiwanese island mole (*Mogera insularis*) most, Masaharu Motokawa and colleagues (2004) argued that both species should in fact belong to the same genus, *Mogera*. The Shenkaku mole has a relatively low number of teeth (only 38), possibly related to its shortened snout. Furthermore, the last lower premolar retains two distinctive cusps, a character not found in any other mole genus. Unfortunately, the Shenkaku mole is classified as endangered, based on its very restricted range and its declining numbers.

Elephants and Stegodons

The reduction in plate number and increase in enamel thickness are often considered an adaptation to a change in diet towards tougher food. This was first observed in the Sicilian dwarf elephant (*Palaeoloxodon falconeri*) with its plate number of 15 and enamel thickness of 1–2 mm. Such a change in molar morphology was also noted for *Mammuthus creticus* (Figure 5.5), the pygmy mammoth of Crete, but here the conclusion was erroneously based on the wrong ancestor (*Palaeoloxodon antiquus*). Compared to *Mammuthus meridionalis*, the most likely ancestor, the number of plates of the third molar is, however, equal, being 10–14 in *meridionalis* and 13 in *creticus*. The same is valid for the enamel thickness, 2.4–4.1 in the third molar of *meridionalis*, and 2.5–3 mm in that of *creticus*. On Flores, on the other hand, a different pattern evolved. The lower second molar of the Late Pleistocene *Stegodon florensis insularis* from Liang Bua cave, the number of plates amounts to 12, two more than in its early Middle Pleistocene large-sized ancestor *Stegodon florensis florensis* from the So'a Basin, whereas the enamel thickness in both subspecies lies around 3 mm. So, here, the number of plates instead increased during dwarfism, with unchanged enamel thickness.

Insular stegodons tend to have higher crowned molars than do mainland stegodons, which have low crowns throughout their evolutionary history. This hypsodonty is best explained as an adaptation towards more abrasive food. An alternative explanation is longevity, as elephants (and most mammals) die when their teeth are worn out. A number of fossil island mammals indeed lived longer, but whether or not this is the case for island proboscideans remains unknown.

Insular proboscideans further often have strongly curved tusks in comparison with mainland forms. This is seen in *Palaeoloxodon falconeri*, *Palaeoloxodon tiliensis*, *Palaeoloxodon cypriotes*, some *Palaeoloxodon mnaidriensis*, and *Stegodon sondaari*. As this is a shared character in unrelated species, it might be simply the result of size reduction (allometry).

Deer and Bovids

Insular dwarf ruminants generally evolved higher molar crowns as well. In the two geologically youngest species of the Balearic mouse-goat (*Myotragus batei* and *Myotragus balearicus*) and in the Late Miocene antelope of Tuscany (*Maremmia lorenzi*), this is accompanied by the presence of ever-growing incisors. These are in fact retained milk incisors with open roots that are not replaced by permanent incisors with closed roots (monophyodonty). Simultaneously, loss of lower premolars is seen in these lineages (the second in *Myotragus* and the second and the third in *Maremmia*). A similar loss of the second lower premolar, but not monophyodonty, is observed in some *Hoplitomeryx*. The Pleistocene deer of Crete (*Candiacervus*), on the other hand, only evolved higher molar crowns, without the loss of dental elements.

Increase in abrasiveness of the food is correlated to an increase in the degree of hypsodonty, as already noted by Alcover and colleagues (1981). Likely, in these insular ruminants, an increased amount of grit intake due to higher population density with higher levels of overgrazing is a more likely driver for the evolution of extreme hypsodonty than abrasiveness of the food proper. Also, in the case of *Myotragus*, a link has been suggested with longevity (Jordana and colleagues, 2012). This is confirmed by histological studies (see below).

Some authors, e.g. Azzaroli and Mazza (1992), consider dwarf deer merely paedomorphic stunted forms. Young, immature deer, however, have less hypsodont teeth than their mature conspecifics, whereas some insular deer have more hypsodont teeth. Another non-paedomorphic feature is a higher degree of foot bone fusions observed in adult island deer. Finally, young ruminants typically have long and slender limbs, whereas insular dwarf forms have proportionally shortened and massive limbs.

Hippos and Pigs

Remarkable is the change in dentition of the world's smallest hippo, the dwarf species from Cyprus, *Hippopotamus minor.* Its molars are more like those of a tapir with ridges (lophodonty), indicating a diet of a leaf browser instead of a grazer like its ancestor, *Hippopotamus amphibius.* This type of molars is a typical feature for more primitive hippos. However, in the case of the Cyprus dwarf hippo, this is not a retention of an ancient character, but instead an adaptation to a new environment and thus secondarily derived. To facilitate a new type of browsing, with a larger transversal component (Sondaar, 1977), this species also lost the fourth premolar in many specimens.

In some pig lineages, a tendency towards hypsodonty, increased enamel thickness, and simplified molar morphology can be observed in the dentition. This is seen in Sondaar's pig (*Sus sondaari*) of Early Pleistocene Sardinia and the Middle Pleistocene Sulawesi pigs (*Celebochoerus*). Primitive molars, either as retention of an ancestral stage of secondarily evolved, are seen in the small Sangiran pig from Early Pleistocene Java (*Sus sangiranensis*).

In addition, two of the three endemic pig species of Sulawesi and neighbouring islands are remarkable for their tusks. The earliest form, the giant Sulawesi pig (*Celebochoerus heekereni*), bears two pairs of impressive tusks. The upper tusks were even larger than those of the living African warthog (*Phacochoerus africanus*). A related dwarf form (*Celebochoerus cagayanensis*) has been described from the Middle Pleistocene of Luzon, but its tusks are much smaller, similar to those of the Middle and Late Pleistocene *Celebochoerus* of Sulawesi. The extant babirusa (*Babyrousa babyrussa*) of Sulawesi sports the largest upper tusks of all wild swine. These upper tusks grow through the skin of the muzzle and then curve backward towards the forehead. The Sulawesi warty pig (*Sus celebensis*), on the other hand, has only small tusks but large bony facial projections instead, being the 'warts'.

Lagomorphs and Mice

Insular pikas (Ochotonidae) of the genus *Prolagus* do not show increased molar crown height, other than due to allometric scaling. Instead, these insular pikas show an increase of enamel complexity in their molars, and an enlargement of the third lower premolars relative to the other lower cheek teeth, as shown by Chiara Angelone (2005). The degree of enamel complexity seems unrelated to the degree of hypsodonty, whereas *Prolagus imperialis* (Late Miocene, Gargano) and *Prolagus figaro* (Early Pleistocene, Sardinia) combined a very complex molar pattern with a low hypsodonty index, *Prolagus sardus* (Late Pleistocene, Sardinia), on the other hand, has a less complex pattern but a higher index. The Late Miocene Tuscany pika (*Paludotona etruria*), on the other hand, preserved a primitive dental morphology, reminding of the Early–Middle Miocene pika lineage *Marcuinomys–Lagopsis*. The same applies to the Amami rabbit (Middle Pleistocene-Holocene; *Pentalagus furnessi*), whose primitive dental pattern resembles that of the Late Miocene hare *Alilepus*. The Sardinian hare (Early Pleistocene, *Sardolagus obscurus*) on the other hand combines an advanced third lower premolar with a primitive second upper premolar.

A progressively more complex molar pattern is also observed in the lineage of Pleistocene Sardinian field mice (*Rhagapodemus-Rhagamys*), as an adaptation to a more abrasive diet together with an increase in molar crown height and in body mass.

Several unrelated insular giant murids evolved hypsodonty along with body size increase. This is the case with the Late Miocene *Anthracomys* of Tuscany, the Late Miocene *Mikrotia* of Gargano, Early Pliocene *Paraethomys balearicus* of Mallorca, the Pleistocene *Raksasamys tikusbesar* of Sumba, and the Late Pleistocene *Rhagamys* of Sardinia and Corsica. Locatelli and colleagues (2012) observed that there is a general trend among insular murids for an increase in size and hypsodonty.

A reduction in the number of molars evolved in a few extant species of shrew-rats of Sulawesi and Luzon and moss-mice of New Guinea. These similar but unrelated groups are all three parts of impressive adaptive radiations of Muridae on their respective very large islands and evolved in parallel towards the ecological niche elsewhere occupied by shrews and hedgehogs. The most spectacular case is doubtlessly the almost toothless *Paucidentomys vermidax* of Sulawesi, a specialist feeder on earthworms and other invertebrates. It lost all molars and retained only the incisors.

Otters

A few unrelated fossil island otter species evolved a more robust dentition with a large talon on the upper carnassial (the last premolar), sometimes coinciding with a change in lifestyle (Box 24.3). These are the otters from Pleistocene Sardinia (*Algarolutra majori*), Late Pleistocene Crete (*Lutrogale cretensis*), Middle Pleistocene Java (*Lutrogale palaeoleptonyx*), Late Miocene Tuscany (*Paludolutra campanii*), and Late Miocene Gargano (*Paralutra garganensis*). The talon in *Lutrogale* is much larger than in most Lutrini, and the dentition is more robust. This is reflected in the diet of the only extant species, the smooth-coated otter (*Lutrogale perspicillata*), which is a shellfish-eater, rather than a fish-eater. Consequently, the fossil *Lutrogale* species of Crete and Java supposedly were shellfish eaters as well. A shellfish diet has also been proposed for the giant Sardinian otter (*Megalenhydris barbaricina*), inferred from its teeth which resemble those of the extant clawless otters (*Aonyx*). The latter genus is, like *Lutrogale*, characterised by a large talon on its upper carnassial. The same is observed in the Miocene Eurasian genus *Paralutra*. The form from the Gargano bears an even larger talon than the type species (*Paralutra jaegeri*), on the basis of which we can assume that it consumed more shellfish than its ancestor.

BOX 24.3

Otters: A Change in Lifestyle

Adaptations other than a change in body size in island otters do not follow a general pattern. The Cretan otter (*Lutrogale cretensis*; Plate 3) seemed to have developed a somewhat more terrestrial way of life, whereas two of the Sardinian otters (*Sardolutra* and *Megalenhydris*) and the Corsican otter (*Lutra castiglionis*), on the other hand, had a more aquatic lifestyle. The Gargano otter was at most semiaquatic, like *Lutra*. Some of the endemic island otters were marine otters, but this is difficult to state with certainty due to insufficient material. *Megalenhydris* and *Sardolutra* are the most likely candidates. The first because of its gigantic size in combination with a much flattened tail and the absence of large freshwater bodies on the island, and the second because of its peculiar large forked penis bone, which has been explained as a feature enabling mating in open sea.

Evolutionary Skeletal Changes

Short and Massive Limbs

Insular megaherbivores generally lost the need for speed as part of their anti-predator strategy and could therefore invest in a higher degree of stability. Stability has a huge advantage in general, as life expectancy for an ungulate is reduced significantly by breaking a leg. This is especially true in a mountainous area, and more so if the body is relatively plump. The increase of stability in insular megaherbivores is mainly achieved by shortening and thickening of the long bones, most markedly the metapodials in ungulates (Figure 24.1) and the femur and humerus in proboscideans and hippos. This was referred to as 'low-gear locomotion' by Sondaar (1977) (Box 24.3), a slow but powerful gait. The observed size reduction of the lower limb bones cannot be explained by a general size reduction of the animal because the relative proportions of limb elements are drastically changed. This inter alia implies that standard equations to predict body mass at the basis of long bone lengths, such as those given by Kathleen Scott (1990) for ungulates, cannot be applied implicitly to insular endemics.

Proportional shortening of the lower limb bones appears to be already present at birth, as shown by van der Geer and colleagues (2006), for the Cretan dwarf deer (*Candiacervus ropalophorus*). This shortening increases further during postnatal ontogeny due to a relative slower growth speed of the distal limb bones compared to the more proximal limb bones. In general, the short metapodials in island ruminants are at the same time more massive than in mainland species of comparable size (Table 24.1). This higher massivity is already observed in neonates of *Myotragus* and *Candiacervus* and increases further during following ontogeny.

Short-leggedness is seen in some endemic insular pigs as well. These are the extant Sulawesi warty pig (*Sus celebensis*) and the two Pleistocene Sulawesi pigs (*Celebochoerus*

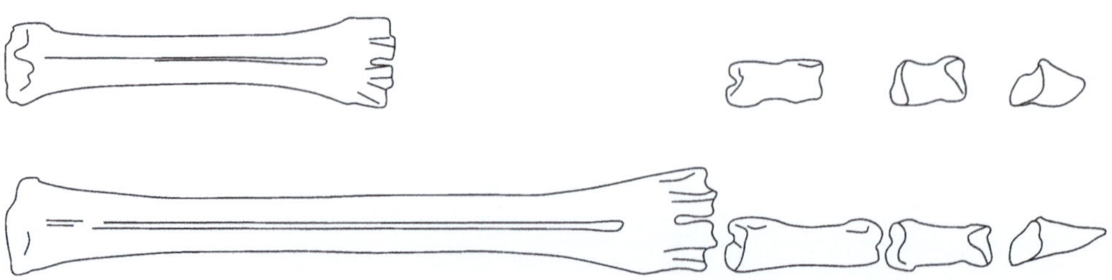

Figure 24.1 Comparison of the metatarsal and phalanges of the insular deer *Candiacervus* size II (top) from Crete with that of Eurasian mainland fallow deer *Dama dama* (bottom).

Table 24.1 Length and massivity (ratio length : distal transversal diameter) of metatarsals of insular deer and continental deer. Measurements in mm.

Deer taxa	*Number of specimens*	*Range of length*	*Mean of Length*	*Massivity*
Crete (Greece)				
Candiacervus ropalophorus (size I)	66	88.4–114.1	98.5	0.23
Candiacervus sp. (size II)	45	99.2–122.2	110.8	0.23
Candiacervus cretensis (size III)	3	131–144.2	138.4	0.20
Candiacervus rethymnensis (size IV)	4	185–206	194.8	0.17
Candiacervus dorothensis (size V)	3	262–284	271.6	0.15
Karpathos (Greece)				
'Candiacervus' cerigensis	3	89.6–110	101.1	0.26
Kume Island (Ryukyu), *Cervus astylodon*				
Kume morph A size 1	4	44.4–48.2	46.3	0.31
Kume morph A size 2	4	54.6–59.8	56.6	0.30
Kume morph B size 3	5	62.6–65.2	64.1	0.26
Kume morph B size 4	16	78.4–89.3	84.6	0.26
Kume morph B size 5	2	105.9–106.3	106.1	0.22
Kume morph B size 6	1		119.2	0.21
Kume morph B size 7	1		141.6	0.18
Gargano (Italy)				
Hoplitomeryx devosi	14	74.4–95	85.5	0.24
Hoplitomeryx matthei	8	139–177	158.9	0.16
Hoplitomeryx macpheei	1		259	0.12
Mainland species				
Rangifer tarandus	10	177.2–205.5	193.6	0.20
Megaloceros giganteus	4	313.6–323.6	319.9	0.21
Cervus elaphus	2	269.4–155.7	276.2	0.15
Axis axis	2	154.4–161.5	155	0.15
Capreolus capreolus	13	150.8–161.5	156.9	0.13
Blastocerus campestris	4	151.5–156	153.4	0.14
Blastocerus bezoartis	2	137.2–137.3	137.2	0.14
Mazama zetta	2	122–122.3	122.1	0.18
Mazama rufa	2	101.9–102	101.9	0.15

heekereni, Celebochoerus sp.), of which the latter has relatively shorter limbs than the first. The Sardinian Early Miocene pig (tentatively ascribed to *Hyotherium insularis*) had only shortened phalanges.

The two living endemic 'rabbits' – *Pentalagus* and *Nesolagus* – are short-legged as well. They are either leftovers of a primitive stock or evolved independently in parallel with each other after isolation from the mainland through vicariance. Until leporid phylogeny is resolved, no firm conclusions can be drawn. It is, however, interesting to note that these two living species, unlike other hares and rabbits, live in dense highland forest. They are strictly nocturnal and shelter in holes during the day. Also, the Sicilian rabbit (*Hypolagus peregrinus*) is considered to have been a relatively slow runner, likely thanks to a decreased predator pressure.

Bony Fusions

Increased stability is further achieved by a restriction of directions of potential movements and the degree thereof. In the most extreme cases, bony fusion can sometimes be observed, such as the navico-cuboid – one of the tarsal bones in ruminants – with the cannon bone in several fossil insular ruminants to form a single unit with the loss of articulation, as first noticed by Leinders and Sondaar (1974). An extremely stable joint permits movement in one direction only, or, in the extreme case of total fusion, no movement at all. Obviously, this seems to pose no problem on islands where there are no fast-running carnivores. Manoeuvrability to escape predators has lost its significance on islands. The trade-offs now differ, and a shift towards higher stability but lower manoeuvrability is seen. The stiff but stable joints with restricted movements in ruminants were explained as 'low-gear system' by Sondaar since 1977.

A progressive increase in the number of fusions, especially that of the cannon bone with the cubonavicular tarsal bone, is seen in several fossil ruminant lineages. The degree to which this particular fusion takes place varies, for example, only 5.9% in the smallest Cretan dwarf deer (*Candiacervus ropalophorus*), against 75% in the Balearic mouse-goat (*Myotragus balearicus*), and 100% in the Gargano five-horned deer (*Hoplitomeryx*, all species). The latter two genera persisted much longer in isolation than the former, ranging to 5 million years or more instead of at most half a million years for *Candiacervus*. Apparently, these processes are slow, in striking contrast to changes in body mass.

In the *Myotragus* lineage, a gradual increase in percentage of fusion between the cubonavicular and the cuneiform tarsal bones took place as well. The percentage is lower in the older species *antiquus* than in the youngest species *balearicus* but still not 100%, indicating that the establishment of such fusions may take a relatively long time as is the case with the fusion between metatarsus and cubonavicular.

The Japanese dwarf *Stegodon aurorae* has not only proportionally short limbs but also a higher degree of synostotic fusion between the radius and ulna. Fusions between the tibia and fibula occur in the Sicilian dwarf elephant *Palaeoloxodon falconeri*. Such fusions have also been observed in old to very old individuals of the woolly mammoth (*Mammuthus primigenius*) of the Eurasian mainland but not in the young individuals, as is the case with the Sicilian pygmy form. Bone fusions in other island proboscideans have not been reported, due to lack of relevant postcranial material, apart from a fused ulna/radius shaft in a juvenile *Stegodon florensis* from Flores.

Apt Climbers

Apart from a smaller body size, the configuration and morphology of the foot of the Cypriot and the Cretan hippos show adaptations to a more terrestrial lifestyle. Insular dwarf hippos seem to have invested both in stability as manoeuvrability with their shorter, stable proximal limb bones but longer phalanges and a more unguligrade walk.

Loss of Skull Pneumatisation

Usually, the lack of strong pneumatisation of the skull in several insular dwarf proboscideans is explained as a response to size reduction. A lighter body has a lighter skull and thus loses the need for strong neck musculature to lift up the heavy head and trunk. The surface area for neck muscle attachment on the skull can thus be reduced. In extreme cases, such as in the Sicilian *Palaeoloxodon falconeri*, the skull becomes globular and parietal swellings are lost. This model was first proposed by Accordi and Palombo (1971) and Sondaar (1977). Reduction of parietal swelling is observed also in the California Channel Islands' dwarf mammoth, *Mammuthus exilis*. The lack of a parietal swelling is also present in the Sulawesi dwarf elephant, *Stegoloxodon celebensis*, an enigmatic species of four-tusked elephant with retention of primitive characters, such as presence of lower tusks in some specimens, and development of premolars. Here, the absence of parietal swellings is sometimes explained as a primitive character as well, instead of what it more likely is: a side-effect of its reduced size.

Evolutionary Brain Changes

Most research on the neurological changes under insular conditions focuses on brain size evolution. Despite the numerous studies on this subject, how the brain responds to phyletic dwarfing is not well known, with sometimes conflicting results. Some insular mammals appear to have a brain of the same relative size as their mainland relatives. According to Palombo and colleagues (2008), the relative brain size of the smaller species of Cretan deer (*Candiacervus*) underwent only minor size changes during its geographic isolation. Similarly, no noticeable change in relative brain size was documented by Schauber and Falk (2008) in the island fox of the California

Channel Islands (*Urocyon littoralis*), relative to its phylogenetically closest mainland ancestor (*Urocyon cinereoargenteus*). Another dwarf canid is the Late Pleistocene Sardinian dog (*Cynotherium sardous*). According to Lyras (2009), its relative brain size is comparable to that of mainland canids of comparable body mass. The same was reported by Montgomery (2013), who, in a study on Holocene insular primates (e.g. species of *Macaca*), did not find a trend of change in relative brain size either.

In contrast, insular hippos seem to have evolved an increasingly larger relative brain size during subsequent stages of dwarfism. The relative brain size of the dwarf hippos of Madagascar and Cyprus (*Hippopotamus madagascariensis*, *Hippopotamus lemerlei*, and *Hippopotamus minor*) is slightly larger than that of the mainland common hippo (*Hippopotamus amphibius*). However, when their brains are examined using other scaling models, opposite patterns emerge. For example, the brains of the two Malagasy hippos were examined by Weston and Lister (2009) from an ontogenetic perspective. They examined whether the brain of these hippos scales allometrically along the ontogenetic curve predicted for the mainland species. In particular, they applied a 'late ontogenetic scaling' (for hippos older than 2 years) and 'ontogenetic scaling' (for all ages of hippos). In both cases, they found that the brain of both Malagasy hippos is smaller than predicted under ontogenetic models. This was confirmed by Lyras (2018) for a larger dataset (Figure 24.2).

Dwarf elephants have particularly high encephalisation values – that is, have large brains relative to the general

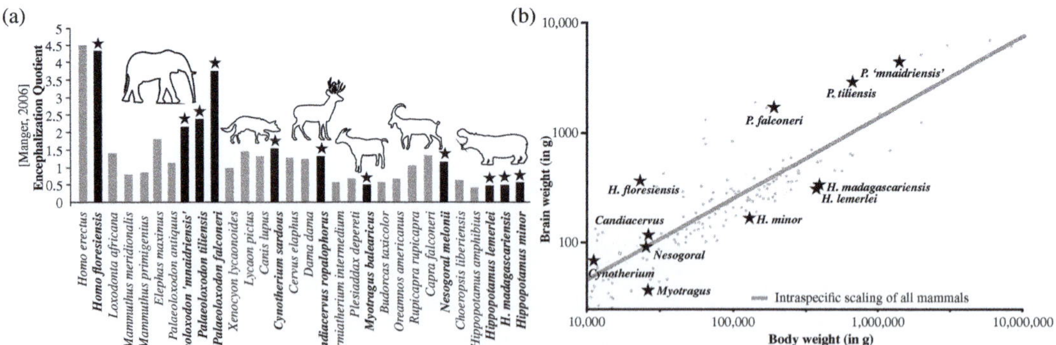

Figure 24.2 Relative brain size of insular phylogenetic dwarfs compared to other mammals. (a) Bar graph of encephalisation quotients of various insular mammals (indicated by asterisks) and of their mainland relatives. (b) Plot of brain weight versus body weight and regressions for *Palaeoloxodon* and *Hippopotamus* as well as for 'all' mammals (excluding Cetacea and Primates). The regression line for 'all' mammals is drawn from Manger, 2006. (Drawing George Lyras.)

mammalian trend, by lack of equally sized relatives. The larger the difference in body mass between insular and mainland ancestral species, the more their encephalisation differs (Figure 24.2). The encephalisation of *Palaeoloxodon falconeri* is significantly higher than that of any mainland elephant. No other insular mammal has a brain that differs so much from that of its mainland relatives. However, as Lyras (2018) indicated, although insular elephants have relatively larger brains than their mainland ancestor, their brains are smaller than what is predicted by the allometric relationship among their continental relatives. The brain of *Palaeoloxodon* aff. *mnaidriensis* is 13%, of *Palaeoloxodon tiliensis* is nearly 30%, and that of *Palaeoloxodon falconeri* is 52% – three dwarf elephant species with increasing body size reduction – smaller than what is predicted by the Elephantidae allometric relationship. To maintain this relative brain size, a proportional increase of brain mass is observed relative to the size of the skull, as shown by Palombo (2001). This, together with a drastically reduced pneumatisation (Figure 24.3), results in a rather round skull with a broad forehead, compared to the skull of adult mainland elephants, as shown by 2018 using 3D geometric morphometrics.

Some insular dwarfs have a smaller brain size than what is predicted both by the general mammalian trend and their closest relatives. *Litovoi tholocephalos* is a multituberculate mammal that was endemic to the Haţeg palaeo-island (part of present-day Romania) during the Late Cretaceous period. According to Zoltán Csiki-Sava and colleagues (2018), *Litovoi* has one of the smallest brains relative to body size of any Mammaliaform. Nevertheless, according to the same authors, despite this small brain, certain parts of the brain – the olfactory bulbs and parafloccular lobes – remained disproportionately well developed, indicating that keen senses and motor control were still important to *Litovoi*, probably essential to its specialised insectivore niche.

The mouse-goat (*Myotragus balearicus*) of Mallorca is another island mammal with a smaller brain. Köhler and Moyà-Solà (2004) considered this a case of extreme brain size reduction. Indeed, its brain is approximately half the size of what is predicted based on the general trend for the contemporary mainland bovids. However, in order to properly evaluate the brain changes during phyletic dwarfing, one should compare it either with its direct late Miocene mainland ancestor or at the least with other late Miocene bovids. Unfortunately, there are only a few publications on the brains of late Miocene bovids, and none on the archaic goat-like ruminants that are phylogenetically related to *Myotragus*. Furthermore, modern bovids display a wide range of relative brain size. Five million years

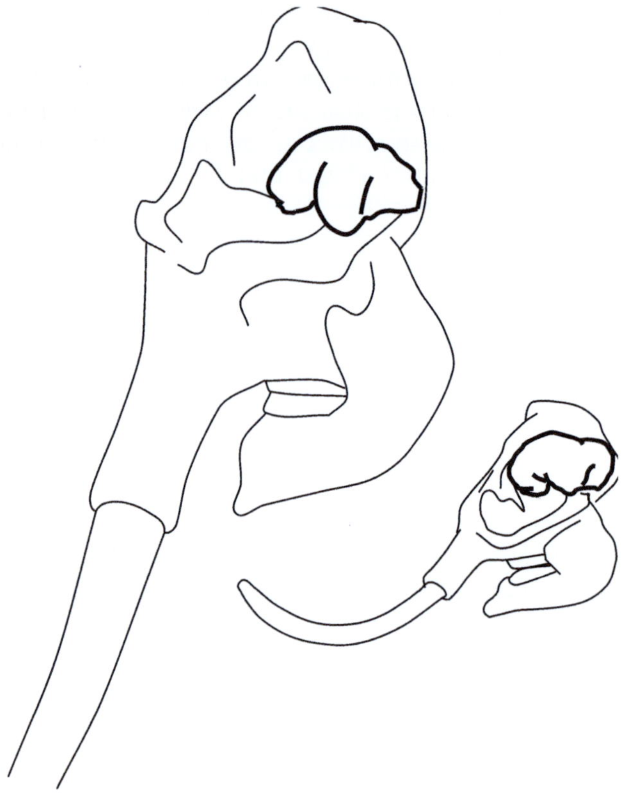

Figure 24.3 Skull and brain of mainland *Palaeoloxodon antiquus* (top) and *Palaeoloxodon falconeri* (Middle Pleistocene, Sicily) drawn to the same scale. (Drawing George Lyras.)

ago, bovids may have had a similarly wide encephalisation range, and the mainland ancestor of *Myotragus* may have had already a low encephalisation value. Therefore, as pointed out by Lyras (2018), without direct information about the relative brain size of its ancestor, *Myotragus* should not be considered as a proven case of extreme brain size reduction.

In addition, the *Myotragus* brain is remarkably convoluted for its size, as already noted by Dechaseaux (1961). This implies significant increase in cortical surface with unchanged volume. Brain studies of *Myotragus balearicus* by Pere Bover and Ferran Tolosa (2003) revealed that the size of the olfactory bulbs is unusually small, and this observation is in accordance with the reduced olfactory ability as inferred from the less perforated criba nasalis. This implies that the need to detect surrounding predators by their smell was substantially reduced. Artiodactyls use their olfactory functions primarily to perceive danger in the form of predators and to a much lesser degree in their social behaviour. For social communication, especially in the breeding season, it is not the olfactory region that is crucial, but Jacobson's organ, or vomeronasal organ, in the rostral base of the nasal septum. The primary olfactory function, to warn against predators, lost its relevance on Mallorca without mammalian carnivores.

The island species which undoubtedly attracted the most attention on this subject is *Homo floresiensis*, the dwarf human from Late Pleistocene Flores. Falk and colleagues (2005) in their study of its brain reported that the anatomical features of its brain resemble those of *Homo erectus*. Furthermore, its relative brain size is comparable to that of the latter species. However, when compared with other hominids, *Homo floresiensis* has an australopithecine-like brain/body size ratio. In addition, according to the same authors, the relative brain size of *Homo floresiensis* does not fall on the projected ontogenetic curve for relative brain size in *Homo erectus*. In accordance with this, Martin and colleagues (2006) claimed that the brain of *Homo floresiensis* was too small to have been a dwarfed descendant of *Homo erectus*. As they pointed out, we would have to go back about 3.5 million years in evolutionary time to find a hominid cranial capacity as small as that of the Flores hominid. That led some researchers to suggest that the Flores hominid is a pathological modern human, resulting in a substantial literature supporting such an interpretation by discussing various syndromes that may be linked to microcephaly. However, these have all been rejected by later works, including Falk and colleagues (e.g., 2007, 2009), on the basis that these syndromes do not account for the total morphological pattern observed in *Homo floresiensis*. An alternative scenario is that *Homo floresiensis* descended from a *Homo habilis*-like ancestor, much reducing the need for such relative brain size reduction. However, there is no fossil evidence yet to suggest that pre-*erectus* hominids were the first to leave Africa. Kubo and colleagues (2013) indicated that *Homo floresiensis* could after all be a dwarfed descendant of *Homo erectus* in terms of brain size allometry. According to them, a significant amount of brain size reduction can be explained by scaling. The remaining amount of brain size reduction remains to be explained by factors other than body size scaling, such as increased cortical folding. As we noted above, brain size reduction has been verified in some island mammals and could, therefore, have been the case for the Flores hominid as well.

Evolutionary Changes in Life History

Life History Shifts: r/K Selection

MacArthur and Wilson (1967) made a distinction between two main types of life history, K and r-strategy, which was elaborated a few years later by Eric R. Pianka (1970). In their view, island colonisers should initially be r-strategists and subsequently evolve the tendency to evolve into K-strategists.

The K-strategists are characterised by a large body size, low reproductive rates, late onset of sexual maturity, low juvenile mortality, high competitive abilities, and a narrow morphological variability, and vice versa for the r-strategists. The first are at advantage in predictable or stable environments where the emphasis is on competition for limited resources, which promotes selection for the better competitor. The latter, on the other hand, are typical for an unstable or changing environment, where such an adaptation loses its advantage and a high generation turnover is essential.

As far as the primary colonisers are concerned, this is not what we see in fossil island faunas. Typically, these founders consist of elephants, hippos, ruminants, and rodents, thus a mixture of K- and r-strategists. Within rodents, voles are successful island colonisers, but these live long and are slow in relation to their body size. Regarding the subsequent evolution into K-strategists, there are indeed several instances of unrelated extant island mammals with a single or twin offspring and increased lifespan. This is the case for, for example, the Philippine giant cloud rat (*Phloeomys pallidus*), several rats and mice of New Guinea (*Hyomys*, *Mallomys*, *Abeomelomys sevia* and *Paramelomys levipes*) (see also Aplin and Helgen, 2010), the Sulawesi babirusa (*Babyrousa celebensis*), and the Hispaniolan hutia (*Plagiodontia aedium*), whereas their mainland relatives have larger litters and a shorter lifespan. The anoa, the dwarf buffalo of Sulawesi (*Bubalus depressicornis*), has a particularly high absolute and relative longevity, as noted by Müller and colleagues (2011).

However, Spinagallo cave on Sicily has been claimed to provide an exception to the rule. Here, remains of more than a hundred individuals of the dwarf elephant (*Palaeoloxodon falconeri*) were retrieved among which were an exceptionally high number of calves and immature individuals, making up almost 60% of the fossil assemblage. This, together with the total lack of tusked females, led Raia and colleagues (2003) to the conclusion that the Spinagallo elephant invested heavily in reproductive efforts because increased reproductive investment and tusklessness are positively correlated in the living African elephants. The survivorship curve for *Palaeoloxodon falconeri* resembles the type II curve of Pianka (2000), whereas fossil assemblages of the woolly mammoth (*Mammuthus primigenius*) follow a type I curve, typical of long-living mammals. According to Raia and colleagues (2003), a selection took place of smaller individuals able to reproduce earlier and at higher rates. Under high mortality rates, litter size tends to increase with lighter offspring, whereas under lower mortality rates, offspring tends to become heavier and litter size decreases, as demonstrated by Daniel Promislow and Paul Harvey (1990).

However, a critical note is appropriate here: first of all, the number of offspring in fossil taxa is, unfortunately, practically impossible to reconstruct, and second, the percentage juveniles in a fossil death assembly are not a reliable proxy for life history strategy.

All other fossil island mammal species that have been investigated regarding life history variables have been invariably shown to indeed have traits that fit K-strategists, such as slow ontogenetic growth and/or longer lifespans. Microscopic studies of fossil bone tissue can reveal information about the onset of skeletal maturation, and thus length of ontogenetic growth, seasonality, and longevity – see for an overview of insular mammals, Kolb and colleagues (2015a). Meike Köhler and Salvador Moyà-Solà (2009) and Köhler (2010) found slow bone growth rates in the Late Pleistocene mouse-goat (*Myotragus balearicus*) of Mallorca, indicating delayed onset of maturity and an extended lifespan. Their conclusion regarding seasonal, reptile-like growth was, however, subsequently refuted by Shai Meiri and Pasquale Raia (2010) in a response: growth lines do occur in mammals. The observation of delayed maturity is in line with the earlier link that Jordana and colleagues (2012) made between extremely increased molar crown height and exceptional longevity in this species. Guillem Orlandi-Oliveras and colleagues (2016) inferred an increased lifespan to over 10 years in the Late Pliocene giant dormouse (*Hypnomys onicensis*) of Mallorca, compared to its closest living relative, the common garden dormouse (*Eliomys quercinus*). Its growth rate, however, had remained unchanged, so gigantism in this case is not achieved by accelerated growth but rather by a delayed maturity. Also, the Ryukyu Late Pleistocene deer (*Cervus astylodon*) evolved increased longevity, as shown by Mugino Kubo and colleagues (2011) based on mortality patterns of fossils of this deer from Okinawa. Christian Kolb and colleagues (2015b) found that the Late Pleistocene dwarf deer of Crete (*Candiacervus*) was characterised by low growth rates and late attainment of skeletal maturity. Moncunill-Solé and colleagues (2013) reported 15 seasonal growth lines in a specimen of the giant burrowing mouse (*Mikrotia magna*) of Gargano, indicating longevity. This was confirmed by Kolb and colleagues (2015b), who observed a high amount of remodelling bone in a sectioned femur, typical for very old individuals. Kolb and colleagues (2015a) found, however, no indication of longevity for the Late Miocene giant gymnures (*Deinogalerix*) or for three insular giant pikas (*Prolagus sardus, Prolagus imperialis, Prolagus apricenicus*).

The fossil cases of insular gigantism contradict McNab's (2012) hypothesis of a faster growth in small mammals in response to increased food levels due to decreased competition on islands. Rather, these fossil insular giants seem to retain an unchanged growth speed, combined with a prolonged growth period, and sometimes evolve longevity (*Hypnomys*), as predicted by MacArthur and Wilson (1967).

Changes in Metabolism

Studies of extant insular taxa indicate that metabolic costs are reduced in some cases. This is, for example, the case for the hutias of the West Indies, where a lower basal metabolism has indeed been measured for *Capromys pilorides*, *Geocapromys browni*, and *Geocapromys ingrahami*. Compared to the values expected from body mass in mammals generally, these rodents appear to have a basal rate of 64, 82, and 67%, respectively, as summarised by McNab (2002). A lower metabolism may also be present in some pteropodid bats. Bonaccorso and McNab (1997) and McNab and Bonaccorso (2001) observed that several small nectar-eating flying foxes (*Macroglossus* and *Syconycteris*), found in Australia, New Guinea, and on large and small islands from the Moluccas to the Solomons, readily enter torpor. This is not observed in other nectar-eating bats of similar or smaller body mass. They suggest that this torpor might facilitate their persistence on small islands.

Information about metabolic rates in fossil species is, unfortunately, still scarce. Justyna Miszkiewicz and colleagues (2020) reconstructed bone metabolic rate by using osteocyte lacunae densities as a proxy for the Pleistocene Timor rats, which range from small to giant. They found that the largest forms indeed seem to have had a relatively slow pace of bone metabolic activity, consistent with predictions of energy conservation in island species. The smaller species from the same assemblage probably had accelerated levels of bone metabolism, as inferred from higher osteocyte lacunae densities. These results indicate that bone metabolic rates in island rodent species scale with body mass and are not necessarily correlated with island conditions but rather an effect of ecological release. Indeed, Bromage and colleagues (2009) found higher osteocyte lacunae densities in the Cypriot dwarf hippo (*Hippopotamus minor*) than in the mainland common hippo (*Hippopotamus amphibius*).

McNab (2002) further proposed that a lower metabolism underlies, at least partly, the loss of flight in birds. This idea is based on his earlier study (2000) on fruit pigeons on small islands in the South Pacific, which have lower basal metabolic rates than their continental relatives. No information for other birds or the fossil record is yet available to confirm this hypothesis.

References

Accordi F.S., Palombo M.R. 1971. Morfologia endocranica degli elefanti nani pleistocenici di Spinagallo (Siracusa) e comparazione con l'endocranio di *Elephas antiquus*. *Rendiconti dell' Accademia Nazionale del Lincei (Series 8)*, 51: 111–124.

Alcover J.A., Moyá-Solá S., Pons-Moyá J. 1981. Les Quimeres dels Passat. Els Vertebrats fóssils del Plio-Quaternari de les Balears i Pitiüses. *Monografíes Científiques*, 1: 1–260.

Angelone C. 2005. Evolutionary trends in dental morphology of the genus *Prolagus* (Ochotonidae Lagomorpha) in the Mediterranean islands. In: Alcover J.A., Bover P. (eds), Proceedings of the International Symposium Insular Vertebrate Evolution: the Palaeontological Approach. *Monografies de la Societat d'História Natural de les Balears*, 12: 17–26.

Aplin K.P., Helgen K.M. 2010. Quaternary murid rodents of Timor Part I: New material of *Coryphomys buehleri* Schaub, 1937, and description of a second species of the genus. *Bulletin of the American Museum of Natural History*, 341: 1–80.

Azzaroli A., Mazza P. 1992. The cervid genus Eucladoceros in the early Pleistocene of Tuscany. *Palaeontographia Italica*, 79: 43–100.

Bonaccorso F.J., McNab B.K. 1997. Plasticity of energetics in blossom bats (Pteropodidae): Impact on distribution. *Journal of Mammalogy*, 78: 1073–1088.

Bover P., Tolosa F. 2003. The olfactory ability of Myotragus balearicus: preliminary notes. In: Alcover J.A., Bover P. (eds), Proceedings of the International Symposium Insular Vertebrate Evolution: The Palaeontological Approach. *Monografies de la Societat d'História Natural de les Balears*, 12: 85–94.

Bromage T.G., Lacruz R.S., Hogg R., Goldman H.M., McFarlin S.C., Warshaw J., Dirks W., Perez-Ochoa A., Smolyar I., Enlow D.H., Boyde A. 2009. Lamellar bone is an incremental tissue reconciling enamel rhythms, body size, and organismal life history. *Calcified Tissue International*, 84: 388–404.

Csiki-Sava Z., Vremir M., Meng J., Brusatte S.L., Norell M.A. 2018. Dome-headed, small-brained island mammal from the Late Cretaceous of Romania. *Proceedings of the National Academy of Sciences of the USA*, 115 (19): 4857–4862.

Dechaseaux C. 1961. Moulages endocraniennes de bovidés fossiles. *Annales de Paléontologie*, 47: 51–73.

Falk D., Hildebolt C., Smith K., Morwood M., Sutikna T., Brown P., Jatmiko E., Saptomo W., Brunsden B., Prior F. 2005. The brain of *Homo floresiensis*. *Science*, 308: 242–245.

Falk D., Hildebolt C., Smith K., Morwood M. J., Sutikna T., Jatmiko, Wahyu Saptomo E., Imhof H., Seidler H., Prior F. 2007. Brain shape in human microcephalics and *Homo floresiensis*. *Proceedings of the*

National Academy of Sciences of the United States of America, 104: 2513–2518.
Falk D., Hildebolt C., Smith K., Jungers W.L., Larson S.G., Morwood M.J., Sutikna T., Jatmiko, Saptomo E.W., Prior F. 2009. The type specimen (LB1) of *Homo floresiensis* did not have Laron Syndrome, *American Journal of Physical Anthropology*, 140: 52–63.
Jordana X., Marín-Moratalla N., DeMiguel D., Kaiser T.M., Köhler M. 2012. Evidence of correlated evolution of hypsodonty and exceptional longevity in endemic insular mammals. *Proceedings of the Royal Society B*, 279: 3339–3346.
Köhler M. 2010. Fast or slow? The evolution of life history traits associated with insular dwarfing. In: Pérez-Mellado V., Ramon C. (eds), *Islands and Evolution*. Mallorca, Institut Menorquí d'Estudis Recerca, pp. 261–280.
Köhler M., Moyà-Solà S. 2004. Reduction of brain size reduction and sense organs in the fossil insular bovid *Myotragus*. *Brain, Behavior and Evolution*, 63: 125–140.
Köhler M., Moyà-Solà S. 2009. Physiological and life history strategies of a fossil large mammal in a resource-limited environment. *Proceedings of the National Academy of Sciences*, 106 (48): 20354–20358
Kolb C., Scheyer T.M., Veitschegger K., Forasiei A.M., Amson E., van der Geer A.A.E., van den Hoek Ostende L.W., Hayashi S., Sánchez-Villagra M.R. 2015a. Mammalian bone paleohistology: A survey and new data with emphasis on island forms. *PeerJ*, 3: e1358.
Kolb C., Scheyer T.M., Lister A.M., Azorit C., de Vos J., Schlingemann M.A.J., Rössner G.E., Monaghan N.T., Sánchez-Villagra M.R. 2015b. Growth in fossil and extant deer and implications for body size and life history evolution. *BMC Evolutionary Biology*, 15: 1–15
Kubo D., Kono R.T., Kaifu Y. 2013. Brain size of *Homo floresiensis* and its evolutionary implications. *Proceedings of the Royal Society B*, 280: 20130338.
Kubo M.O., Fujita M., Matsu'ura S., Kondo M., Suwa G. 2011. Mortality profiles of late Pleistocene deer remains of Okinawa island: Evidence from the Hananda-Gama cave and Yamashita-cho cave I sites. *Anthropological Science*, 119: 18–201.
Leinders J.J.M., Sondaar P.Y. 1974. On functional fusions in footbones of Ungulates. *Zeitschrift für Säugetierkunde*, 39 (2): 109–115.
Locatelli E., Awe Due R., van den Bergh G.D., van den Hoek Ostende L.W. 2012. Pleistocene survivors and Holocene extinctions: The giant rats from Liang Bua (Flores, Indonesia). *Quaternary International*, 281: 47–57.
Lyras G.A. 2009. The evolution of the brain in Canidae (Mammalia: carnivora). *Scripta Geologica*, 139:1–93.
Lyras G.A. 2018. Brain changes during phyletic dwarfing in elephants and hippos. *Brain Behaviour and Evolution*, 92: 167–181.
MacArthur R.H., Wilson E.O. 1967. *The Theory of Island Biogeography*. Monographs on Population Biology, Princeton University Press, Princeton.
Manger P.R. 2006. An examination of cetacean brain structure with a novel hypothesis correlating thermogenesis to the evolution of a big brain. *Biological Reviews*, 81 (2): 293–338.

Martin R.D., MacLarnon A.M., Philips J.L., Dobyns W.B. 2006. Flores hominid: new species or microcephalic dwarf? *The Anatomical Record* (Part A), 288A (11): 1123–1145.
McNab B.K. 2000. The influence of body mass, climate, and distribution on the energetics of South Pacific pigeons. *Comparative Biochemistry and Physiology (Part A: Molecular & Integrative Physiology)*, 127: 309–329.
McNab B.K. 2002. Minimizing energy expenditure facilitates vertebrate persistence on oceanic islands. *Ecology Letters*, 5: 693–704.
McNab B.K. 2012. *Extreme Measures: The Ecological Energetics of Birds and Mammals*. University of Chicago Press, Chicago.
McNab B.K., Bonaccorso F.J. 2001. The metabolism of New Guinean pteropodid bats. *Journal of Comparative Physiology (B Biochemical Systemic and Environmental Physiology)*, 171: 201–214.
Meiri S., Raia P. 2010. Reptilian all the way? *Proceedings of the National Academy of Sciences*, 107 (8): E27.
Miszkiewicz J.J., Louys J., Beck R.M.D., Mahoney P., Aplin K., O'Connor S. 2020. Island rule and bone metabolism in fossil murines from Timor. *Biological Journal of the Linnean Society*, 129: 570–586.
Moncunill-Solé B., Marín-Moratalla N., Jordana X., Casanovas-Vilar I., Rook L., Köhler M. 2013. Preliminary results on life history traits of the insular fossil rodent *Mikrotia* (Muridae, Rodentia) from Gargano archipelago (Apulia, Italy). In: Lamm E.T., Bailleul A., Flynn K. (eds), *Book of Abstracts of the Second International Symposium on Paleohistology. Bozeman, Montana, July 18–20, 2013*: 68. Museum of the Rockies, Bozeman.
Montgomery S.H. 2013. Primate brains, the "island rule" and the evolution of *Homo floresiensis*. *Journal of Human Evolution*, 65: 750–760.
Motokawa M., Haradab M., Lin L.-K. 2004. Variation in the Y chromosome of *Crocidura tadae kurodai* (Insectivora, Soricidae). *Mammalian Biology – Zeitschrift für Säugetierkunde*, 69 (4): 273–276.
Müller D.W.H., Zerbe P., Codron D., Clauss M., Hatt J.-M. 2011. A long life among ruminants: Giraffids and other special cases. *Schweizer Archiv für Tierheilkunde*, 153: 515–519.
Orlandi-Oliveras G., Jordana X., Moncunill-Solé B., Köhler M. 2016. Bone histology of the giant fossil dormouse *Hypnomys onicensis* (Gliridae, Rodentia) from Balearic Islands. *Comptes Rendus Palevol*, 15: 238–244.
Palombo M.R. 2001. Paedomorphic features and allometric growth in the skull of *Elephas falconeri* from Spinagallo (Middle Pleistocene, Sicily). In: Cavaretta G., Gioia P., Mussi M., Palombo M.R. (eds), *The World of Elephants. Proceedings of the 1st International Congress, Roma, 16–20 October 2001*, pp. 492–496.
Palombo M.R., Köhler M., Moyà-Solà S., Giovinazzo C. 2008. Brain versus body mass in endemic ruminant artiodactyls: A case studied of *Myotragus balearicus* and smallest *Candiacervus* species from Mediterranean Islands. *Quaternary International*, 182: 160–183.

Pianka E. 1970. On r and k selection. *American Naturalist*, 104: 592–597.
Pianka E. 2000. *Evolutionary Ecology*. Addison Wesley Inc., San Francisco.
Promislow D.E.L., Harvey P.H. 1990. Living fast and dying young: A comparative analysis of life history variation among mammals. *Journal of Zoology*, 220: 417–437.
Radinsky L.B. 1987 *The Evolution of Vertebrate Design*. University of Chicago Press, Chicago.
Raia P., Barbera C., Conte M. 2003. The fast life of a dwarfed giant. *Evolutionary Ecology* 17: 293–312.
Schauber A., Falk D. 2008. Proportional dwarfism in foxes, mice, and humans: Implications for relative brain size in *Homo floresiensis*. *American Journal of Physical Anthropology*, 46: 185–186.
Scott K. 1990. Postcranial dimensions of ungulates as predictors of body mass. In: Damuth J., MacFadden B.J. (eds), *Body Size in Mammalian Paleobiology. Estimation and Biological Implications*. Cambridge University Press, Cambridge, pp. 301–336.
Sondaar P.Y. 1977. Insularity and its effect on mammal evolution. In: Hecht M.N., Goody P.L., Hecht B.M. (eds), *Major Patterns in Vertebrate Evolution*. Plenum, New York, pp. 671–707.
van der Geer A.A.E. 2005. The postcranial of the deer *Hoplitomeryx* (Mio-Pliocene; Italy): Another example of adaptive radiation on Eastern Mediterranean Islands. In: Alcover J.A., Bover P. (eds), *Proceedings of the International Symposium Insular Vertebrate Evolution: The Palaeontological Approach. Monografies de la Societat d'Història Natural de les Balears*, 12: 325–336.
van der Geer A.A.E. 2014. Parallel patterns and trends in functional structures in island mammals. *Integrative Zoology*, 9: 167–182.
van der Geer A.A.E., de Vos J., Lyras G., Dermitzakis M. 2006. New data on the Pleistocene Cretan deer *Candiacervus* sp. II (Mammalia, Cervinae). *Courier Forschungsinstitut Senckenberg*, 256: 131–137.
van der Geer A.A.E., Lyras G., MacPhee R.D.E., Mitteroecker P. 2018 (a). From Jumbo to Dumbo: Cranial shape changes in elephants and hippos during phyletic dwarfing. *Evolutionary Biology*, 45 (3): 303–317.
Weston E.M., Lister A.M. 2009. Insular dwarfism in hippos and a model for brain size reduction in *Homo floresiensis*. *Nature*, 459: 85–88.

CHAPTER TWENTY FIVE

Extinction of Island Mammals

Evolution of Island Mammals: Adaptation and Extinction of Placental Mammals on Islands, Second Edition.
Alexandra van der Geer, George Lyras and John de Vos.

© 2021 John Wiley & Sons Ltd. Published 2021 by John Wiley & Sons Ltd.

The timing and cause of large-scale extinction of endemic fossil faunas, or of a single endemic taxon – termed *faunal turnover* and *species extinction*, respectively – is often poorly known and understood due to the incomplete fossil record and lack of absolute datings. The subject is, however, of great importance mainly because of the extensive and ongoing extinctions of islanders caused by us and our ancestors – directly as well as indirectly, introduced species, and climatic changes, including sea level rise. Edward O. Wilson (1992) calculated that about 30,000 species will go extinct each year since the start of the twentieth century. This Late Pleistocene–Holocene mass extinction, or the sixth mass extinction, had and still has a decisive impact on especially the megafauna and explains why so few large mammal species survived until today. But, also, smaller sized species have been disappearing, often silently.

There are two main classes of causes for extinctions of insular endemics. First, those that are part of the natural dynamics of the island system: These include competition with new species – new arrivals or locally evolved species – natural changes of habitat – e.g. due to climate change – natural disasters and the end of the island – either by total submersion or the connection to the mainland. Evidence of these factors is extremely limited in the fossil record; yet, they play a crucial role in island biogeography, as extinction, speciation, and immigration are regarded as the three main factors shaping biodiversity patterns on islands, with species interactions as major feedback system (see also Figure 22.4b). Second, those that are related to human activity: The impact of humans is manifold – hunting, active and passive introductions of alien taxa, spread of pathogens, habitat destruction, etc. – and acts on levels from the individual taxon to entire faunas. The timing of the colonisation of islands by Palaeolithic and Neolithic humans is relatively well documented by archaeological findings – e.g. Flores, Ryukyu Islands, Cyprus, and Caribbean Islands – but much remains to be understood about the degree of such impact. Most is known, of course, about the impact of colonisations during historical times, but this does not provide a reliable proxy for the impact of Palaeolithic nor of Neolithic humans. It is merely a rough indication of the potential dangers that our species can bring to an endemic island fauna.

Anthropogenic factors are relatively well documented in the fossil record compared to the 'natural' factors. Extinction events, however, are often caused by a combination of factors and rarely by a single factor alone, as demonstrated by Ross MacPhee (2018). In addition, natural factors and anthropogenic factors are often difficult to separate, for example, the colonisation by humans may coincide with drastic sea level drops due to a colder and drier climate, both potentially fatal to the fragile island ecology. Also, the combination of natural factors and anthropogenic factors may give the fatal blow to a species, for example, climate change and extensive hunting. One should keep in mind, though, that island endemics survived throughout the entire Pleistocene, a geological epoch with drastic climatic changes, often abrupt, with glacial and interglacial stadia. Thus, to point to climate change during the Holocene as major cause of extinction is at its best an oversimplification.

The Island Prison

Endemic insular species are particularly prone to extinctions in general, independent of the cause, and often the entire species or population may be wiped out, as they cannot migrate elsewhere as their mainland relatives can. In addition, to make things worse, many if not most insular forms lost the dispersal abilities of their mainland ancestors. For birds, this is attested by flightless insular birds, such as the dodo (*Raphus cucullatus*) of Mauritius and the giant pigeon (*Natunaornis gigoura*) of Fiji. That this is valid for terrestrial mammals as well is corroborated by the fact that insular species have hardly ever managed to colonise mainlands. For example, the skulls of dwarf elephants lack the necessary large sinuses that help their continental nephews to float. Another problem is integration into local continental environments, with their much higher species (including competitor) loads. As a result, the vast majority of insular vertebrates are imprisoned on their own island or at most to nearby islands.

In general, small islands are more at risk of losing species than are large islands, as predicted by the species–area relationship models. Although, on average, the number of extinctions is highest on larger islands, simply because there are more species on larger islands, the percentage of extinctions is the lowest, likely because of the greater ecological diversity of these islands and the potentially larger population sizes, reducing the risk of inbreeding depression and increasing the chance of finding a suitable partner.

The Burden of Body Mass

It is widely known that extinctions on islands, and elsewhere, had a disproportionate effect on the larger species, that is, species over 45 kg or for herbivores, over a ton. Extinction risk is thus correlated with body mass of insular species. Indeed, the Late Pleistocene and early Holocene extinctions of the West Indies primarily consisted of large-sized mammals,

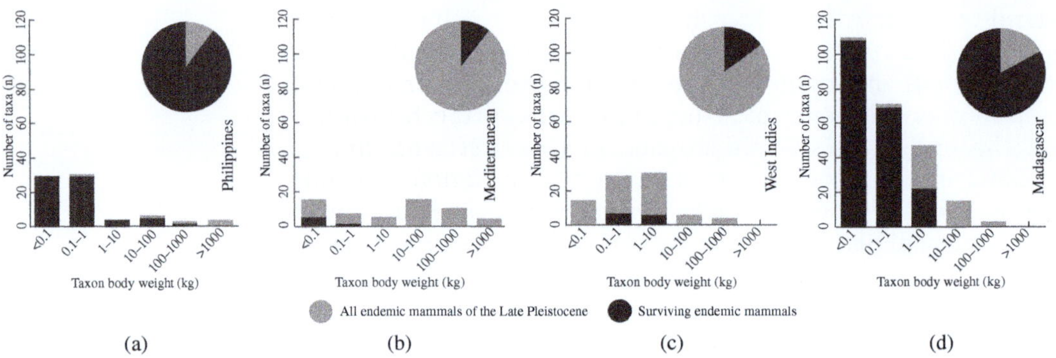

Figure 25.1 Mammalian extinction patterns relative to body size classes for the largest archipelagos and Madagascar. (From Kouvari and van der Geer, 2019; correction of Kouvari and van der Geer, 2018.)

as summarised by Ross MacPhee (2009), hence the popular term *megafauna extinction*. Based on a global dataset, Miranta Kouvari and van der Geer (2018) confirmed that all islands with a fossil record lost their heaviest mammal species indeed, whereas the maximum surviving body size class differs per archipelago, ranging from heavier than 100 kg for the Philippines to well below 100 g for Crete (Figure 25.1). As a side effect, this unbalanced extinction has partially obscured the effect of the island rule because the dwarfed megafauna – dwarf elephants, hippos, and deer – is practically missing today, leaving only giant small mammals as survivors, as noticed by Søren Faurby and Jens-Christian Svenning (2016).

Extinction Debt

The predictive model developed by David Tilman and colleagues (1994) applied to a hitherto overlooked phenomenon, that of extinction debt after severe habitat destruction. They observed that the surviving, dominant species in the few undisturbed fragments that remain after destruction sooner or later drop in numbers and go extinct. Their model shows that this time-delayed extinction is unavoidable even in moderate habitat destruction. With increasing habitat destruction, the more species are predicted to become extinct. Also, the number of extinctions increases further with increasing habitat fragmentation.

Extinction debt then should also occur naturally, for example, when a peninsula becomes an island after a sea level rise. The remaining, vicariant species, part of a fauna with a mainland composition, have now to compete for less resources. In time, the poorest competitors and predators will go extinct. Examples presented by the fossil record, for example, in the Mediterranean. At the start of its isolation in the earliest Pliocene, Mallorca was inhabited by a richer fauna than later on. The fauna had arrived over land during or around the Messinian Salinity Crisis and included a hamster (*Apocricetus darderi*), a murid (*Paraethomys balearicus*), a hare (*Hypolagus*), a bovid, a dormouse, and a shrew. Only the latter three species would survive into the Holocene in the form of the three Pleistocene lineages of *Myotragus*, *Hypnomys*, and *Nesiotites*. Apparently, the hamster, the murid, and the hare were poorer competitors and went extinct early.

The ecological meltdown that took place in Barro Colorado, Central America – which became an island when in the early twentieth century Charges River was dammed for the construction of the Panama Canal – provides a present-day example of extinction debt and subsequent evolution of the 'winners' on a suddenly isolated, continental island.

Natural Disasters

Whittaker and Fernández-Palacios (2007) considered hurricanes, volcanic eruptions, and mega-landslides as the most important natural disasters that might precipitate species extinctions. Of these, only volcanic eruptions may leave a reliable and datable trace in the lithostratigraphic record. The Indonesian island of Flores was thought to provide such a case, where a volcanic eruption marked the end of the Late Pleistocene megafauna, including *Homo floresiensis* and dwarf stegodon. Fossils of the Pleistocene endemic megafauna are found only below the black tuffaceous silts, deposited around 17,000 years, while subfossils of newcomers, including modern humans, occur only in deposits overlying the white tuffaceous silts. There is no overlap in megafauna between the two biozones. A revised stratigraphy of the cave and new datings (see Chapter 13), however, revealed that this is not the case and that there is a depositional hiatus between about 60,000 years ago (the last evidence of hominids and stegodon) and about 12,000 years ago, when the modern fauna arrived.

Tectonics and Sea Level

The total erosion and submerging of an island leads, naturally, to the total extinction of its fauna. The fossil bearing terrestrial deposits of the palaeo-island Gargano, for example, are covered by marine sediments. The submerging of the Gargano during the Early Pliocene inevitably marked the end of the latest

Miocene endemic fauna. In other cases, the submerging is only partially, yet dramatic enough to lead to the extinction of the island fauna, or at least contribute to it. This is the case, for example, in Naxos and Paros, islands in the Mediterranean Sea. During the Late Pleistocene, they were connected to each other by a large, fluvial plain, which provided an ideal habitat for the dwarf elephants living there (*Palaeoloxodon lomolinoi*). With the rising sea level, this plain disappeared below the water, and only mountainous terrain with steep peaks remained. A similar scenario has been suggested in the case of the extinction of the Channel Island dwarf mammoths (*Mammuthus exilis*). The sea level rise at the start of the Holocene caused drowning of significant part of the island and with it low lying areas suitable for mammoths. However, Muhs and colleagues (2015) challenged this view. In case the ancestor arrived during a colder stage during the Middle Pleistocene, then it had already experienced a similar warming of the climate and concurrent sea level rise. An additional factor, however, was present at the end of the Late Pleistocene in the form of humans – represented by the Arlington Springs Man – who arrived, probably not coincidental, around the extinction event.

Another way in which an island stops existing is when it gets connected to the mainland through sea level drop or tectonic activity. This merging to a larger land mass generally results in an invasion by mainland species and the subsequent complete disappearance of the insular fauna. The continentalisation of an island is generally the result of tectonic processes, low sea levels, or both. The three Indonesian Greater Sunda Islands Java, Sumatra, and Borneo, for example, were at intervals connected to the mainland during the Pleistocene, enabling the immigration of balanced mainland faunas, such as the tropical rainforest fauna that spread over the three islands around 128,000 years ago and which is represented by similar fossils found in caves on Java (Punung) and Sumatra (Lida Ayer), as well as in Vietnam (Lang Trang), Malaysia (Niah), Cambodia (Phnom Loang), Laos (Tam Hang), and South China (Ho-shang-tung).

In the fossil record, some extinction and invasion events occurring at the same time on different islands are explained as being related to worldwide (eustatic) sea level changes. These faunal turnovers are particularly evident at roughly 700,000 –800,000 years ago (Early to Middle Pleistocene transition), 243,000 –300,000 years ago (early Riss glaciation of the Middle Pleistocene), and at 20,000 –30,000 years ago (Last Glacial Maximum), when the sea level was indeed very low. Continental faunas invaded the former islands and led to the extinction of the endemic fauna.

Impact of Exotic Competitors

New colonisers may reach the islands via natural means or as part of human activities. However, just a fraction of the colonisers manage to establish a viable population. Mark Williamson (1996) reviewed several cases of species introduced to islands by humans. He concluded that only 10% of these species had been successful. The percentage of unsuccessful colonisers among the species that arrived by natural means is unknown, because a fossil record of failures simply does not exist. However, it is reasonable to assume that in the deep past, a roughly similar percentage of newcomers could establish itself.

However, in the unfortunate cases of a successful establishment, the impact of these alien competitors is dramatic. Insular species often evolved loss of competitive and anti-predator skills, termed *ecological naiveté* by Mark Lomolino and colleagues (2017). Thus, many island species went extinct after the successful establishment of a new species. This is evident from the fossil record, as well as from the history of introductions of alien species by humans. But why are insular species so vulnerable? After all, island species evolved to fit the particular conditions of each island, so it would be expected that they have an advantage over any invasive species. It appears, however, that typical insular adaptations make specialised island species vulnerable to intrusions by alien species whose adaptive success was shaped elsewhere.

Quentin Cronk and Janice Fuller (1995) indicated some reasons why the successful establishment of a colonist has an impact on the competitive native species. They focused on endemic plants, but the same principles can be applied to any insular species. One reason is related to island size. Many islands are small and often have limited geomorphologic variation. This means that the new species may spread quickly over the island. Another reason lies in the species poverty of islands. Islands typically contain many potential ecological niches because they harbour only a few species. Thus, at the very beginning, the new coloniser may fit to one of them without having the need to compete with endemic forms. In due time, the new colonisers will experience an ecological release from their mainland predators and competitors, giving the species the possibility to radiate into several ecomorphs. Without natural enemies, they only have to compete with a few insular species, which they eventually may do with great success.

According to a model proposed by Robert Ricklefs and George Cox (1972), the arrival of new colonists restricts the geographic range and narrows the niche breadth of the already established species. Their observations were based on the recent avifauna of the Lesser Antilles, but similar processes have been confirmed for endemic rats in Japan. For example, the spiny rats (*Tokudaia*) are today restricted to the northern part of Okinawa, Tokuno, and Amami, but lived during the Late Pleistocene also in the southern part of Okinawa and on the nearby islet of Ie. Presumably, they were outcompeted in their original range by the introduction of common rats and mice (*Rattus norvegicus*, *Rattus tanezumi*, and *Mus musculus*).

In some cases, the success of the new colonists might be related to their superior design. On islands, mammals may evolve new anatomical or behavioural features suitable for their particular island habitat. These adaptations often parallel those of mainland species belonging to a different taxon. The parallel evolution of adaptive strategies though may be incomplete because of phylogenetic constraints. For example, according to van der Geer and colleagues (2006), the two smallest sizes of the Late Pleistocene Cretan deer (*Candiacervus ropalophorus* and *Candiacervus* size II) were adapted to the mountainous environments of Crete and therefore had evolved limbs similar to those of goats. However, deer generally are forest or swamp dwellers, and because of this limitation, the Cretan deer could not reach the level seen in the wild goat (*Capra aegagrus*).

In a few cases, the resident species may evolve in a different direction after the successful establishment of an ecologically competitive newcomer. Van der Geer and colleagues (2013) showed that the giant dormouse of the Balearics (*Hypnomys morpheus*) evolved a smaller body size after a competitor belonging to the same genus as its ancestor had arrived. This was also observed in the Cretan giant rat (*Kritimys catreus*) after the arrival of the mouse (*Mus bateae*) and the giant hamster (*Hattomys gargantua*) of Gargano at the time truly giant rats appeared (*Mikrotia magna*) in the fossil record.

Impact of Exotic Predators

> *. . . what havoc the introduction of any new beast of prey must cause in a country, before the instincts of the indigenous inhabitants have become adapted to the stranger's craft of power.* (Charles Darwin, 1839, p. 401)

The introduction and successful establishment of a non-native predator can have a catastrophic impact on the insular fauna. Many cases have been recorded of insular species driven to extinction due to hyperpredation – enhanced predation pressure on a secondary prey – by predators that were introduced by humans, accidentally or deliberately. These predators are usually cats, dogs, mongooses, pigs, and rats. Many insular vertebrate species have evolved on islands devoid of such predators and gradually lost the behavioural and morphofunctional mechanisms to defend themselves against them.

Island species with very small numbers are particularly vulnerable to hyperpredation. Perhaps, the best-known story is that about the extinction of the flightless wren (*Traversia lyalli*) of Stephens Island, New Zealand. The first specimen was caught in 1894, and a year later, it was reported as extinct. The popular account, mentioned at a 1905 conference by Walter Rothschild (1907), holds that the population of the species was so small that it was wiped out completely by a single cat, the pet of the lighthouse keeper. The cat obviously caught the birds whole, as the birds were collected by the keeper's assistant, David Lyall, whose name is honoured in the species name. Although Ross Galbreath and Derek Brown (2004) weakened the story somewhat – by adding that it was not just one cat and that the extermination did not happen within only one year – it still remains a fact that this tiny songbird became extinct within a few years, apparently due to predation by introduced cats. A similar case has been reported the same year by Ella Vázquez-Domínguez and colleagues (2004). According to them, a single introduced cat caused the extirpation of an insular subspecies of the deer mouse *Peromyscus guardia* from Estanque Island of Mexico.

The above-mentioned examples are predators directly introduced by humans. There are also cases of predators invading an island secondarily. A typical example is that of the golden eagle (*Aquila chrysaetos*) on the northern California Channel Islands. It was attracted to the islands after introduction of feral pigs by humans. These pigs provided abundant food for the golden eagles. Before the establishment of the golden eagles, the only carnivores were the endemic island fox (*Urocyon littoralis*), the skunk (*Spilogale gracilis amphiala*), and the fish-eating bald eagle (*Haliaeetus leucocephalus*). After the disappearance of bald eagles due to toxic DDT levels in fish, golden eagles could colonise the islands, increase their population size, and start preying on not only the pigs, but also on the foxes. As a result, the foxes experienced a catastrophic decline in numbers in only a few years. Today, conservation measures have enabled the recovery of the fox population, but, most probably, a similar scenario in the past would have resulted in a rapid extinction of the foxes. The speed of the phenomenon is so fast that in the fossil record we would only see a faunal turnover, with foxes and bald eagles being replaced by pigs and golden eagles. This is a general problem of the fossil record. Its resolution is not fine enough to give us evidence of the predator in its mainland form and its impact on the insular fauna. We only have the end result, which often includes the predator in its fully adapted insular form. The insular carnivore with the best fossil record is the Pleistocene Sardinian dog (*Cynotherium sardous*). We know that this species became progressively smaller during its evolutionary history, but, even in this case, fossils of its mainland ancestor have never been found and the earliest *Cynotherium* fossil belongs already to an endemic form, being much smaller than its mainland ancestor, *Xenocyon*.

Genetic Disorders and Infections

With small population numbers, mammals suffer an increasing risk of the loss of genetic variation. This inevitably leads to a higher chance for deleterious genes to become dominant or the loss of genes that deal with the immune system through inbreeding depression. Such processes are difficult to prove for fossil species, yet, DNA analysis of the Wrangel Island mammoth population (*Mammuthus primigenius*), which survived on the Siberian island until about 3,700 years ago, has shown this to be the case. Eleftheria Palkopoulou and colleagues (2015), Amongst others, Rebekah Rogers and Montgomery Slatkin (2017) found that these dwarfed mammoths had accumulated a set of damaging mutations. Erin Fry and colleagues (2020) elaborated on these earlier findings and tested the genes in vitro. They discovered that the dwarf mammoths suffered from increased risk of diabetes, developmental defects, low sperm count, and even a loss of smell. Presumably, this was the result of the small area of the island and the isolation of this last surviving mammoth population, and their extinction was associated with a 'mutational meltdown'.

The importance of novel infectious diseases as a cause of extinction was stressed by Ross MacPhee and Preston Marx (1997). Contact with introduced mainland organisms may expose native species to exotic pathogens, often resulting in severe disease and death, simply because the insular species had no opportunity to develop resistance to alien pathogens. As David Duffy and Hernan Vargas (2018) wrote, the loss of genes and adaptation to novel environments may over time reduce the capacity of island mammals to deal with new pathogens.

Although persuasive evidence is still lacking in the fossil record, it is very reasonable to assume that species extinctions due to the introduction of exotic pathogens have occurred in the prehistoric past and still do today. Indeed, historical accounts of cases of exotic diseases devastating native populations of humans as well as animals are well attested. Perhaps, the most cited example is that of avian malaria, which caused population collapses of several species of Hawaiian honeycreeper birds, according to Richard Warner (1968). The

Hawaiian birds had not been exposed to avian malaria since their isolation and therefore had not developed any resistance in the way that mainland birds had. Although recent studies have shown that additional factors played a role as well, it is clear that the disease played an important role in these extinctions. Another documented example of a pathogen as (partial) cause of extinction is provided by the Japanese or Honshu wolf (*Canis lupus hodophilax*). The first case of rabies in this wolf was reported in Kyushu and Shikoku in 1732. The disease increased the wolf's aggressive interaction with humans, and together with extensive deforestation by farmers, led to active eradication of the endemic wolf. The last known specimen died in 1905 in Nara Prefecture, southern Honshu.

Kelly B. Wyatt and colleagues (2008) studied museum specimens of the extinct endemic arboreal rat (*Rattus macleari*) from Christmas Island in the Indian Ocean. With the use of ancient DNA analysis, they suggested that the rat had become extinct due to a pathogenic trypanosome carried by fleas hosted on black rats that had arrived through human agency. At the end of the nineteenth century, Maclear's rat was abundant on the island, and Charles Andrews (1900) reported that it was 'by far the commonest of the mammals found in the island; in every part I visited, it occurred in swarms'. Nine years later, he returned to the island and reported that they had gone extinct. He speculated that black rats had brought a fatal epidemic disease to the island. This was the only reasonable explanation for the disappearance of such an abundant animal in such a short time. Black rats could not have outcompeted Maclear's rat because the former were not present over the entire island. In addition, the latter was an arboreal species eating fruits and shoots from the trees. The black rats are reported to have arrived in 1899 by means of the S.S. Hindustan, while Maclear's rat was seen alive for the last time between 1902 and 1904. This means that the extinction took place in less than 5 years. No data of this type has been published so far concerning fossil or subfossil species.

Habitat Loss

The destruction, alteration, including fire, and over-exploitation of natural habitats by humans are among the most important causes of extinction of endemic species. Under the term habitat alteration also falls habitat reduction and fragmentation.

Although today habitat destruction and alteration proceed at an alarming rate, islands have been deforested and over-exploited since millennia. As noted by Whittaker and Fernández-Palacios (2007), the popular image of the pre-industrial hunter and gatherer living in total harmony with the natural environment is rather misleading. There is ample evidence that humans contributed to the degradation of the island ecosystems prior to European colonisation. One of the best-known examples is the case of the Polynesian Islands, which were over-exploited by humans between 3,200 and 1,200 years before present. In several instances, the Polynesians had to leave the island after depletion of natural resources and soil degradation. These so-called Mystery Islands bear witness to a once flourishing culture in the form of archaeological remains or the presence of the Polynesian rat (*Rattus exulans*), which they had intentionally introduced. The smaller Polynesian islands, such as Rapa Nui or Easter Island, were completely cleared of forest, likely driving animal as well as plant extinctions.

In Madagascar, habitat alteration was not solely caused by deliberate human actions but was also due to the local elimination of the large grass-eating vertebrates (lemurs, elephant birds, hippos, and giant tortoises). According to David Burney and colleagues (2003), the first settlers of Madagascar hunted the large grass-eating herbivores to local extinction. As a result, plant biomasses began to form increasingly inflammable accumulations, which caused huge fires that fragmentised the habitats. As evidence Burney presented data from sediment cores, which show a fast decline in the quantity of a dung fungus (*Sporormiella*) and a simultaneous peak in the charcoal content. They suggested that many now totally extinct species could have persisted for more than a thousand years in more remote places, such as the humid forests, explaining in this way the large extinction window estimated for the Malagasy megafauna.

An indirect way in which humans may alternate habitats is via the introduction of domestic goats. These animals, often living as feral goats on islands, are notorious ecosystem transformers. Wherever humans brought goats (Box 25.1), either to claim land or as a source of food for seafarers, they increased incredibly in number and reduced the original vegetation to a sparse cover of thorny bushes and other sclerophytic plants.

BOX 25.1

Goats versus the Senkaku Mole

The tiny island Uotsuri is the largest of the Senkaku or Pinnacle Islands, north-west of Taiwan, and measures only 4.3 km^2. A pair of domestic goats was introduced to this uninhabited rocky island in 1978, and due to the total absence of natural predators, all their offspring survived and the population increased rapidly. At present, there are probably more than 300 goats on this miniature island. Naturally, goats destroy the native vegetation through grazing and trampling. A severe side effect of this damage to the fragile island ecosystem is its negative – if not disastrous – effect on the rare Senkaku mole (*Nesoscaptor uchidai*), which is now endangered and on the Red List of Threatened Species of the International Union for Conservation of Nature (IUCN). The mole has an extremely limited range, as it occurs only on Uotsuri Island and thus runs a high risk to get extinct. The ongoing dispute concerning the governmental territoriality of the islands among Japan, China, and Taiwan is a severe obstacle for conservation work, such as the removal of the goats. At present, no actions have been carried out to date.

Hunting to Extinction

Humans hunt in search for food and for various cultural reasons, such as the collecting of traditional medicines and amulets (e.g. tiger and rhinoceros parts), trophy hunting (e.g. deer and big cats), protection of themselves and their fields (e.g. big cats and large bovids), extirpation of inauspicious creatures (e.g. aye-ayes), collecting for scientific purposes, and pet trade. There is but sparse and often inconclusive evidence of humans hunting insular mammals before the time of European colonisation of the islands. Some of the most interesting examples include cut marks on dwarf stegodon bones dated to before 50,000–60,000 years ago from Liang Bua on Flores and butchery marks on rhino bones dated to about 631,000–777,000 years ago from Luzon. At Liang Bua, the stegodon bones here are found in association with stone tools, distributed on occupation floors, indicating that the stegodons had been butchered there. Much younger evidence at the same cave is provided by burnt bones of large-bodied rats, fruit-eating bats, and monitor lizards from early Holocene archaeological deposits. On Flores, stone tools in association with dwarf *Stegodon florensis* fossils date back to at least about 700,000 year ago in the So'a Basin, indirect evidence that early humans ate them, but whether they were scavengers or active hunters remains elusive. Van den Bergh and colleagues (2014) preliminarily reported at a conference that the population structure – corresponding with that of a living population – of the So'a Basin stegodons shows that they died by some catastrophic event and were not selectively accumulated by humans.

In any case, humans and dwarf megafauna lived together on the island for thousands of years, apparently without extinction of the latter. Another direct evidence for hunting by humans are the butchery cut marks on hippo and lemur bones from

Madagascar dated to about 2,000 years ago by MacPhee and Burney (1991). These bones not only show that hunting of the Malagasy hippos indeed took place but also that the impact of humans on the hippo population was not a fast event as explained by the Blitzkrieg ('lightning war'; see Box 25.2) theory, but instead lasted for at least a thousand years. After all, one of the dwarf species, *Hippopotamus lemerlei*, survived until at least about 980. They thus co-occurred with humans for centuries, or in case an Early Holocene colonisation of the island is accepted – as postulated by Hansford and colleagues (2018), but dismissed by most scholars – for thousands of years.

Following the European colonisation of islands, hunting, including that for zoological collecting purposes, constituted the final blow for many insular species. One of these unfortunate species was the Falkland Islands wolf (Figure 25.2) (*Dusicyon australis*). This species is a good example of deliberate eradication by humans. The first settlers, who came to the islands in 1692, regarded the wolf as a threat to their sheep and thus a pest that had to be controlled. They poisoned and shot the animals on a massive scale, which was immensely facilitated by the wolf's tameness, as described by Charles Darwin (1839). Darwin reported that the wolf, referred by him as *Canis antarcticus*, could be lured with a piece of meat held out in one hand and subsequently be killed with a knife held in the other hand. At the time of Darwin's visit in 1834, the species had already disappeared from the east part of the

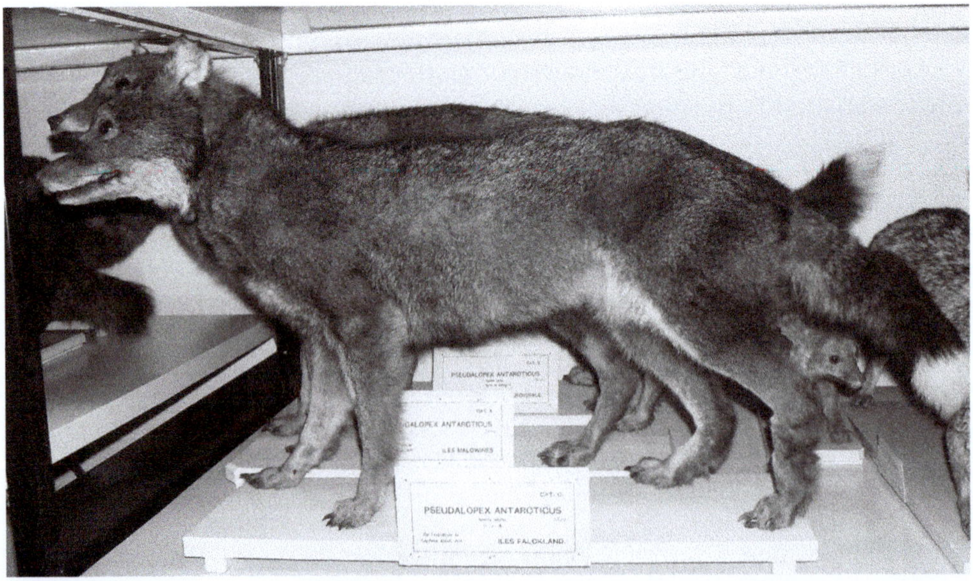

Figure 25.2 Three taxidermy specimens of the Falkland wolf or fox (*Dusicyon australis*). National Museum of Natural History, Leiden. (Photograph: George Lyras.)

island. He foresaw its total extinction, analogue to that of the dodo of Mauritius, which indeed happened 40 years later. The usual reason ascribed for the loss of the Falklands Island wolf is fur trade – Astor and others collected a huge number of furs in the late 1830s – after which the wolf population collapsed. The final blow though may have been given by Scottish settlers in the 1860s who considered the wolves a threat to their sheep.

The impact of hunting is related to island size and characteristics of the species itself, such as population size and reproduction rate. Species with low population numbers and a low reproduction rate on a small island run a higher risk to decrease significantly in numbers when hunted by humans than those with a high rate of reproduction on large islands. Although hunting certainly plays a role in the decrease of insular endemics, it is not clear how many species actually went extinct because of this hunting. According to MacPhee (2009, 2018), recent models suggest that though hunting may result in the rarity of a species, it rarely causes its total loss. More likely, hunting in combination with other factors, such as habitat destruction and competition with introduced species, may lead to extinction.

BOX 25.2

Blitzkrieg or Sitzkrieg

The dramatic and devastating impact of hunting by first settlers is generally referred to as Blitzkrieg extinction: fast, overwhelming, and with no chance for defence. The theory of Blitzkrieg extinction has been used to explain the extinction of the megafauna of New Zealand, Australia, Madagascar, and the Americas practically immediately after the arrival of humans. The basic assumption here is that the endemic fauna is naive by nature and thus easy to hunt down. According to Steve Wroe and colleagues (2004), the model is an oversimplification. First of all, the fauna was not that naive and contained large predatory animals. Second, the theory entirely overlooks the effects of climate change. Australia's climate became hotter and drier and large parts lost their original vegetation cover. It might be that the model works for some extinction events, such as that of the moas of New Zealand, but not for all. The theory that the moas went extinct practically overnight after the Polynesians set food on the island in the thirteenth century is widely accepted, though MacPhee (2009) puts a question mark here. As the moas became rarer, it took considerably more time to find them, and after a while, with just a few birds left, the hunt would not have been worth the effort anymore, and the Polynesians would have turned to easier game, like seals and seafood. The giant birds would have survived in low numbers in remote or inaccessible places, as did the still extant kiwi, which has an equally low reproduction rate. Therefore, even in the moa case, another factor must have contributed to their extinction, apart from hunting.

A problem in most extinction studies is the absence of reliable data concerning the timing of the extinctions. MacPhee (2009) assembled data concerning all known mammalian extinction events in the West Indies. The diagram (Figure Box 25.2) is a simplified version of the one published by MacPhee. It shows that the extinctions started after the arrival of the first Amerindians, some 6,000 years ago. However, the Amerindians had no devastating impact on the fauna and many taxa continued to exist after their arrival. Thus, in the case of West Indies, the Blitzkrieg scenario is not supported by facts. On the contrary, the extinction caused by the Amerindians and later by the Europeans presumably took place within a rather broad time frame, sometimes referred to as *Sitzkrieg*, as opposed to Blitzkrieg. MacPhee's results give support to Samuel Turvey and colleagues (2007), who explained the extinction of West Indian mammalian taxa by the culminating effects of centuries of human-induced distress.

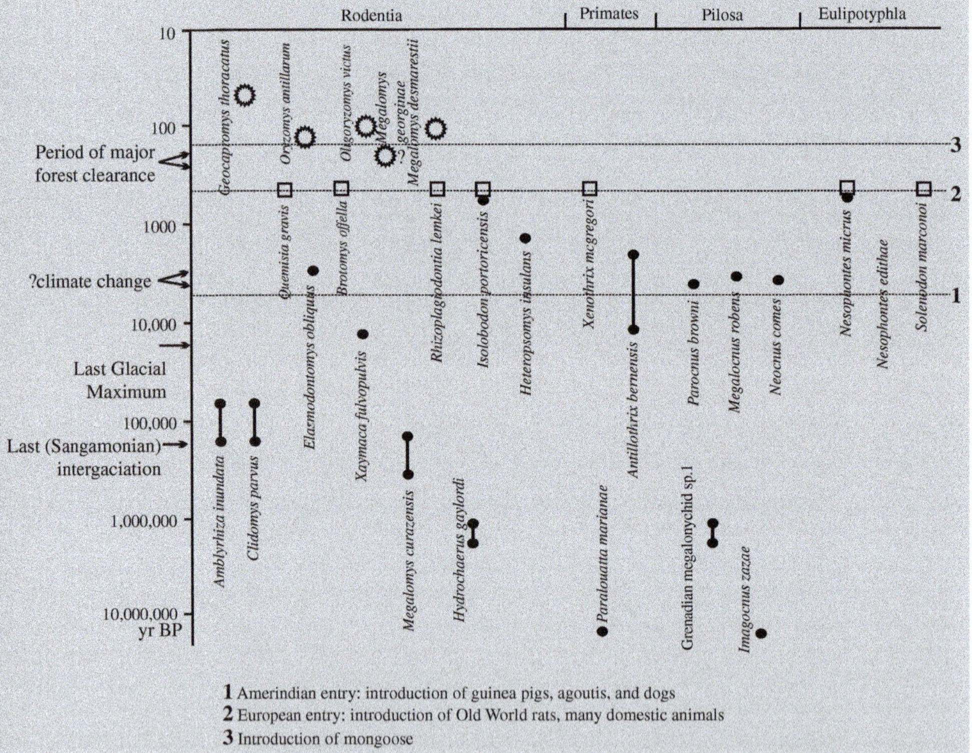

Figure Box 25.2 Extinction schedule of West Indian taxa, summarised from MacPhee, 2009. Symbols indicate the method of estimating the last appearance date (LAD) of each taxon: Squares and oblongs: the date is based on direct radiometric and/or a good stratigraphic record (the length of the interval is represented by the height of the oblong). Stars: the date is based on dated observation or collection. Solid circles: the date is based on indirect evidence (e.g. co-occurrence with *Rattus* remains). Note that the vertical axis is logarithmic.

Kouvari and van der Geer (2018, 2019) assembled a larger dataset, encompassing all islands with a fossil record, and found that the major part – almost 80% – of island mammal extinctions occurred after the first human arrival, with the highest percentages during the Late Pleistocene and the Modern Era. Considering the substantially longer time span of the Late Pleistocene (about 113,000 years) than that of the Modern Era (about five centuries), increased rate of extinctions in the latter period can be observed. Likely, increased rates of introductions of alien species worldwide in the wake of global trade in combination with substantial anthropogenic habitat alteration underlies this pattern. However, it remains unclear whether these extinction waves follow a fast or slow scenario, although the gradual increase in extinctions through the Holocene, with a peak (31%) in the last 500 years, is suggestive of a slow scenario.

References

Andrews C.W. 1900. *A Monograph of Christmas Island (Indian Ocean).* British Museum of Natural History, London.

Burney D.A., Robinson G.S., Burney L.P. 2003. *Sporormiella* and the late Holocene extinctions in Madagascar. *Proceedings of the National Academy of Sciences of the United States of America*, 100: 10800–10805.

Cronk Q.C.B., Fuller J.L. 1995. *Plant Invaders.* Chapman & Hall, London.

Darwin Ch. 1839. *Narrative of the Surveying Voyages of His Majesty's Ships Adventure and Beagle between the years 1826 and 1836, Describing Their Examination of the Southern Shores of South America, and the Beagle's Circumnavigation of the Globe. Journal and Remarks 1832–1836.* London: Henry Colburn.

Duffy D.C., Vargas F.H. 2018. From the vagile to the sedentary: Disease implications and new host relationships on islands. In: Parker P.G. (ed.), *Disease Ecology, Social and Ecological Interactions in the Galapagos Islands.* Springer, pp. 113–135.

Faurby S., Svenning J.C. 2016. Resurrection of the island rule: Human-driven extinctions have obscured a basic evolutionary pattern. *American Naturalist*, 187: 812–820.

Fry E., Kim S.K., Chigurapti S., Mika K.M., Ratan A., Dammermann A., Mitchell B.J., Miller W., Lynch V.J. 2020. Functional architecture of deleterious genetic variants in the genome of a Wrangel Island mammoth. *Genome Biology and Evolution*, 12 (3): 48–58.

Galbreath R., Brown D. 2004. The tale of the lighthouse-keeper's cat: Discovery and extinction of the Stephens Island wren (*Traversia lyalli*). *Notornis*, 51: 193–200.

Hansford J.P., Wright P.C., Rasoamiaramanana A., Pérez V.R., Godfrey L.R. Errickson D., Thompson T., Turvey S.T. 2018. Early Holocene human presence in Madagascar evidenced by exploitation of avian megafauna. *Science Advances*, 4 (9): 6925.

Kouvari M., van der Geer A.A.E. 2018. Biogeography of extinction: The demise of insular mammals from the Late Pleistocene till today. *Palaeogeography, Palaeoclimatology, Palaeoecology*, 505: 295–304.

Kouvari M., van der Geer A.A.E. 2019. Corrigendum to "Biogeography of Extinction: The demise of insular mammals from the Late Pleistocene till today" [PALAEO 505 (15 September 2018) pages 295–304]. *Palaeogeography, Palaeoclimatology, Palaeoecology*, 518, 232–233.

Lomolino M.V., Riddle B.R., Whittaker R.J., 2017. *Biogeography: Biological Diversity Across Space and Time*, 5th edn. Sinauer Associates, Sunderland, MA.

MacPhee R.D.E. 2009. *Insulae infortunatae*: Establishing a chronology for Late Quaternary mammal extinctions in the West Indies. In: Haynes, G. (ed.), *American Megafaunal Extinctions at the End of the Pleistocene.* Springer Netherlands, Dordrecht, pp. 169–193.

MacPhee R.D.E. 2018. *End of the Megafauna: The Fate of the World's Hugest, Fiercest, and Strangest Animals.* W. W. Norton.

MacPhee R.D.E., Burney D. 1991. Dating of modified femora of extinct dwarf *Hippopotamus* from southern Madagascar: Implications for constraining human colonization and vertebrate extinction events. *Journal of Archaeological Science*, 18: 695–706.

MacPhee R.D.E., Marx P.A. 1997. The 40,000-year plague: Humans, hyperdisease, and first-contact extinctions. In: Goodman S., Patterson B. (eds), *Natural Change and Human Impact in Madagascar.* Smithsonian Institution Press, Washington, DC, pp. 169–217.

Muhs D.R., Simmons K.R., Groves L.T., McGeehin J.P., Schumann R.R., Agenbroad L.D. 2015. Late Quaternary sea-level history and the antiquity of mammoths (*Mammuthus exilis* and *Mammuthus columbi*), Channel Islands National Park, California, USA. *Quaternary Research*, 83: 502–521.

Palkopoulou E., Mallick S., Skoglund P., Enk J., Rohland N., Li H., Omrak A., Vartanyan S., Poinar H., Götherström A., Reich D., Dalén L. 2015. Complete genomes reveal signatures of demographic and genetic declines in the woolly mammoth. *Current Biology*, 25: 1395–1400.

Ricklefs R.E., Cox G.W. 1972. Taxon cycles in West Indian avifauna. *American Naturalist*, 106: 195–219.

Rogers R.L., Slatkin M. 2017. Excess of genomic defects in a woolly mammoth on Wrangel island. *PLoS Genetics*, 13: e1006601.

Rothschild W. 1907. On extinct and vanishing birds. *Proceedings of the 4th International Ornithological Congress, London* 1905: 191–217.

Tilman D., May R.M., Lehman C.L., Nowak M.A. 1994. Habitat destruction and the extinction debt. *Nature*, 371: 65–66.

Turvey S.T., Oliver J.R., Narganes Storde Y.M., Rye P. 2007. Late Holocene extinction of Puerto Rican native land mammals. *Biology Letters*, 3: 193–196.

van den Bergh G.D., Brumm A., Puspaningrum M.R., Setiabudi E., Kurniawan I., Prasetyo U.W. 2014. Taphonomy of *Stegodon florensis* remains from the early Middle Pleistocene archaeological site Mata Menge, Flores, Indonesia. In: Kostopoulos D.S., Vlachos E., Tsoukala E. (eds), *Scientific Annals, School of Geology, Aristotle University of Thessaloniki, Special Volume*, 102: 207–208.

van der Geer A.A.E., de Vos J., Lyras G., Dermitzakis M. 2006. New data on the Pleistocene Cretan deer *Candiacervus* sp. II (Mammalia, Cervinae). *Courier Forschungsinstitut Senckenberg*, 256: 131–137.
van der Geer A.A.E., Lyras G.A., Lomolino M., Palombo M.R., Sax D. 2013. Body size evolution of palaeo-insular mammals: Temporal variations and interspecific interactions. *Journal of Biogeography*, 40: 1440–1450.
Vázquez-Domínguez E., Ceballos G., Cruzado J. 2004. Extirpation of an insular subspecies by a single introduced cat: The case of the endemic deer mouse *Peromyscus guardia* on Estanque Island, Mexico. *Oryx*, 38 (3): 347–350.
Warner R.E. 1968. The role of introduced diseases in the extinction of the endemic Hawaiian avifauna. *Condor*, 70: 101–120.
Whittaker R.J., Fernández-Palacios J.M. 2007. *Island Biogeography: Ecology, Evolution, and Conservation*, 2nd edn. Oxford University Press, Oxford.
Williamson M.H. 1996. *Biological Invasions*. Population and Community Biology Series 15. Chapman & Hall, London.
Wilson E.O. 1992. *The Diversity of Life*. Belknap Press, Cambridge, MA.
Wroe S., Field J., Fullagar R., Jermiin L. 2004. Late Quaternary extinctions of megafauna and the global overkill hypothesis. *Alcheringa*, 28: 291–331.
Wyatt K.B., Campos P.F., Gilbert M.T.P., Kolokotronis S.-O., Hynes W.H., DeSalle R., Daszak P., MacPhee R.D.E., Greenwood A.D. 2008. Historical mammal extinction on Christmas Island (Indian Ocean) correlates with introduced infectious disease. *PLoS ONE*, 3 (11): e3602.

INDEX

Page numbers in *italics* represent plates; those in **bold** represent figures.

Plate 1 Reconstruction of four insular dwarf proboscideans with their respective mainland ancestors. Mainland proboscideans: 1, *Palaeoloxodon antiquus*; 2, *Mammuthus columbi*; 3, *Stegodon zdanskyi*. Insular proboscideans: 4, *Palaeoloxodon* aff. *mnaidriensis*; 5, *Palaeoloxodon falconeri*; 6, *Mammuthus exilis*; 7, *Stegodon aurorae*. Source: Adapted from van der Geer A.A.E., van den Bergh G., Lyras G.A., Prasetyo U.W., Due R.A., Setiyabudi E., Drinia H. 2016. The effect of area and isolation on insular dwarf proboscideans. Journal of Biogeography, 43: 1656–66.

(a) (a)

Plate 2 Partial skull (a) and mandible (b) of the dwarf hippopotamus (*Hippopotamus minor*) of Late Pleistocene Cyprus. Scale bar: 5 cm. University of Utrecht, the Netherland. (Photograph: George Lyras.)

Evolution of Island Mammals: Adaptation and Extinction of Placental Mammals on Islands, Second Edition.
Alexandra van der Geer, George Lyras and John de Vos.
© 2021 John Wiley & Sons Ltd. Published 2021 by John Wiley & Sons Ltd.

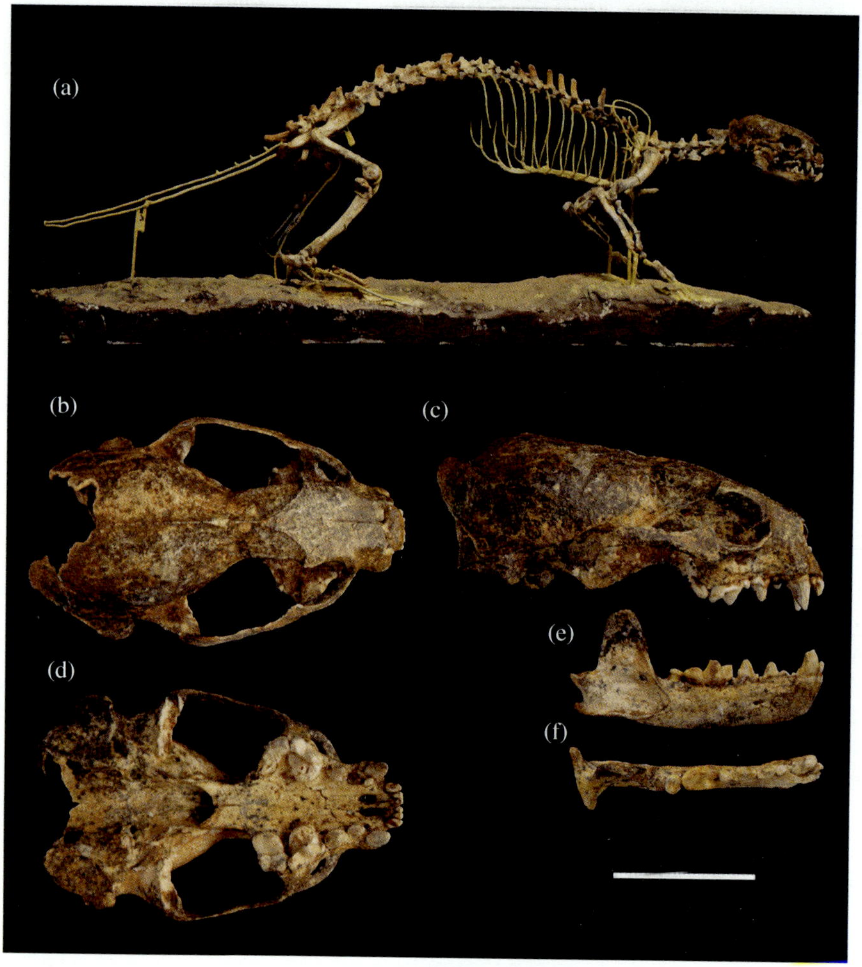

Plate 3 Holotype of the Cretan otter (*Lutrogale cretensis*) found at Liko cave, Crete, late Middle Pleistocene. (a) Mounted skeleton; (b) dorsal, (c) lateral, and (d) ventral views of the skull; (e) lateral and (f) occlusal views of the mandible. Scale bar: 5 cm. Museum of Geology and Palaeontology, University of Athens, Greece. (Photograph: George Lyras.).

Plate 4 Maxilla of the Cretan dwarf hippo (*Hippopotamus creutzburgi*), occlusal view. Katharo, Middle Pleistocene. Scale bar: 5 cm. Natural History Museum, London. (Photograph: George Lyras.)

Plate 5 Mounted skeleton of Cretan deer (*Candiacervus ropalophorus*), Late Pleistocene, and a taxidermy specimen of the Sulawesi anoa (*Bubalus depressicornis*), Recent. National Museum of Natural History, Leiden. (Photograph: Eelco Kruidenier.)

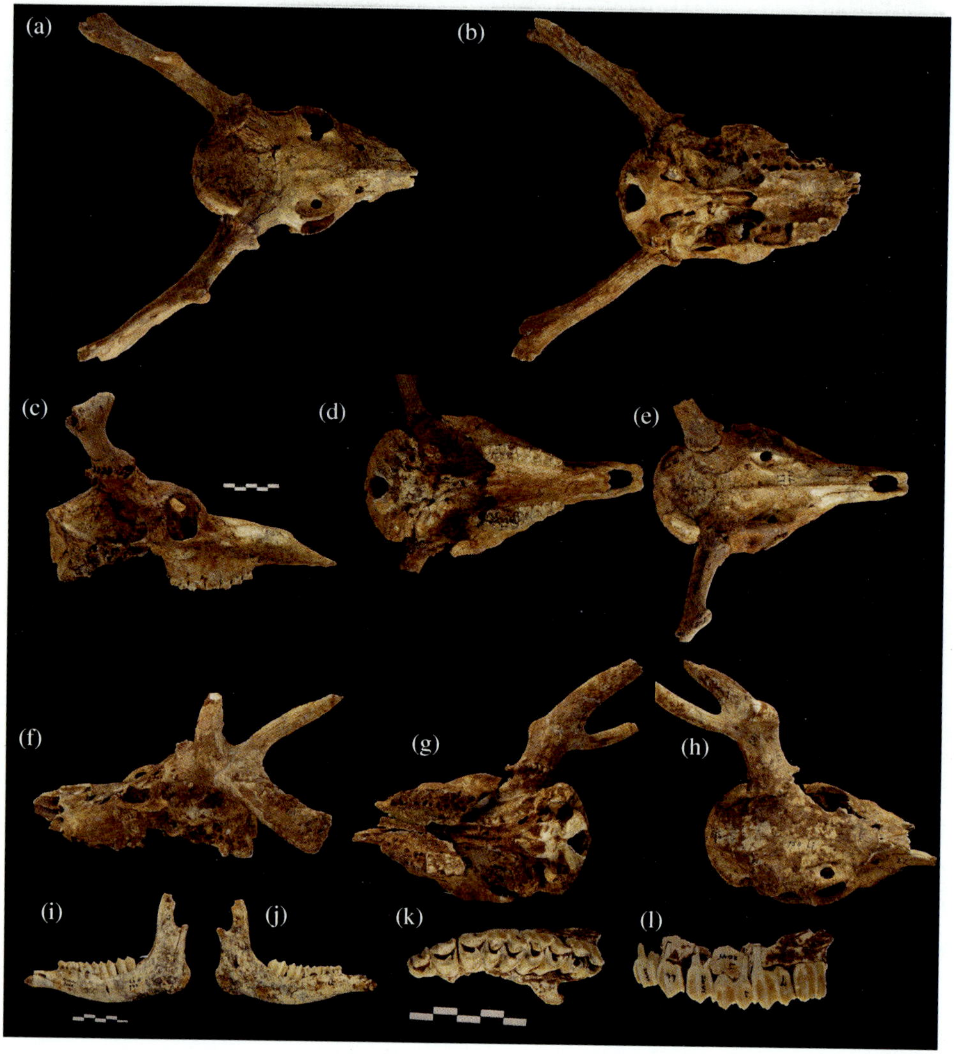

Plate 6 Cranial material of *Candiacervus* sp. II., Crete, Late Pleistocene. (a) Dorsal and (b) ventral views of the holotype of *Candiacervus reumeri*. (c) Lateral, (d) ventral, and (e) dorsal views of the holotype of *Candiacervus listeri*. (f) Lateral, (g) ventral, and (h) dorsal views of the holotype of *Candiacervus devosi*. (i, j) Lateral view of mandible of *Candiacervus* size 2. (k) Occlusal and (l) buccal views of upper toothrow of *Candiacervus* size 2. Museum of Geology and Palaeontology, University of Athens.

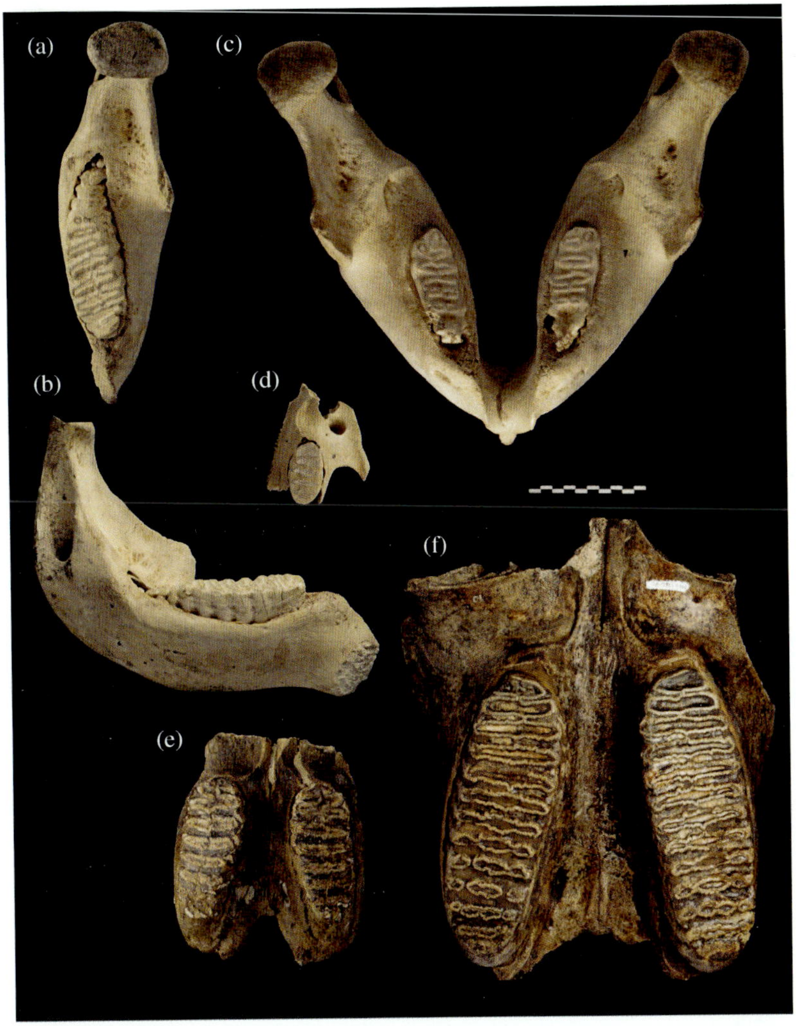

Plate 7 Dwarf elephant remains in comparison to their mainland relative. *Palaeoloxodon tiliensis* from Tilos, Late Pleistocene: mandibles (a–c) and maxilla fragment with milk molar (d). *Palaeoloxodon lomolinoi* from Naxos, Late Pleistocene: maxilla (e). Mainland *Palaeoloxodon* from Peloponnesus, Middle Pleistocene: maxilla (f). Occlusal (a, c–f) and lateral (b) views. Scale bar: 10 cm. Museum of Geology and Palaeontology, University of Athens.

Plate 8 Overview (left) and detail (right) of fossiliferous fissures, Gargano, Italy, Late Miocene–Early Pliocene. (Photographs: John de Vos.)

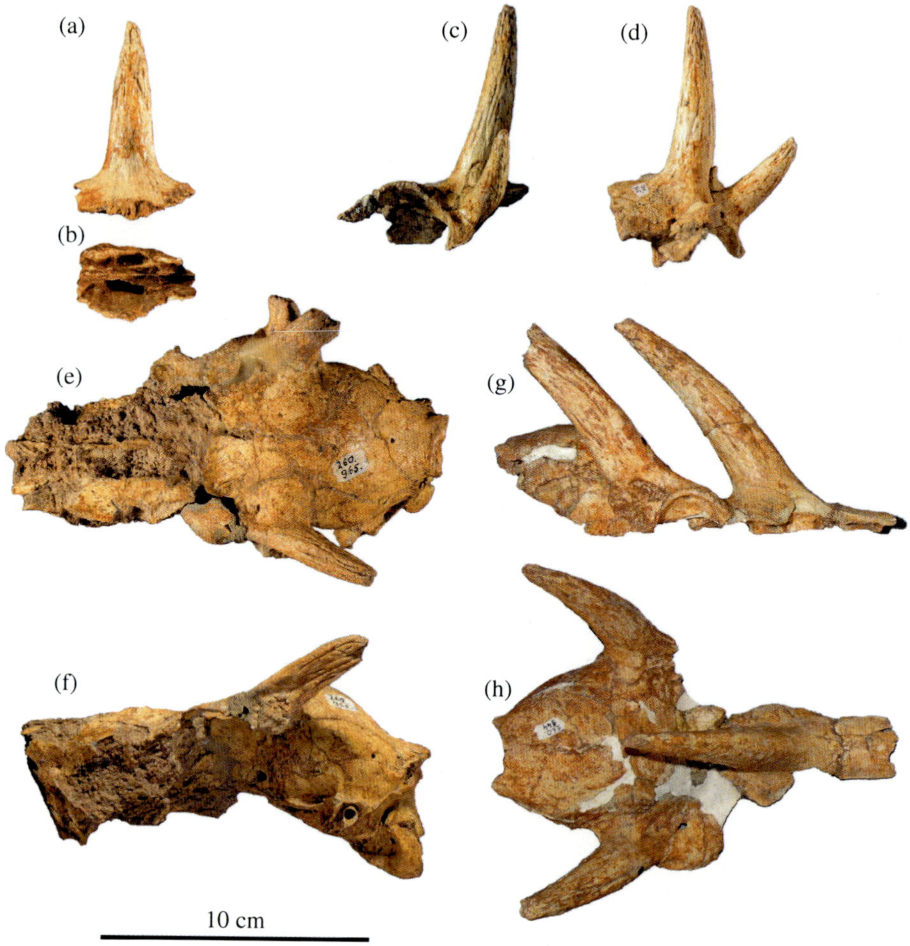

Plate 9 Skulls and horn cores of *Hoplitomeryx matthei*, Gargano, Late Miocene. (a) Nasal horn, lateral view; (b) same, ventral view; (c) orbital horn, lateral view; (d) same, frontal view; (e) skull, holotype, dorsal view; (f) same, lateral view; (g) skull roof and nasals, right side; (h) same, dorsal view. National Museum of Natural History, Leiden. (Photographs: George Lyras.)

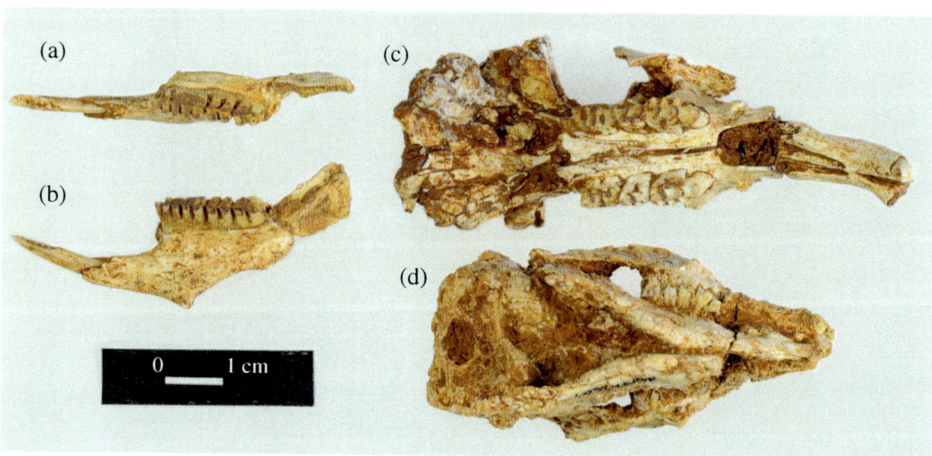

Plate 10 Mandible (a, b) and skull (c) of the burrowing mouse *Mikrotia magna* and skull with associated mandible of the giant pika *Prolagus apricenicus*, Gargano, Late Miocene. Occlusal (a), lateral (b), and ventral (c, d) views. National Museum of Natural History, Leiden. (Photograph: Eelco Kruidenier.)

Plate 11 Mounted skeleton (cast) of the Sicilian dwarf hippo (*Hippopotamus pentlandi*), late Middle Pleistocene. Museum of Geology G.G. Gemmelaro, Palermo. Original described and figured in Accordi, 1955. (Photograph: George Lyras.)

Plate 12 Mounted skeleton (cast) of the Sicilian pygmy elephant (*Palaeoloxodon* aff. *mnaidriensisi*, late Middle Pleistocene). Museum of Geology G.G. Gemmelaro, Palermo. (Photograph: George Lyras.)

Plate 13 Detail of the head of the Sardinian pika (*Prolagus sardus*), Late Pleistocene. Museo dei Palaeoambienti Sulcitani 'E.A. Martel', Carbonia, Sardinia. (Photograph: Alexandra van der Geer.)

Plate 14 Excavations at Corbeddu Cave, Oliena, Sardinia, Late Pleistocene. (Photograph: Alexandra van der Geer.)

Plate 15 Overview of limestone quarry X at Monte Tuttavista, Orosei, Sardinia, Italy, Late Pliocene–terminal Pleistocene. (Photograph: Alexandra van der Geer.)

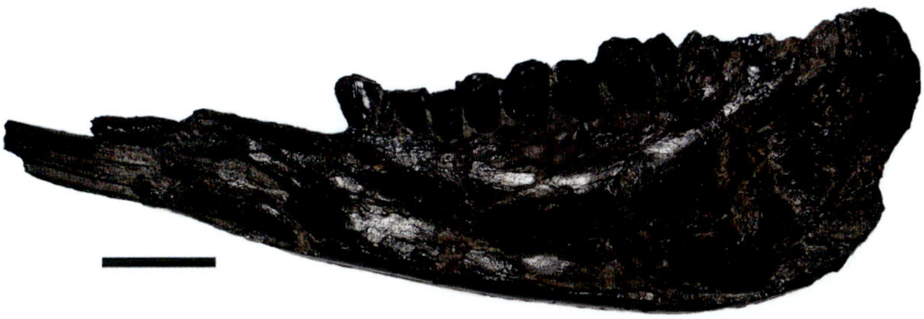

Plate 16 Left lower jaw of the Late Miocene bovid (*Maremmia haupti*) of Tuscany with ever-growing incisors. Scale bar: 2 cm. Natural History Museum, Basel. (Photograph: George Lyras.)

(a) (b)

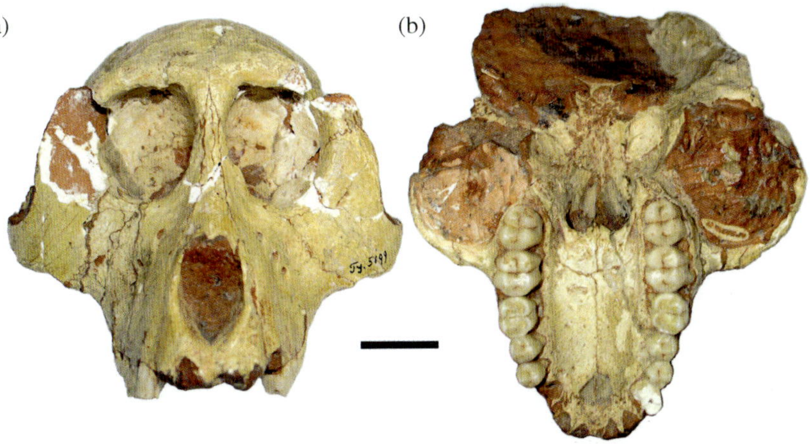

Plate 17 Skull of the Early Pleistocene macaque (*Macaca majori*) of Sardinia, frontal and ventral views. Scale bar: 2 cm. Natural History Museum, Basel. (Photograph: George Lyras.)

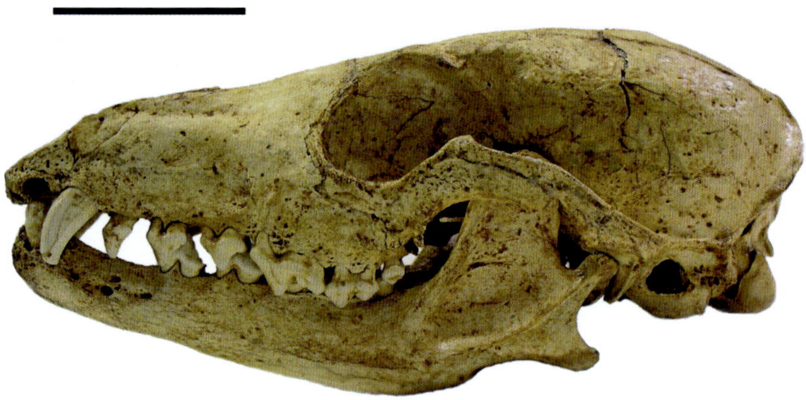

Plate 18 Skull and associated mandible of *Cynotherium sardous* from Corbeddu Cave, Sardinia, Late Pleistocene; left side. Scale bar: 4 cm. Archaeological Museum at Nuoro, Sardinia. (Photograph: George Lyras.)

Plate 19 Skull and mandible of *Myotragus balearicus*, Muleta Cave, Mallorca, Late Pleistocene. Skull in lateral (right side) and frontal view, mandible in medial (interior) view. Scale bar 5 cm. National Museum of Natural History, Leiden. (Photograph Eelco Kruidenier.)

Plate 20 Extinct Madagascar lemurs with profile of *Indri indri*, the largest living lemur, for comparison. Late Pleistocene–Holocene. (Reconstructions by Stephen D. Nash.)

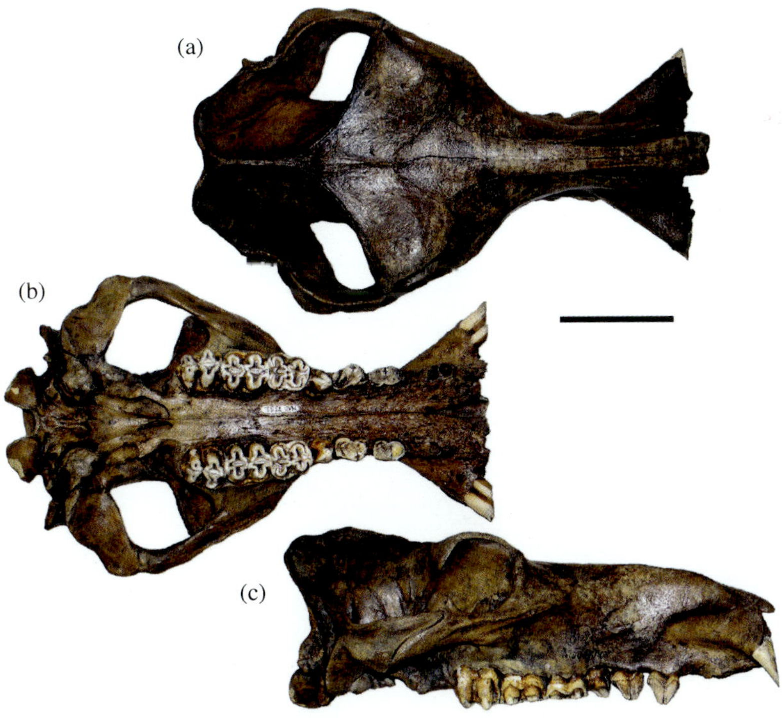

Plate 21 Skull of the Malagasy dwarf hippo (*Hippopotamus madagascariensis*), Late Pleistocene–Holocene. (a) Dorsal, (b) lateral, and (c) ventral views. Scale bar: 10 cm. Museum National d'Histoire Naturelle, Paris. (Photograph: George Lyras.)

Plate 22 Skull of the Malagasy dwarf hippo (*Hippopotamus lemerlei*), Late Pleistocene–Holocene. (a) Dorsal, (b) lateral, and (c) ventral views. Scale bar: 10 cm. Museum of Geology and Palaeontology, University of Athens. (Photograph: George Lyras.)

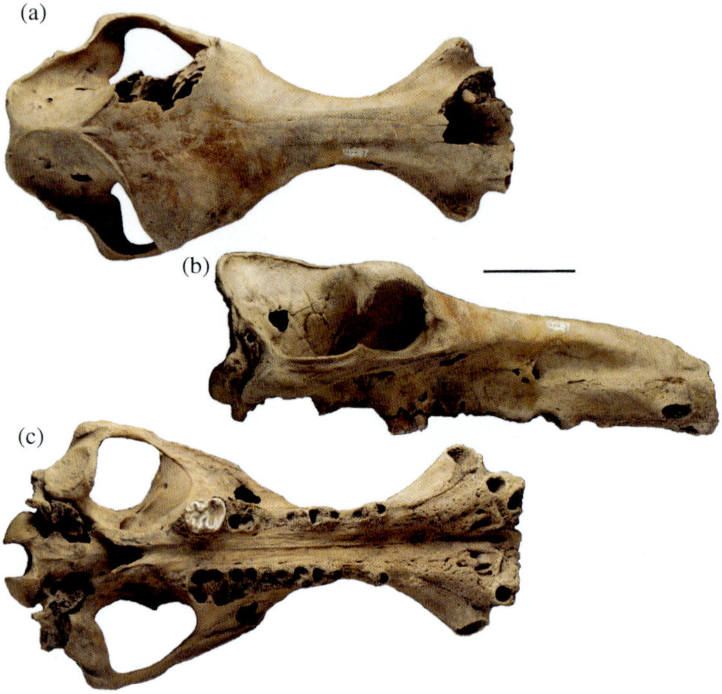

Plate 23 Skull part of the giant lemur *Palaeopropithecus* sp., Late Pleistocene–Holocene. (a) Dorsal, (b) lateral, and (c) ventral views. Scale bar: 10 cm. Museum National d'Histoire Naturelle, Paris. (Photograph: George Lyras.)

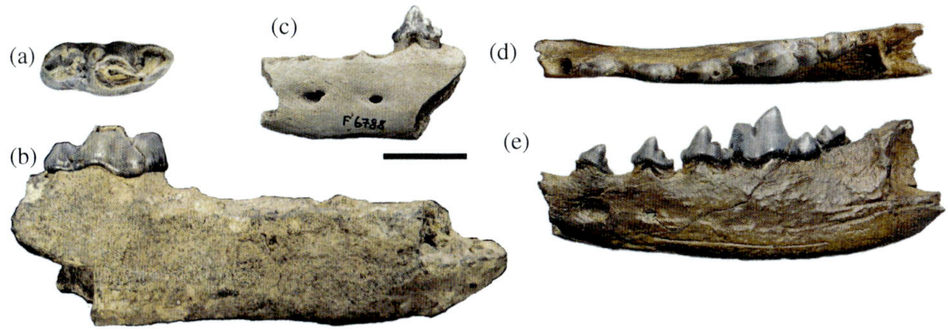

Plate 24 Scale bar: 2 cm. Early-early Middle Pleistocene canid material from Java, the large *Xenocyon merriami* (a, b) and the small *Xenocyon trinilensis* (c–f). (a) Lower carnassial, occlusal view. (b) Right mandible fragment with carnassial and partially filled alveoli of second incisor to fourth premolar, occlusal view. (c, d) Upper right first molars, occlusal view. (e) Left mandible fragment with second premolar to third molar, occlusal view. (a, b) Senckenberg Research Institute, Frankfurt am Main, Germany (Von Koenigswald Collection). (Photographs: Rebekka Volmer.) (c–e) Museum für Naturkunde, Berlin, Germany. (Photograph: Alexandra van der Geer.)

Plate 25 Detail of the 2015 excavation at the Middle Pleistocene locality Mata Menge, Flores, west part. A *Stegodon* pelvis is visible at the boundary between Layer I (yellow mudflow; above) and Layer II (grey sandstone; below). Presumably, the bones lay in the sandy riverbed and were subsequently covered by a mudflow after a volcanic eruption. All hominin fossils were retrieved from the top of Layer II. (Photograph: Gert van den Bergh.)

Plate 26 Upper third molar of the dwarf stegodon, *Stegodon florensis*, Flores, Middle Pleistocene; latero-occlusal view. Scale bar: 5 cm. National Museum of Natural History, Leiden. (Photograph: Eelco Kruidenier.)

Plate 27 Mandibles of giant Flores rats, Late Pleistocene. *Papagomys armandvillei*, lateral (a, b) and occlusal (c) views. *Papagomys theodorverhoeveni*, occlusal view (d). *Speleomys florensis*, occlusal (e) and lateral (f) views. Scale bar: 1 cm. National Museum of Natural History, Leiden. (Photograph: Eelco Kruidenier.)

Plate 28 Mandible fragment with molar fragment of the dwarf stegodon of Timor, *Stegodon timorensis*. Late Pleistocene, Raehio, Atambua, West Timor. Scale bar: 10 cm. Mau Buti, Sadi Laun, West Timor. (Photograph: Bartholomeus van der Geer.)

Plate 29 Left mandibles of the four-tusked Sulawesi dwarf elephant *Stegoloxodon celebensis*, Late Pliocene–Early Pleistocene. Top: medial view, showing part of the lower tusk. Below: occlusal view. Scale bar: 5 cm. National Museum of Natural History, Leiden. (Photograph: Eelco Kruidenier.)

Plate 30 Skull and associated mandible of *Rattus* cf. *everetti*, Masbate, Philippines, Late Pleistocene. Skull length: 5.9 cm. National Museum, Manila. (Photograph: Eelco Kruidenier.)

Plate 31 Excavation at Cagayan Valley, detail of the rhinoceros site at Kalinga, Cagayan Valley. Middle Pleistocene. (Photograph: Thomas Ingicco.)

Plate 32 Jaw fragment of *Stegodon luzonensis* with fragments of the second and third molars. Scale bar: 5 cm. National Museum of the Philippines, Manila. (Photograph: Alexandra van der Geer.)

Plate 33 Fossil trackways of dwarf stegodon and deer, dated to about 1.8 Ma. Kobiwako Group, Hino-cho, Shiga Prefecture, Japan. Right: trackway of *Stegodon aurorae*. Top, centre: petrified giant Chinese sequoia (*Metasequoia*). Top, left: detail of footprints of deer (*Cervus kyushuensis*). Below, left: overview of the site. (Photographs: John de Vos.)

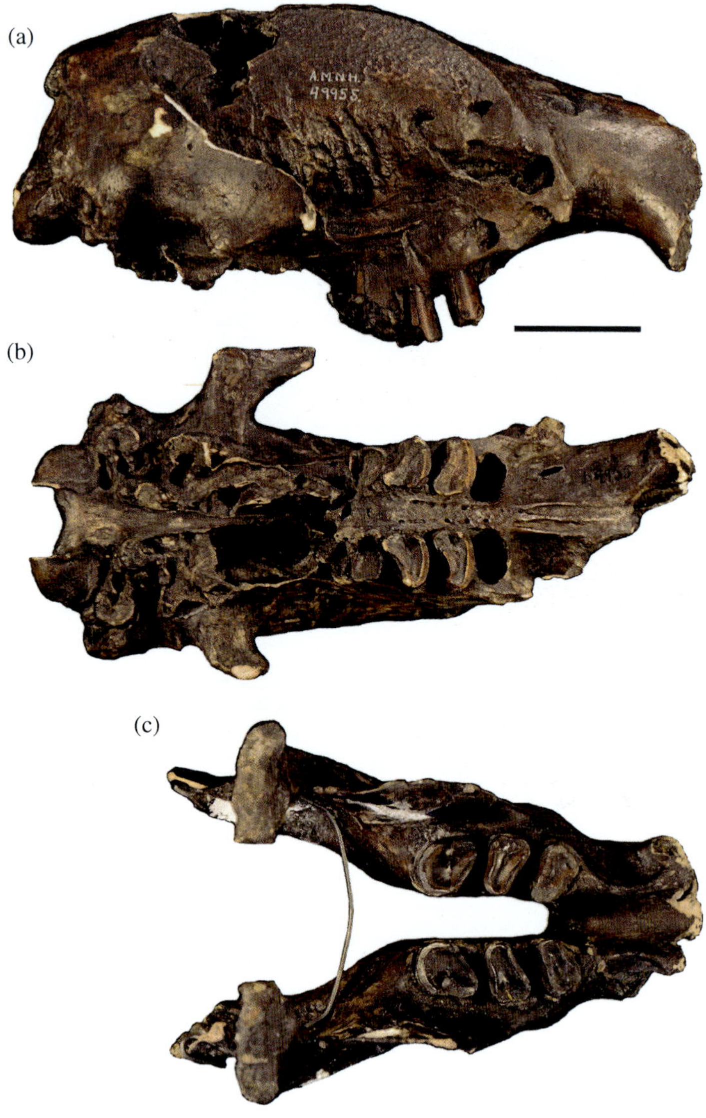

Plate 34 Skull and mandible of the giant Cuban ground sloth, *Megalocnus rodens*, excavated by Barnum Brown in 1918. (a) Lateral and (b) ventral views of the skull, and (c) occlusal view of the mandible. Scale bar: 5 cm. American Museum of Natural History, New York. (Photograph: Bartholomeus van der Geer.)

Plate 35 Ventral view of a maxilla of *Amblyrhiza inundata*, St Martin, Late Pleistocene. Scale bar: 1 cm. National Museum of Natural History, Leiden. (Photograph: Eelco Kruidenier.)

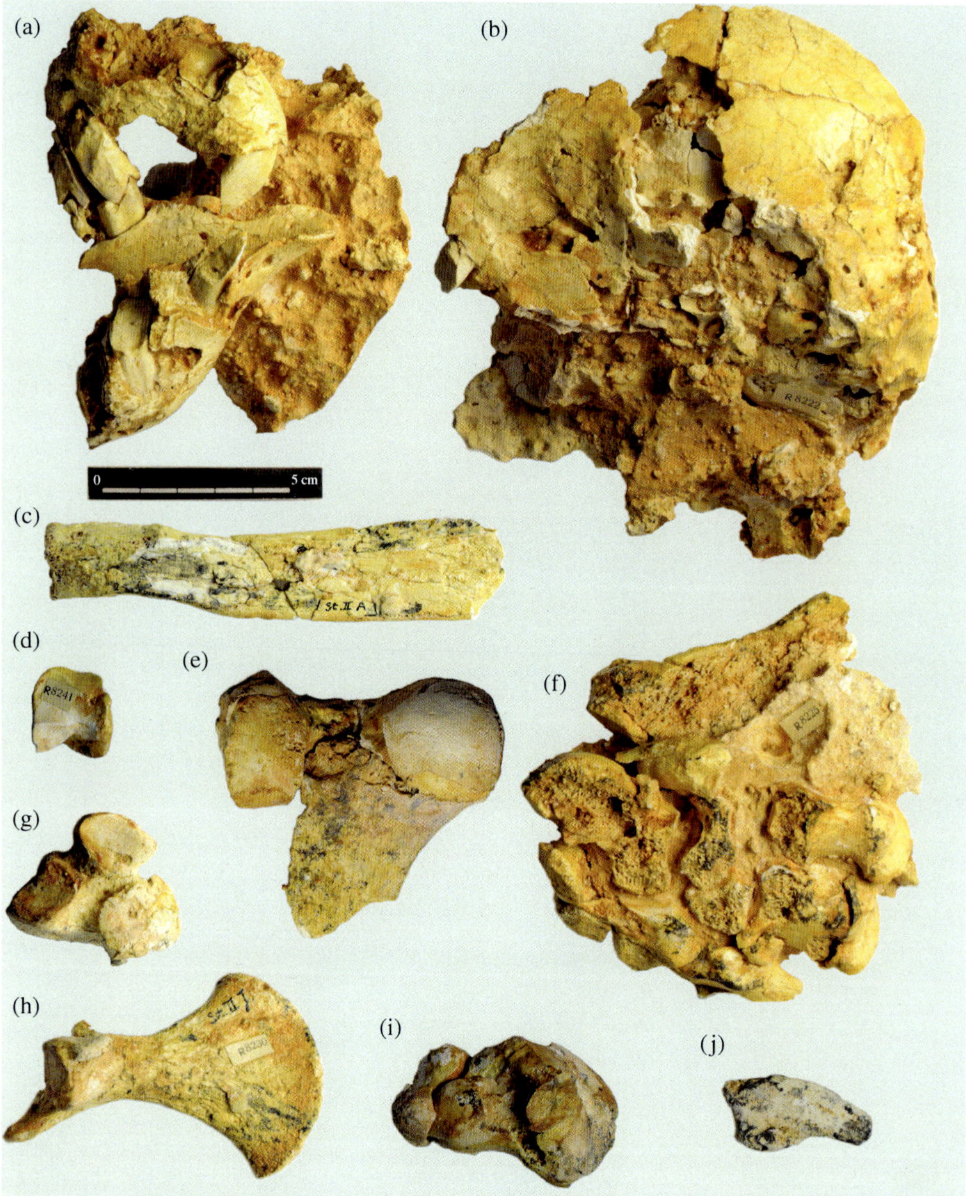

Plate 36 Skeletal elements of *Paulocnus petrifactus,* Curacao, Late Pleistocene. (a) Frontal part of the mandible and left side of the rostrum, (b) parieto-occipital part of cranium (holotype), (c) shaft of right radius, (d) right scaphoid, (e) distal end of right tibia, (f) bones of left manus, (g) distal view and (h) external view of left calcaneum, and (i) left astragalus and navicular, associated, (j) side view of ungular phalanx. National Museum of Natural History, Leiden. (Photographs: Eelco Kruidenier.)

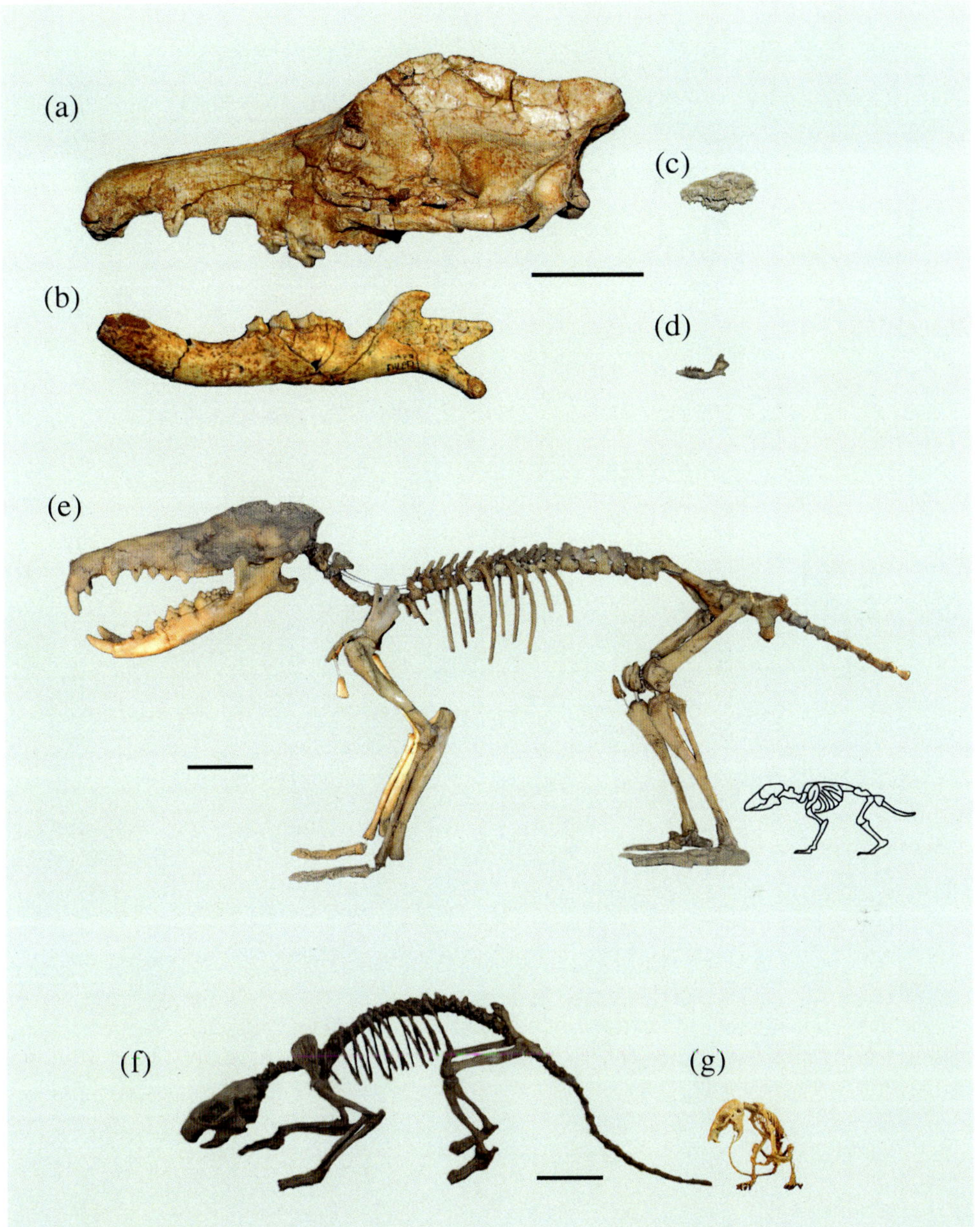

Plate 37 Examples of extreme insular gigantism: (a) skull and (b) mandible of the giant gymnure *Deinogalerix koenigswaldi* (Late Miocene, Gargano) compared to the skull (c) and mandible (d) of its mainland ancestor *Parasorex socialis*; (e) cast of a mounted skeleton of *Deinogalerix koenigswaldi* and hypothetical outline of *Parasorex socialis*; (f) cast of a mounted skeleton of the giant dormouse *Leithia melitensis* (Middle Pleistocene, Sicily) compared to (g) a mounted skeleton of its mainland ancestor, the common garden dormouse *Eliomys quercinus*. Scale bar: 5 cm. National Museum of Natural History, Leiden (a, b, and e); State Museum of Natural History Stuttgart (c and d); Museo Geologico G.G. Gemmelaro, Palermo, Sicily (f and g). (Photographs: Eelco Kruidenier (a, b, and e) and George Lyras (c, d, f, and g).)